Improving Animal Welfare, 2nd Edition

A Practical Approach

D1082476

Improving Animal Welfare, 2nd Edition

A Practical Approach

Edited by

Temple Grandin

Professor of Animal Science
Colorado State University
USA

www.cabi.org

CABI is a trading name of CAB International

CABI	CABI
Nosworthy Way	38 Chauncy Street
Wallingford	Suite 1002
Oxfordshire OX10 8DE	Boston, MA 02111
UK	USA
Tel: +44 (0)1491 832111	Tel: +1 800 552 3083 (toll free)
Fax: +44 (0)1491 833508	E-mail: cabi-nao@cabi.org
E-mail: info@cabi.org	
Website: www.cabi.org	

A catalogue record for this book is available from the British Library, London, UK.

Library of Congress Cataloging-in-Publication Data

Improving animal welfare : a practical approach / edited by Temple Grandin, Professor of Animal Science, Colorado State University, USA. -- 2nd edition.
     pages cm
   Includes bibliographical references and index.
   ISBN 978-1-78064-467-7 (hardback : alk. paper) -- ISBN 978-1-78064-468-4 (pbk. : alk.
paper) 1. Animal welfare. 2. Animal industry--Moral and ethical aspects. I. Grandin, Temple, editor.
HV4708.I47 2015
179'.3--dc23

                                    2015011828

ISBN-13: 978 1 78064 467 7 (hbk)
         978 1 78064 468 4 (pbk)

Commissioning editor: Caroline Makepeace
Assistant editor: Alexandra Lainsbury
Production editor: James Bishop

Typeset by SPi, Pondicherry, India
Printed and bound by Gutenberg Press Ltd, Tarxien, Malta

Contents

Contributors

Anne Marie de Passillé, University of British Columbia, British Columbia, Canada. E-mail: passille@mail.ubc.ca

Lily N. Edwards-Callaway, Animal Welfare Specialist, JBS, Greeley, Colorado, USA. E-mail: Lily.Edwards-Callaway@jbssa.com

Wendy Fulwider, Global Animal Partnership, 107 S West Street, PMB #771, Alexandria, Virginia 22314, USA. E-mail: wfulwider@globalanimalpartnership.org

Temple Grandin, Department of Animal Science, Colorado State University, Fort Collins, Colorado, USA. E-mail: cheryl.miller@colostate.edu

Camie R. Heleski, Michigan State University, East Lansing, Michigan, USA. E-mail: heleski@msu.edu

Hubert J. Karreman, VMD, Rodale Institute, 611 Siegfriedale Road, Kutztown, Pennsylvania 19530, USA. E-mail: penndutch@earthlink.com

David C.J. Main, University of Bristol, Langford, Bristol, UK. E-mail: main@bristol.ac.uk

Amy K. McLean, American Quarter Horse Association, Amarillo, Texas, USA. E-mail: amclean@aqha.org

David J. Mellor, Massey University, Palmerston North, New Zealand. E-mail: D.J.Mellor@Massey.ac.nz

Bernard Rollin, Colorado State University, Fort Collins, Colorado, USA. E-mail: bernard.rollin@colostate.edu

Jeffrey Rushen, University of British Columbia, British Columbia, Canada. E-mail: rushenj@mail.ubc.ca

Jan K. Shearer, Iowa State University, Ames, Iowa, USA. E-mail: jks@iastate.edu

Kevin J. Stafford, Massey University, Palmerston North, New Zealand. E-mail: K.J.Stafford@massey.ac.nz

Janice C. Swanson, Michigan State University, East Lansing, Michigan, USA. E-mail: swansoj@msu.edu

Helen R. Whay, University of Bristol, Langford, Bristol, UK. E-mail: Bec.Whay@bristol.ac.uk

Tina Widowski, University of Guelph, Guelph, Ontario, Canada. E-mail: twidowsk@uoguelph.ca

Jennifer Woods, J. Woods Livestock Services, Blackie, Alberta, Canada. E-mail: livestockhandling@mac.com

Preface

Animal welfare is an increasing concern all over the world. The second Edition provides updates on the latest research and still retains its practical emphasis. The World Organization for Animal Health (OIE) now has recommendations in the *Terrestrial Animal Health Code* for transport, slaughter, killing animals for disease control, broiler chickens, beef cattle and fish. An additional introductory chapter covers the new OIE Guiding Principles of Animal Welfare. In some countries, animal welfare is a new concept and this book will provide practical information which will enable veterinarians, managers and animal scientists to implement effective practical programs to improve animal welfare. It will be especially useful for both students and training of animal welfare specialists. A new chapter on organic agriculture has been added that explains differences between European and North American systems. Emphasis is placed on an international approach. Two of the authors worked on OIE animal welfare committees, and the most important parts of these guidelines are reviewed. The authors are from the USA, Canada, the UK and New Zealand, and they have extensive experience improving animal welfare in both the developed and the developing world. In addition to their work in North America and Europe, they have also worked in Brazil, Mali, western Africa, Uruguay, Chile, Australia, the Philippines, Mexico, China, Thailand, Argentina and New Zealand. *Improving Animal Welfare: A Practical Approach* 2nd Edition covers how to both measure and assess welfare, plus 'how to' instructions on methods to improve practices in areas of major welfare concern such as animal handling, euthanasia, painful surgical procedures, transport, slaughter and treatment of draught animals. Another major area of emphasis is how to use animal-based outcome measures such as scoring of body condition, lameness, lesions, behavior and coat/feather condition. The use of numerical scoring to measure handling and stunning practices will also be covered. Measurement is essential because people manage the things that they measure. Key references to important scientific papers are included but this book is not intended to be a complete review of the literature. Additional chapters on the benefits of good stockmanship, economic factors, ethics and proven methods for motivating producers will also be helpful to bring about improvements. This book is aimed at the people who will be putting animal welfare program into practice to improve conditions for animals all over the world.

1 An Introduction to Implementing an Effective Animal Welfare Program

TEMPLE GRANDIN

Colorado State University, Fort Collins, Colorado, USA

Summary

There has been an exponential increase in the publication of guidelines and scientific papers on animal welfare. Students, farm managers, veterinarians, and others who work in the field might ask themselves, "Animal welfare is so complex, where do I start?" The first step is to improve management, breeding, housing, and handling to reduce conditions that cause poor welfare. Some examples are lameness (difficulty in walking), bruises from rough handling, leg injuries, poor body condition, high ammonia levels in the air, dirty animals, tail biting, or breast blisters on poultry. There are three components of an effective welfare program – a guidance document, assessment tools, and standard operating procedures. There are four internationally recognized conceptual frameworks for assessing welfare. They are *Understanding Animal Welfare: The Science in its Cultural Context* (Fraser, 2008), The Five Freedoms (FAWC, 1992), Welfare Quality Network (2009) and the World Organisation for Animal Health (OIE) Guiding Principles on Animal Welfare. Recommendations on how to interpret the OIE principles are included in this chapter. Studies have shown that in addition to preventing suffering, a higher level of animal welfare can be attained if environments are provided that enable animals to experience positive emotions. Converting multiple welfare measures to a single score is risky because it may mask a serious welfare problem such as lameness.

Learning Objectives

- How to implement animal welfare assessment programs
- Identify the most severe animal welfare problems that must be corrected immediately
- How to measure bruises, injuries, and other animal welfare problems at the slaughter plant
- Understand the conceptual frameworks for forming animal welfare guidance documents
- How to apply the OIE (World Organisation for Animal Health) general principles of animal welfare

Introduction

Since the publication of the first edition of this book in 2010, there has been a worldwide exponential increase in the number of research studies, books, new animal welfare guidelines, legislation, and websites about animal welfare (Walker *et al.*, 2014). Government regulations in different countries vary greatly, from very strict to none at all. A person who is new to the field of animal welfare is likely to become overwhelmed by the massive amounts of new information. Students, farm managers, veterinarians, and others who are in the field may say to themselves, "Animal welfare has become so complex; where do I start?" This book will not attempt to review all the regulations and new literature but, instead, it will include the most practical scientific information and combine it with recommendations for use on commercial operations. Animal welfare is a global issue (Fraser *et al.*, 2013), and this book contains information that can be put to everyday use in both the developed and the developing world. There is certain basic information that can be used to improve animal welfare anywhere in the world. Quick Access Information boxes have been placed throughout the text to allow the reader easy access to summarized basic information.

There is also a need for information on how to implement assessment programs, SOPs (standard operating procedures), and other strategies that will improve animal welfare and prevent animal abuse. This is a hands-on, "how-to" guide that provides practical information for veterinarians, animal scientists, producers, transporters, auditors, government agencies, quality assurance managers, and others who work in the field with animals. Too often, legislation will be passed to ban some terrible practice, but it still continues because little is done in the field to implement change. Another problem is that people who draft legislation may have little knowledge of conditions in the field.

Recommendations on implementing animal welfare programs are based on over 15 years of the editor's experience in developing and implementing welfare auditing systems for major retailers and restaurants (Grandin, 2000, 2005). The author has visited over 500 farms and slaughter plants in 26 different countries. This book will also help the reader use more effectively the knowledge that can be obtained from many other sources. It will make it possible to bring about real changes that will improve the treatment of livestock, poultry, and fish on farms, in transport vehicles, and in slaughter plants. The principles of implementing an effective animal welfare program are the same for all species.

The World Organisation for Animal Health (OIE, 2014a) has published specific welfare guidelines for the slaughter of livestock and poultry, the transport of livestock, depopulation of farms after a disease outbreak, beef cattle production, and broiler chicken production. Additional guidelines are available for farmed fish slaughter, transport, and depopulation (OIE, 2014a). Guidelines for dairy farms, pork production, and laying hens are under development. OIE guidelines are minimum, basic international standards for the entire world. Many countries and large meat-buying customers will often have more strict standards.

Large food retailers and restaurant chains are now requiring that their suppliers comply with their own animal welfare standards. The economic incentive provided by these large buyers is a major force for improving animal welfare in both the developed and the developing world (Webster, 2012). Non-governmental organization (NGO) animal advocacy groups are also influential in promoting the development of animal welfare standards and legislation.

Goals of the book

This book has five main goals:

1. To help managers, veterinarians, and policy makers to implement effective practical auditing, regulatory, and assessment programs that will improve the welfare and treatment of livestock, poultry, and farmed fish.
2. To provide practical information that will directly improve welfare in critical areas such as slaughter, transport, handling, euthanasia, care of working animals, and painful surgical procedures.
3. To help the reader understand the importance of animal behavior in assessing animal welfare and its role in the design of production housing and handling systems.
4. To discuss the role of ethics in animal welfare in an easy-to-understand manner.
5. To understand how economic incentives can be used both to improve welfare and reduce economic losses in farm animals. Improvements in husbandry, handling, stockmanship, and transport will improve animal productivity, and reduce losses due to bruises, sickness, mortalities, lameness, and other problems (Hemsworth and Coleman, 2010; Huertas *et al.*, 2010).

Effects of electronic media on animal welfare

Since the first edition of this book, there has also been a massive increase in the number of mobile telephones that can take high-quality video. Videos of atrocious animal abuse have caused trade disruptions and public outcry, and have forced companies to improve practices. The first step in creating an effective animal welfare program is to prevent animal abuse. Compliance with legislative paperwork is not going to protect a farm when animal abuse is caught on video that has been shown around the world. Some videos have gone viral and have been seen by millions of viewers. There is a tendency in many welfare guidelines and legislation to turn compliance into keeping records and filling out forms instead of monitoring actual practices in the field. The author's emphasis is going to be on what is actually happening on the farm or at the slaughter plant.

The first step in implementing an animal welfare program is stopping obvious abusive practices. Some examples of overt abuse that is not permitted by the OIE (2014a) are beating animals, deliberately breaking legs, cutting tendons, dragging nonambulatory downed animals, or driving an animal by poking sensitive areas such as the rectum. Overloading a truck that causes animals to die or get trampled under other animals is another practice that must be stopped. Some examples of production conditions on a farm that are totally abusive are animals housed in filth, where manure is not removed, or high ammonia levels or dust concentrations in intensive housing. Neglected health problems, such as necrotic prolapses, advanced cancer eye, starvation, or overcrowding to the point that animals have to lie on top of each other are also not acceptable. The first step in any program is to eliminate all of the above problems. There is further information in Quick Access Information 1.1.

Quick Access Information 1.1. The most severe animal welfare problems caused by abuse, neglect, or bad management, that cause obvious suffering. These conditions must be corrected immediately.

Handling and transport – prohibited practices and conditions[a]	Welfare problems caused by poor housing, environmental conditions, nutrition, or neglected health problems	Slaughter – prohibited practices[a]
• Beating, throwing, or kicking animals. • Poking out eyes or cutting tendons to restrain an animal. • Dragging and dropping animals. • Overloading trucks so tightly that a downed animal is trampled. • Deliberately driving animals over the top of other animals. • Poking animals in sensitive areas such as the eyes, anus, or mouth. • Breaking tails or legs. • Overloading a draft animal and working it to exhaustion. • Poking animals with pointed sticks. • Conditions that cause animals to fall frequently or become injured or bruised during handling.	• Starvation or allowing animals to become severely dehydrated. Many animals that are skinny or have poor body condition. • High ammonia levels that cause eye or lung damage. • Death or severe stress from extreme heat or cold. • Large swellings or other injuries caused by either a lack of bedding or poorly designed housing. • Dirty animals or birds covered with manure, with no dry place to lie down. • Failure to treat obvious health problems. • Nutritional problems that compromise the animal's health. • Any condition that causes many animals to become lame (difficulty in walking). • Saddle or harness sores on a working animal. • Failure to euthanize animals with either severe injuries or disease that will not be able to recover. • Neglected health problems such as necrotic prolapses or advanced cancer eye.	• Scalding, skinning, leg removal or other carcass-dressing procedures performed on sensible, conscious animals. • Immobilizing animals with an electrical current (Grandin et al., 1986; Pascoe, 1986), not to be confused with effective electrical stunning. • Puntilla method of immobilizing animals before slaughter by severing the spinal cord, which does not cause instantaneous insensibility (Limon et al., 2010). • Highly stressful methods of restraining conscious animals. One example is hoisting cattle by one leg.

[a]The items in the handling and transport, and slaughter columns would be in violation of OIE (2014 a,b) codes for slaughter and transport. The OIE standards for animal welfare are the most basic standards that everybody in both developed and developing countries should follow. To achieve a higher level of welfare will require some additional standards. Many countries have several additional standards. Disease control standards between different countries are easier to make uniform between countries than welfare standards, which have more complex ethical considerations.

Steps for the Effective Implementation of an Animal Welfare Program

Many experts on animal welfare agree that preventing overt abuse is not the only thing that guarantees good welfare (Broom, 2011; Keeling *et al.*, 2011; Mellor and Webster, 2014). The author agrees with these concerns. To simplify understanding of welfare issues, the author has listed four types of welfare problems in order of priority. This will be especially useful in developing countries when a program is being implemented initially.

Step 1

Eliminate abuse, neglect, and cruelty (Quick Access Information 1.1). This requires management supervision, and fancy new facilities will not fix these problems.

Step 2

Initiate the measurement systems and assessment tools outlined in the rest of this book to reduce the percentage of animals that have a serious welfare problem, such as poor body condition (skinny, ribs showing), lameness (difficulty in walking), dirty, or sores and injuries (Quick Access Information 1.1). Lameness is a major problem (Von Keyserlingh *et al.*, 2012). Programs should also be implemented for the assessment and continuous monitoring of handling, transport, and slaughter practices. Many serious welfare problems can be identified easily at the slaughter plant (Quick Access Information 1.2). The advantage of assessment at the slaughter plant is that large numbers of animals from many farms can be viewed in a short period. The disadvantages of slaughter-plant assessment are that the welfare issues in Steps 3 and 4 cannot be assessed easily.

Step 3

Provide animal housing that satisfies basic behavioral needs, such as being able to turn around and stand in a normal standing posture. Provide enrichments for highly motivated behavioral needs, such as a secluded nest box for laying hens, or rooting substrate or devices for pigs to chew. Train producers in good stockmanship skills that reduce an animal's fear of humans and improve productivity (Hemsworth and Coleman, 2010). Methods to

---

Quick Access Information 1.2. Indicators of animal welfare problems that can be measured easily at the slaughter plant.

Outcomes of rough handling, transport problems, and abuse	Outcomes of housing problems	Outcomes of poor management genetic conditions or neglected health problems
• Bruises • Broken wings on poultry • Number dead on arrival • Number of nonambulatory animals or injuries such as broken horns or legs • Broken tails • Injuries from specific abusive practices such as poked out eyes, puncture wounds from nails, or cut tendons	• Animal or birds covered with manure • Hock burn on poultry • Swellings on the legs of dairy cattle • Pressure sores on sows • Serious wounds from fights • Foot pad lesions on poultry • Eye and lung pathology due to high ammonia levels • Breast blisters on poultry • Hoof problems • Fin erosion on fish	• Poor body condition • Lameness (difficulty in walking) • Poor coat/feather condition • Difficult to handle wild animals • Weak animals • Cancer eye • External parasites • Internal parasites • Tail biting • Twisted legs on poultry • Poor leg conformation in cattle and pigs • Prohibited practices and mutilations • Disease conditions • Dehydration • Horns cut on mature cattle • Liver abscesses – grain-fed cattle

T. Grandin

reduce pain during castration or dehorning should also be implemented. There are chapters in this book on pain mitigation after painful procedures, stockmanship, and behavioral needs.

Step 4

Since the first edition of this book, there has been a huge amount of research on providing environments that enable an animal to have truly positive emotions (Boissey *et al.*, 2007; Rutherford *et al.*, 2012). Some critics of the first edition stated that the book did not have enough emphasis on facilitating an animal's positive emotions. Some new assessment tools for positive emotions are QBA (qualitative behavioral assessment) (Rutherford *et al.*, 2012) and cognitive bias (Douglas *et al.*, 2012). These assessments will be discussed in Chapters 2, 12, and 16, this volume. Scientific research clearly shows that animals have emotions (Morris *et al.*, 2011; Panksepp, 2011). Some simple assessments of positive emotions are: animals have low fear and will approach people, play behavior, social grooming where animals lick each other, and willingness to approach a novel object.

The Four Components of an Effective Welfare Program

A guidance document, legislative code, or conceptual framework

Examples of guidance documents would be the OIE Guiding Principles on Animal Welfare (OIE, 2014a), the Welfare Quality Network (2009), the five freedoms (FAWC, 1992), Fraser's (2008) four guiding principles, government legislation, or a private corporate guidance document. Guidance documents range in detail from complex to simple anti-cruelty legislation. These documents specify the overall principles.

Assessment tools

Assessment tools are used to assess the conditions of the animals so that they comply with the guidance documents, standards, and regulations. They range from complete programs such as the welfare quality protocols for each species or the American Meat Institute (2013) Audit Guide, to single assessment tools for particular welfare problems such as lameness, animal handling, body condition, or a behavior problem such as tail biting. Assessment tools can be used internally on a farm to determine if conditions are improving or deteriorating. They are also used by government inspectors, corporate quality assurance people, or auditors from major buyers to determine compliance with guidelines and legislation. Chapter 4, this volume, contains a discussion of the practical application of assessment tools. Other chapters will provide specific information on handling, euthanasia, transport, slaughter, painful procedures, and behavior. These chapters will be useful for training stockpeople and writing SOPs.

Regulations or standards

The guidance documents contain overall principles, but more specific standards are still needed. For example, the OIE has general principles for animal welfare, which is a guidance document (OIE, 2014a). They then have separate standards that are more specific for slaughter, transport, beef and broiler chicken production. Another example would be a standard written by either a private corporation or a certifying organization. Some examples would be McDonald's Corporation, Global Animal Partnership, or an organic standards organization. These standards may require the use of a particular assessment tool. For example, 95% of cattle must remain silent and not vocalize (moo or bellow) in the stun box or while entering it. If more than 5% vocalize, the plant is in non-compliance with the corporate standard of McDonalds. Another example would be a requirement that animals are grazed on pasture during the growing season.

A well-written private corporate standard is compatible with the OIE. They should be written to avoid direct conflict with international OIE standards. For example, numerical scoring of cattle vocalization during handling provides guidance for enforcing the OIE standard that states, "Animals should be handled in such a way to avoid harm, distress or injury" (OIE, 2014b).

Regulations written by the government in your country would also be included in this category. You are required by law to obey the government regulations in your country. OIE standards are compatible with animal welfare regulations in most countries. Developed countries may have regulations that are stricter, and the implementation of OIE standards in less developed areas will improve animal welfare. Some countries may write OIE standards into their own legislation.

Standard operating procedures (SOPs) and animal welfare officers

Each company or individual farm should develop their own SOPs for animal welfare, which cover both "how to do" different procedures and methods for the use of assessment tools. Large slaughter plants (European Union, 2012) and corporate farming operations should have an animal welfare officer to write SOPs. The welfare officer is the key person who ensures that a company complies with welfare regulations. Another essential part of the welfare officer's job is to implement a welfare program that stops obvious abuses and corrects the problems outlined in Quick Access Information 1.1 and 1.2.

SOPs are used in both food safety and animal welfare to provide guidance for the specific conditions on a particular farm or slaughter plant. They are not assessment tools. SOPs provide detailed guidance on how to conduct different procedures such as lameness scoring, castration, euthanasia, truck loading, or stunning at slaughter. The EU requires specific stunning parameters, such as amperage and application time, to be specified for each abattoir in an SOP (European Union, 2012). Many companies will use published training and educational materials as SOPs. Some examples would be the euthanasia guidelines published by the American Veterinary Medical Association (AVMA, 2013), guidelines published by the cattle association in your country, or information from this book. An SOP for lameness scoring of dairy cows might state that lameness scoring should be conducted monthly when the cows exit the milking facility. Since each dairy is different, the SOP should state where the observer should stand. It should also specify which lameness assessment tool should be used. An SOP for euthanasia could specify either the methods shown in this book, the guidelines from a veterinary association, the OIE guideline, or the requirements of a major retailer. It would instruct farm employees and managers on how to carry out different procedures.

Conditions vary greatly around the world. For example, an environmentally controlled building where the animals are dependent on the ventilation system to prevent death should have an SOP stating that the backup generator must be tested and run monthly. A building with completely open sides that has natural ventilation would not have a generator, or an SOP about a generator. SOPs vary in the level of detail. A very detailed set of SOPs would be a complete employee training book, which would have detailed instructions on how to detect disease, treat disease, handle animals, euthanasia, transport, and slaughter. A less detailed SOP might specify the following of an existing published guideline.

Four Conceptual Frameworks to Form Guidance Documents for Assessing Welfare

The four internationally recognized conceptual frameworks are David Fraser's four guiding principles (Fraser, 2008), the Welfare Quality Network (2009), the five freedoms (FAWC, 1992), and the OIE Guiding Principles on Animal Welfare (OIE, 2014a).

Four guiding principles

David Fraser (2008), at the University of British Columbia in Canada, states that from both an ethical and a scientific viewpoint, there are four guiding principles for good welfare.

1. Maintain basic health – examples: provide sufficient feed, water, vaccinations, housing, and air quality to prevent disease and reduce death losses. Maintain the body condition and productivity of the animals or birds. Health is a major component of animal welfare but it is not the only factor.
2. Reduce pain and distress – examples: use of anesthetics for dehorning, preventing lameness, reducing bruises, preventing injuries, and elimination of rough, stressful handling methods that cause fear or pain. Preventing hunger, thirst, heat stress, and cold stress are also covered by the second principle.
3. Accommodate natural behaviors and affective states – examples: provide a nest box for hens, and straw for pigs to root in. The affective state is the animal's emotional state (see Chapter 8, this volume; Duncan, 1998).
4. Natural elements in the environment – examples: outdoor access or natural sunlight.

Welfare Quality System

The Welfare Quality System developed in Europe assesses animal welfare within the framework of four factors (Welfare Quality Network, 2009; Blokhuis et al., 2010; Rushen et al., 2011).

1. Good feeding – absence of prolonged hunger and thirst. Some examples of assessments for good feeding would be: body condition score to locate skinny animals, access to drinking water, signs of dehydration, and feeder space.

2. Good housing – includes thermal comfort, ease of movement, resting comfort. Some examples of assessments for good housing would be: animal cleanliness, air quality, heat stress symptoms such as panting, stocking density, condition of litter or bedding, and cold stress symptoms such as shivering. Handling scoring can also be induced in this section under ease of movement.

3. Good health – includes injuries, disease, and controlling pain during surgical procedures. Some examples of assessment for good health would be: lameness scoring, death losses, wounds such as swollen hocks on dairy cows, hock burn in poultry, breast blisters on poultry, foot pad lesions in poultry, vulva bites on sows, hernias on pigs, somatic cell count in a dairy, internal and external parasites, dystocia, and pain relief for painful procedures. All clinical signs of disease would also be included in this section.

4. Appropriate behavior – includes expression of social behavior and good human–animal relationship. Some examples of assessment for appropriate behavior would be: measures of flight zone to assess fearfulness (avoidance) of people, Qualitative Behavioral Assessment, scoring of wounds caused by aggression, and scoring of stereotypies and other abnormal behavior. Assessments that show the signs of positive emotions would also be included. Some examples are play, approach to a novel object, and social grooming (Rushen *et al.*, 2011).

The five freedoms

The OIE (2014a) recognizes the five freedoms listed below as part of a guidance document for evaluating animal welfare. The five freedoms were developed originally by the Brambell Committee in the UK (FAWC, 1992). The five freedoms are:

1. Freedom from hunger, thirst, and malnutrition.
2. Freedom from physical and thermal discomfort.
3. Freedom from pain, injury, and disease.
4. Freedom to express normal behavior.
5. Freedom from fear and distress.

OIE General Principles of Animal Welfare

Since the first edition of this book, the OIE (2014a) has published 'General Principles for the Welfare of Animals in Livestock Production Systems' in Article

7.1.4. This guidance document covers farm production units (Fraser *et al.*, 2013). There are other, older OIE guidelines that cover transport, slaughter, and depopulation for disease control. Below, the author will provide specific examples of how the OIE (2014a) guidelines for animal production on the farm can be used. At that time of writing, the OIE still had no specific guidelines and assessments for the production of pigs, dairy cattle, and laying hens. These three types of animals have more complex and contentious welfare issues related to the design of housing systems than fish, beef cattle, or broiler chickens. Since the issues are more complex, it will probably be more difficult for all the countries in the OIE to agree on an international guideline. This may explain why guidelines for raising beef cattle, broiler chickens, and fish were developed first. The author will provide opinions on how the OIE guidelines could be interpreted. The reader must remember that the OIE guidelines are designed to provide guidance in many different countries where there are a great variety of conditions.

"Genetic selection should always take into account the health and welfare of animals" (OIE, 2014a)

This is an area of concern to many scientists (Rodenburg and Turner, 2012). Some of the severe welfare problems caused by indiscriminant selection for productivity are: high percentages of broken bones due to osteoporosis in laying hens (Wilkins *et al.*, 2011), lameness and leg abnormalities in rapidly growing broiler chickens (Caplen *et al.*, 2012), decreased disease resistance in modern hybrid pigs (Jiang *et al.*, 2013), loss of parasite resistance in sheep (Greer, 2008), and reproduction problems in dairy cows that are correlated with increased milk production (Spencer, 2013). Further research has shown that high-producing cows have lower body condition and thinner fat pads in their feet (Green *et al.*, 2014). Walsh *et al.* (2011) reported that negative energy balance was a major factor in low dairy cow fertility. Negative energy balance occurs when a high-producing dairy cow is not able to eat enough nutrients to prevent loss of body condition. The author refers to these problems as "biological system overload". In the future, overloading of the animal's biological system may cause some of the most severe animal welfare problems (Grandin and Deesing, 2013). It is interesting that the OIE listed the genetic problems first. People from developing countries who are on OIE committees may have observed both welfare and production problems when modern, high-producing livestock and poultry were first introduced to their countries. These high-producing animals were more likely to get sick or die under more primitive conditions. Rodenburg and Turner (2012) are concerned about loss of functionality and welfare problems in animals genetically selected to produce more eggs, milk, or meat. Both scientists and some producers have stated that we need to select animals for optimum production instead of maximum production. To maintain a reasonable level of disease resistance, reproductive ability, and structural integrity of the animal, managers may need to select for slightly less productivity.

"Animals chosen for introduction into new environments should be suited to the local climate and able to adapt to diseases, parasites, and nutrition" (OIE, 2014a)

Some problem areas would be high-producing Holstein dairy cows housed in a very hot climate and not provided with artificial cooling. A good alternative would be crossing hardy local cattle with Holsteins to increase milk production but still retain sufficient heat tolerance to prevent heat stress (Galukande et al., 2013). It is important that indigenous, disease-resistant local breeds are preserved (Yilmaz et al., 2013). Vordermeier et al. (2012) reports that Ethiopian native Zebu cattle are more resistant to tuberculosis than imported Holstein-Friesian dairy cows. To avoid either welfare or productivity problems, animals bred for high meat, egg or milk production usually require higher inputs of both expensive feeds (Thatcher et al., 2011) and environmentally controlled facilities.

"The physical environment, including the substrate (walking surface, resting surface, etc.) should be suited to the species so as to minimize risk of injury and transmission of diseases and parasites to animals" (OIE, 2014a)

The trend in animal welfare assessments is to use animal-based outcome measures (Whay et al., 2003; Velarde and Dalmau, 2012). The main outcome measures for welfare problems under this guideline would be a high percentage of animals with the following problems – lameness (difficulty in walking), dirty animals, leg injuries/swellings, hoof disease, foot pad lesions, hock burn, and breast blisters in poultry and areas of worn-off fur or feathers (see Chapters 2 and 4, this volume).

Some examples of poor housing and physical environments that may cause injuries or disease are:

- Slatted floors with the wrong slat spacing that causes lameness and leg injuries (Kilbride et al., 2009)
- Cubicles (freestalls) for dairy cows that are too small and cause hock swelling (Fulwider et al., 2007)
- Housing heavy cattle for long periods on bare concrete that causes lameness
- Wet, dirty litter in poultry barns that causes hock burn and breast blisters
- Muddy feedlots that cause damage to the hide surface and foot rot.

Some examples of good environments are:

- Providing sufficient bedding to prevent soil and manure from transferring on to the animals
- Rubber or plastic flooring that reduces lameness or injuries
- Providing cows with daily access to dry dirt lots or pasture to reduce lameness (Flowers et al., 2007)
- Rotating pastures to reduce parasite transmission
- Providing sufficient ventilation in totally enclosed housing to keep ammonia levels below 10 ppm.

"The physical environment should allow comfortable resting, safe and comfortable movement including normal postural changes and the opportunity to perform types of natural behaviors that animals are motivated to perform" (OIE, 2014a)

The first requirement to be in compliance would be that the animals should be able to turn around and stand and lie in normal positions. This guideline covers the most controversial animal welfare issues. Fraser et al. (2013), in their discussion of the application of this section, did not address the issue of sow gestation stalls where sows are not able to turn around, and small hen cages that prevent a hen from standing at full posture. Fraser et al. (2013) did present strong evidence on the necessity to provide for behavioral needs such as nest building in hens. They made it very clear that hens need a place to perch and have a nesting area. It is the author's opinion that the following systems would be in noncompliance with this guideline – sow gestation stalls that prevent a sow from turning around, and small battery cages for laying hens that prevent a hen from standing in a normal upright position with her head up. Group sow housing and enriched furnished cages (colony housing) would be in compliance. Figure 1.1 shows a hen in an enriched housing system that has sufficient space so she can stand at full posture and flap her wings. A well-designed furnished cage also provides both a secluded nest box and a perch (Tactacan *et al.*, 2009). Hens are highly motivated to use a nest box. It also has a sufficient ceiling to allow normal walking posture. Pigs are motivated to manipulate and root things with their nose (van de Weerd and Day, 2009). They should be given either straw, other fibrous bedding material, or specially designed rooting and chewable objects. Dairy cubicles should be sufficiently wide so that a cow is able to lie with her head in a normal curled back position (uses her body as a pillow) and not damage her legs on the rear concrete curb (Fulwider *et al.*, 2007). Tethered cattle should be able to turn around. Cattle kept in tie-stall housing should have exercise time where they are released from the tether. There are increased risks for compromising welfare in tethered cattle (Loberg *et al.*, 2004; Tucker *et al.*, 2009). Intensive housing facilities for all species should be stocked so that all animals or birds have the ability to sleep at the same time without being on top of each other. All of the above are bare minimum requirements.

Fig. 1.1. Enriched colony housing provides a hen with sufficient space that enables her to stand at full posture and spread her wings. (Photo courtesy of Big Dutchman.)

"Social regrouping of animals should be managed to allow positive social behavior and minimize injury distress and chronic fear" (OIE, 2014a)

Farm animals and poultry are social, and being reared alone or in individual stalls does not allow normal social behavior. On dairies, young calves are often reared in individual stalls for disease control purposes. It is the author's opinion that for a minimum compliance with this guideline, dairy calves should be housed in groups after 6 weeks. Sows in gestation stalls would not comply with this standard. There are differences in pig genetics in aggression (D'Eath *et al.*, 2009). Practical experience has shown that certain hybrid lines of domestic pigs will fight aggressively when mixed. To make a successful switch from gestation stalls to group housing may require a change in pig genetics. For all species, when strange animals are mixed, they should be placed simultaneously in a new pen or pasture to avoid the problem of having another animal invading a resident animal's territory. Small groups of five or six pigs mixed in a small pen may fight more than mixing larger groups. Therefore, small groups of five or six sows are usually kept in the same group for their entire productive life. In larger pens filled with many more animals, an animal that is being attacked has room to escape. Another contentious issue is tie stalls for dairy cattle. These systems are common in many countries, and dairy cows housed in these facilities usually do not have abnormal behavior. It is the author's opinion that dairy cows housed in tie stalls should be let out every day to either pasture or an outdoor lot. The only exception to this recommendation would be severe seasonal weather conditions.

"For housed animals, air quality, temperature and humidity should support good animal health and not be aversive. Where extreme conditions occur, animals should not be prevented from using natural methods of thermo-regulation" (OIE, 2014a)

Outcome-based indicators of environmental problems would be high percentages of animals with eye problems such as conjunctivitis (OIE, 2014c), open mouth panting in response to heat stress (Mader *et al.*, 2006), shivering or huddling in response to cold stress, or death losses. Research shows that, during the summer, providing shade for feedlot cattle improves productivity (Barajas *et al.*, 2013). Black Angus cattle had higher reductions in respiration rate when provided with shade than light tan Charolais cattle (Brown-Brandl *et al.*, 2013). Black cattle have hotter surface temperature compared to breeds with lighter colored coats. High ammonia levels are aversive to poultry and sheep (Phillips *et al.*, 2012). For both human and animal welfare, ammonia levels should not exceed 25 ppm in any type of animal facility. People reported more symptoms of headaches or eye discomfort at 25 ppm compared to 5 ppm or 0 ppm of ammonia (Sundblad *et al.*, 2014). Ammonia should be measured at the level of the animals. On farms where intensively housed animals are completely dependent on mechanical ventilation to prevent heat stress or suffocation deaths, either a backup generator or a method for opening the building sides is required. In buildings with natural ventilation, this is not required. In cold climates, cold stress can often be prevented with windbreaks, deep straw, or shelter (Dronen, 1988; Anderson *et al.*, 2011). Heat stress or cold stress may also be deadly when unacclimated animals are brought to an area with a different climate. An example of death losses due to a lack of acclimation would be bringing cattle with slick summer coats to a cold climate. If these same cattle had been brought in earlier in the season, they would have grown a winter coat before the weather got really cold, and they probably would have been fine. Breeds of animals that originate from the tropics may have problems with cold stress in colder climates. Animal breeds (Holstein dairy cows) developed in colder climates may have more problems with heat stress in hot climates. Breeds that have been developed in the tropics have greater ability to lose heat. Several days or weeks is required for the animal's metabolism to adapt and acclimatize to a change in temperature (Roy and Collier, 2012).

"Animals should have access to sufficient feed and water suited to the animal's age and needs to maintain normal health and productivity and to prevent prolonged hunger, thirst or dehydration" (OIE, 2014a)

This is one area where exclusive use of outcome measurements would be detrimental to welfare because a severe lack of feed or water may result in a slow, stressful death. Ideally, there should be sufficient feeder space so all birds or animals can all eat at the same time. Less feeder space can be provided if feed is always available. Outcome measures of problems with access to feed may be a few animals that are thin, with poor body condition. These animals became skinny because they were pushed away from the feeder by dominant aggressive animals. Sufficient water supplies must be available to supply all animals on the hottest days. In many animals, water requirements may double during hot weather (Arias and Mader, 2011). Beta-agonists fed at high doses may also violate the above principle. Longeragan *et al.* (2014) found that beta-agonists fed during the summer increased death losses in cattle.

"Diseases and parasites should be prevented and controlled as much as possible through good management practices. Animals with serious health problems should be isolated and treated promptly or killed humanely if treatment is not feasible or recovery is unlikely" (OIE, 2014a)

Examples of outcome measure for this guideline would be health records that keep track of the percentages of sick animals, poor body condition, or animals that have high levels of internal or external parasites. Poor condition of the hair coat or areas of hair loss are often signs of external parasites. In some organic or natural programs that forbid the use of antibiotics or other medications, the producer may be tempted to withhold treatment of a sick animal. This is not acceptable from an animal welfare standpoint. Animal production is expanding rapidly in the developing world. There is a great need to train animal caretakers in basic husbandry because some people do not have basic knowledge of simple clinical signs of disease. In a recent bad case, the caretakers did not know that coughing cattle were sick. Bald spots on animals due to hair loss from external parasites or scratching are also not acceptable.

"When painful procedures cannot be avoided, the resulting pain should be managed to the extent possible that available methods allow" (OIE, 2014a)

See Chapters 2 and 6, this volume, for detecting behaviors that may indicate pain. Animals definitely perceive pain, and research supports the use of anesthetics and analgesics (Chapter 6, this volume; Coetzee, 2011, 2013; Stafford and Mellor, 2011). In areas of the world where pain relief is not available, castration and dehorning should be done at a young age. There are some difficult ethical questions in poor countries where antibiotics and anesthetics are not available for people. If I were the mother of a sick child, I would probably steal antibiotics from a pig farm to save my child. When a big corporation starts a pig or poultry farm in a developing country, the animals may have better access to health care than the people. It is the author's opinion that in countries with poor access to health care, large corporations should ensure that their employees also have access to essential medications.

"Owners and handlers should have sufficient skill and knowledge to ensure that animals are treated in accordance with these principles" (OIE, 2014a)

The importance of management commitment to good animal welfare cannot be overemphasized. Often, people want to buy the magic new technology because they mistakenly believe that it will solve all their problems. Technology is NEVER a substitute for good management. It is possible to have a high level of animal welfare in very simple facilities. Training of employees in animal behavior and basic husbandry is essential.

Combining Numerous Welfare Measures to a Single Score is Not Recommended

Newer assessment tools, such as Welfare Quality Network (2009) and Blokhuis et al. (2010), have attempted to combine research on positive emotions with everything listed in Steps 1, 2, and 3 earlier in this chapter. The author commends the inclusion of positive welfare indicators in this assessment, but unfortunately it is too complex for routine commercial use. Welfare Quality Network (2009) is a good example of a valuable scientific research tool. It needs to be simplified for commercial use (Andreasen et al., 2014) because it takes too long to perform. The use of a single aggregate score that combines all the data on both negative and positive welfare factors is not recommended. DeVries et al. (2013) found that when dairies were assessed with the Welfare Quality® Assessment Protocol, a dairy with 47% lame cows had an acceptable rating and another dairy with 25% lame cows got an enhanced rating. Most people would agree that a dairy with 47% lame cows does not have acceptable welfare. Lameness causes pain (Flowers et al., 2007; Rushen et al., 2007). Andreasen et al. (2013) stated that the complex method for aggregating all the scores lacked transparency and was difficult for consumers to understand. Vessier et al. (2011) explained how the aggregate scoring for the welfare quality system was developed based on ethical concerns.

Critical Control Points to Prevent Poor Welfare

The author suggests that a more practical approach is to have certain critical control points (core criteria) where an acceptable score is required on ALL of them to achieve an acceptable welfare score. Another name for this method is critical noncompliances. A farm that has an acceptable score on the critical noncompliance should then be able to get extra points for the factors that indicate positive emotions such as low fear, presence of play, or social grooming. Some of the obvious welfare basics are: no acts of abuse and low percentages of lame, sick, skinny, dirty, or injured animals. Scoring of animal handling (Chapter 5, this volume) for low percentages of animals falling or vocalizing during movement and restraint for veterinary procedures are additional basics required for a minimum acceptable level of animal welfare. In order to get an overall acceptable score, a farm should have an acceptable score on all the basic critical control points. They could then get extra points for high scores on assessments of positive emotions. To have a minimum acceptable level of animal welfare, a farm should have low percentages of affected animals on the following critical control points:

- lameness (difficulty in walking)
- no acts of abuse by stockpeople
- dirty animals
- animals with injuries
- sick animals
- emaciated animals
- injuries and lesions on both production and work animals
- poor air quality that causes eye or lung irritation
- abnormal stereotypic behavior
- tail biting or feather picking.

Chapters in this book will also discuss how to measure welfare issues on the farm. There are also chapters on both assessment and how to perform critical procedures such as euthanasia, transport, slaughter, and pain mitigation for surgical procedures such as castration. Additional chapters cover the importance of ethics, good stockmanship, understanding behavioral needs, and care of working animals.

References

American Meat Institute (2013) Recommended Animal Handling Guidelines and Audit Guide: A Systematic Approach to Animal Welfare. Available at: http://www.animalhandling.org/ht/d/sp/i/26752/pid/26752 (accessed 9 May 2015).

Anderson, V., Ilse, B., Stoltenow, C., Burr, D., Schroeder, T., et al. (2011) Winter Management of Feedlot Cattle. North Dakota State University, Extension Service, Fargo, North Dakota.

Andreasen, S.N., Wernelsfelder, F., Sandøe, P. and Forkman, B. (2013) The correlation of qualitative behavior assessments with Welfare Quality® protocol outcomes in on-farm welfare assessment of dairy cattle. Applied Animal Behaviour Science 143, 9–17.

Andreasen, S.N., Sandøe, P. and Forkman, B. (2014) Can animal-based welfare assessment be simplified? A comparison of the Welfare Quality® protocol for dairy cattle and the simpler less time-consuming protocol developed by the Danish Cattle Federation. Animal Welfare 23, 81–94.

Arias, R.A. and Mader, T.L. (2011) Environmental factors affecting daily water intake on cattle finished in feedlots. Journal of Animal Science 89, 245–251.

AVMA (2013) AVMA Guidelines for the Euthanasia of Animals: 2013 Edition. American Veterinary Medical Association, Schaumberg, Illinois.

Barajas, R., Garces, P. and Zinn, R.A. (2013) Interactions of shade and feeding management on feedlot performance of crossbred steers during seasonal periods of high ambient temperatures. The Professional Animal Scientist 29, 645–651.

Blokhuis, H.J., Veissier, I., Miele, M. and Jones, B. (2010) The Welfare Quality® project and beyond: Safeguarding farm animal well-being. Acta Agriculturae Scandinavica 60, 129–140.

Boissy, A., Manteuffel, G., Jensen, M.B., Oppermann Moe, R., Spruijt, B., et al. (2007) Assessment of positive emotions in animals to improve their welfare. Physiology & Behavior 92, 375–397.

Broom, D.M. (2011) A history of animal welfare science. Acta Biotheoretica 59, 121–137.

Brown-Brandl, T.M., Eigenberg, R.A. and Nienaber, J.A. (2013) Benefits of providing shade to feedlot cattle of different breeds. Transactions of the ASABE 56, 1563–1570.

Caplen, G., Hothersall, B., Murrell, J.C., Nicol, C.J., Waterman-Pearson, A.E., et al. (2012) Kinematic analysis quantifies gait abnormalities associated with lameness in broiler chickens and identifies evolutionary gait differences. PLoS One doi: 10.1371/journal.pone.0040800.

Coetzee, J.F. (2011) A review of pain assessment techniques and pharmacological approaches to pain relief after bovine castration: Practical implications for cattle production within the United States. Applied Animal Behaviour Science 135, 192–213.

Coetzee, J.F. (2013) Assessment and management of pain associated with castration in cattle. Veterinary Clinics of North America Food Animal Practice 29, 75–101.

D'Eath, R.B., Roehe, R., Turner, S.P., Ison, S.H., Farish, M., et al. (2009) Genetics of animal temperament: Aggressive behaviour at mixing is genetically associated with response to handling in pigs. Animal 3, 1544–1554.

De Vries, M., Bokkers, E.A.M., van Schaik, G., Botreau, R., Engel, B., et al. (2013) Evaluating the results of the welfare quality multi-criteria evaluation model for classification of dairy cattle welfare at the herd level. Journal of Dairy Science 96, 6264–6273.

Douglas, C., Bateson, M., Walsh, C., Bédué, A. and Edwards, S.A. (2012) Environmental enrichment induces optimistic cognitive biases in pigs. Applied Animal Behaviour Science 139, 65–73.

Dronen, S.I. (1988) Layout and design criteria for livestock windbreaks. In: Brandle, J.R., Hintz, D.L. and Sturrock, J.W. (eds) Windbreak Technology. Elsevier, Amsterdam, pp. 231–240.

Duncan, I.J. (1998) Behavior and behavioral needs. Poultry Science 77, 1766–1772.

European Union (2012) The Animal Welfare Officer in the European Union. Regulation 1099/2009, Directorate General for Health and Consumers European Commission – 3-1048, ISBN 979-92-79-26103-9. Brussels.

FAWC (Farm Animal Welfare Council) (1992) FAWC updates – the five freedoms. Veterinary Record 131, 357.

Flower, F.C., de Passillé, A.M., Weary, D.M., Sanderson, D.J. and Rushen, J. (2007) Softer, higher-friction flooring improves gait of cows with and without sole ulcers. Journal of Dairy Science 90, 1235–1242.

Fraser, D. (2008) Understanding Animal Welfare: The Science in its Cultural Context. Wiley Blackwell, USA.

Fraser, D., Duncan, I.J., Edwards, S.A., Grandin, T., Gregory, N.G., et al. (2013) General principles for the welfare of animals in production systems: The underlying science and its application. The Veterinary Journal 198, 19–27.

Fulwider, W.K., Grandin, T., Garrick, D.J., Engle, T.E., Lamm, W.D., et al. (2007) Influence of free-stall base on tarsal joint lesions and hygiene in dairy cows. Journal of Dairy Science 90, 3559–3566.

Galukande, E., Mulindwa, H., Wurzinger, M., Roschinsky, R., Mwai, A.O., *et al*. (2013) Cross-breeding cattle for milk production in the tropics: Achievements, challenges and opportunities. *Animal Genetic Resources* 52, 111–125.

Grandin, T. (2000) Effect of animal welfare audits of slaughter plants by a major fast food company on cattle handling and stunning practices. *Journal of the American Veterinary Medical Association* 216, 848–851.

Grandin, T. (2005) Special Report: Maintenance of good animal welfare standards in beef slaughter plants by use of auditing programs. *Journal of the American Veterinary Medical Association* 226, 370–373.

Grandin, T. and Deesing, M.J. (2013) Genetics and animal welfare. In: Grandin, T. and Deesing, M.J. (eds) *Genetics and the Behavior of Domestic Animals* (2nd edition). Academic Press, San Diego, California, pp. 435–472.

Grandin, T., Curtis, S.E., Widowski, T.M. and Thurmon, J.C. (1986) Electro-immobilization versus mechanical restraint in an avoid–avoid choice test for ewes. *Journal of Animal Sciences* 62, 1469–1480.

Green, L.E., Huxley, J.N., Banks, C. and Green, M.J. (2014) Temporal associations between low body condition lameness and milk yield in a UK dairy herd. *Preventive Veterinary Medicine* 113, 63–71.

Greer, A.W. (2008) Trade-offs and benefits: Implications of promoting a strong immunity to gastrointestinal parasites in sheep. *Parasite Immunology* 30, 123–132.

Hemsworth, P.H. and Coleman, G.J. (2010) *Human–Livestock Interactions, The Stockperson and the Productivity and Welfare of Intensively Farmed Animals* (2nd edn). CAB International, Wallingford, UK.

Huertas, S.M., Gil, A.D., Piaggio, J.M. and van Eerdenburg, F.J.C.M. (2010) Transportation of beef cattle to slaughterhouses and how this relates to animal welfare and carcase bruising in an extensive production system. *Animal Welfare* 19, 281–285.

Jiang, C., Xing, F., Xing, J., Jiang, Y. and Zhou, E. (2013) Different expression patterns of PRRSV mediator genes in the lung tissues of PRRSV resistant and susceptible pigs. *Developmental and Comparative Immunology* 39, 127–131.

Keeling, L., Rushen, J. and Duncan, I.J.H. (2011) Understanding animal welfare. In: Appleby, M.C., Mench, J.A., Olsson, I.A.S. and Hughes, B.O. (eds) *Animal Welfare*. CAB International, Wallingford, UK, pp. 13–26.

Kilbride, A.L., Gillman, C.E., Ossent, P. and Green, L.E. (2009) Across sectional study of the prevalence of lameness in finishing pigs gilts and pregnant cows, and associations with limb lesions, and floor types on commercial farms in England. *Animal Welfare* 18, 215–224.

Limon, G., Guitian, J. and Gregory, N.G. (2010) An evaluation of the humaneness of puntilla in cattle. *Meat Science* 84, 352–355.

Loberg, J., Telezhenko, E., Bergsten, C. and Lidfors, L. (2004) Behavior and claw health in tied dairy cows with varying access to exercise in an outdoor paddock. *Applied Animal Behaviour Science* 89, 1–16.

Longeragan, G.H., Thomson, D.U. and Scott, H.M. (2014) Increased mortality in groups of cattle administered B-adrenergic agonists ractopamine hydrochloride and zilpaterol hydrochloride. *PLoS One* 9(3), e9117, doi:10.1371/journal.pone.0091177.

Mader, T.L., Davis, M.S. and Brown-Brandl, T. (2006) Environmental factors influencing heat stress in feedlot cattle. *Journal of Animal Science* 84, 712–719.

Mellor, D.J. and Webster, J.R. (2014) Development of animal welfare understanding drives change in minimum welfare standards. *Review Scientific Technical Office of International Epizootics* 33, 121–130.

Morris, C.L., Grandin, T. and Irlbeck, N.A. (2011) Companion Animals Symposium: Environmental enrichment for companion, exotic, and laboratory animals. *Journal of Animal Science* 89, 42237–4238.

OIE (World Organisation for Animal Health) (2014a) Article 7.1.4. General principles for the welfare of animals in livestock production systems. In: Chapter 7.1 Introduction to the Recommendations for Animal Welfare. *Terrestrial Animal Health Code*. OIE, Paris.

OIE (2014b) Article 7.5.2. Moving and handling animals. In: Chapter 7.5 Slaughter of Animals. *Terrestrial Animal Health Code*. OIE, Paris.

OIE (2014c) Chapter 7.10. Animal welfare and broiler chicken production systems. In: Section 7 Animal Welfare. *Animal Health Code*. OIE, Paris.

Panksepp, J. (2011) The basic emotional circuits of mammalian brains, Do animals have affective lives? *Neuroscience and Biobehavioral Reviews* 35, 1791–1804.

Pascoe, P.J. (1986) Humaneness of an electro-immobilization unit for cattle. *American Journal of Veterinary Research* 47, 2252–2256.

Phillips, C.J., Pines, M.K., Latter, M., Muller, T., Petherick, J.C., *et al*. (2012) Physiological and behavioral responses of sheep to gaseous ammonia. *Journal of Animal Science* 90, 1562–1569.

Rodenburg, T.B. and Turner, S.P. (2012) The role of breeding and genetics in the welfare of farm animals. *Animal Frontiers* 2, 16–21.

Roy, K.S. and Collier, R.J. (2012) Regulation of acclimation to environmental stress. In: Collier, R.J. and Collier, J.L. (eds) *Environmental Physiology of Livestock*. Wiley Blackwell, USA, pp. 49–64.

Rushen, J., Pombourcq, E. and de Passillé, A.M. (2007) Validation of two measures of lameness in dairy cows. *Applied Animal Behaviour Science* 106, 173–177.

Rushen, J., Butterworth, A. and Swanson, J.C. (2011) Animal behavior and well-being symposium: Farm animal welfare assurance: science and application. *Journal of Animal Science* 89, 1219–1228.

Rutherford, K.M., Donald, R.D., Lawrence, A.B. and Wemelsfelder, F. (2012) Qualitative behavioural assessment of emotionality in pigs. *Applied Animal Behavior Science* 139, 218–224.

Spencer, T.E. (2013) Early pregnancy: Concepts, challenges, and potential solutions. *Animal Frontiers* 3, 48–55.

Stafford, K.J. and Mellor, D.J. (2011) Addressing the pain associated with disbudding and dehorning in cattle. *Applied Animal Behaviour Science* 135, 226–231.

Sundblad, B.M., Larssen, B.M., Acevedo, F., Ernstgard, L., Johanson, G., *et al.* (2004) Acute respiratory effects of exposure to ammonia on healthy persons. *Scandinavian Journal of Work Environmental Health* 30, 313–321.

Tactacan, G.B., Guenter, W., Lewis, N.J., Rodriguez-Lecompte, J.C. and House, J.D. (2009) Performance and welfare of laying hens in conventional and enriched cages. *Poultry Science* 88, 698–707.

Thatcher, W., Santos, J.E. and Staples, C.R. (2011) Dietary manipulations to improve embryonic survival in cattle. *Theriogenology* 76, 1619–1631.

Tucker, C.B., Weary, D.M., von Keyserlingk, M.A. and Beauchemin, K.A. (2009) Cow comfort in tie-stalls: Increased depth of shavings or straw bedding increases lying time. *Journal of Dairy Science* 92, 2684–2690.

van de Weerd, H.A. and Day, J.E.L. (2009) A review of environmental enrichment for pigs housed in intensive housing systems. *Applied Animal Behaviour Science* 116, 1–20.

Velarde, A. and Dalmau, A. (2012) Animal welfare assessments at slaughter in Europe: Moving from inputs to outputs. *Meat Science* 92, 244–251.

Veissier, I., Jensen, K.K., Botreau, R. and Sandøe, P. (2011) Highlighting ethical decisions underlying the scoring of animal welfare in the Welfare Quality® scheme. *Animal Welfare* 20, 89–101.

von Keyserlingk, M.A.G., Barrientos, A., Ito, K., Galo, E. and Weary, D.M. (2012) Benchmarking cow comfort on North American freestall dairies: Lameness, leg injuries, lying time, facility design, and management for high-producing Holstein dairy cows. *Journal of Dairy Science* 95, 7399–7408.

Vordermeier, M., Ameni, G., Berg, S., Bishop, R., Robertson, B.D., *et al.* (2012) The influence of cattle breed on susceptibility to bovine tuberculosis in Ethiopia. *Comparative Immunology, Microbiology and Infectious Diseases* 35, 227–232.

Walker, M., Díez-León, M. and Mason, G. (2014) Animal welfare science: Recent publication trends and future research priorities. *International Journal of Comparative Psychology* 27, 80–100.

Walsh, S.W., Williams, E.J. and Evans, A.C. (2011) A review of the causes of poor fertility in high milk producing dairy cows. *Animal Reproduction Science* 123, 127–138.

Webster, J. (2012) Critical control points in the delivery of improved animal welfare. *Animal Welfare* 21, 117–123.

Welfare Quality Network (2009) Available at: http://www.welfarequalitynetwork.net/network (accessed 22 April 2014).

Whay, H.R., Main, D.C.J., Green, L.E. and Webster, A.J.F. (2003) Assessment of the welfare of dairy cattle using animal-based measurements: Direct observations and investigation of farm records. *Veterinary Record* 153, 197–202.

Wilkins, L.J., McKinstry, J.L., Avery, N.C., Knowles, T.G., Brown, S.N., *et al.* (2011) Influence of housing system and design on bone strength and keel bone fractures in laying hens. *Veterinary Record* 169, 414.

Yilmaz, O., Cengiz, F., Ertugrul, M. and Wilson, R.T. (2013) The domestic livestock resources of Turkey: Sheep breeds and cross-breeds and their conservation status. *Animal Genetic Resources* 52, 147–163.

2 The Importance of Measurement to Improve the Welfare of Livestock, Poultry, and Fish

TEMPLE GRANDIN

Department of Animal Science, Colorado State University, Fort Collins, Colorado, USA

Summary

It is essential to measure serious animal welfare problems such as lameness or poor handling practices continuously. When numerical scoring is used, managers can determine if conditions are improving or becoming worse. Surveys indicate that there are big differences between the very best and worst producers. For example, the best 20% of dairy farms had 0% cows with swollen legs and the worst 20% had 7.4–12.5%. Measurement can help prevent bad practices from becoming routine. Animal handling should be measured continuously to maintain high standards. Training people and simple improvements in equipment can reduce bruises, animals falling during handling, and electric goad usage greatly. Vocalization (moos and bellows) scoring in cattle or squeals in pigs is a useful measure of severe stress during handling and restraint. Improvements in equipment and handling can reduce greatly the percentage of cattle or pigs that vocalize during restraint. The chapter also contains behavioral measures for pain that can be performed easily on the farm. It also discusses the importance of reducing fear stress.

Learning Objectives

- Learn that measurement indicates big differences between the best and poorest producers
- Good health alone does not guarantee good welfare
- Learn about problems with biological overload
- Measurement of pain and distress
- Use of measurement to show either improvement or worsening of practices

Introduction – Manage Things You Measure

Livestock producers measure weight gain, death losses, and sickness routinely, but they may not be measuring painful or distressing conditions such as lameness (difficulty in walking), falling during handling, bruises, or electric goad use, which severely compromises an animal's welfare. Measuring these indicators on a regular basis makes it possible to determine if conditions are improving or becoming worse. Lameness, which causes an abnormal walking gait, is one of the most serious welfare problems in many species of livestock and poultry. Lameness definitely causes pain, because giving dairy cows the anesthetic, lidacaine, reduces it (Flower *et al.*, 2007; Rushen *et al.*, 2007). People often fail to be effective managers of conditions that they do not measure. Lameness in intensively housed dairy cattle is a good example.

Dairy cows and sow lameness

Over a period of many years, lameness in dairy cows housed on concrete has become steadily worse. One of the reasons why this has happened is that relatively few dairies measure walking ability in dairy cows until lameness becomes really bad. A British study showed that 16.2% of the dairy cows were lame (Rutherford *et al.*, 2009). Cows housed in freestall (cubicle) barns had an average of 24.6% of clinically lame cows (Espejo *et al.*, 2006). However, lameness was only 5.4% in the top 10% of dairies (Espejo *et al.*, 2006). Von Keyserlingk *et al.* (2012) found in a more recent study that the average percentage of lame dairy cows had not improved. In Canada and the northern USA, it was over 27%. In England and Wales, the average percentage of lame dairy cows was 36.8% (Barker *et al.*, 2010). A British survey of 53 dairies indicated that in the best 20% of the dairies, only 0–6% of the cows were lame and in the worst 20% of the dairies, 33–62% of the cows were lame (Webster, 2005a,b). This shows that good management can reduce lameness.

The author observed a big increase in lame slaughter-weight pigs between 1995 and 2008. A major breeder of lean, rapidly growing pigs in the USA did nothing about it until 50% of the slaughter-weight pigs in some herds were clinically lame with very poor leg conformation. This breeder was selecting for leanness, loin eye size, and rapid growth, and over a 10-year period they did not notice that there were more and more lame pigs. The increase in lameness was mainly genetic, because the pigs were all housed on the same concrete slats that had been used for years. Some breeders are now selecting breeding stock with better leg conformation, but since the first edition of this book, the author is still observing lame market-weight pigs with structural leg problems. A US study indicated that 21% of the sows were lame (Vansickle, 2008), and in the UK, 8.1% were lame after weaning (Pluym *et al.*, 2013). A study of sows in Minnesota indicated that risk of removal from the breeding herd increased when leg conformation was poor. Culling of breeding sows that was attributable to poor legs was 16.37% for the forelimbs and 12.90% for the back legs (Tiranti and Morrison, 2006). Lameness was a major reason for culling (Kirk *et al.*, 2005; Pluym *et al.*, 2013). A study carried out in Spain showed that poor leg conformation was associated with higher sow culling rates (deSeville *et al.*, 2008). Selecting breeding gilts with structurally correct feet and legs will provide both better welfare and productivity.

Preventing "bad from becoming normal"

How could dairy cows and pigs become so severely lame before anything was done to correct it? Surveys on dairies show that producers often underestimate the percentage of lame cows, and may not view lameness as a major welfare problem (Leach *et al.*, 2010; Bennett *et al.*, 2014). The people who were breeding the pigs or managing the dairies were only looking at their own animals. They seldom saw other groups of pigs or cows with which to compare their animals. Since the increase in lameness occurred slowly over a period of years, they failed to notice it because they were not measuring it. This is a good example of a situation the author calls "bad becoming normal". The reason why the author noticed the increase in lame pigs and the pig breeders failed to see it was due to the author's observations of thousands of pigs in many different slaughter plants. Observations of slaughter-weight pigs from many different breeders but raised in similar buildings indicated that there were large differences between breeders on the percentage of lame pigs and leg conformation defects. To correct this problem, progressive dairy producers and pig breeders have incorporated formal lameness measurements and leg conformation evaluations into their programs.

Leg problems in beef cattle

Since the first edition of this book, the author has started observing similar structural lameness problems in Angus beef cattle. Beef cattle breeders are repeating the mistakes that were made by the pork industry. Data collected in 2013 by a US packer indicated that up to 10% of fed beef cattle were lame. These animals were kept in a dry feedlot with a dirt surface that remained dry. They were never fed beta-agonists. Lameness is one of the most serious animal welfare problems. Figure 2.1 contains drawings of both correct leg conformation and different types of structurally poor legs in pigs and cattle. Charts like this can be obtained easily for all livestock species and poultry from the Internet. Cattle, sheep, and other mammals have similar problems with leg conformation. They should be used when breeding stock is being selected.

Handling Practices Deteriorate Unless Measured

On large farms and slaughter plants where many different people handle animals, the author has observed that animal handling practices may deteriorate slowly and become rougher and rougher without anyone realizing it. Over the years, the author has given many seminars on low stress handling and quiet movement of pigs and cattle through corral systems on ranches, feedlots, and slaughter plants. Many managers were eager to implement the new methods. Employees were taught to use behavioral principles of animal movement, stop yelling and greatly reduce the use of electric goads. A year later,

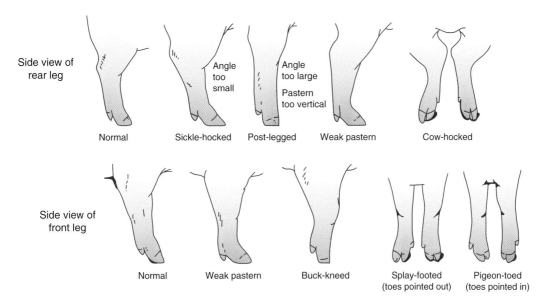

Fig. 2.1. Leg conformation chart – showing poor conformation. Diagrams of common poor conformation of the feet and legs of cattle, pigs, and other livestock. Poor leg and hoof structure is a major cause of lameness (difficulty in walking). Breeders should select animals with the normal correct hoof and lower leg angle. An animal's hoof should point straight forward. The two diagrams on the bottom right show poor conformation with toes pointed either in or out.

when the author returned to re-evaluate the handling practices, it was discouraging to observe that many employees had reverted back to their old rough ways. When the manager was informed that the animal handling methods were bad, he was surprised and upset. Since the regression back into old rough ways had happened slowly over the course of a year, the manager did not notice the slow deterioration of handling practices. Bad practices had become normal because the manager had not measured handling practices in an objective manner. In Chapters 4 and 5 of this volume, an easy method to measure livestock handling objectively will be covered.

The Power of Comparison Measures

When large numbers of farms are surveyed, practices and conditions can range from excellent to atrocious. Fulwider *et al.* (2007) surveyed 113 freestall (cubicle) dairies on the incidence of lesions and swellings on the dairy cow's legs. Figure 2.2 shows a cow that would be scored for a severe swelling. Table 2.1 shows the big differences between the best 20% of the farms and the worst 20%. The best 20% had 0% of the cows with swollen hocks and on the worst 20% of the dairies, 12.5–47% of the cows had swollen hocks. In a Canadian study of 317 tie-stall

Fig. 2.2. Dairy cow with a severe leg lesion that has a diameter greater than 7.4 cm (size of a baseball). Photographs like this should be put on laminated cards for scoring leg lesions on cows. This cow is also splay footed, where the toes point outward. The toes should point straight ahead. See Quick Access Information 2.1 (Photograph courtesy of Wendy Fulwider).

dairies, 26% of the dairies were well managed and had no open wounds on the hocks. However, 16% of the dairies were really bad and 15% or more of the cows had open hock wounds (Zurbrigg *et al.*, 2005a,b). In another study of 53 dairy farms in the UK, the best 20% had 0–13.6% lame cows and the

Table 2.1. Freestall (cubicle) dairies sorted by the best 20% to the worst 20% of farms for each welfare issue in 113 dairies. (Adapted from Fulwider *et al.*, 2007.)

Well-being issue	Percentage of cows on				
	Best 20% of farms	Second best 20% of farms	Middle 20% of farms	Second worst 20% of farms	Worst 20% of farms
Hock hair loss only	0–10	10.6–20	20.8–35.8	36.2–54.4	56–96.1
Hock swelling	0	0.7–1.7	1.9–4.2	4.2–11.9	7.4–12.5
Severe swelling[a]	0	0	0	0–1.5	1.8–10.7
Dirty cows[b]	0–5	5.3–9.8	10.3–15.4	16.8–28.9	29.4–100
Thigh lesions	0	0	0	0	0–28.8

Notes: [a]Cows were rated on having a severe swelling if the worst leg had a swelling more than 7.4 cm (size of a baseball) in diameter, or open or oozing injuries; [b]Cows were rated as dirty if there was dried or wet manure on their body, belly, udder, or upper portions of the leg.

worst 20% had 34.9–54.4% lame (Whay *et al.*, 2003). A panel of dairy veterinarians agreed that lameness and swollen or ulcerated hocks were the most serious problems that needed to be corrected (Whay *et al.*, 2003). It is likely that many of the dairy producers in the worst groups did not realize how bad they were compared to the other 80% of their colleagues. Research undertaken in Ontario, Canada, showed that on 40% of the farms, there was 0% broken tails and on the worst 20%, there was 5–50% of the cows with broken tails (OMAFRA, 2005).

A huge survey conducted by Knowles *et al.* (2008) in the UK showed that 27.6% of chickens were lame, with a score of 3 or higher on a 6-point scale. The scores ranged from normal walking gait to down and unable to walk. A score 3 bird is mobile, but obviously lame. There was a big difference between five different chicken companies and there was also a lot of variation between the best and the worst flocks. The best flock had 0% score of 3 or worse lame birds, and the worst flock had 83.7% obviously lame birds. The standard deviation for data collected on 176 flocks was 24.3% for score 3 lameness. A high standard deviation indicates that there were huge differences between the best and the worst farms. Unpublished industry data in the USA from chicken farms that were being audited by major customers had only 2% obviously lame 3-kg chickens. Lameness causes pain. Giving broilers NSAID pain medication improved walking ability (Caplen *et al.*, 2013).

Numerical Scoring Enables Quantification of Improved Practices

Grandin (2005, 2010, 2012) discussed the implementation of measurements of practices in slaughter plants as a method for auditing animal welfare.

This measurement system is now used by McDonald's Corporation, Wendy's, Tesco International, Burger King, Whole Foods, and many other large meat-buying customers. Instead of an auditor subjectively determining whether or not a plant has good or bad practices, stunning and handling is evaluated with objective numerical scoring. Before the audits began, it was routine to use an electric prod multiple times on every animal, and in some plants, the stun guns were not kept repaired.

Baseline data prior to the audits indicated that only 30% of the plants could stun 95% of the cattle with a single shot from a captive bolt. After the audits began, and the plants became concerned about losing a major customer, the percentage of plants that could achieve this rose to over 90% (Fig. 2.3) (Grandin, 1998a, 2005). Both audit and survey data indicated that a lack of maintenance of the captive bolt was a major cause of poor stunning.

Similar dramatic results were obtained with the reduction of electric goad use. Usage of electric goads dropped from multiple shocks administered to every animal to more than 75% of the cattle and pigs moving through the entire plant with no shocks (Grandin, 2005). Similar studies have been done in South America. Training employees in animal handling will reduce bruises and damage to carcasses (Pilecco *et al.*, 2013; Paranhos de Costa *et al.*, 2014). The education of handlers and improved procedures have resulted in a huge reduction both in cattle trampling on top of another animal and in lost vaccine. Carefully restraining each animal one at a time in the head stanchion and squeeze was compared to vaccinating a long line of animals held in a race. Restraining each animal individually reduced trampling from 10% to 0%, and lost vaccine dropped from 7% to 1% (Chiquitelli Neto *et al.*, 2002; Paranhos de Costa, 2008).

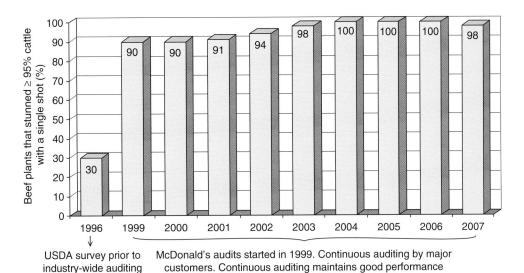

Fig. 2.3. Percentage of beef plants that stunned 95% or more cattle with the first shot. The baseline scores were measured in 1996 in ten US beef slaughter plants. After the restaurant audits started in 1999, the number of plants audited each year varied from 41 to 59. After the first 4 years of auditing, the scores further improved because the plants now had documented stunner maintenance programs and test stands that measured the bolt velocity of the stunner.

Measuring Improvements within a Farm or Plant

Objective outcome measurements can also be used by the managers of farms and plants to quantify improvements, and continuous internal audits prevent practices from slipping back into old sloppy or rough ways. Measurements done every week or month make it easy to determine if handling practices are improving, staying the same, or becoming slowly worse. Measurements on a regular basis of welfare-related issues such as lameness or leg lesions will enable farm managers to determine if their veterinary care, bedding, and husbandry programs are improving or becoming worse.

Measurements can also be used to determine if a new piece of equipment, a new procedure, or a repair has made an improvement. Figure 2.4 illustrates that a simple modification such as adding a light to the entrance of a race makes it possible to reduce greatly the use of electric prods due to pigs balking and refusing to enter the race. The addition of a lamp reduces balking and fewer pigs have to be shocked. Pigs have a natural tendency to approach illuminated areas (Van Putten and Elshof, 1978; Grandin, 1982; Tanida *et al.*, 1996). Measurement also makes it possible to locate producers who have animals that are difficult to handle. Figure 2.5

shows that some groups of pigs are more difficult to drive, so there is increased electric goad use and squealing. In Chapter 4, this volume, the measuring tools that can be used to assess welfare will be covered in more detail.

Good Health Does Not Guarantee Good Welfare

Many people mistakenly assume that good health automatically means that an animal has satisfactory welfare. Good health is a key component for an animal to have good welfare. The World Organisation for Animal Health (OIE, 2014a,b) code states, "Good welfare requires disease prevention and veterinary treatment". However, there are some situations where an animal can be healthy but its welfare may be poor. For example, a dairy cow that is healthy, disease free and providing lots of milk could have painful lesions on her legs from lying in a poorly bedded stall that did not provide enough cushioning to prevent injuries. Fulwider *et al.* (2007) and Green *et al.* (2014) found that high-producing dairy cows had more leg lesions, a shorter productive life, and poorer body condition. Some genetic lines of chickens that have been bred to grow quickly have a high rate of lameness and leg abnormalities (Knowles *et al.*, 2008). Another example of poor welfare would

Fig. 2.4. Electric prod use on pigs was reduced by adding lighting at the restrainer entrance. Simple changes such as placing a light on the entrance of a stunning race greatly reduced balking and refusal to enter the races, so less use of an electric goad was required. All handlers were well trained and only pigs that balked or backed up were prodded.

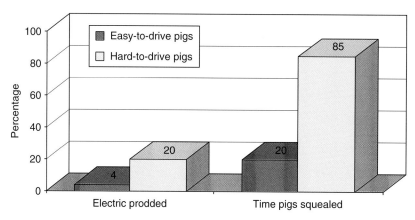

Fig. 2.5. Comparison of electric goad use and squealing between easy-to-drive pigs and hard-to-drive pigs when moved by well-trained people who only used the electric goad on pigs that refused to move. Some pigs are more excitable and difficult to handle compared to other pigs. Measurement of each producer's pigs can be used to locate problem pigs that balk more and are more difficult to move through races. Squealing can be measured in small plants by counting the number of pigs that squeal. In large plants, this is difficult, so the percentage of time the room is quiet is measured.

be healthy laying hens with high levels of osteoporosis and bone fractures (Sherwin *et al.*, 2010; Wilkins *et al.*, 2011). They keep laying eggs, even though the cages are so tightly stocked that the hens cannot all lie down at the same time without being on top of each other.

Some animals that have been bred for rapid weight gain are difficult to transport and handle in a manner that ensures good welfare. Certain genetic lines of pigs that have been bred for rapid weight gain become easily fatigued and weak when grown to heavy

weights of 130 kg. The author has also observed healthy pigs that had been fed high doses of Paylean® (ractopamine) that were too weak to walk from one end of a lairage to the other. This is another example of bad conditions that some people perceive as normal. The author observed pigs in the 1980s that were strong enough to walk up long, steep ramps.

Healthy animals may also have abnormal behavior if they are housed in an environment that does not allow them to express normal social and species-specific behaviors. Some examples of abnormal

T. Grandin

behavior are pacing in a circle, bar biting in sows housed in gestation stalls, tail biting, and pulling out feathers or hair. The behavioral needs of animals will be discussed in Chapter 8, this volume. There are also many excellent books on farm animal behavior and ethology that review research on behavioral needs (Broom and Fraser, 2007; Fraser, 2008b; Blokhuis *et al.*, 2013).

Reducing Pain, Fear, and Distress

Many welfare problems cause pain and distress, and pain can be measured. There are numerous scientific studies, review papers, and books which document that livestock and poultry feel both pain and fear (Gentle *et al.*, 1990; Rogan and LeDoux, 1996; Grandin, 1997; Grandin and Johnson, 2005, 2009; Panksepp, 1998, 2011). Both rats and chickens will self-medicate for pain (Danbury *et al.*, 2000; Colpaert *et al.*, 2001; Pham *et al.*, 2010). They will eat or drink bitter-tasting food or water containing painkillers when they are lame and have sore leg joints. To make it easier to implement a welfare program, pain and distress items can be grouped into four main categories. They are: (i) routine painful procedures such as castration or beak trimming; (ii) situations such as handling and restraint that cause fear; (iii) separation distress that occurs when animals are weaned or separated from their herdmates; and (iv) chronic long-term painful conditions such as lameness. Pain, fear, and separation distress are processed in separate brain systems (Panksepp, 1998, 2011).

Painful routine procedures in livestock and poultry

The pain and distress category includes all of the painful procedures that producers routinely perform on animals (Quick Access Information 2.1). Many of these procedures are done with no anesthetics or painkillers. Numerous research studies on dehorning of livestock clearly indicate the need for pain relief (Faulkner and Weary, 2000; Stafford and Mellor, 2005a; Coetzee *et al.*, 2012). Cautery disbudding young calves is strongly recommended. Stafford and Mellor (2005b) report that removing the horns on an older animal causes greater increases in cortisol compared to disbudding calves. Other research has shown the welfare benefits of providing painkillers for castration. There has been much controversy

---

Quick Access Information 2.1. Routine painful management procedures.

- Beak trimming in poultry
- Spaying female animals
- Castrating male animals
- Dehorning
- Notching ears for identification
- Removing tusks on boars
- Hot iron brand
- Clipping needle teeth on piglets
- Docking dairy cow and pig tails
- Mutilating and making big cuts in ears to identify animals
- Mulesing sheep
- Wattling – cutting flaps of skin for identification
- Tail docking
- Hot iron branding

---

and discussion about whether or not certain procedures should be done at all. For example, selecting beef cattle to be polled (hornless) eliminates the need for dehorning. Further discussion of painful management procedures is in Chapter 6, this volume. Animal welfare legislation and animal welfare standards will vary greatly in different countries. However, most researchers and veterinary organizations agree that major surgery where the main body cavity is opened, such as flank spaying a heifer, requires the use of an anesthetic.

Quick Access Information 2.2 shows behaviors that are associated with pain that are easy for people to score and quantify. The behaviors in Quick Access Information 2.2 occur AFTER the procedure has been done in cattle, calves, pigs, and lambs. Animals should be scored immediately after a procedure is done, for a minimum of 1 h, to detect signs of acute pain.

To detect signs of long-term pain, scoring can also be done over a period of days. The behaviors listed in Quick Access Information 2.2 are associated with physiological measures of pain and stress (Molony and Kent, 1997; Eicher and Daly, 2002; Sylvester *et al.*, 2004; Stafford and Mellor, 2005b; Viñuela-Fernández *et al.*, 2007; Stillwell *et al.*, 2012). The behaviors associated with different painful surgeries will vary depending on the procedure and the species. Quantifying pain-related behavior provides an easy economical way to evaluate welfare in large numbers of animals. Animals often conceal pain-related behavior when they see a person watching. This is a natural behavior for cattle and sheep, which helps to protect them

> **Quick Access Information 2.2. Behaviors associated with pain that are easy to quantify numerically in lambs, cattle, calves, piglets, and other animals. Evaluate these behaviors AFTER the painful procedure has been done.**
> **(From Molony and Kent, 1997; Eicher and Daly, 2002; Sylvester *et al.*, 2004; Stafford and Mellor, 2005a; Viñuela-Fernández *et al.*, 2007; Stillwell *et al.*, 2012).**
>
> | Behaviors associated with pain | Species[a] |
> |---|---|
> | Time in contorted abnormal lateral or ventral recumbency | Lambs, calves, cattle |
> | Time in lateral recumbency | Lambs, calves, cattle |
> | Number of times foot stamping | Lambs, calves, cattle |
> | Number of kicks | Lambs |
> | Number of lip curls | Lambs |
> | Number of ear flicks | Calves, cattle |
> | Number of tail switches (wags) | Cattle, calves |
> | Time standing still like a statue | Cattle, calves |
> | Time walking (restless) | Cattle, calves |
> | Time trembling | Calves |
> | Time lying down in all positions | All species |
> | Time huddling | Piglets |
> | Number of times rubs or shakes head | Calves dehorning |
> | Time kneeling | Piglets |
>
> *Note*: [a]The scientific studies were done in calves, lambs, and piglets. Many of these behaviors may also occur in other species.

from predators. To assess accurately the occurrence of pain-related behaviors, the observer must either be hidden from the animal's view or a remote video camera should be used.

Vocalization scoring of painful, stressful procedures

Vocalization scoring of squeals, moos, and bellows in cattle and pigs is a useful indicator that a procedure is stressful due to either pain, fear, or separation distress. Watts and Stookey (1999) recommend that vocalization scoring should be used to assess the stressfulness of a procedure across a group of animals and not for assessing the welfare of each individual. Watts and Stookey (1999) found that hot iron branding caused 23% of the cattle to vocalize, while freeze branding vocalizations were only 3%. In cattle, vocalization may be especially useful for evaluating severe stress (Watts and Stookey, 1999), and high-pitched vocalizations are associated with higher stress.

In another study, 98% of the cattle that vocalized during handling and stunning at a slaughter plant had been subjected to an obvious painful event such as being poked with an electric goad, poor stuns, excessive pressure from a restraint device, or sharp edges on a restraint device (Grandin, 1998b, 2001).

Excessive pressure from a restraining device resulted in 23–25% of the cattle vocalizing (Grandin, 2001; Bourguet *et al.*, 2011). Loosening a neck stanchion reduced to zero the percentage of cattle that vocalized (Grandin, 2001). Vocalization scoring works well for showing that improvements in equipment and handling procedures lower the vocalization score (Grandin, 2001). Table 2.2 shows some data where the variable of restraining the animal in the squeeze chute was separated from the variable of the stressful, painful, or frightening procedure. It also shows how a less severe electro-ejaculation method reduced the number of vocalizations in Angus beef cattle. Velarde *et al.* (2014) reported that cattle vocalized less when restrained in an upright position compared to inversion on their backs.

Vocalization scoring should be done DURING painful procedures such as branding, castration, weaning, or frightening practices such as restraint or handling. Vocalization in cattle and pigs during painful or stressful procedures is correlated with physiological measures of stress (Dunn, 1990; Warriss *et al.*, 1994; White *et al.*, 1995; Hemsworth *et al.*, 2011).

Vocalization scoring must not be used in sheep for scoring reactions to painful procedures or the fear of being restrained or handled. Cattle and pigs will vocalize when they are hurt or frightened, but sheep

Table 2.2. Average number of vocalizations per bull in response to restraint only or restraint plus electro-ejaculation. (From B.D. Voisinet and T. Grandin, 1997, unpublished data.)

	Controls restraint in a squeeze chute with a stanchion head gate	High voltage electro-ejaculation machine	Low voltage electro-ejaculation machine
Average number of vocalizations per bull[a]	0.15 ± 0.1	8.9 ± 1.1[b]	3.9 ± 1.0[b]

Notes: [a]The total number of bellows for each treatment was tabulated and divided by the number of bulls; [b]the difference between the high voltage and low voltage was significant at the $P < 0.001$ level.

will usually remain silent. Sheep are the ultimate defenceless prey species. They have evolved to remain silent when they are hurt so that they do not advertise their vulnerability to predators. However, lambs will vocalize loudly when they are separated from either the mother at weaning or a lone lamb is separated from the flock. This is the only time that distressed sheep will vocalize. Vocalization scoring cannot be used if an animal is immobilized with electricity. Immobilization prevents vocalization. These devices are highly stressful and should not be used (see Chapter 5, this volume).

How to score vocalizations

There are two ways for scoring vocalizations that work really well in cattle and pigs. These methods can be used easily on a farm, ranch, or slaughter plant. They are:

1. Score each animal as either silent or vocal to determine the percentages of animals that vocalized. This simple method works really well for assessing problems with excessive electric prod use and other handling problems such as excessive pressure or sharp edges on a restraint device (Grandin, 1998a, 2001).
2. Score the total number of vocalizations in a group of animals. The number of vocalizations are counted and divided by the number of animals to determine an average vocalization score per animal.

Behavioral versus physiological measures of pain and stress

Many scientists and veterinarians would prefer to use physiological assessments instead of behavioral measures. For research studies, it is recommended to use physiological and production measures combined with behavioral measures. The problem with physiological measures such as cortisol is that the laboratory tests are too expensive for routine farm use. Hand-held lactate meters have been validated as a relatively inexpensive way to assess stress in cattle and pigs. Burfeind and Heuwiesser (2012) reported that the Lactate Scout from Leipzig, Germany, was comparable to a commercial laboratory, when used with lithium heparin plasma or sodium fluoride plasma anticoagulants. Short-term stresses such as electric prod use or jamming in a race 5 min before slaughter raise lactate, and longer-term stresses will lower lactate (Edwards et al., 2010). To assess an acute stressor such as restraint, the lactate measurement should be taken within 5 min. Behavioral agitation in cattle shortly before stunning at a slaughter plant was associated with higher lactate (Gruber et al., 2010). The major advantage of behavioral measurements is that they can be implemented easily on the farm. Researchers should work to validate easily observable behavioral measures. There is a huge need for research in species such as goats, camels, donkeys, and many other animals.

Fear stress

Fear stress occurs when animals become agitated and excited during restraint and handling. Rough handling and multiple shocks from electric prods will increase fear stress. Stress from poor handling methods will elevate stress hormones greatly (Grandin, 1997; Edwards at al., 2010a). Pearson et al. (1977) found that sheep slaughtered in a small, quiet research abattoir had lower cortisol levels compared to sheep in a large, noisy commercial abattoir. Animals that are fearful of people or become agitated during handling will have lower weight gains and be less productive (Hemsworth and Coleman, 1994; Voisinet et al., 1997). The fear circuits in animal's brains have been completely mapped. The literature on fear is reviewed by Rogan and LeDoux (1996), Grandin (1997), Morris *et al.* (2011) and Panksepp (1998, 2011). Scientists have been able to determine that the brain contains a fear center called the amygdala. Destroying the amygdala will eliminate both learned and unlearned fear responses (Davis, 1992). Electrical stimulation of the amygdala in rats and cats raises corticosterone levels (Setekleiv *et al.*, 1961; Matheson *et al.*, 1971; Redgate and Fabringer, 1973). The amygdala is also the fear center in humans (Rogan and LeDoux, 1996). Animals that are fearful will have poor welfare. There is a further discussion of fear stress during the handling and restraint of both livestock and poultry in the chapters on handling and slaughter, this volume. Chapter 4 reviews many research studies that clearly show the benefits of good stockmanship on reducing fear stress.

Tonic immobility looks calm but is really frightened

Many animals and birds will become highly agitated when they are frightened. However, others such as Zebu cattle may lie down and go into tonic immobility. They remain completely still and look calm. This phenomenon has been researched extensively in poultry. Tonic immobility can be induced by placing a chicken on its back in a U-shaped trough. It is held down lightly by a person for 10 s (Faure and Mills, 1998). If tonic immobility is induced, it will make no attempt to get up for at least 10 s (Jones, 1987). Poultry from genetic lines that are highly fearful will stay motionless for longer periods (Jones and Mills, 1982). The length of time a bird will remain motionless in a state of tonic immobility is used as an index of fear in chickens (Jones, 1984). Stimuli that are stressful to poultry, such as electrical shocks or being kept in continuous light with no period of darkness, increase the duration of tonic immobility (Hughes, 1979; Campo *et al.*, 2007). Many people mistakenly assume that chickens that go into tonic immobility are calm and relaxed. Tonic immobility testing can be used as one indicator of conditions that are stressful to poultry.

Do Fish and Invertebrates Suffer?

Since the early 2000s, there has been an explosion of research on fish welfare. A search of the literature located many new welfare papers and lots of recent patents for equipment for stunning fish at the processing plant. The vast majority of this research has been on the teleost fish such as salmon, trout, tilapia, and other finned fish that are farmed. Scientists in England, Canada, Norway, Brazil, and other countries have written on the welfare of teleost fish (Chandroo *et al.*, 2004; Braithwaite and Boulcott, 2007; Lund *et al.*, 2007; Volpato *et al.*, 2007; Branson, 2008). Lund *et al.* (2007) state that fish can detect noxious stimuli, and these authors conclude that "farmed fish should be given the benefit of the doubt and we should make efforts to make sure that their welfare needs are met as well as possible". Since the first edition of this book, there are still no definitive conclusions concerning the perception of pain by fish. Rose *et al.* (2014) state there is still no clear evidence that fish experience pain. However, there is evidence that fish may experience fear.

Research shows that fish respond to painful stimuli, in a manner that is not just a simple reflex. The most convincing evidence that fish feel pain is the studies by Sneddon (2003), Sneddon *et al.* (2003a,b) and Reilly *et al.* (2008). Acetic acid was injected into the lips of fish to create a painful stimulus. Some of the fish engaged in weird rocking back and forth and rubbing the injected lip against the tank walls. Some individuals exhibited the behavior and others did not. It is common in pain studies in all species to have big differences in the reaction between different individuals. There were also species differences in the occurrence of this behavior. Zebra fish did not react in this way (Reilly *et al.*, 2008).

A study by Laming *et al.* (2006) showed that fish could be fear conditioned and that their reactions were affected in a complex manner by the presence of other fish. There is also evidence that fish react to handling stress or restraint stress with increases in cortisol (Wolkers *et al.*, 2013). This would be similar to the increase in cortisol after stressful handling in mammals. Recent evidence shows that a noxious stimulus is processed in multiple brain regimes (Ludvigsen *et al.*, 2014). The final experiment that needs to be done to verify that finned fish suffer from pain is the self-medication experiment that has clearly shown that rats and chickens will self-medicate for pain (Danbury *et al.*, 2000; Colpaert *et al.*, 2001).

From a practical standpoint, this research indicates that equipment such as a stunner should be used to render farmed fish insensible to external stimuli at the slaughter plant. Some of the behavioral indicators of stress that could be quantified easily on a fish farm are loss of equilibrium (fish is belly up), high respiration rate, and agitated swimming (Newby and Stevens, 2008). Other researchers scored fish for fast swimming escape responses and a tail flip behavior called the Mauthner-initiated startle response (Eaton et al., 1977). Fish is one area where more research will be needed to develop simple on-farm assessments. There are significant species differences in a fish's reaction to stress. Specific behavioral assessments will have to be developed for each species of farmed fish.

Invertebrates

During the literature search on this section of fish welfare, the author found only one paper on possible pain perception in invertebrates such as prawns. In this experiment, acetic acid was applied to the antenna. The prawns reacted by rubbing their antenna on the sides of the tank. The research was done by Stuart Barr at the Queen's University in Belfast, Northern Ireland (Barrett, 2008). Much more research will need to be done. At the time of writing this chapter, the author recommends that welfare programs should be initiated for farmed teleost fish, and further research is required on invertebrates and the cartilaginous elasmobranches such as sharks and rays.

Overloaded Biology and Animal Welfare

Overloading the animal's biological system by either genetic selection for more and more production or performance-enhancing substances may also cause pain and distress. This includes welfare problems caused by overuse of beta-agonists, rBST growth hormone, and other performance-enhancing substances. Many of these conditions may cause serious suffering. Quick Access Information 2.3 lists conditions caused by either genetic selection for a narrow range of production traits or the indiscriminant use of substances such as hormones or beta-agonists such as ractopamine. All of the conditions listed in Quick Access Information 2.3 are caused by pushing the animal's biology too hard to keep producing more and more meat, eggs, or milk. It is the author's opinion that overloading of the biological system is the cause of many serious

---

Quick Access Information 2.3. Welfare problems caused by overloading the animal's biological system.

- Lameness or leg abnormalities in rapidly growing pigs and poultry (de Sevilla et al., 2008; Knowles et al., 2008; Caplen et al., 2012; Dinev et al., 2012).
- Increased aggression in some genetic lines of pigs or chickens (Craig and Muir, 1998).
- Future problems with animals that have been genetically modified (OIE, 2006).
- Increased excitability in some genetic lines of lean pigs bred for rapid growth.
- High rates of calving problems in cattle bred for large muscle mass (Webster, 2005a,b).
- Heat stress or death losses in cattle due to beta-agonists such as ractopamine or Zilpaterol (Grandin, 2007; Longeragan et al., 2014).
- Health problems caused by rBST in dairy cows (Willeberg, 1993; Kronfield, 1994; Collier et al., 2001).
- Stress gene in pigs that causes porcine stress syndrome, which increases death losses (Murray and Johnson, 1998).
- Loss of disease or parasite resistance (Greer, 2008; Vordermeier et al., 2012; Jiang et al., 2013).

- Weak, heavily muscled pigs due to either genetics or the use of beta-agonists such as ractopamine (Marchant-Forde et al., 2003; Grandin, 2007). These pigs may be reluctant to move.
- Cracks and hoof lesions in pigs fed ractopamine (Poletto et al., 2009).
- High appetite drive and frustration in breeding animals selected for high weight gain when they are fed a restricted diet to prevent obesity.
- Metabolic problems that may increase death losses in poultry (Pakdel et al., 2005).
- Lameness in cattle due to excessive use of beta-agonists such as ractopamine or Zilpaterol.
- Increased bites from fighting in pigs caused by the beta-agonist ractopamine (Poletto et al., 2008).
- Lameness caused by poor leg conformation and overselection for a narrow range of production traits.
- Dairy cows that last only two lactations.
- Decline in dairy cow reproduction and fertility (Walsh et al., 2011).
- High percentages of laying hens with bone fracture due to osteoporosis (Wilkins et al., 2011).

---

welfare problems. Overloading the animal's biology is a matter of degree. Selection for higher production will usually have no detrimental effects on welfare if it is done in moderation. Recent research shows that animals selected for high production have less disease resistance, and laying hens have more broken bones (Wilkins *et al.*, 2011; Jiang *et al.*, 2013). Careful use of low amounts of performance-enhancing substances is probably not detrimental. It is just like revving up the RPMs in a car engine. A little revving up does no harm, but if the engine RPMs get too high, the engine will be ruined. The problem is that breeders and producers often fail to see problems until they become very serious. There are trade-offs in animal production. Overselection for meat, milk, or egg production may result in lowered fertility or lower disease resistance. Research also indicates that intensive selection for high production and less backfat may be related to increased tail biting in pigs (Brumberg *et al.*, 2013).

Assessment of Behavioral Needs

Fraser (2008b) maintains that affective states are a central animal welfare principle. An affective state is the animal's emotional state. The animal's affective state provides the motivation for many natural behaviors. Scientific research clearly shows that animals are highly motivated to perform certain species-typical behavior. Pigs are highly motivated to explore and root in soft fibrous materials such as straw, maize stalks, wood chips, or other bedding materials (Van de Weerd *et al.*, 2003; Studnitz *et al.*, 2007; Day *et al.*, 2008; see Chapter 8, this volume). The author has observed that fresh straw is rooted more by pigs than old straw. After the straw is chewed up into little pieces, the pigs lose interest in it. Fraser (1975) found that providing small amounts of straw to tethered sows

prevented abnormal stereotypic behavior. Together, these studies show that pigs should receive a daily ration of hay, straw, or maize stalks to satisfy their rooting and chewing needs. Other behavioral needs that are supported strongly by scientific research are providing nest boxes and perches for laying hens (Duncan and Kite, 1989; Hughes *et al.*, 1993; Freire *et al.*, 1997; Olsson and Keeling, 2000; Cordiner and Savory, 2001). A hen seeks a nest box to hide in so that she has freedom from fear. Finding a secluded place to lay her eggs is an instinctual behavior that prevented the wild ancestors of domestic hens from being eaten by predators.

Motivation strength can be measured in a very objective manner. Some of the methods that can be used to measure the strength of an animal's motivation to perform natural behaviors are: (i) the amount of time an animal is willing to go without feed so that it can perform a behavior; (ii) the number of times it will push a switch to get to something it wants; and (iii) weighted doors that become increasingly heavy (Widowski and Duncan, 2000).

Scientific research clearly shows that to give an animal a high level of welfare, the most highly motivated behaviors should be accommodated (O'Hara and O'Connor *et al.*, 2007). Behavioral needs are important, but in places where conditions are really poor, the very serious animal welfare problems listed in Quick Access Information 1.1, Chapter 1, this volume, should be corrected first. Quick Access Information 2.4 lists the most important behavioral needs.

Natural Elements and Ethical Considerations

The fourth principle in Dave Fraer's conceptual framework (Fraser, 2008a,b) about providing natural elements has little scientific basis compared to the

---

Quick Access Information 2.4. Basic behavioral needs that should be accommodated.

- Roughage feed for ruminant animals and equines.
- Animals should be able to turn around, stand up and lie down in a natural position.
- Secluded nest boxes for poultry.
- Perches for poultry.
- Straw or other fibrous materials for pigs to root and chew on or specialized enrichment devices.
- Repetitive stereotyped behavior in a barren cage or pen is an indicator of a poor environment that does

not satisfy behavioral needs. Environmental enrichment should be provided to prevent abnormal repetitive behavior (see Chapter 8, this volume).
- Opportunities for social interaction with other animals.
- An environment that helps prevent damaging abnormal behavior such as feather pecking, wool pulling, or tail biting. Damage on animals can be quantified and measured easily.

T. Grandin

first three principles of health, pain and distress and natural behavior. The first three principles are backed by many scientific studies. Providing natural elements is mainly an ethical concern. Ethical concerns must not be ignored by the veterinarians, managers, and other people who are in charge of implementing an animal welfare program.

Different stakeholders have different priorities. Consumers usually put more ethical emphasis on the natural elements and having cage-free chickens (de Jonge and van Trijp, 2013; Van Loo et al., 2014). Farmers put more ethical emphasis on the animals health and biology (Tuyttens et al., 2010; Hansson and Lagerkvist, 2012). Many organic programs and large meat buyers specify that animals should be able to go outside or have daylight. A mistake that some people have made is to provide the natural elements but the animals have poor welfare due to neglect of health. Health is an essential component of good welfare. The author has seen disgusting outdoor pig units full of sick pigs and excellent outdoor units with lots of healthy pigs. It is much easier to reach a consensus on welfare issues where science can provide clear answers compared to ethical concerns that have no clear answer. Ethical concerns are part of the decision-making process of legislators, animal advocacy groups, and others who make policies. Lassen et al. (2006) provide a good summary: "The main message of this chapter for those who are professionally involved in animal production is that ethnical assumptions and potential conflicts of view should be recognized and brought into the discussion of animal welfare." Science cannot provide all the answers to ethnical concerns. In some cases, ethics will overrule science. Sow gestation stalls are a good example. Research shows that sows can be highly productive in stalls, but confining a pig in a box where she cannot turn around for most of her life is not acceptable to two-thirds of the public. The author showed pictures of sow gestation stalls to passengers she sat next to on many flights. One-third of the passengers had no opinion, one-third said, "this does not seem right", and one-third hated them. One person said, "I would not keep my dogs in that." Roughly two-thirds of the public disliked sow stalls. In 2008 in California, 63% of the voters decided to ban sow gestation stalls. Sow stalls are being phased out in both Europe and the USA. Farm animals are sentient beings and it is the author's opinion that people should provide both food animals and working animals with a decent life that is worth living. Confining an animal in a box for most of its life where it is not able to turn around does not provide a decent life.

Assessment of Trade-Offs Between Different Groups of Housing Systems for Sows and Layer Hen Housing

The author is asked all the time which group housing system should be used for sows or which is the best alternative for replacing small battery cages for hens. The industry should be allowed to innovate and measure outcome variables such as number of bites, piglets born, injuries, body condition, lameness scoring, and other measures. Group housing will require more space (Wildman et al., 1982; Hemsworth et al., 2013). D'Eath et al. (2009) report that some genetic lines of pigs are more aggressive. Many producers have switched to group housing successfully by using less aggressive genetics. There are four basic group housing systems: (i) electronic sow feeders; (ii) free access stalls; (iii) small groups of sows that are never mixed; and (iv) feeding a large group of sows from a trough. Each system has different advantages and disadvantages. Electronic cow feeders require operators with computer skills. The free access stalls are simple to operate but may be more expensive to build due to great amounts of welded steel. The system with small groups of sows that are never mixed may be a good choice if a barn with gestation stalls is being converted. The back half of the stall is cut off and the same feeding system can be used. More information on these different group housing systems is in Spoolder et al. (2009), Hemsworth et al. (2013) and National Pork Board (2013). In properly managed group housing, sow performance can be excellent (Li and Gonyou, 2013). Stockpeople need to be observant and make sure that all sows are getting access to feed. If one or two sows lose body condition, this is an indicator that they are being pushed away from the feeder by the dominant animals.

Different types of hen housing systems

For laying hens, there are three basic systems: (i) conventional battery cages; (ii) cage free; and (iii) enriched colony housing, also called furnished cages. Cage-free systems vary greatly from multi-tier indoor aviary, barn eggs with a single layer of freely moving chickens, and free range where hens can go outside. The enriched colony cage provides some basic amenities for the hen, such as an enclosed nest box, a higher ceiling that allows a hen to walk upright, perches, and space for full wing flapping. Both the conventional battery cage and the enriched cage will provide clean eggs (Gast *et al.*, 2014). People must be careful not to overcrowd the hens in enriched colony housing. Overcrowding may increase agonistic behavior (Hunniford *et al.*, 2014). Aviary systems with multiple tiers of hens have poorer air quality and higher microbiological contamination of eggs (Laywel, Feedstuffs, 2014; Zhao *et al.*, 2014; Parisi *et al.*, 2015). The cage-free systems provide the birds with more space and opportunities to forage in litter. Possibly, cage-free systems without litter could improve air quality and reduce egg contamination. There are always trade-offs. Removing litter prevents the natural behavior of foraging.

Making Ethical Decisions

Scientific research shows that some behavioral needs are more important to the animal than others. For example, providing a secluded nest box is more important for a hen than providing a place to dust bathe (Widowski and Duncan, 2000). O'Hara and O'Connor (2007) state that there are priority behaviors that must be provided for in order to satisfy the minimum requirements for behavioral needs. The author recommends that the list in Quick Access Information 2.4 would satisfy the minimum behavioral requirements for livestock and poultry. Higher welfare systems would provide for additional behaviors such as dust bathing in poultry or mud wallows for pigs.

Science can provide information to help people make good decisions about animal welfare. However, there are some ethical concerns that science cannot answer (see Chapter 3, this volume). To help make good decisions, many governments and large meat buyers have animal welfare advisory councils. The author has served on these councils for livestock industry associations, major retailers, and restaurant chains. Most councils consist of scientific researchers in the field of welfare, animal advocacy groups, and lay people. They provide advice and guidance. In Europe, advisory councils make recommendations on legislation. In England, the Farm Animal Welfare Council (FAWC) has been advising the government for many years. Another example is the Council on Animal Ethics at the National Veterinary Institute in Norway. It has both expert and lay members (Mejdell, 2006).

Measurements and ethics

Numerical quantification of lameness, electric goad use, and feather pecking or other areas of welfare concern is a powerful tool for showing that practices or conditions have either improved or deteriorated. Some atrocious practices such as poking out an animal's eye or cutting leg tendons to restrain cattle should be banned. However, it is impossible to eliminate lameness totally. Chapter 4 discusses practical ways to develop reasonable limits on the percentage of lame animals that would be permitted to pass a welfare audit. With good management, very low levels of lameness are possible.

From an ethical standpoint, interpretation of physiological measures such as cortisol levels or heart rate is more difficult. What level of cortisol should be permitted? The most practical way to help people make ethical decisions about physiological measurements is to compare the stressful or painful treatment to a control condition that most people find acceptable, such as restraining the animal. It is best to evaluate physiological data by comparing it to a control condition within the same study with the same type of animals.

There are extreme levels of physiological measures that most scientific experts on an advisory council would be able to say, "This is absolutely not acceptable"; for example, extremely high average levels of cortisol such as the 93 ng/ml average level reported by Dunn (1990) in cattle. The level is 30 units higher than acceptable cortisol levels, due to poor handling. It is important to use the AVERAGE level of a physiological measure in a group of animals. Individual animals can have great variation in stress levels. More information on assessing stress has been reviewed in Grandin (1997, 2014) and Knowles *et al.* (2014). Another example of conditions that are absolutely not acceptable is the capture myopathy cases that are described in Chapter 5, this volume.

The ethics of animal treatment versus the treatment of people

The author went to Mexico and saw a man with a skinny, sickly donkey. When my host asked him about the poor condition of his donkey, he pulled up his shirt to show us his skinny, bony chest with ribs that showed. He said, "I suffer too". Obviously, he cannot afford to feed his donkey because he cannot afford to feed himself. It would be unethical to ask the man to feed all his family's food to the donkey.

The most constructive way to implement improving animal welfare in this situation would be to show the man simple ways to help his donkey, to help him survive. Simple changes in the harness may prevent saddle sores, show the person how to take care of the animal's feet, and work with people in the community on donkey husbandry (see Chapter 17, this volume). He cannot afford to give the donkey more food, but some simple improvements in husbandry, such as providing plenty of water, could be taught to help the donkey live longer and be a more useful working animal. Numerical scoring of lameness, injuries, and deaths in many donkeys would help show the entire community that when you take care of your donkey's welfare and you do not overload it, he will last longer. Even in the poorest country, there is never any justification for beating up animals or torturing them. Quick Access Information 2.5 shows where in this book to locate important welfare information.

Quick Access Information 2.5. Where in the book to find important animal-based measures for assessing welfare.

- Body condition scoring – Chapters 4 and 15
- Lameness scoring – Chapters 4 and 11
- Condition of coat and feathers scoring – Chapter 4
- Lesion and injury scoring – Chapters 2, 4, and 17
- Handling scoring – Chapters 4, 5, and 9
- Transport losses scoring – Chapter 11
- Scoring of stunning at slaughter – Chapter 9
- Animal cleanliness scoring – Chapter 4
- Behavioral measurements – Chapters 2, 8, and 12
- Assessing pain – Chapters 2 and 6
- Vocalization scoring – Chapters 2, 5, and 9
- Panting scoring for heat stress – Chapter 11
- Ethical issues – Chapters 1, 2, 3, 8, and 12
- Condition of pastures – Chapter 4
- Lists of practices that should be banned – Chapter 1
- List of welfare problems easily assessed at slaughter – Chapter 1

References

Barker, Z.E., Leach, K.A., Whay, H.R., Bell, N.J. and Main, D.C.J. (2010) Assessment of lameness prevalence and associated risk factors in dairy herds in England and Wales. *Journal of Dairy Science* 93, 932–941.

Barrett, L. (2008) Decapods feel the pinch. *Animal Behavior* 75, 743–744.

Bennett, R., Barker, Z.E., Main, D.C.J., Whay, H.R. and Leach, K.A. (2014) Investigating the value dairy farmers place on a reduction of lameness in their herds using a willingness to pay approach. *The Veterinary Journal* 199, 72–75.

Blokhuis, H., Miele, M., Veissier, I. and Jones, B (2013) *Improving Farm Animal Welfare. Science and Society Working Together: The Welfare Quality Approach.* Wageningen Academic Publishers, the Netherlands.

Bourguet, C., Deiss, V., Tannugi, C.C. and Terlouw, E.M. (2011) Behavioral and physiological reactions of cattle in a commercial abattoir: relationships with organisational aspects of the abattoir and animal characteristics *Meat Science* 88, 158–168.

Braithwaite, V.A. and Boulcott, P. (2007) Pain perception, aversion and fear in fish. *Diseases of Aquatic Organisms* 75, 131–138.

Branson, E. (2008) *Fish Welfare.* Blackwell Publishing, Oxford, UK.

Broom, D.M. and Fraser, A.F. (2007) *Domestic Animal Behaviour and Welfare.* CAB International, Wallingford, UK.

Brunberg, E., Jensen, P., Isaksson, A. and Keeling, L.J. (2013) Brain gene expression differences are associated with abnormal tail biting behavior in pigs. *Genes, Brain, and Behavior* 12, 275–281.

Burfeind, O. and Heuwieser, W. (2012) Validation of handheld meters to measure blood L-lactate concentration in dairy cows and calves. *Journal of Dairy Science* 95, 6449–6456.

Campo, J.L., Gil, M.G., Dávila, S.G. and Muñoz, I. (2007) Effect of lighting stress on fluctuating asymmetry, heterophil-to-lymphocyte ratio, and tonic immobility duration in eleven breeds of chickens. *Poultry Science* 86, 37–45.

Caplen, G., Colborne, G.R., Hothersall, B., Nicol, C.J., Waterman-Pearson, A.E., *et al.* (2013) Lame broiler chickens respond to non-steroidal anti-inflammatory drugs with objective changes in gain function: A controlled clinical trial. *The Veterinary Journal* 196, 477–482.

Chandroo, K.P., Duncan, I.J.H. and Moccia, R.D. (2004) Can fish suffer? Perspectives on sentience, pain, fear and stress. *Applied Animal Behaviour Science* 86, 225–250.

Chiquitelli, Neto, M., Paranhos de Costa, M.J.R., Piacoa, A.G. and Wolf, V. (2002) Manejo racional na vacinacio de bovinos Nelore: uma availacao preliminar da eficiencia e qualidade do trabalho. In: Josahkian, L.A. (ed.) 5th Congresso dos Racos Zebuinas, ABCZ, UberabaMG, pp. 361–362.

Coetzee, J.F., Mosher, R.A., Kukanich, B., Gehring, R., Robert, B., *et al*. (2012) Pharmokinetics and effect of intravenous meloxicam in weaned Holstein calves following scoop dehorning without local anesthesia. *BMC Veterinary Research* 8, 153.

Collier, R.J., Byatt, J.C., Denham, S.C., Eppard, P.J., Fabellar, A.C., *et al*. (2001) Effects of sustained release bovine somatotropin (sometribove) on animal health in commercial dairy herds. *Journal of Dairy Science* 84, 1098–1108.

Colpaert, F.C., Tarayre, J.P., Alliaga, M., Bruins Slot, L.A., Attal, N., *et al*. (2001) Opiate self-administration as a measure of chronic nociceptive pain in arthritic rats. *Pain* 91, 33–45.

Cordiner, L.S. and Savory, C.J. (2001) Use of perches and nest boxes by laying hens in relation to social status, based on examination of consistency of ranking orders and frequency of interaction. *Applied Animal Behaviour Science* 71, 305–317.

Craig, J.C. and Muir, W. (1998) Genetics and the behavior of chickens. In: Grandin, T. (ed.) *Genetics and the Behavior of Domestic Animals*. Academic Press, San Diego, California, pp. 265–298.

Danbury, T.C., Weeks, C.A., Chambers, J.P., Waterman-Pearson, A.E. and Kestin, S.C. (2000) Self-selection of the analgesic drug carprofen by lame broiler chickens. *Veterinary Research* 146, 307–311.

Davis, M. (1992) The role of the amygdala in fear and anxiety. *Annual Review of Neuroscience* 15, 353–375.

Day, J.E.L., Van de Weerd, H.A. and Edwards, S.A. (2008) The effect of varying lengths of straw bedding on the behaviour of growing pigs. *Applied Animal Behaviour Science* 109, 249–260.

D'Eath, R.B., Roehe, R., Turner, S.P., Ison, S.H., Farish, M., *et al*. (2009) Genetics of animal temperament: Aggressive behaviour at mixing is genetically associated with the response to handling in pigs. *Animal* 3, 1544–1554.

de Jonge, J. and van Trijp, J.C.M. (2013) The impact of broiler production system practices on consumer perceptions of animal welfare. *Poultry Science* 92, 3080–3095.

de Sevilla, X.F., Fàbrega, E., Tibau, J. and Casellas, J. (2008) Effect of leg conformation on survivability of Duroc, Landrace, and Large White sows. *Journal of Animal Science* 86, 2392–2400.

Dinev, I., Denev, S.A. and Edens, F.W. (2012) Comparative clinical and morphological studies on the incidence of tibial dyschondroplasia as a cause of lameness in three commercial lines of broiler chickens. *Journal Applied Poultry Research* 21, 637–644.

Duncan, I.J. (1998) Behavior and behavioral needs. *Poultry Science* 77, 1766–1772.

Duncan, I.J.H. and Kite, V.G. (1989) Nest site selection and nest-building behavior in domestic fowl. *Animal Behavior* 37, 215–231.

Dunn, C.S. (1990) Stress reactions of cattle undergoing ritual slaughter using two methods of restraint. *Veterinary Record* 126, 522–525.

Eaton, R.C., Bombardieri, R.A. and Meyer, D.L. (1977) The Mauthner-initiated startle response in teleost fish. *The Journal of Experimental Biology* 66, 65–81.

Edwards, L.N., Grandin, T., Engle, T.E., Porter, S.P., Ritter, M.J., *et al*. (2010) Use of exsanguination blood lactate to assess the quality of pre-slaughter pig handling. *Meat Science* 86, 384–390.

Eicher, S.D. and Dailey, J.W. (2002) Indicators of acute pain and fly avoidance behaviors in Holstein calves following tail-docking. *Journal of Dairy Science* 85, 2850–2858.

Espejo, L.A., Endres, M.I. and Salfer, J.A. (2006) Prevalence of lameness in high-producing Holstein cows housed in freestall barns in Minnesota. *Journal of Dairy Science* 89, 3052–3058.

Faulkner, P.M. and Weary, D.M. (2000) Reducing pain after dehorning in dairy calves. *Journal of Dairy Science* 83(9), 2037–2041.

Faure, J.M. and Mills, A.D. (1998) Improving the adaptability of animals by selection. In: Grandin, T. (ed.) *Genetics and the Behavior of Domestic Animals*. Academic Press, Elsevier, San Diego, California, pp. 233–265.

Feedstuffs (2014) Laying hen housing trade-offs. *Feedstuffs Food Link Connecting Farm to Fork*, 12 August 2014, p. 9.

Flower, F.C., de Passillé, A.M., Weary, D.M., Sanderson, D.J. and Rushen, J. (2007) Softer, higher-friction flooring improves gait of cows with and without sole ulcers. *Journal of Dairy Science* 90, 1235–1242.

Fraser, D. (1975) The effect of straw on the behaviour of sows in tether stalls. *Animal Production* 21, 59–68.

Fraser, D. (2008a) Toward a global perspective on farm animal welfare. *Applied Animal Behaviour Science* 113, 330–339.

Fraser, D. (2008b) *Understanding Animal Welfare*. Wiley-Blackwell, Oxford, UK.

Freire, R., Appleby, M.C. and Hughes, B.O. (1997) Assessment of pre-laying motivation in the domestic hen using social interaction. *Animal Behaviour* 54, 313–319.

Fulwider, W.K., Grandin, T., Garrick, D.J., Engle, T.E., Lamm, W.D., *et al*. (2007) Influence of free-stall base on tarsal joint lesions and hygiene in dairy cows. *Journal of Dairy Science* 90, 3559–3566.

Gast, R.K., Guraya, R., Jones, D.R. and Anderson, K.E. (2014) Contamination of eggs by *Salmonella enteritidis* in experimentally infected laying hens housed in conventional or enriched cages. *Poultry Science* 93, 728–733.

Gentle, M.J., Waddington, D., Hunter, L.N. and Jones, R.B. (1990) Behavioral evidence for persistent pain following partial beak amputation in chickens. *Applied Animal Behaviour Science* 27, 149–157.

Grandin, T. (1982) Pig behavior studies applied to slaughter-plant design. *Applied Animal Behaviour Science* 9, 141–151.

Grandin, T. (1997) Assessment of stress during handling and transport. *Journal of Animal Science* 75, 249–257.

Grandin, T. (1998a) Objective scoring of animal handling and stunning practices at slaughter plants. *Journal of the American Veterinary Medical Association* 212, 36–39.

Grandin, T. (1998b) The feasibility of using vocalization scoring as an indicator of poor welfare during cattle slaughter. *Applied Animal Behaviour Science* 56, 121–128.

Grandin, T. (2001) Cattle vocalizations are associated with handling and equipment problems at beef slaughter plants. *Applied Animal Behaviour Science* 71, 191–201.

Grandin, T. (2005) Special Report: Maintenance of good animal welfare standards in beef slaughter plants by use of auditing programs. *Journal of the American Veterinary Medical Association* 226, 370–373.

Grandin, T. (2007) *Livestock Handling and Transport.* CAB International, Wallingford, UK.

Grandin, T. (2010) Auditing animal welfare at slaughter plants. *Meat Science* 86, 56–65.

Grandin, T. (2012) Developing measures to audit animal welfare of cattle and pigs at slaughter. *Animal Welfare* 21, 351–356.

Grandin, T. (2014) Improving welfare and reducing stress on animals at slaughter plants. In: Grandin, T. (ed.) *Livestock Handling and Transport*, 4th edn. CAB International, Wallingford, UK, pp. 421–450.

Grandin, T. and Johnson, C. (2005) *Animals in Translation.* Scribner, New York.

Grandin, T. and Johnson, C. (2009) *Animals Make Us Human.* Houghton Mifflin Harcourt, Boston, Massachusetts.

Green, L.E., Huxley, J.N., Banks, C. and Green, M. (2014) Temporal associations between low body condition, lameness and milk yield in a UK dairy herd. *Preventive Veterinary Medicine* 113, 63–71.

Greer, A.W. (2008) Trade-offs and benefits: Implications of promoting strong immunity to gastrointestinal parasites in sheep. *Parasite Immunology* 30, 123–132.

Gruber, S.L., Tatum, J.D., Engle, T.E., Chapman, P.L., Belk, K.E., et al. (2010) Relationships of behavioral and physiological symptoms of pre-slaughter stress to beef longissimus muscle tenderness. *Journal of Animal Science* 88, 1148–1159.

Hansson, H. and Lagerkvist, C.J. (2012) Measuring farmers' attitudes to animal welfare and health. *British Food Journal* 114, 840–852.

Hemsworth, P.H. and Coleman, G.J. (1994) *Human–Livestock Interactions.* CAB International, Wallingford, UK.

Hemsworth, P.H., Rice, M., Karlen, G.A., Calleja, L.J., Barnett, J., et al. (2011) Human–animal interactions at abattoirs: Relationships between handling and animal stress in sheep and cattle. *Applied Animal Behaviour Science* 91, 4953–4964.

Hemsworth, P.H., Rice, M., Nash, J., Giri, K., Butler, K.L., et al. (2013) Effects of group size and floor space allowance on grouped sows: Aggression, stress, skin injuries, and reproductive performance. *Journal of Animal Science* 91, 4953–4964.

Hughes, B.O., Wilson, S., Appleby, M.C. and Smith, S.F. (1993) Comparison of bone volume and strength as measures of skeletal integrity in caged laying hens with access to perches. *Research in Veterinary Medicine* 54, 202–206.

Hughes, R.A. (1979) Shock induced tonic immobility in chickens as a function of post hatch age. *Animal Behavior* 27, 782–785.

Hunniford, M.E., Torrey, S., Bedecarrats, G., Duncan, I.J.H. and Widowski, T. (2014) Evidence of competition for nest sites in large furnished cages. *Applied Animal Behaviour* 161, 95–104.

Jiang, C., Xing, F., Xing, J., Jiang, Y. and Zhou, E. (2013) Different expression patterns of PRRSV medicator genes in lung tissues PRRSV resistant and susceptible pigs. *Developmental and Comparative Immunology* 39, 127–131.

Jones, R.B. (1984) Experimental novelty and tonic immobility in chickens (*Gallas domesticus*). *Behavioral Processes* 9, 255–260.

Jones, R.B. (1987) The assessment of fear in the domestic fowl. In: Zayan, R. and Duncan, I.J.H. (eds) *Cognitive Aspects of Social Behaviour in the Domestic Fowl.* Elsevier, Amsterdam, pp. 40–81.

Jones, R.B. and Mills, A.D. (1982) Estimation of fear in two lines of the domestic chick. Correlations between various methods. *Behavioral Processes* 8, 243–253.

Kirk, R.K., Svensmark, B., Elegaard, L.P. and Jensen, H.D. (2005) Locomotion disorders associated with sow mortality in Danish pig herds. *Journal of Veterinary Medicine and Physiological and Pathology Clinical Medicine* 52, 423–428.

Knowles, T.G., Kestin, S.C., Hasslam, S.M., Brown, S.N., Green, L.E., et al. (2008) Leg disorders in broiler chickens: Prevalence, risk factors and prevention. *PLOS One*, doi:10.1371/journal.pone.0001545.

Knowles, T.G., Warriss, P.D. and Vogel, K. (2014) Stress physiology in animals during transport. In: Grandin, T. (ed.) *Livestock Handling and Transport*, 4th edn. CAB International, Wallingford, UK, pp. 399–419.

Kronfield, D.S. (1994) Health management of dairy herds treated with bovine somatotropin. *Journal of American Veterinary Medical Association* 204, 116–130.

Laming, P., Dunlop, R. and Millsopp, S. (2006) Avoidance learning in goldfish (*Carassius auratus*) and trout (*Oncorhynchus mykiss*) and implications for pain perception. *Applied Animal Behaviour Science* 97, 255–271.

Lassen, J., Sandoe, P. and Forkman, B. (2006) Happy pigs are dirty! – conflicting perspectives on animal welfare. *Livestock Science* 103, 221–230.

Leach, K.A., Whay, H.R., Maggs, C.M., Barker, Z.E., Paul, E., et al. (2010) Working towards a reduction in cattle lameness: 1 Understanding barriers to lameness control on dairy farms. *Research in Veterinary Science* 89, 311–317.

Li, Y.Z. and Gonyou, H.W. (2013) Comparison of management options for sows kept in pens with electronic feeding stations. *Canadian Journal of Animal Science* 93, 445–452.

Longeragen, G.H., Thomson, D.U. and Scott, H.M. (2014) Increased mortality in groups of cattle administered the b-adrenergic agonists ractopamine hydrochloride and zilpaterol hydrochloride. *PLos One*, doi. org/10.1371/journalpone0091177.

Ludvigsen, S., Stenklev, N.C., Johnson, H.K., Laukli, N.C., Johnsen, H.K., et al. (2014) Evoked potentials in the Atlantic cod following putatively innocuous and putatively noxious electrical stimulation: a minimally invasive approach. *Fish Physiology and Biochemistry* 40, 173–181.

Lund, V., Mejdell, C.M., Röcklinsberg, H., Anthony, R. and Håstein, T. (2007) Expanding the moral circle: farmed fish as objects of moral concerns. *Diseases of Aquatic Organisms* 75, 109–118.

Marchant-Forde, J.N., Lay, D.C., Pajor, E.A., Richert, B.T. and Schinckel, A.P. (2003) The effects of ractopamine on the behavior and physiology of finishing pigs. *Journal of Animal Science* 81, 416–422.

Matheson, G.K., Branch, B.J. and Taylor, A.N. (1971) Effects of amygdaloid stimulation on pituitary-adrenal activity in conscious cats. *Brain Research* 32, 151–167.

Mejdell, C.M. (2006) The role of councils on animal ethics in assessing acceptable welfare standards in agriculture. *Livestock Science* 103, 292–296.

Molony, V. and Kent, J.E. (1997) Assessment of acute pain in farm animals using behavioral and physiological measurements. *Journal of Animal Science* 75, 266–272.

Morris, C.L., Grandin, T. and Irlbeck, N.A. (2011) Companion Animals Symposium: Environmental enrichment for companion, exotic, and laboratory animals. *Journal of Animal Science* 89, 4227–4238.

Murray, A.C. and Johnson, C.P. (1998) Impact of the halothane gene on muscle quality and pre-slaughter death in western Canadian pigs. *Canadian Journal of Animal Science* 78, 543–548.

National Pork Board (2013) Group housing systems, choices and designs. Available at: http://www.Pork.org/.

Newby, N.C. and Stevens, E.D. (2008) The effects of the acetic acid "pain" test on feeding, swimming, and respiratory responses of rainbow trout (*Oncorhynchus mykiss*). *Applied Animal Behaviour Science* 114, 260–269.

O'Hara, P. and O'Connor, C. (2007) Challenge of developing regulations for production animals that produce the welfare outcomes we want. *Journal of Veterinary Behavior, Clinical Applications and Research* 2, 205–212.

OIE (2006) *Introduction to the Recommendations for Animal Welfare. Terrestrial Animal Health Code*. World Animal Health Organization, Paris.

OIE (2014a) *Transport of Animals by Land. Terrestrial Animal Health Code*. World Organisation for Animal Health, Paris.

OIE (2014b) *Slaughter of Animals. Terrestrial Animal Health Code*. World Health Organization, Paris.

Olsson, I.A.S. and Keeling, L.J. (2000) Night-time roosting in laying hens and the effect of thwarting access to perches. *Applied Animal Behaviour Science* 68, 243–256.

OMAFRA (2005) Score Your Farm or Cow Comfort. Ontario Ministry of Agricultural Food and Rural Affairs, Ontario, Canada. Available at: www.gov.on.ca/ OMAFRA/english/livestock/dairy/facts/info_tsdimen.htm

Panksepp, J. (1998) *Affective Neuroscience*. Oxford University Press, New York.

Panksepp, J. (2011) The basic emotional circuits of mammalian brains: Do animals have affective lives. *Neuroscience and Biobehavioral Reviews* 35, 1791–1804.

Paranhos de Costa, M. (2008) Improving the welfare of cattle, practical experience in Brazil. In: Dawkins, M.S. and Bonney, R. (eds) *The Future of Animal Farming*. Blackwell Publishing, Oxford, UK, pp. 145–152.

Paranhos de Costa, M.J.R., Huertas, S.M., Strappins, A.C. and Gallo, C. (2014) Handling and transport of cattle and pigs in South America. In: Grandin, T. (ed.) *Livestock Handling and Transport*, 4th edn. CAB International, Wallingford, UK, pp. 174–192.

Parisi, M.A., Northcutt, J.K., Smith, D.P., Steinberg, E.L. and Dawson, P.L. (2015) Microbiological contamination of shell eggs produced in conventional and free-range housing systems. *Food Control* 47, 161–165.

Pakdel, A., Van Arendonk, J.A.M., Vereijken, A.L. and Bovenhuis, H. (2005) Genetic parameters of ascites-related traits in broilers: correlations with feed efficiency and carcase traits. *British Poultry Science* 46, 43–51.

Pearson, A.J., Kilgour, R., de Langen, H. and Payne, E. (1977) Hormonal responses of lambs to trucking, handling and electric stunning. *Proceedings of the New Zealand Society for Animal Production* 37, 243–248.

Pham, T.M., Hagman, B., Codita, A., Van Loo, P.L., Strömmer, L., et al. (2010) Housing environment influences the need for pain relief during post-operative recovery in mice. *Physiology and Behavior* 99, 663–668.

Pilecco, M., Almeida Paz, I.C.L., Tabaldi, L.A., Nääs, I.A., Garcia, R.G., et al. (2013) Training of catching teams and reduction of back scratches in broilers. *Revista Brasileira de Ciência Avícola* 15, doi.org/10.1590/ S1516-635X2013000300017.

Pluym, L.M., Van Nuffel, A., Van Weyenberg, S. and Maes, D. (2013) Prevalence of lameness and claw lesions during different stages in the reproductive cycle of sows and the impact on reproduction results. *Animal* 7, 1174–1181.

Poletto, R., Garner, J., Cheng, H.W. and Marchant-Forde, J. (2008) The effects of ractopamine, gender and social rank on aggression and peripheral monoamine levels in finishing pigs. *Journal of Animal Science* 86(E.Suppl 2) (Abstract 352).

Poletto, R., Rostagno, M.H., Richert, B.T. and Marchant-Forde, J.N. (2009) Effects of a "step up" ractopamine feeding program, sex and social rank on growth performance, hoof lesions, and Enterobacteriaceae shedding in finishing pigs. *Journal of Animal Science* 87, 304–313.

Redgate, E.S. and Fahringer, E.E. (1973) A comparison of the pituitary adrenal activity elicited by electrical stimulation of preoptic, amygdaloid and hypothalamic sites in the rat brain. *Neuroendocrinology* 12, 334–343.

Reilly, S.C., Quinn, J.P., Cossins, A.R. and Sneddon, L.U. (2008) Behavioral analysis of a nociceptive event in fish: Comparisons between three species demonstrate specific responses. *Applied Animal Behaviour Science* 114, 248–259.

Rogan, M.T. and LeDoux, J.E. (1996) Emotion: systems, cells, synaptic plasticity. *Cell* 85, 469–475.

Rose, J.D., Arlinghaus, R., Cooke, S.J., Diggles, B.K., Sawynok, W., *et al.* (2014) Can fish really feel pain? *Fish and Fisheries* 15, 97–133.

Rushen, J., Pombourcq, E. and de Passillé, A.M. (2007) Validation of two measures of lameness in dairy cows. *Applied Animal Behaviour Science* 106, 173–177.

Rutherford, K.M., Langford, F.M., Jack, M.C., Sherwood, L., Lawrence, A.B., *et al.* (2009) Lameness prevalence and risk factors in organic and non-organic dairy herds in the United Kingdom. *Veterinary Journal* 180, 95–105.

Setekleiv, J., Skaug, O.E. and Kaada, B.R. (1961) Increase in plasma 17-hydroxycorticasteroids by cerebral cortical and amygdaloid stimulation in the cat. *Journal of Endocrinology* 22, 119–127.

Sherwin, C.M., Richards, G.J. and Nicol, C.J. (2010) Comparison of the welfare of layer hens in 4 housing systems in the UK. *British Poultry Science* 51, 488–499.

Sneddon, L.U. (2003) The evidence for pain in fish: The use of morphine as an analgesic. *Applied Animal Behaviour Science* 83, 153–162.

Sneddon, L.U., Braithwaite, V.A. and Gentle, M.J. (2003a) Do fish have nociceptors: Evidence for the evolution of a vertebrate sensory system. *Proceedings of the Royal Society B Biological Sciences* 270, 1115–1121.

Sneddon, L.U., Braithwaite, V.A. and Gentle, M.J. (2003b) Novel object test: Examining nociception and fear in the rainbow trout. *Journal of Pain* 4, 431–440.

Spoolder, H.A.M., Geudeke, M.J., Van der Peet-Schwering, C.M.C. and Soede, N.W. (2009) Group housing of sows in early pregnancy: A review of success and risk factors. *Livestock Science* 125, 1–14.

Stafford, K.J. and Mellor, D.J. (2005a) Dehorning and disbudding distress and its alleviation in calves. *Veterinary Journal* 169, 337–349.

Stafford, K.J. and Mellor, D.J. (2005b) The welfare significance of the castration of cattle: A review. *New Zealand Veterinary Journal* 53, 271–278.

Stillwell, G., Lima, M.G., Carvalho, R.C. and Broom, D.M. (2012) Effects of hot-iron disbudding, using regional anaesthesia with and without carprofen, on cortisol and behaviour of calves. *Research in Veterinary Science*, 92, 338–341.

Studnitz, M., Jensen, M.B. and Pedersen, L.J. (2007) Why do pigs root and in what will they root? A review on the exploratory behaviour of pigs in relation to environmental enrichment. *Applied Animal Behaviour Science* 107, 183–197.

Sylvester, S.P., Stafford, K.J., Mellor, D.J., Bruce, R.A. and Ward, R.N. (2004) Behavioural responses of calves to amputation dehorning with and without local anaesthesia. *Australian Veterinary Journal* 82, 697–700.

Tanida, H., Miura, A., Tanaka, T. and Yoshimoto, T. (1996) Behavioral responses of piglets to darkness and shadows. *Applied Animal Behaviour Science* 49, 173–183.

Tiranti, K.I. and Morrison, R.B. (2006) Association between limb conformation and retention of sows through the second parity. *American Journal of Veterinary Research* 67, 505–509.

Tuyttens, F.A.W., Vanhonacker, F., Poucke, E.V. and Verbeke, W. (2010) Quantitative verification of the correspondence between the Welfare Quality® operational definition of farm animal welfare and the opinions of Flemish farmers, citizens and vegetarians. *Livestock Science* 131, 108–114.

Van de Weerd, H.A., Docking, C.M., Day, J.E.L., Avery, P.J. and Edwards, S.A. (2003) A systematic approach towards developing environmental enrichment for pigs. *Applied Animal Behaviour Science* 84, 101–118.

Van Loo, E.J., Caputo, V., Nayga, R.M. and Verbeke, W. (2014) Consumers' valuation of sustainability labels on meat. *Food Policy* 49, 137–150.

Van Putten, G. and Elshof, W.J. (1978) Observations of the effects of transport on the well being and lean quality of slaughter pigs. *Animal Regulation Studies* 1, 247–271.

Vansickle, J. (2008) Sow lameness underrated. *National Hog Farmer*, 15 June 2008, p. 34.

Velarde, A., Rodriguez, P., Dalmau, A., Fuentes, C., Llonch, P., *et al.* (2014) Religious slaughter: evaluation of current practices in selected countries. *Meat Science* 96, 278–287.

Viñuela-Fernández, I., Jones, E., Welsh, E.M. and Fleetwood-Walker, S.M. (2007) Pain mechanisms and their implication for the management of pain in

---

farm and companion animals. *Veterinary Journal* 174, 227–239.

Voisinet, B.D., Grandin, T., Tatum, J.D., O'Connor, S.F. and Struthers, J.J. (1997) Feedlot cattle with calm temperaments have higher average daily gains than cattle with excitable temperaments. *Journal of Animal Science* 75, 892–896.

Volpato, G.L., Gonçalves-de-Freitas, E. and Fernandes-deCastilho, M. (2007) Insights into the concept of fish welfare. *Diseases of Aquatic Organisms* 75, 165–171.

von Keyserlingk, M.A.G., Barrientos, A., Ito, K., Galo, E. and Weary, D.M. (2012) Benchmarking cow comfort on North American freestall dairies: lameness, leg injuries, lying time, facility design, and management for high producing Holstein dairy cows. *Journal of Dairy Science* 95, 7399–7408.

Vordermeier, M., Ameni, G., Berg, S., Bishop, R., Robertson, B.D., *et al.* (2012) The influence of cattle breed on susceptibility to bovine tuberculosis in Ethiopia. *Comparative Immunology, Microbiology and Infectious Diseases* 35, 227–232.

Walsh, S.W., Williams, E.J. and Evans, A.C. (2011) A review of the causes of poor fertility in high milk producing dairy cows. *Animal Reproduction Science* 123, 127–138.

Warriss, P.D., Brown, S.N., Adams, S.J. and Corlett, I.K. (1994) Relationship between subjective and objective assessments of stress at slaughter and meat quality in pigs. *Meat Science* 38, 329–340.

Watts, J.M. and Stookey, J.M. (1999) Effects of restraint and branding on rates and acoustic parameters of vocalization in beef cattle. *Applied Animal Behaviour Science* 62, 125–135.

Webster, J. (2005a) The assessment and implementation of animal welfare: Theory into practice. *Review Science Technology, Office International Epizootics* 24, 723–734.

Webster, J. (2005b) *Animal Welfare, Limping Towards Eden*. Blackwell Publishing, Oxford, UK.

White, R.G., DeShazer, J.A., Tressler, C.J., Borcher, G.M., Davey, S., *et al.* (1995) Vocalization and physiological response of pigs during castration with or without a local anesthetic. *Journal of Animal Science* 73, 381–386.

Widowski, T.M. and Duncan, I.J.H. (2000) Working for a dust bath: Are hens increasing pleasure rather than relieving suffering? *Applied Animal Behaviour Science* 68, 39–53.

Wildman, E.E., Jones, G.M., Wagner, P.E., Boman, R.L., Troutt, H.F., *et al.* (1982) A dairy cow body condition scoring system and its relationship to selected production characteristics. *Journal of Dairy Science* 65, 495–501.

Wilkins, L.J., McKinstry, J.L., Avery, N.C., Knowles, T.G., Brown, S.N., *et al.* (2011) Influence of housing system and design on bone strength and keel bone fractures in laying hens. *Veterinary Record* 169, 414.

Willeberg, P. (1993) Bovine somatotropin and clinical mastitis: epidemiological assessment of the welfare risk. *Livestock Science* 36, 55–66.

Wolkers, C.P.B., Barbosa, A., Menescal-de-Oliveira, L. and Hoffmann, A. (2013) Stress-induced antinociception in fish reversed by Naloxone. *PLoS One*, doi.10.1371/journal.pone.0071175.

Whay, H.R., Main, D.C.J., Green, L.E. and Webster, A.J.F. (2003) Assessment of the welfare of dairy cattle using animal-based measurements: direct observations and investigation of farm records. *Veterinary Record* 153, 197–202.

www.welfarequalitynetwork.net (2009) Accessed 19 April 2014.

Zhao, Y., Shepherd, H. and Xin, H. (2014) Environmental assessment of three egg production systems – (Part II) Ammonia, greenhouse gas and particulate matter emissions. *Poultry science* 94, 534–543.

Zurbrigg, K., Kelton, D., Anderson, N. and Millman, S. (2005a) Stall dimensions and the prevalence of lameness, injury, and cleanliness on 317 tie-stall dairy farms in Ontario. *The Canadian Veterinary Journal* 46, 902–909.

Zurbrigg, K., Kelton, D., Anderson, N. and Millman, S. (2005b) Tie-stall design and its relationship to lameness, injury, and cleanliness on 317 Ontario dairy farms. *Journal of Dairy Science* 88, 3201–3210.

3 Why is Agricultural Animal Welfare Important? The Social and Ethical Context

BERNARD ROLLIN

Colorado State University, Fort Collins, Colorado, USA

Summary

Your concept of what animal welfare is will influence your views on what is really good, sound science. Science should not be used solely to increase productivity. Throughout the world, the ethics of animal treatment is switching from personal ethics to the ethics of society. The great increase in animal welfare legislation reflects this trend. The use of industrial animal production has had the unfortunate effect of a loss of emphasis on husbandry and care of the individual animal. A major problem is diseases and pathologies associated with ever-increasing production. Some examples are shorter productive lifespan, lower reproduction, lameness, liver abscesses, and osteoporosis and broken bones in old laying hens. Along with cost to the individual animals, there are also costs to the environment and society. Animal science must address both values and ethics. Consumer demand has resulted in better treatment of animals.

Learning Objectives

- Understand different concepts of animal welfare
- Learn about welfare problems caused by a loss of animal husbandry principles
- Overview of production diseases and pathologies
- The importance of the animal's innate nature (telos)
- Address the issues that animal scientists and veterinarians must address: welfare and ethics

Introduction – How Should We Raise Animals?

There is one monumental conceptual error that is omnipresent in the agricultural industry's discussions of animal welfare – an error of such magnitude that it trivializes the industry's responses to ever-increasing societal concerns about the treatment of agricultural animals. These concerns are emerging as non-negotiable demands by consumers. Failure to respect such concerns can essentially destroy the economic base for intensive animal agriculture. When one discusses farm animal welfare with industry groups or with the American Veterinary Medical Association, one finds the same response – animal welfare is solely a matter of "sound science".

Those of us serving on the Pew Commission, better known as the National Commission on Industrial Farm Animal Production, encountered this response regularly during our dealings with industry representatives. This commission studied intensive animal agriculture in the USA (Pew Commission, 2008). For example, one representative of the Pork Producers, testifying before the Commission, answered that while people in her industry were quite "nervous" about the Commission, their anxiety would be allayed were we to base all of our conclusions and recommendations on "sound science". Hoping to rectify the error in that comment, as well as educate the numerous industry representatives present, I responded to her as follows:

> Madame, if we on the Commission were asking the question of how to raise swine in confinement, science could certainly answer that question for us. But that is not the question the Commission, or

society, is asking. What we are asking is, ought we to raise swine in confinement? And to this question, science is not relevant.

Judging by her "huh", I assume I did not make my point.

Questions of animal welfare are at least partly "ought" questions, questions of ethical obligation. The concept of animal welfare is an ethical concept to which, once understood, science brings relevant data. When we ask about an animal's welfare, or about a person's welfare, we are asking about *what* we owe the animal, and to *what extent*. A document referred to as the Council for Agricultural Science and Technology (CAST) report, first published by US agricultural scientists in the early 1980s, discussed animal welfare. It affirmed that the necessary and sufficient conditions for attributing positive welfare to an animal were represented by the animal's productivity. A productive animal enjoyed positive welfare; a nonproductive animal enjoyed poor welfare (CAST, 1997).

This notion was fraught with many difficulties. First of all, productivity is an economic notion predicated of a whole operation; welfare is predicated of individual animals. An operation such as caged laying hens may be quite profitable if the cages are severely overcrowded, yet the individual hens do not enjoy good welfare. Second, as we shall see, equating productivity and welfare is, to some significant extent, legitimate under husbandry conditions, where the producer does well if, and only if, the animals do well, and square pegs, as it were, are fitted into square holes with as little friction as possible. Under industrial conditions, however, animals do not naturally fit in the niche or environment in which they are kept, and are subjected to "technological sanders" that allow for producers to force square pegs into round holes – antibiotics, feed additives, hormones, air handling systems – so the animals do not die but produce more and more kilograms of meat or milk. Without these technologies, the animals could not be productive. We will return to the contrast between husbandry and industrial approaches to animal agriculture.

The key point to recall here is that even if the CAST report definition of animal welfare does not suffer from the difficulties we outline, it is still an ethical concept. It essentially says 'what we owe animals and to what extent is simply what it takes to get them to create profit'. This in turn would imply that the animals are well off if they have only food, water, and shelter, something the industry has sometimes asserted. Even in the early 1980s, however, there were animal advocates and others who would take a very different ethical stance on what we owed farm animals. Indeed, the famous five freedoms articulated in the UK by the Farm Animal Welfare Council (FAWC) during the 1970s (even before the CAST report) represents quite a different ethical view of what we owe animals, when it affirms that:

> The welfare of an animal includes its physical and mental state, and we consider that good animal welfare implies both fitness and a sense of well-being. Any animal kept by man, must at least, be protected from unnecessary suffering.
>
> We believe that an animal's welfare, whether on farm, in transit, at market or at a place of slaughter should be considered in terms of 'five freedoms' (see www.fawc.org.uk).
>
> 1. **Freedom from Hunger and Thirst** – by ready access to fresh water and a diet to maintain full health and vigour.
> 2. **Freedom from Discomfort** – by providing an appropriate environment including shelter and a comfortable resting area.
> 3. **Freedom from Pain, Injury or Disease** – by prevention or rapid diagnosis and treatment.
> 4. **Freedom to Express Normal Behavior** – by providing sufficient space, proper facilities and company of the animal's own kind.
> 5. **Freedom from Fear and Distress** – by ensuring conditions and treatment which avoid mental suffering.
> (FAWC, 2009)

Clearly, the two definitions contain very different notions of our moral obligation to animals (and there is an indefinite number of other definitions). Which one is correct, of course, cannot be decided by gathering facts or doing experiments – indeed, which ethical framework one adopts will, in fact, determine the shape of the science of studying animal welfare!

Your Concept of Welfare Determines What is Sound Science

To clarify: suppose you hold the view that an animal is well off when it is productive, as per the CAST report. The role of your welfare science in this case will be to study what feed, bedding, temperature, etc.. are most efficient at producing the most meat, milk, or eggs for the least money – much what animal and veterinary science does today. On the other hand, if you take the FAWC

B. Rollin

view of welfare, your efficiency will be constrained by the need to acknowledge the animal's natural behavior and mental state, and to assure that there is minimal pain, fear, distress, or discomfort – not factors in the CAST view of welfare, unless they have a negative impact on economic productivity. Thus, in a real sense, sound science does not determine your concept of welfare; rather, your concept of welfare determines what counts as sound science!

The failure to recognize the inescapable ethical component in the concept of animal welfare leads inexorably to those holding different ethical views talking past each other. Thus, producers ignore questions of animal pain, fear, distress, confinement, truncated mobility, bad air quality, social isolation, and impoverished environment unless any of these factors impacts negatively on the "bottom line". Animal advocates, on the other hand, give such factors primacy, and are totally unimpressed with how efficient or productive the system may be.

A major question obviously arises here. If the notion of animal welfare is inseparable from ethical components, and people's ethical stance on obligations to farm animals differs markedly across a highly diverse spectrum, whose ethic is to predominate and define, in law or regulation, what counts as "animal welfare"? This is of great concern to the agriculture industry, worrying as they do about "vegetarian activists hell-bent on abolishing meat". In actual fact, of course, such concern is misplaced, for the chance of such an extremely radical thing happening is vanishingly small. By and large, however, the ethic adopted in society reflects a *societal consensus*, what most people either believe to be right and wrong or are willing to accept upon reflection.

All of us have our own personal ethics, which rule a goodly portion of our lives. Fundamental questions as to what we read, what we eat, to whom we give charity, what political and religious beliefs we hold, and myriad others are answered by our personal ethics. These derive from many sources – parents, religious institutions, friends, reading books, and watching movies and television. One is certainly entitled to believe ethically, as do some People for the Ethical Treatment of Animals (PETA) members, that "meat is murder", that one should be a vegan, it is immoral to use products derived from animal research, and so on.

Clearly, a society, particularly a free society, contains a bewildering array of such personal ethics, with the potential for significant clashes between them. If my personal ethic is based in fundamentalist religious beliefs and yours is based in celebrating the pleasures of the flesh, we are destined to clash, perhaps violently. For this reason, social life cannot function simply by relying on an individual's personal ethics, except perhaps in singularly monolithic cultures where all members overwhelmingly share the same values. One can find examples of something resembling this in small towns in rural farming areas, where there is no need to lock one's doors, remove one's keys from the car, or fear for one's personal safety. But of course such places are few, and probably decreasing in number. In larger communities, of course, the extreme case being New York City or London, one finds a welter of diverse cultures and corresponding personal ethics crammed into a small geographical locus. For this reason alone, as well as to control those whose personal ethic may entail taking advantage of others, a *social consensus ethic* is required, one that transcends personal ethics. This social consensus ethic is invariably articulated in law, with manifest sanctions for its violation. As societies evolve, different issues emerge, leading to changes in the social ethic.

Ethics of Animal Treatment Switches from Personal Ethics to Societal Ethics

My claim then, is that in the USA and Europe, beginning roughly in the late 1960s, the treatment of animals has moved from being a paradigmatic example of personal ethics to ever-increasingly falling within the purview of societal ethics and law. How and why has this occurred, and to what extent?

If one looks to the history of animal use in society back to the beginning of domestication some 11,000 years ago, one finds very little social ethics dictating animal treatment. The one exception to this generalization is the prohibition against deliberate, purposeless cruelty; that is, needless infliction of pain and suffering or outrageous neglect, such as failing to provide food or water. This mandate is well illustrated in the Old Testament, where many injunctions illustrate its presence. For example, one is told that when collecting eggs from a bird's nest, one should leave some eggs so as not to distress the animal. Kosher and halal slaughter accomplished by a trained person using a very sharp knife were clearly intended as a viable alternative to the much more traumatic bludgeoning. (That is not, of course, to suggest that such slaughter remains welfare friendly in high-throughput industrialized slaughterhouses.) The rule of Kashrut, prohibiting the eating of milk and meat – "do not seethe a calf in its mother's milk" – seems to be aimed at avoiding loss of sensitivity to animal suffering. Raza Gharebaghi and colleagues in Iran (2007) have reviewed Islamic prohibitions that forbid separating baby birds or animals from their mother. There are also many passages in the Koran that sustenance and water must be provided to working animals (Al Masri, 2007). Rahman and Aidaros (2012) members of the World Organisation for Animal Health (OIE) animal welfare working group, cite many Koranic passages that clearly emphasize kindness to animals.

In the middle ages, St Thomas Aquinas provided a more anthropocentric reason for prohibiting cruelty, based in the prescient psychological insight that those who would abuse animals will inexorably progress to abusing humans. Aquinas does not see animals as direct objects of moral concern but, none the less, strongly prohibits their abuse.

In the late 18th century in Britain, and in subsequent years elsewhere, the prohibition against deliberate, sadistic, deviant, wilful, malicious cruelty, that is inflicting pain and suffering on animals to no reasonable purpose, or outrageous neglect such as not providing food or water, were encoded in the anti-cruelty laws of all civilized societies. While adopted in part out of a moral notion of limiting animal suffering, an equally important reason was the Thomistic one – to ferret out individuals who might graduate to harming humans; case law in the USA and elsewhere makes this manifest.

In one revealing case in the 19th century, a man was charged with cruelty after throwing pigeons into the air and shooting them to demonstrate his skill. After killing the birds, he ate them. The court ruled that the pigeons were not "needlessly or unnecessarily killed" because the killing was done "in the indulgence of a healthful recreating during an exercise tending to promote strength, bodily agility and courage". In discussing a similar 19th-century case of a tame pigeon shoot in Colorado, the court affirmed that

> every act that causes pain and suffering to animals is not prohibited. Where the end or object in view is reasonable and adequate, the act resulting in pain is ... necessary and justifiable, as ... where the act is done to protect life or property, or to minister to the necessities of man.

To the credit of the Colorado court, it did not find that such tame pigeon shoots met the test of "worthy motive" or "reasonable object". Even today, however, there are jurisdictions where tame pigeon shoots and "canned hunts" do not violate the anti-cruelty laws.

It is certainly true that cruelty to animals is closely linked to psychopathic behavior – animal cruelty, along with fire starting, are signs of future psychopaths. The majority of children who shoot up their schools have early histories of animal abuse, as do 80% of the violent offenders in Leavenworth Prison, Kansas, USA, and most serial killers. Animal abusers often abuse wives and children (Ascione *et al.*, 2007; Volant *et al.*, 2008). Most battered women's shelters must make provisions for keeping the family pet, as the abuser will hurt the animal to hurt the woman. Several studies have shown a relationship between childhood animal cruelty and violence towards people (Miller, 2001). But these laws conceptually provide little protection for animals. Animal cruelty accounts for only a tiny fraction of the suffering that animals undergo at human hands. For example, the USA produces 9 billion broiler chickens a year, and many have bruises and fractures or other skeletomuscular injuries that occur during catching and transport. Before restaurant companies started doing animal welfare audits, careless, rough handling of chickens resulted in 5% of the birds suffering broken wings, which is a shocking 450,000,000 birds with an injury as severe as a broken arm. If even 1% of chickens are so injured (a ridiculously low proportion), then we have 90,000,000 suffering animals there alone – there is nothing like 90,000,000 incidents of cruelty, and those chickens are legally unprotected. In the USA, they are not even subject to humane slaughter law! In Europe and Canada, humane slaughter laws include poultry.

The End of Animal Husbandry

Why was the law or societal ethic protecting animals so minimalistic historically? When I first began to work in animal ethics, I thought the answer lay in cynical exploitation of the powerless. I later realized that the answer was much more subtle, and lay in what I have called "the end of husbandry".

The traditional account of the growth of human civilization out of a hunter-gatherer society invariably invokes the rise of agriculture, that is the domestication of animals and the cultivation of crops. This, of course, allowed for as predictable a food supply as humans could create in the vagaries of the natural world – floods, droughts, hurricanes, typhoons, extremes of heat and cold, fires, etc. Indeed, the use of animals enabled the development of successful crop agriculture, with the animals providing labour and locomotion, as well as food and fibre.

This eventuated in what Dr Temple Grandin has called the "ancient contract" with animals, a highly symbiotic relationship that has endured essentially unchanged for thousands of years. Humans selected animals congenial to human management, and further shaped them in terms of temperament and production traits by breeding and artificial selection. These animals included cattle – dubbed by Calvin Schwabe (1978), a famous veterinarian, the "mother of the human race" – sheep, goats, horses, dogs, poultry and other birds, swine, ungulates, and other animals capable of domestication. The animals provided food and fibre – meat, milk, wool, leather; power to haul and plough; transportation; and served as weaponry – for example, horses and elephants. As people grew more effective at breeding and managing the animals, productivity was increased.

As humans benefited, so simultaneously did the animals. They were provided with the necessities of life in a predictable way. And thus was born the concept of husbandry, the remarkable practice and articulation of the symbiotic contract.

"Husbandry" is derived from the Old Norse words "hus" and "bond"; the animals were bonded to one's household. The essence of husbandry was care. Humans put animals into the most ideal environment possible for the animals to survive and thrive, the environment for which they had evolved and been selected. In addition, humans provided them with sustenance, water, shelter, protection from predation, such medical attention as was available, help in birthing, food during famine, water during drought, safe surroundings, and comfortable appointments.

Eventually, what was borne of necessity and common sense became articulated in terms of a moral obligation inextricably bound up with self-interest. In the Noah story, we learn that even as God preserves humans, humans preserve animals. The ethic of husbandry is, in fact, taught throughout the Bible; the animals must rest on the Sabbath even as we do, one is not to seethe a calf in its mother's milk (so we do not grow insensitive to animals' needs and natures); we can violate the Sabbath to save an animal. The Islamic tradition has similar instructions on animal use, "do not ride an animal as much, as it can no longer bear your load, and be fair to animals … If an animal is exhausted, it must take some rest" (Shahidi, 1996).

Proverbs tells us that "the wise man cares for his animals". The Old Testament is replete with injunctions against inflicting unnecessary pain and suffering on animals, as exemplified in the strange story of Balaam, who beats his ass and is reprimanded by the animal's speaking through the grace of God.

The true power of the husbandry ethic is best expressed in the 23rd Psalm. There, in searching for an apt metaphor for God's ideal relationship to humans, the Psalmist invokes the good shepherd:

> The Lord is my shepherd; I shall not want. He maketh me to lie down in green pastures; He leadeth me beside still waters. He restoreth my soul.

We want no more from God than what the good shepherd provides to his animals. Indeed, consider a lamb in ancient Judaea. Without a shepherd, the animal would not find forage or water easily, would not survive the multitude of predators the Bible tells us prowled the land – lions, jackals, hyenas, birds of prey, and wild dogs. Under the aegis of the shepherd, the lamb lives well and safely. Shepherds and herders from many different cultures ranging from the Nues in Africa, the Incas, the Basques in Spain, and Indian herders all composed poems and songs about being a shepherd (Kessler, 2009). All of these traditions promoted good husbandry and a bond with the animals and the land.

In return, the animals provide their products, and sometimes their lives, but while they live, they live well. As we saw, even slaughter, the taking of the animal's life, must be as painless as possible, performed with a sharp knife by a trained person to avoid unnecessary pain. Ritual slaughter, both kosher and halal, was, in antiquity, a far kinder death than bludgeoning; most importantly, it was the most humane modality available at the time.

When Plato discusses the ideal political ruler in *The Republic*, he deploys the shepherd–sheep metaphor: the ruler is to his people as the shepherd is to his flock. Qua shepherd, the shepherd exists to protect, preserve and improve the sheep; any payment tendered to him is in his capacity as wage earner. So, too, it is for the ruler, again illustrating the power of the concept of husbandry on our psyches. To this day, ministers are called shepherds of their congregation, and "pastor" derives from "pastoral".

The singular beauty of husbandry is that it is at once an ethical and prudential doctrine (Rollin, 2003). It was prudential in that failure to observe husbandry inexorably led to ruination of the person keeping animals. Not feeding, not watering, not protecting from predators, not respecting the animals' physical, biological, psychological needs and natures, what Aristotle called their *telos* – the "cowness of the cow", the "sheepness of the sheep" – meant your animals did not survive and thrive, and thus neither did you. The ultimate sanction of failing at husbandry – erosion of self-interest – obviated the need for any detailed ethical exposition of moral rules for husbandry: anyone unmoved by self-interest is unlikely to be moved by moral or legal injunctions! And thus one finds little written about animal ethics and little codification of that ethic in law before the 20th century, with the bulk of what is articulated aimed at the pathological cruelty we have discussed. Thus, lack of animal ethics and law was explained by the nature of the predominant use of animals – agriculture – and what was needed for its success – good husbandry.

The Rise of Industrial Agriculture

In the mid-20th century, husbandry agriculture was supplanted by industrial agriculture in many countries. Ruth Harrison (1964), in her famous book, *Animal Machines*, described the horrid conditions that she observed on intensive farms in the UK. Chickens crammed into tiny cages, tethered veal calves, and sows that were not able to turn around. Her book was translated into seven languages, and it served as an impetus for the formation of the Brambell Committee that wrote the five freedoms (McKenna, 2000; Van de Weerd and Sandilands, 2008). Her book caused the public to become concerned about the intensification and lack of husbandry that was occurring on British poultry, veal, and pig farms. Industrialization of animal agriculture in the USA has occurred for a variety of understandable and even laudable reasons that are worth recounting.

1. When industrial agriculture began, roughly in the 1940s, the USA was confronted with a variety of new challenges related to food. In the first place, the economic Great Depression and Dust Bowl (severe drought) had soured many people on farming and, even more dramatically, had raised the spectre of starvation for the American public for the first time in US history. Vivid images of breadlines and soup kitchens drove the desire to assure a plentitude of cheap food. By the late 1960s and 1970s, the USA had large-scale industrialized animal agriculture with much bigger units compared to Europe.

2. Better jobs were to be found in cities, and rural people flocked to them in the hope of a better life, creating a potential shortage in agricultural labour.

3. Correlative with the growth of cities and suburbs came encroachment on agricultural land for various forms of development, raising land prices and moving acreage once available for agriculture out of that pool.

4. Many people who would otherwise have been happy with a slow, rural way of life were exposed to greater sophistication by virtue of military service in World Wars I and II, and thus were dissatisfied with an agrarian existence. Recall the song popular after World War I, "How ya gonna keep 'em down on the farm (after they've seen Paree?)". Ironically, by the late 1960s, many urban people were yearning for the "simple life" represented by the small farm. Today in developed countries, there is a huge interest in buying animal products from small, local family farms. Urban people are yearning for a return to animal husbandry.

5. Demographers predicted a precipitous and dramatic increase in population, which turned out to be accurate.

6. With the success of industrialization in new areas, notably Henry Ford's application of the concept to the automobile, it was probably inevitable that the concepts of industrialization would be applied to agriculture. (Ford himself had already characterized slaughterhouses as "disassembly lines".)

Thus was born an industrial approach to agriculture, with machines taking the place of labour. The traditional departments of animal husbandry in agricultural schools symbolically marked this transition by changing their names to departments of animal sciences, a field defined in textbooks as "application of industrial methods to the production of animals".

In this transition, the traditional bedrock values of agriculture, husbandry, sustainability, agriculture as a way of life not only a way of making a living, were transmuted into values of efficiency and productivity. With human labour replaced by machinery, in turn requiring large amounts of capital, farm units grew larger, eventuating in the mantra of the 1970s "get big or get out". Agricultural research stressed producing cheap and plentiful food, and moved in unprecedented directions. With animals confined for efficiency and away from forage, much research was directed towards finding cheap sources of nutrition, in turn leading to feeding such deviant items to animals as poultry and cattle manure. Animals were kept under conditions alien to their natural needs for the sake of productivity. Whereas husbandry animal agriculture stressed putting square pegs into square holes, round pegs into round holes, while producing as little friction as possible, industrialized animal agriculture forced square pegs into round holes by utilizing what I have called "technological sanders", such as antibiotics, hormones, extreme genetic selection, air handling systems, artificial cooling systems, and artificial insemination to force animals into unnatural conditions while they none the less remained productive.

Consider, for example, the egg industry, one of the first areas of agriculture to experience industrialization. Traditionally, chickens ran free in barnyards, able to live off the land by foraging and expressing their natural behaviors of moving freely, nest building, dust bathing, escaping from more aggressive animals, defecating away from their nests, and, in general, fulfilling their natures as chickens. Industrialization of the egg industry, on the other hand, meant placing the chickens in small cages, in some systems with six birds in a tiny wire cage, so that one animal may stand on top of the others and none can perform any of their inherent behaviors, unable even to stretch their wings. In the absence of space to establish a dominance hierarchy or pecking order, they cannibalize each other, and must be "debeaked", producing painful neuromas because the beak is innervated (Gentle *et al.*, 1990). The biology of the laying hen has been pushed to the point where a high percentages of old layers have osteoporosis and broken keel bones (Sherwin *et al.*, 2010; Wilkins *et al.*, 2011). Keel bone fractures are painful (Nasr *et al.*, 2012). The animal is now an inexpensive cog in a machine, part of a factory, and the cheapest part at that, and thus totally expendable. If a 19th-century farmer had attempted such a system, he would have gone broke, with the animals dead of disease in a few weeks. Some genetic lines of pigs and chickens are so highly selected for egg and meat production that they have less disease resistance. Pigs have become so susceptible to disease that some farmers have installed antibacterial filters to take germs out of the air that enters the building (Vansickle, 2008). This is a technological sander taken to the extreme.

Problems with Industrial Agriculture

The steady state, assuring the balance of humans, animals, and land, is lost. Putting chickens in cages and cages in an environmentally controlled building requires large amounts of capital and energy, and technological "fixes"; for example, to run the exhaust fans to prevent lethal build-up of ammonia. The value of each chicken is negligible, so one needs more chickens; chickens are cheap, cages are expensive, so one crowds as many chickens into each cage as is physically possible. The vast concentration of chickens requires huge amounts of antibiotics and other drugs to prevent wildfire spread of disease in overcrowded conditions. Breeding of animals is oriented solely towards productivity, and genetic diversity – a safety net allowing response to unforeseen changes – is lost. Bill Muir, a genetics specialist at Purdue University, Indiana, USA, found that commercial lines of poultry had lost 90% of their genetic diversity compared to noncommercial poultry (Lundeen, 2008). Dr Muir is extremely concerned about the lack of genetic diversity. Small poultry producers are lost, unable to afford the capital requirements; agriculture as a way of life as well as a way of making a living is lost; small farmers, who Thomas Jefferson argued are the backbone of society (Torgerson, 1997), are superseded by large corporate aggregates. Giant corporate entities, vertically integrated, are favored. Manure becomes a problem for disposal, and a pollutant, instead of fertilizer for pastures. Local wisdom and know-how essential to husbandry is lost; what "intelligence" there is is hard-wired into "the system". Food safety suffers from the proliferation of drugs and chemicals, and widespread use of antimicrobials to control pathogens in effect serves to breed – select for – antibiotic-resistant pathogens as susceptible ones are killed off. Above all, the system is not balanced – constant inputs are needed to keep it running and to manage the wastes it produces, and create the drugs and chemicals it consumes. And the animals live miserable lives, for productivity has been severed from well-being.

One encounters the same dismal situation for animals in all areas of industrialized animal agriculture. Consider, for example, the dairy industry, once viewed as the paradigm case of bucolic, sustainable animal agriculture, with animals grazing on pasture, giving milk and fertilizing the soil for continued pasture with their manure. Though the industry wishes consumers to believe that this situation still exists – the California dairy industry ran advertisements proclaiming that California cheese came from "happy cows", and showing the cows on pastures – the truth is radically different. The vast majority of California dairy cattle spend their lives on dirt and concrete, and in fact never see a blade of pasture grass, let alone consume it. So outrageous is this duplicity that the dairy association was sued for false advertising and a friend of mine, a dairy practitioner for 35 years, was very outspoken against such an "outrageous lie".

In actual fact, the life of dairy cattle is not a pleasant one. In a problem ubiquitous across contemporary agriculture, animals have been bred single-mindedly for productivity; in the case of dairy cattle, for milk production. Today's dairy cow produces three to four times more milk than 60 years ago. In 1950, the average dairy cow annually produced 2410 l (5314 lb) of milk per lactation. Fifty years later, it was close to 9072 l (20,000 lb) (Blaney, 2002). From 1995 to 2004 alone, milk production per cow increased 16%. Today, the dairy cow has had further increases in milk production. This has resulted in an animal that is having greater and greater reproductive problems (Rodenburg and Turner, 2012). The result is a milkbag on legs, and unstable legs at that. A high percentage of the US dairy herd is chronically lame, and these cows suffer serious reproductive problems. Espejo *et al.* (2006) found that an average of 24.6% of dairy cows housed in cubicle (freestall) barns were lame. A more recent study showed that the average percentage of lame dairy cows was 28% (von Keyserlingk *et al.*, 2012). Whereas, in traditional agriculture, a milk cow could remain productive for 10 and even 15 years, today's cow lasts slightly longer than two lactations, a result of metabolic burnout and the quest for ever-increasingly productive animals, hastened in the USA by the use of the hormone, bovine somatotropin (BST), to increase production further. Such unnaturally productive animals naturally suffer from mastitis, and the industry's response to mastitis in portions of the USA has created a new welfare problem by the docking of cow tails without anesthesia in a futile effort to minimize teat contamination by manure. Still practiced, this procedure has been definitively demonstrated not to be relevant to mastitis control or lowering somatic cell count (Stull *et al.*, 2002). (In my view, the stress and pain of tail amputation coupled with the concomitant inability to chase away flies may well dispose to more mastitis.) Calves are removed from mothers shortly after birth, before receiving colostrum, creating significant distress in both mothers and infants. Bull calves may be shipped to slaughter or a feedlot immediately after birth, generating stress and fear.

The intensive swine industry, which, through a handful of companies, is responsible for 85% of the pork produced in the USA, is also responsible for significant suffering that did not affect husbandry-reared swine. Certainly, the most egregious practice in the confinement swine industry, and possibly, given the intelligence of pigs, in all of animal agriculture, is the housing of pregnant sows in gestation crates or stalls – essentially small cages. The *recommended* size for such stalls, in which the sow spends her entire productive life of about 4 years, with a brief exception we will detail shortly, according to the industry is 0.9 m high by 0.64 m wide by 2.2 m long – this for an animal that may weigh 275 kg or more. (In reality, many stalls are smaller.) The sow cannot turn around, walk, or even scratch her rump. In the case of large sows, she cannot even lie flat, but must remain lying on her sternum. The exception alluded to is the period of farrowing – approximately 3 weeks – when she is transferred to a "farrowing crate" to give birth and nurse her piglets. The space for her is not greater, but there is a "creep rail" surrounding her, so the piglets can nurse without being crushed by her postural adjustments.

Under extensive conditions, a sow will build a nest on a hillside so excrement runs off; forage an area covering 2 km a day; and take turns with other sows to watch piglets, so allowing all sows to forage. With the animal's nature thus aborted, she goes mad, exhibits bizarre and deviant behavior such as compulsively chewing the bars of the cage; she also endures foot and leg problems, and lesions from lying on concrete in her own excrement (see Chapters 8 and 15, this volume).

These examples suffice to illustrate the absence of good welfare in confinement. Rest assured that a long litany of issues could be addressed. In general, all animals in confinement agriculture (with the sometime exception of beef cattle who live most of their lives on pasture and are "finished" on grain in dirt feedlots, where they can actualize much of their nature) suffer from the same generic set of affronts to their welfare absent in husbandry agriculture.

Quick Access Information 3.1. Problems caused by loss of husbandry.

1. Production diseases – by definition a production disease is a disease that would not exist or would not be of serious epidemic import were it not for the method of production. Examples are liver and rumenal abscesses resulting from feeding cattle too much grain, rather than roughage. In fed beef, 24% had ruminitis lesions and only 68% had normal livers (Rezac *et al.*, 2014). The animals that get sick are more than balanced out economically by the remaining animals' weight gain. Other examples are confinement-induced environmental mastitis in dairy cattle, weakness caused by beta-agonists to increase muscle mass in pigs, and "shipping fever" (bovine respiratory disease) in beef cattle. There are textbooks of production diseases, and one of my veterinarian colleagues calls such disease "the shame of veterinary medicine", because veterinary medicine should be working to eliminate such pathogenic conditions, rather than treating the symptoms.

2. Loss of workers who are "animal smart" – in large industrial operations such as swine factories, the workers are minimum-wage, sometimes illegal, often migratory, workers with little animal knowledge. Confinement agriculturalists will boast that "the intelligence is in the system", and thus the historically collective wisdom of husbandry is lost, as is the concept of the historical shepherd, now transmuted into rote, cheap, labour.

3. Lack of individual attention – under husbandry systems, each animal is valuable. In intensive swine operations, the individuals are worth little. When this is coupled with the fact that workers are no longer caretakers, the result is obvious.

4. The lack of attention to animal needs determined by their physiological and psychological natures – as mentioned earlier, technological sanders allow us to keep animals under conditions that violate their natures, thus severing productivity from assured well-being.

It is absolutely essential to stress here that there was no malicious attempt to ride roughshod over animal welfare by those who created and perpetuated these systems. (Almost all animal scientists and producers in industrial agriculture came from a husbandry background.) The failure to account for animal welfare in these new systems basically arose from a conceptual error.

Let us recall that traditional agricultural success under husbandry depended on the proper treatment of animals – generally, if the animals suffered, the producer lost out. Under such circumstances, it was perfectly reasonable (subject to some refinements) to use animal productivity as a good (rough) criterion for welfare. The people who developed confinement agriculture continued to use this criterion, failing to mark the singular difference between industry and husbandry. In husbandry conditions, animal suffering would *usually* lead to loss of productivity. Under industry, however, the technological sanders mentioned sever the close connection between productivity and welfare, in so far as the animals may suffer harm in ways that do not impinge on economic productivity. In some cases, such as the modern dairy cow, great productivity has been univocally selected into the cow's genetics, in ways that actually create new welfare problems, such as metabolic "burnout", foot and leg problems, and reproductive problems (Rodenburg and Turner, 2012; Spencer, 2013). In one study, the most productive cows had a higher percentage of swollen legs (Fulwider *et al.*, 2007). On many intensive dairies, a cow lasts for only two lactations. We saw this conceptual error expressed in the CAST report of 1982 (CAST, 1982) mentioned earlier.

Environmental and Societal Cost of Intensive Agriculture

Animal welfare is not the only problem inadvertently occasioned by the industrialization of animal agriculture. The Pew Commission on Industrial Farm Animal Production in America, on which I was privileged to serve, recently issued a report on all of the issues after over 2 years of intensive study. (The report may be found at www.ncifap.org.) We conclude that although industrial agriculture does indeed yield food that is cheap at the cash register, there are many significant costs of such agriculture that are borne by society (the consumers) that do not appear at the cash register. (Economists call this *externalization* of costs.) These costs include:

1. Environmental despoliation – in highly concentrated industrial agriculture, animal waste disposal, for example, becomes a major avenue to pollution. Air pollution is a related problem.

2. Loss of small farmers and robust rural communities to consolidation – the USA, for example, lost 85% of the swine producers it had in the early 1980s.

3. Human health danger costs – these include incubation of pathogens in high confinement; respiratory problems among those working in some high animal confinement units; low-level antibiotic nontherapeutic use in confined animals, driving antibiotic resistance (Marshall and Levy, 2011); and other issues of food safety.

4. Animal welfare – as discussed.

At any rate, let us return to our main topic: the notion of animal welfare implicit in emerging social ethics for animals that must be understood for animal agriculture to meet societal demands.

Increased Public Concern for Animals

It is manifest first of all that public concern for animal treatment has increased greatly over the past few years, particularly as people have become aware that much modern animal use is not a fair contract with the animals, but is patently exploitative, particularly regarding animal agriculture and animal research and testing. This has, in turn, led globally to a proliferation of laws attempting to address animal suffering not covered by the cruelty laws, even as the cruelty laws themselves have been strengthened and elevated to felony status in 40 US states. Over 2100 laws relevant to animal welfare were proposed in state legislatures in the USA in 2004, and laboratory animal legislation has proliferated throughout Europe and other countries. The OIE welfare codes on slaughter, transportation, and animal production have now brought animal welfare issues to the attention of the developing world. Many abusive practices have been eliminated by public pressure, and consumers increasingly express their buying preferences for humanely raised foodstuffs and cosmetics derived without additional animal testing.

As media coverage of animal issues increases, the social paradigm for animals moves towards the companion animal "member of the family", prominent people speak out for animals, and books on animal ethics proliferate, people want assurance of proper animal treatment. And, as I predicted 30 years ago, with husbandry gone and most animal suffering invisible to the anti-cruelty ethic and laws,

society will require a new ethic for animals and will look to extant ethics for humans, appropriately modified, to do the job. As Plato remarked, ethics proceeds from pre-existing ethics, and one develops new ethical coverage, not by creating new principles (what he calls "teaching") but rather by *reminding* people of the logical implications of their current ethical beliefs.

The portion of our ethical beliefs being so extended to animals is patent. Every society faces a fundamental conflict of two goods – the good of the group and the good of the individual. It is, for example, in the group interest to tax the wealthy, but not in the interest of a wealthy individual. It is in the group interest to draft young men to fight wars, but not in the interest of those individuals. Many societies, totalitarian societies, resolve this conflict by positing the primacy of the corporate entity.

With regards to animal agriculture, the pastoral images of animals grazing on pasture and moving freely are iconic. As the 23rd Psalm indicates, people who consume animals wish to see the animals live decent lives, not lives of pain, distress, and frustration. It is for this reason in part that industrial agriculture conceals the reality of its practices from a naïve public – witness Purdue's advertisements about raising "happy chickens", or the California "happy cow" advertisements. As ordinary people discover the truth, they are shocked. When I served on the Pew Commission and other commissioners had their first view of sow stalls, many were in tears, and all were outraged.

The Importance of the Animal's Innate Nature (Telos)

Just as our use of people is constrained by respect for the basic elements of human nature, people wish to see a similar notion applied to animals. Animals, too, have natures; what I call *telos*, following Aristotle – the "pigness of the pig", the "cowness of the cow". Pigs are "designed" to move about on soft loam, not to be in gestation crates. If this no longer occurs naturally, as it did in husbandry, people wish to see it legislated. This is the mainstream sense of "animal rights".

As property, strictly speaking, animals cannot have legal rights. But a functional equivalent to rights can be achieved by limiting property rights. When I and others drafted the US federal laws for laboratory animals, we did not deny that research animals were the property of researchers. We merely placed limits on their use of their property. I may own my car, but that does not mean I can drive it on the sidewalk, or at any speed I choose. Similarly, our law states that if one hurts an animal in research, one must control pain and distress. Thus, research animals can be said to have the *right* to have their pain controlled.

In the case of farm animals, people wish to see their basic needs and nature, *teloi*, respected in the systems in which they are raised. Since this no longer occurs naturally, as it did in husbandry, it must be imposed by legislation or regulation. A Gallup poll conducted in 2003 showed that 62% of the public wanted legislated guarantees of farm animal welfare. This is what I call "animal rights as a mainstream phenomenon". Legal codification of rules of animal care respecting animal *telos* is thus the form animal welfare takes where husbandry has been abandoned.

Thus, in today's world, the ethical component of animal welfare prescribes that the way we raise and use animals must embody respect and provision for their psychological needs and natures. It is therefore essential that industrial agriculture phases out those systems that cause animal suffering by violating animals' natures and replaces them with systems respecting their natures.

Animal Science and Veterinary Medicine Must Address Values and Ethics

As I have told many animal agriculture and veterinary groups, this does not mean the end of animal science. What it does mean is that while the fundamental values driving animal science have been efficiency and productivity, animal science must vector into the design of its systems the other values we have discussed – respect for animal nature and minimizing animal pain and distress, control of environmental degradation growing out of production systems, concern for the effect of animal production on rural communities and human health, and concern for animals' needs and natures. In short, animal science and production veterinary medicine must become what the 18th century called a "moral science". The irreducibly moral component of the concept of animal welfare, long ignored, must be embraced and engaged by the industry.

This is why highly restrictive systems such as sow gestation stalls are being phased out. They have already been eliminated in the UK, and the EU countries in Europe banned sow gestation stalls in 2013. In the USA and Canada, pressure from consumers and animal advocacy groups has been instrumental in getting large corporations such as Smithfield and Maple Leaf to phase out sow stalls. It is being done out of response to the ethical concerns of consumers.

Europe continues to pursue legislative remedies to farm animal welfare issues. For example, the European Union has banned piglet castration without anesthesia, effective 2012, and projects a total ban on such castration by 2018. In general, the European public is considerably more knowledgeable about issues of farm animal welfare than is the US public. For example, supermarket chains such as Tesco have long distributed information to customers on the standards they employ when purchasing animal products. While speciality markets in the USA such as Whole Foods do publicize their standards, the information disseminated tends to be significantly less detailed. It is difficult to know why Europeans are more knowledgeable about animal welfare than US citizens, but it is probably due to the fact that the USA is so geographically expansive that confinement systems remain hidden from public view, and unapproachable, for alleged reasons of "biosecurity". (On one occasion, my son and I were on a motorcycle trip in a very remote area of southern Utah. We stopped to climb a hill. At the top, we could see the sun reflecting off low buildings with metal roofs. He asked me what those buildings were. I assumed they represented some secret military facility such as Area 51, but suggested we look for a side road off the main road. Driving slowly, we found a road that was virtually invisible from the highway, and took it for 5 miles until we reached a guardhouse. We asked the guard if this was a military installation. "No," he replied proudly. "This is the Acme Swine Company and we produce more fecal waste than the city of Los Angeles.")

At any rate, in the USA, there has been far less legislation directed against factory farming than in Europe. In fact, a major concern the Pew Commission put at the top of our list of action items was feeding antibiotics to farm animals, both to promote growth and to blunt the negative effects of how the animals were kept, resulting in breeding pathogens impervious to antibiotics and dangers to humans. Despite this concern being brought forward to Congress by private and governmental health authorities, the influence of the agricultural lobby is so great that absolutely no action has been taken. The most influential political action has been at the state level, as a result of public referenda and public initiatives placed on state legislative ballots over the past 15 years by the

Humane Society of the United States, aimed at eliminating sow stalls, high confinement crates for veal calves, preventing these animals from developing muscle that would make the meat less tender, and grossly overpopulated and impoverished "battery cages" for laying hens. Tellingly, these initiatives passed by 2 to 1 in all the states they were attempted.

Consumers Demand Good Animal Welfare

The most progress in animal welfare has occurred as a result of consumer demand. Beginning with pressure on such corporate giants as McDonald's, Burger King, and Wendy's, these entities have been forced to buy from producers utilizing humane practices. In my view, such an approach may, in the long run, actually be preferable to legislative action, in a way illustrated in the following anecdote. In 2008, when I was working with Chipotle to help develop their animal welfare commitment as emblematic of the way they do business, I was asked if I would be willing to have two Smithfield (the world's large pork producer) executives attend a speech that I was giving to a division of Chipotle. I said, "Of course!" The talk was to be held in Vail, Colorado, and we arrived the evening before. During dinner, I laid out the social-ethical reasons for them to abandon gestation crates voluntarily. They attended my talk about emerging social animal ethics the next day and we continued our conversation. I urged them not just to take my word regarding animal ethics, but to ask their consumers. Six months later, I received a call from them telling me they had done that. "And what did you find out?", I queried. "Well, we used questionnaires and focus groups to determine what our customers thought of sow stalls. Actually we found out you were wrong. You had told us that 75% of the public did not accept gestation crates. In fact, it was 78%!" With that basis, they proceeded to phase out gestation crates in their own facilities and recently moved to eliminate them from their contract growers as well. They key point of this story is to illustrate that by talking to consumers, these people got a rich picture of animal ethics among consumers. If the ban on sow stalls had come through external government pressure, nothing like the same level of understanding would have been accomplished. Recently, Smithfield has moved to expand the end of sow stalls to its contract suppliers.

In January of 2014, corporate giant, Tyson, made an announcement relevant to new animal welfare suggestions addressed to its suppliers. Those suggestions include the use of video monitoring in sow farms, discontinuation of manual blunt-force trauma as a primary method of euthanasia, use of pain mitigation for tail docking and castration, and the requirement for sow housing built or renovated in 2014 to provide adequate quality and quantity of space for gestating sows. Although these are not mandated requirements, they represent a step in the right direction, again presumably based on consumer concerns.

There are currently more than 60 companies that have rejected pork from suppliers who use gestation crates. These companies include:

McDonald's
DineEquity (IHOP and Applebees)
Chipotle
Campbell Soup
Atlantic Premium
Wendy's
Costco
Sysco
Bon Appetit
Williams Sausage
Burger King
Smithfield (the world's largest pork producer)
Kraft Foods (Oscar Mayer)
Quiznos
Harris Teeter
Subway
Cracker Barrel
Compass Group
Einstein Noah Restaurant Group
Kmart
Denny's
Hormel Foods (leading pork producer)
Target Corp.
Carnival Cruise Lines
TrustHouse
Dunkin' Brands
Maple Leaf (top Canadian pork producer)
Jack in the Box/Qdoba
Brinker (Chili's, Romano's, Maggiano's)
Whole Foods
Sonic Drive-In
Kroeger
ConAgra Foods
Wienerschnitzel
Heinz

Arby's
Safeway
Carl's Jr/Hardees
Hillshire Brands (Jimmy Dean, Ballpark)
Bruegger's Bagels
General Mills
SuperValu
The Cheesecake Factory
Royal Caribbean Cruise Lines
Baja Fresh Aramark
Sodexo
Wolfgang Puck
Tim Hortons
Bob Evans Farms
TGI Fridays
Olymel (top Canadian pork producer)
Metz Culinary
Tyson Foods
Johnsonville Sausage
Roundy's
The Retail Council of Canada (Walmart, Safeway, Loblaw, Metro, Costco, Metro and the other largest Canadian retailers)

(List courtesy of Paul Schapiro, Vice President, Humane Society of the United States)

The list eloquently illustrates the degree to which attention to social ethics can generate societal change without legislative intervention. Reference to ethics appears in virtually all of the corporate statements explaining the abandonment of sow stalls. Though the change may take somewhat longer, it is likely to be better understood.

References

Al Masri, H. (2007) *Animal Welfare in Islam*. Islamic Foundation, London.

Ascione, F.R., Weber, C.V., Thompson, T.M., Heath, J., Maruyama, M., et al. (2007) Battered pets and domestic violence: Animal abuse reported by women experiencing intimate violence and by nonabused women. *Violence Against Women* 13, 354–373.

Blaney, D.P. (2002) The changing landscape of U.S. milk production. United States Department of Agriculture (USDA), Statistical Bulletin Number 978. USDA Agricultural Research Service (ARS), Washington, DC.

Council for Agricultural Science and Technology (CAST) (1982) *Scientific Aspects of the Welfare of Food Animals*. Report No. 91. CAST, Ames, Iowa.

CAST (1997) *Wellbeing of Agricultural Animals Task Force Report, Chaired by Stanley Curtis*. CAST, Ames, Iowa.

Espejo, L.A., Endres, M.I. and Salfer, J.A. (2006) Prevalence of lameness in high producing Holstein cows housed in freestall barns in Minnesota. *Journal of Dairy Science* 89, 3052–3058.

Farm Animal Welfare Council (FAWC) (2009) Available at:http://webarchive.nationalarchives.gov.uk/20121007104210/http://www.fawc.org.uk/freedoms.htm(accessed 13 April 2015).

Fulwider, W.K., Grandin, T., Garrick, D.J., Engle, T.E., Lamm, W.D., et al. (2007) Influence of free stall base on tarsal joint lesions and hygiene in dairy cows. *Journal of Dairy Science* 90, 3559–3566.

Gallup (2003) Available at: www.gallup.com (accessed 18 June 2009).

Gentle, M.J., Waddington, D., Hunter, L.N. and Jones, R.B. (1990) Behavioral evidence for persistent pain following partial beak amputation in chickens. *Applied Animal Behavioural Science* 27, 149–157.

Gharebaghi, R., Reza Vaez Mahdavi, M., Ghasemi, H., Dibaei, A. and Heidary, F. (2007) Animal rights in Islam. In: *Proceedings of the Sixth World Congress in Alternatives and Animal Use in the Life Sciences*, 21–25 August 2007, Tokyo, Japan, pp. 61–63.

Harrison, R. (1964) (Reprinted 1966) *Animal Machines*. Ballantine Books, New York.

Kessler, B. (2009) *Goat Song*. Scribner, New York.

Lundeen, T. (2008) Poultry missing genetic diversity. *Feedstuffs*, 1 December, p. 11.

McKenna, C. (2000) Ruth Harrison. *Guardian*, 6 July.

Marshall, B.M. and Levy, S.B. (2011) Food animals and antimicrobials: Impacts on human health. *Clinical Microbiology Review* 24, 718–733.

Miller, C. (2001) Childhood animal cruelty and interpersonal violence. *Clinical Psychology Review* 21, 735–749.

Nasr, M.A.F., Nico, C.J. and Murrell, J.C. (2012) Do laying hens with keel bone fractures experience pain? *PLoS One*, doi:10.1371/journal.pone.0042420.

Pew Commission (2008) Putting Meat on the Table: Industrial Farm Production in America. Pew Commission on Industrial Farm Animal Production in America. Available at: http://www.ncifap.org/_images/pcifapfin.pdf (accessed 13 April 2015).

Rahman, S.A. and Aidaros, H. (2012) Islam and animal welfare with special reference to cruelty to animals during transport and slaughter. Available at: http://www.australiananimalwelfare.com.au/article-m/islam-and-animal-welfare-with-special-reference-to-cruelty-to-animals-during-transport-and-slaughter (accessed 30 March 2015).

Rezac, D.J., Thomson, D.U., Bartle, S.J., Osterstock, J.B., Prouty, F.L., et al. (2014) Prevalence, severity and relationships of lung lesions, liver abnormalities, and rumen health scores measured at slaughter of beef cattle. *Journal of Animal Science* 92, 2595–2602.

Rodenburg, T.B. and Turner, S.P. (2012) The role of breeding and genetics in the welfare of farm animals. *Animal Frontiers* 2, 16–21.

Rollin, B. (2003) *Farm Animal Welfare, Social, Bioethical, and Research Issues*. Wiley, Chichester, UK.

Schwabe, C. (1978) *Cattle, Priests, and the Progress of Medicine*. University of Minnesota Press, Minneapolis, Minnesota.

Shahidi, S.J. (1996) *Translation of Nahj-albalagheh by Iman Ali enbe Abu Talih*. Cultural and Scientific Publishing, Iran, 286 pp.

Sherwin, C.M., Richards, G.J. and Nicol, C.J. (2010) Comparison of the welfare of layer hens in 4 housing systems in the UK. *British Poultry Science* 51, 488–499.

Spencer, T.E. (2013) Early pregnancy: Concepts, challenges and potential solutions. *Animal Frontiers* 3, 48–55.

Stull, C.L., Payne, M.A., Berry, S.L. and Hullinger, P.J. (2002) Evaluation of the scientific justification for tail docking in dairy cattle. *Journal of the American Veterinary Medical Association* 220, 1298–1303.

Torgerson, R. (1997) Jeffersonian democracy and the role of cooperatives. *Rural Cooperatives* 64, 2.

Van de Weerd, H. and Sandilands, V. (2008) Bringing the issue of animal welfare to the public: a biography of Ruth Harrison (1920–2000). *Applied Animal Behaviour Science* 113, 404–410.

Vansickle, J. (2008) Filtration works at a cost. *National Hog Farmer,* 15 November. Available at: http://national-hogfarmer.com/health-diseases/1115-filtration-works-cost (accessed 13 April 2015).

Volant, A.M., Johnson, J.A., Guillone, E. and Coleman, G.J. (2008) The relationship between domestic violence and animal abuse: an Australian Study. *Journal of Interpersonal Violence* 23, 1277–1295.

von Keyserlingk, M.A., Barrientos, A., Ito, K., Galo, E. and Weary, D.M. (2012) Benchmarking cow comfort on North American freestall dairies: Lameness, leg injuries, laying time, facility design, and management for high producing Holstein dairy cows. *Journal of Dairy Science* 95, 7399–7408.

Wilkins, L.J., McKinstry, J.L., Avery, N.C., Knowles, T.G., Brown, S.N., *et al*. (2011) Influence of housing system and design on bone strength and keel bone fractures in laying hens. *Veterinary Record* 169, 414–417.

4 Implementing Effective Standards and Scoring Systems for Assessing Animal Welfare on Farms and Slaughter Plants

Temple Grandin

Department of Animal Science, Colorado State University, Fort Collins, Colorado, USA

Summary

When standards and guidelines are being written for animal welfare, vague wording should be avoided. Avoid terminology such as "proper handling" or "adequate veterinary care". What one person thinks is proper handling, another person may consider the handling practice as abusive. Both the World Organisation for Animal Health (OIE) and many animal specialists recommend the use of animal-based outcome measures. Some of the commonly used outcome measures are body condition score, percentage of lame animals that have difficulty in walking, eye damage from high ammonia levels, animal cleanliness, sores/lesions, and feather or coat condition. There is less emphasis on resource-based input standards that specify how to build animal housing. Resource-based standards are still required for feeder space, water access space, and truck loading densities. Both legislative and private standards will have acts of abuse or housing methods that are prohibited. Some examples would be sow gestation stalls or cruel acts such as dragging downed, nonambulatory animals. The most effective audit and inspection programs have three components: (i) internal (first-party) audits done by a farm's own stockpeople or veterinarian; (ii) third-party audits done by an independent auditing organization; and (iii) managers or inspectors (second-party audits) from either the corporate or government agencies who check random farms to ensure that the auditors are doing their job.

Learning Objectives

- Learn how to write clear standards and guidelines that will be interpreted the same way by different people
- Know the difference between animal-based outcome measures and input resource-based standards
- How to determine the most important core criteria or critical points to prevent abuse or neglect
- Provide easy-to-use measures for assessing body condition, lameness, injuries, condition of haircoat/feathers/animal handling, hygiene, heat and cold stress, and presence of abnormal behavior
- How to set up effective animal welfare auditing programs

Introduction: The Need for Clear Standards and Methods to Improve Consistency Between Different Auditors and Inspectors

Many standards and regulations for animal welfare, disease control, food safety, and many other important areas are too vague and subjective. When vague standards are used, there will be great variation in how they are interpreted. One inspector may interpret the guidelines in a very strict manner and another may allow abuse to occur and have a lax interpretation. The author has trained many auditors and inspectors to assess animal welfare in slaughter plants and on farms. These people need specific information on what conditions are acceptable and what conditions would either be a failed audit or a violation of the law. They ask very specific questions. For example, what is the size of a bruise that counts as a bruised carcass? When does nudging an animal with your foot become kicking the animal, which would be considered an act of abuse? In the poultry barn, they asked for a very specific description of acceptable and nonacceptable litter condition. A good example of a clear description of good litter condition would be: the litter does not transfer soil on to the animals and there is some loose, dry material for chickens to scratch in. Vague terminology such as, handle animals properly or provide adequate veterinary care, should be avoided. What one person may consider as proper handling, another person may think the methods are abusive.

Improving Inter-Observer Reliability

A good auditor training program is essential to make an auditing program really effective. Good training will improve consistency between different auditors. Variation in judgements between different auditors can be reduced through good training (Webster, 2005).

Inter-observer reliability of a welfare assessment can be improved greatly when the wording of the standard is clear. A measure with good inter-observer reliability will produce similar scores when it is done by different people. A study by Phythian et al. (2013) reported good inter-observer agreement for body condition, lameness, and eye damage. High inter-observer agreement and repeatability is possible for the body condition of dairy cows and lameness (difficulty in walking) in dairy cows (Thomsen et al., 2008; D'Eath, 2012; Vasseur et al., 2013).

Data collected by the author in 66 US slaughter plants indicated that there were no significant differences between three different restaurant auditors when they had to count the percentage of cattle rendered insensible with a single shot from a captive bolt gun ($P = 0.529$), and count the percentage of cattle that vocalized (moo or bellow) in the race and stun box ($P = 0.22$). A possible reason for high inter-observer agreement was use of simple yes/no scoring. For example, each animal was scored as silent or vocal. Unfortunately, there were significant differences between auditors on the percentage of cattle moved with an electric goad ($P = 0.004$). The reason for this was that the standards for stunning and vocalization were clearly written and the

standards for electric goad use were not clear. Some auditors counted all touches with the electric goad and others did not. Some auditors did not know that all touches with an electric goad should count, because it is impossible to determine accurately if the electric button has been pushed.

A comparison of three commonly used dairy welfare evaluation assessment tools indicated that they all picked out the 20% of the worst dairies accurately, but they differed greatly on other measures (Stull et al., 2005). Nicole (2014) compared the Welfare Quality System to two industry assessments. All three assessments identified the farms with the poorest welfare.

Research done by Smulders et al. (2006) with pigs showed that high inter-observer reliability could be attained for evaluating welfare and behavior on the farm when simple yes/no scoring was used. These researchers developed an easy-to-administer behavior test, and the results correlated highly with physiological measures of stress, such as salivary cortisol, urinary epinephrine, norepinephrine, and production traits. On the startle test, a pen of pigs was scored as either fearful or not fearful. A 21-cm-diameter yellow ball was tossed into the pen. The pen of animals was rated as fearful if more than half the animals initially ran away. Startling is influenced by both environmental enrichment and the attitude of the stockperson (Grandin et al., 1987; Hemsworth et al., 1989; Beattie et al., 2000). Animals that have both environmental enrichment and a good stockperson caring for them will have a weaker, less fearful startle response. One of the reasons this test had such good inter-observer reliability was because it was simple.

Assessment tools for commercial use should be simpler than assessment used for research

It is the author's opinion, based on in-the-field training of over 400 auditors and inspectors, that a student auditor needs to be accompanied by a highly experienced person on visits to five operations before they carry out audits on their own. If they are going to audit or inspect dairies, they need to visit five dairies, and if they are going to assess poultry slaughter, they need to visit five poultry plants.

There is a difference between an audit or assessment tool that is used commercially to assess animal welfare and measurements used for detailed scientific study or veterinary diagnosis. A screening tool used

T. Grandin

for commercial certifications or legislative compliance has to be much simpler, to make the training of large numbers of auditors easier. The purpose of a screening tool is to determine that there is a problem. The purpose of more sophisticated measures is to diagnose and fix a problem or to carry out scientific research. An example of a good research tool is the Welfare Quality Network (2009) dairy assessment. Unfortunately, it is too time-consuming for routine use as a certification tool (Heath *et al.*, 2014). The Welfare Quality Network contains good assessment tools for pigs, poultry, and beef. Parts of these assessments can be incorporated into simpler audits. Andreasen *et al.* (2014) developed a simpler dairy cow audit that correlated well with Welfare Quality.

Vague wording should be eliminated to improve consistency

There are several vague words that should be eliminated from all standards and guidelines: they are "adequate", "proper", and "sufficient". One person's interpretation of proper handling may be totally different from another person's. For example, if a standard states "minimize the use of electric goads", one person may interpret this as almost never using it and another may think that using it once on every animal is in compliance with this standard.

Some examples of vague wording in United States Department of Agriculture (USDA) standards in the USA are to avoid "excessive electric goad use", or avoid "unnecessary pain and suffering". It is impossible to train an auditor or an inspector on what is excessive use of an electric goad and have good inter-observer reliability between different inspectors. One inspector may suspend meat inspection and shut down a plant for using an electric goad once on every animal, and another may think this is perfectly normal and does not penalize the plant. The author has repeatedly observed great inconsistency in the enforcement of vague USDA directives and standards. The use of objective numerical scoring can improve agreement greatly between different auditors or government inspectors.

In another example, a standard states that pigs should have adequate space on a truck or in a housing system. This is too vague. An example of a more effective way to write the standard is that the pigs must have enough space so they can all lie down at the same time without being on top of each other. Two other examples of clearly written standards

are: ammonia levels in an animal house must not exceed 25 ppm, and 10 ppm is the goal, or tail docking of dairy cows is prohibited (Kristensen and Wathes, 2000; Kristensen *et al.*, 2000; Jones *et al.*, 2005). Some welfare standards require that animals must be housed on pasture. There should be a good definition of the minimum requirements for pasture. At what point does barren, overgrazed ground switch from being a pasture to a dirt lot? The author suggests that to qualify as pasture, 75% or more of the area occupied by the animals must have vegetation with a root system (Fig. 4.1).

Traffic safety standards are a good example of clearly written standards that work effectively in many countries. People who write welfare standards should use traffic laws as a model. When a car is speeding, the police officer pulls it over after he has measured its speed. He does not stop cars that he thinks are speeding; he measures the car's speed. Both the drivers and the police officers know the numerical speed limits and that rules such as a stop sign mean stop. In most countries, enforcement of traffic laws is effective and fairly uniform.

Unfortunately, there are some politicians and policy makers who deliberately make standards vague. A political appointee who was in charge of food inspection in a major developed country resisted the author's suggestions to make standards less vague. He admitted that his agency wanted enforcement flexibility. The reason he wanted this was so enforcement could be strengthened or weakened depending on the political conditions. His vague standards resulted in one inspector being super strict and others very lax. There was no consistency between inspectors.

Three Types of Animal Welfare Standards: Animal Based, Prohibited Practices, and Input Standards

Animal-based measures of welfare problems that can be directly observed by an auditor when he or she visits a farm can be very effective for improving welfare. Some common examples are body condition score (BCS), lameness (difficulty in walking), coat or feather condition, animal cleanliness, and sores, swellings, and injuries. They are outcomes of poor management practices. The World Organisation for Animal Health (OIE) has moved towards the use of more animal-based outcome standards (OIE, 2014a,b). The use of outcome-based standards is recommended by many animal welfare researchers

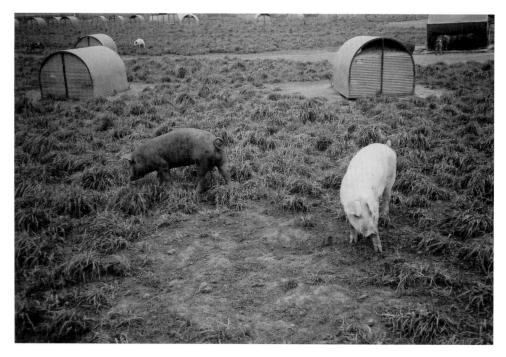

Fig. 4.1. Pigs housed outdoors on pasture. For the ground to qualify as being pasture, 75% or more of the paddock where the animals are housed must have vegetation with a root system. The pasture in this photograph has some bare spots, but it would definitely be in compliance.

Quick Access Information 4.1. Different types of welfare standards.

There are five basic types of standards for assessing animal welfare:

1. Animal-based measures, also called performance standards or outcome criteria. Animal-based measures can be used both as assessment tools and for setting standards. When they are used as a standard, a certain level of performance is required. An example of "excellent" would be a farm with only 5% lame, lactating cows (most emphasis).

2. Practices that are prohibited (most emphasis).

3. Input-based requirements, also called engineering resource, or design standards (less emphasis).

4. Subjective evaluations (less emphasis) for improving actual conditions in the field.

5. Record-keeping, stockperson training documents and paperwork requirements. Documentation of management procedures and standard operating procedures (SOPs) (less emphasis) for improving actual conditions in the field. Regulators may require more emphasis.

(Hewson, 2003; Whay *et al.*, 2003, 2007; Webster, 2005; Velarde and Dalmau, 2012). The large European animal welfare assessment project is also emphasizing outcome-based measures (Welfare Quality Network, 2009; EFSA, 2012; Linda Keeling, 2008, personal communication). Linda Keeling states that measures should be: (i) science based; (ii) reliable and repeatable; and (iii) feasible and practical to be implemented in the field. Work on animal-based

assessment has been done by Whay *et al.* (2003, 2007) and LayWel (2006). Directly observable conditions are easy for auditors to score and quantify. Some important examples are body condition scoring (BCS), lameness scoring, falling during handling, and animal cleanliness scoring. There are many published scoring systems which have pictures and diagrams that make it easy to train auditors (Laywel, 2006; Welfare Quality Network, 2009). Edmonson *et al.*

T. Grandin

(1989) is one example. For a welfare audit, scoring will be easier because the assessor only has to identify animals that are too skinny (see Chapters 1 and 2, this volume, for further information on measurements).

For assessing animal handling, scoring systems where handling faults are scored numerically are easy to implement and are very effective (Maria *et al.*, 2004; Grandin, 2005, 2010; Edge and Barnett, 2008). Some of the items that are measured are the percentage of animals that fall, the percentage poked with an electric goad, the percentage that vocalize (moo, bellow, squeal), or the percentage that move faster than a trot or walk. Lameness scoring of dairy cows has high inter-observer repeatability of locomotion scores of individual cows, and the variation between observers is low (Winckler and Willen, 2001). This indicates that a five-point lameness scale will provide reliable data. A study with broiler chickens indicated that a three-point lameness scoring system had better inter-observer agreement than a six-point system (Webster *et al.*, 2008). To improve inter-observer reliability, scoring systems with more than five categories should be avoided. In another study, observers who visited seven dairy farms on three occasions had high inter-observer repeatability on the variables of lameness scoring, kicking and stepping during milking, cow cleanliness, and avoidance (flight) distance (DeRosa *et al.*, 2003). They had similar results in water buffaloes for kicking and stepping and avoidance distance (DeRosa *et al.*, 2003). Lameness was nonexistent in buffaloes, and cleanliness scores are meaningless because buffaloes wallow.

The OIE codes (2014a,b) also support the use of numerical scoring – Chapter 7.3 states:

> Performance standards should be established in which numerical scoring is used to evaluate the use of such instruments, and to measure the percentage of animals moved with an electric instrument and the percentage of animals slipping or falling as a result of their usage.
> (OIE, 2014d)

At the slaughter plant, directly observable animal-based conditions that would be detrimental to welfare can be assessed easily. A few examples are bruises, death losses during transport, broken wings on poultry, hock burn on poultry, disease conditions, poor body condition, lameness, injuries, and animals covered with manure (see Chapter 1, this volume). On the farm, abnormal behavior such as stereotypical pacing, bar biting in sows, cannibalism in poultry, excessive startle response, and tail biting are also easy to quantify and observe numerically.

Animal-based welfare criteria that is not directly observable

Animal health measures that can be obtained from producers' records are important indicators of welfare problems. Some common examples are death losses, culling rates, animal treatment records, and health records. These measures are useful, but directly observable measures should be weighted more heavily, because records can be falsified. Paperwork and records are more important for tracing disease outbreaks than for assessing welfare.

Animal-based measures are continuous

It is impossible to never have a sick animal or never have a lame animal. When handling animals, doing it perfectly is not attainable. All of the animal-based measures are *continuous variables*. When a standard is being written, a decision has to be made on the acceptable level of faults. These decisions should be based on a combination of scientific data, ethical concerns, and field data on levels that can actually be attained.

The percentage of faults that is considered acceptable may vary between different countries or customer specifications. Many people who are concerned about animal welfare have difficulty writing a standard that allows for some faults to occur. Unless a numerical limit is placed on an animal-based measure, it is impossible to enforce it in an objective manner. An example would be 5% as the maximum acceptable percentage of lame animals. Data presented from the studies reviewed in Chapter 2, this volume, show that well-managed dairies can achieve this easily. Vague terms such as "minimizing lameness" should not be used, because one auditor may think 50% is acceptable and another may consider 5% lame animals as a failed audit.

For example, the scoring system used in slaughter plants (Grandin, 1998a, 2012) allows 1% of the animals to fall during handling. Some people thought that allowing this level of falling was abusive. Data from audits done in many beef and pork plants indicated that for a plant to reliably pass an audit of 100 animals at the 1% level, their actual falling percentages dropped to less than one in 1000. The reason why the standard has remained at 1% is that during an audit of 100 animals, a farm or plant should not be penalized for a single animal that may have jumped and fallen or was spooked by additional people getting too close to it. Putting a hard number on the allowable percentage of animals that fall has brought about remarkable improvements. The use of hard numbers also prevents the gradual deterioration of practices. The OIE code states:

> Animals should not be forced to move at a speed greater than their normal walking pace, in order to minimize injury through falling or slipping. Performance standards should be established with numerical scoring of the percentage of animals slipping or falling used to evaluate whether animal moving practices and/or facilities should be improved. In properly designed and constructed facilities with competent animal handlers, it should be possible to move 99% of animals without their falling.
>
> (OIE, 2014c)

Many people are reluctant to assign hard numbers, or they make the allowable numbers of bruised, injured, or lame animals or birds so high that the worst operations can pass. For example, the National Chicken Council in the USA originally set the limit for broken wings during catching and transport at 5% of the chickens. When progressive managers improved their catching practices, broken wings dropped to 1% or less. The standard is now much stricter.

All scores simple yes or no scoring per animal

To simplify auditing, each animal-based welfare criterion should be scored on a yes or no, pass or fail basis per animal. For example, each animal is scored as either lame or not lame, or fell or did not fall. The number of animals that have a fault is used to determine the percentage of lame animals or the percentage of animals falling during handling. Lameness and BCS methods used for auditing must be simpler than scoring systems used in research or veterinary diagnosis. The author has learned from training many auditors that measurements must be simple to use. For welfare audits, the measure for body condition score or lameness score should be divided into two categories: pass or fail. Each animal is scored and the percentages of animals that comply are tabulated for each variable. The use of simple yes or no scoring will help avoid some of the observer's bias problems discussed by Tuyttens *et al.* (2014).

Prohibited practices that are either severe animal abuse or practices that are banned in a particular country or market

These standards are much easier to write because they are discrete variables. Certain practices such

as sow gestation stalls or docking tails on dairy cows are prohibited. There is no room for interpretation. Certain practices are banned. The farm either has sow gestation stalls or it has group housing. Another example is specific guidelines on surgical procedures. For example, after a certain age, anesthetics would be required for castration. Severely abusive practices such as throwing animals, puntilla, poking out eyes, breaking tails, and dumping or dropping animals off large trucks must be banned. The OIE codes (2014c) for slaughter have lists of practices that should not be used (see Chapter 9, this volume). These types of standards are easy to write, and they usually do not have the interpretation problems that occur with vague wording. Below are OIE (2014c) codes for handling animals which specify practices that should not be used:

- The use of such devices (electric goads) should be limited to battery-powered goads on the hindquarters of pigs and large ruminants, and never on sensitive areas such as the eyes, mouth, ears, anogenital region, or belly. Such instruments should not be used on horses, sheep, or goats of any age, or on calves or piglets.
- Conscious animals should not be thrown, dragged, or dropped.
- Animals for slaughter should not be forced to walk over the top of other animals.
- Animals should be handled in such a way as to avoid harm, distress, or injury. Under no circumstances should animal handlers resort to violent acts, such as crushing or breaking tails of animals, grasping their eyes, or pulling them by the ears, to move animals. Animal handlers should never apply an injurious object or irritant substance to animals, and especially not to sensitive areas such as eyes, mouth, ears, anogenital region, or belly. The throwing or dropping of animals, or their lifting or dragging by body parts such as their tail, head, horns, ears, limbs, wool, hair, or feathers, should not be permitted. The manual lifting of small animals is permissible.

Input-based engineering resource or design criteria standards

Input-based standards may also be called engineering resource or design standards. These standards tell producers how to build housing or specify space requirements. Input-based standards are easy to write clearly. They may work well with one breed of animal and poorly for another. For example, a small hybrid hen needs less space than a large hybrid hen. A single space guideline would not work for both large and small chickens. In many cases, input-based standards should be replaced with animal-based ones. However, there are other situations where input-based standards are recommended.

Useful input standards

Input-based standards work well for specifying baseline minimum conditions for acceptable levels of welfare. Some examples are minimum space requirements on transport vehicles, space in housing, feed troughs, and water-trough space. For nipple waterer systems, specify the number of animals per nipple. Another important input standard is limiting ammonia levels in animal houses to 25 ppm, with 10 ppm being the goal. The above items can all be quantified with numbers. For these standards, charts with animal species and weights can easily be made to fit local conditions. Animal weight is often a better variable to use because breed characteristics change. For example, Holstein and Angus cattle were much smaller in the 1960s and 1970s compared to 2015. Input standards work poorly when they specify exactly how specific details of animal housing should be designed. For example, detailed specifications on how to build dividers and neck rails on dairy cow stalls (cubicles) is not recommended because an outdated specification may block innovation and new designs. To detect a problem with poorly designed or poorly maintained stalls, animal-based measures such as injuries, swellings, sleeping posture, or cow cleanliness should be used. If the stalls are either badly designed or poorly bedded, the cows will have higher percentages of swollen hocks and lameness.

To have a bare minimum acceptable level of welfare, animals must NEVER be jammed into a crate or pen so tightly that they have to sleep on top of each other. The author has observed some caged layer farms where hens have had to walk on top of each other to reach the feeder. An input standard should be implemented to ban this type of abuse. However, on broiler chicken farms, both Dawkins *et al.* (2004) and Meluzzi *et al.* (2008) found that stocking density was not a straightforward indicator

of welfare. Severe welfare problems such as foot-pad lesions or mortality were more related to the poor condition of the litter (Meluzzi *et al.*, 2008). The problems were worse in the winter, when airflow through the building was reduced. There have been serious problems with conducting audits when an input standard is vague. One example was a standard for pasture; it stated that animals must have access to pasture. There was no stipulation on how much time the cows had to be on pasture.

Training of Auditors and Assessors to Improve Consistency of Audits

Unfortunately, it is impossible to eliminate all subjectivity from an animal welfare audit. Some variables that will require subjectivity to evaluate are the overall maintenance of a facility and the attitude of the staff. One of the best ways for an auditor or inspector to become skilled in subjective measures is to visit many different places. This will provide a range of farms or plants ranging from excellent to bad.

Photographs and videos of abusive practices on animals in poor conditions or poorly maintained equipment should be used for training auditors and inspectors. Photographs of both poorly maintained and well-maintained equipment can be used. Common faults in equipment such as broken gates, dirty ventilation fans, dirty water troughs, and worn-out milking machine equipment are shown alongside properly maintained equipment. People working with animals should never kick or beat animals. A question an inspector may ask is, "When does tapping become abusive beating?" A video has been made that shows proper handling, where an animal is being moved correctly by tapping it. The video also demonstrates crushing a small empty corrugated cardboard box by beating it. When enough hitting force has been applied to the box to start crushing it, tapping has turned into beating. An animal must never be beaten to make a training video. Download videos of animals being beaten from the Internet or use the video of cardboard boxes being beaten. The title, "Proper Use of Livestock Driving Tools with Temple Grandin", can be typed into a search engine. There are also excellent photographs on the European Welfare Quality Network website (www.welfarequalitynetwork.net). They can be downloaded easily from the Internet for dairy cattle, beef, pigs, and poultry.

Clear Comments

Auditors and inspectors must write clear comments. The author has reviewed many audits and inspection reports, and there are too many vague comments, or not enough comments. There should be well-written comments to describe conditions on failed audit items and areas of noncompliance. Observations of both really good and really bad practices should be recorded. Well-written comments will help both customers and regulatory officials to make wise decisions after an audit or inspection is failed. Some examples of vague and well-written comments are shown in Quick Access Information 4.2.

Record-keeping and paperwork requirements

Well-kept records are essential for an efficient farm that has high standards. They are also essential for identification and trace-back for disease control purposes. However, there is an unfortunate tendency for some regulatory agencies to turn the entire audit into a paperwork audit. This may create a situation where the paperwork is correct but the conditions out on the farm may be poor. Records are valuable for looking at culling rates and the longevity of animals such as dairy cows and breeding sows. The longevity of an animal is one important indicator of animal welfare (Barnett *et al.*, 2001; Engblom *et al.*, 2007). On some farms, a dairy cow lasts for only two lactations.

Quick Access Information 4.2. Examples of poor and well-written comments.

Vague comments	Well-written comments that explain the actual problem
Rough handling of pigs	The handler kicked piglets and threw them across the room
Poor stunning	The broken stun gun misfired and failed to render the animal insensible about one-third of the time
Poor litter in poultry barn	The litter was clumped, chopped up newspaper. It was wet and transferred soil on to the birds

---

Core Criteria and Critical Control Points to Prevent Poor Welfare

Some items on an audit are much more important than others. The situation must be avoided where the paperwork has been inspected and it passes but the farm is full of skinny, emaciated, lame animals. There are certain major core standards or critical noncompliances that must be met in order to pass an audit. Many of the most important core standards are directly observable animal-based measures that are the outcomes of many bad practices or conditions. The severe welfare problems listed in Quick Access Information 1.1 in Chapter 1, this volume, are all core criteria. Some examples are moving animals by poking sensitive areas such as the rectum, high ammonia levels in a building that cause eye damage, or dirty animals covered with manure with no dry place to lie down, or emaciated animals. The principle of Hazard Analysis Critical Control Points (HACCP) is used in many audit systems. The principle is to have relatively few critical control points (CCPs) or core standards that must ALL be in compliance to pass an audit. An effective CCP is an outcome measure of many poor practices. HACCP systems were originally used for food safety. When the first HACCP programs were developed, they were often too complicated, with too many CCPs. Later versions of the programs were simplified. The numerical scoring system the author developed for slaughter plants has five core standards or CCPs that are continuous outcome-based measures that are scored numerically, one engineering- or input-based standard on watering animals, and one discreet core criterion that prohibits any act of abusive handling (Grandin, 1998a, 2005, 2010). A plant has to pass on ALL seven core standards to pass a welfare audit (see Chapter 9, this volume). The HACCP principles are being applied increasingly to animal welfare audits (von Borell *et al.*, 2001; Edge and Barnett, 2008). There are a number of different ways that core standards or CCPs can be classified. The rationale behind the author's classification is for easier implementation in the field.

The core standards in Quick Access Information 1.1 (Chapter 1, this volume) can be used for welfare standards for all types of systems, ranging from completely free-range to intensive large-scale farms. To pass a welfare audit, the farm should have an acceptable score on all these core standards. The items in Quick Access Information 1.1 and 1.2 in

---

T. Grandin

Chapter 1, this volume, are all things that an auditor can observe without having to look at records. For each species of animal, training materials will have to be developed for local conditions. The reason for placing less emphasis on examining records is because the author has observed that some records are faked. It is the author's opinion that auditing paperwork is an important part of a welfare audit, but it should be weighted less heavily than the directly observable items in Quick Access Information 4.3, 4.4, 4.5, 4.6, and 4.7.

Problems with converting multiple measures to a single score

Busy managers want data that are easy to evaluate, and they like to have all the welfare and food safety data combined into single scores. The danger of converting multiple measures to a single score is that serious problems may be masked. For example, when multiple welfare indicators on a European Welfare Quality Network dairy audit were combined into a single score, a dairy with 47% lame cows received an acceptable rating (deVries et al., 2013). A dairy with 47% lame cows would have very poor welfare. If a single welfare score is used, there must be a provision for an automatic audit failure for deliberate acts of abuse. On animal-based measures, such as lameness, a few percentage points over the 5% level would be points off, but if 25% of the cows were lame, the audit would be failed. Therefore, each animal-based measure could have two levels of failure – points off and audit failure.

The Most Important Animal-based Measures (CCPs) that Assess Many Welfare Problems

Body condition scoring (BCS)

There are many different published scoring systems. It is often best to use charts that have been developed in your country. For American Holstein and European dairy cows, there are good pictures and charts in Wildman et al. (1982), Ferguson et al. (1994), University of Wisconsin (2005), and Welfare Quality Network (2009). All the articles are free on the Internet. The BCS pictures should be on a plastic-laminated card that assessors can always have with them. Body condition scoring systems range from three scores to nine scores. The most practical systems use four or five scores. It is recommended to

> **Quick Access Information 4.3. Body condition scoring system.**
>
> **1.** Emaciated – ribs and spine showing. Not acceptable.
> **2.** Thin.
> **3.** Normal.
> **4.** Overcondition.
> **5.** Obese – extremely fat – sometimes is a welfare problem.
>
> Conditions vary greatly in different parts of the world. Body condition scoring charts should be made to fit local conditions.

use a scoring system that producers in the local area are familiar with. For example, cattle that had starved to death in extremely cold conditions, such as −18°C, often did not have an emaciated body condition score of 1 (Terry Whiting, 2008, personal communication). Autopsy revealed that there was no fat around the heart and kidney. To have adequate welfare, cattle that live in very cold climates must have better (fatter) body condition than cattle that live in warmer climates. Laminated cards with photographs of a rear view and side view of animals that are acceptable and are not acceptable are recommended. Poor body condition can be caused by a lack of food, parasites, disease, or poor management of rBST (rBST is a growth hormone given to dairy cows to increase milk production). It is an important outcome measure of poor management, nutrition, and health. Lactating animals will usually be thinner, with a poorer BCS than nonlactating animals.

Lameness (difficulty in walking) scoring

Lameness is caused by many different conditions. Measuring the percentage of lame animals is an outcome of a variety of factors that cause animals to become lame. Some examples of factors that may increase the percentage of lame animals are:

- rapid growth in poultry and pigs genetically selected for growth
- wet bedding or muddy lots
- standing on wet concrete
- lack of hoof trimming or foot care
- disease conditions – for example, hoof rot
- poor leg conformation

- founder (laminitis) due to feeding high levels of concentrates
- lameness due to feeding high levels of beta-agonists, such as ractopamine or zilpaterol
- poor handling that causes animals to slip and fall
- poorly designed cubicles (freestalls) in dairies
- genetic selection for high milk production.

Lameness in chickens can be measured with a simple three-point scoring system (Dawkins *et al.*, 2004; Webster *et al.*, 2008). The scores are: 1 – walks normally for ten paces; 2 – walks crooked and obviously lame for ten paces; and 3 – downer or not able to walk ten paces. The paper by Knowles *et al.* (2008) contains videos for training auditors that can be accessed online. Zinpro.com/lameness has excellent videos for scoring lameness in cattle with a five-point system. Cattle with scores of 3, 4, or 5 on a five-point scale would be classified as lame. On this scale, a 1 is completely normal and a 5 can barely stand and walk.

Since the first edition of this book, a simpler four-point scoring system has become popular. Lameness is classified as mild or severe, depending on whether or not the animal can keep up with its herdmates when the group is walking. A three-point lameness scoring system works well in poultry, but it may underestimate the percentage of lame cattle. There are many variations of lameness scoring systems with slight differences in wording. When scoring systems are used for research, it is important to specify in the methods section the lameness scoring tool that has been used. Some systems designate zero as normal and others designate one as normal.

Ideally, all lactating cows should be scored for lameness. On very large dairies where this is not possible, research by Hoffman *et al.* (2013) indicates that a sample of lactating cows should be taken in the middle of each group of cows exiting the milking parlor. New methods for measuring lameness with electronic devices are also being developed. In dairies and other operations where animals are kept tied, lameness can be assessed by watching the cow to see if she shifts weight from side to side or shows uneven weight bearing (Gibbons *et al.*, 2014). There are differences in numbering in some systems. One popular four-point lameness scoring system uses a zero for normal and then 1,2,3. When lameness scores are being compared between farms, ALL data MUST use the same numbering system.

Quick Access Information 4.4. Simple four-point lameness (difficulty in walking) scoring of cattle, pigs, and other mammals.

1. Walks normally with smooth, even steps.
2. Walks with a limp or has a stiff gait, with head down or bobbing head. The animal keeps up when a group of animals are walking (classify as mild lameness).
3. Walks with difficulty but still fully mobile. It cannot keep up and is left behind when a group of animals is walking (classify as severe lameness).
4. Can barely stand and walk, and may become nonambulatory. Same as a 5 on the Zinpro dairy cow scales (classify as severe lameness and not fit for transport). On the New Zealand four-point scale, the most severe score is assigned to a fully mobile animal that cannot keep up with the walking herd. This animal would be fit for transport.

Note: Some guidelines will label normal walking as 0 and renumber degrees of lameness with 1,2,3. A detailed photographic guide to identify hoof lesions that cause lameness can be found in Mulling *et al.* (2014).

Quick Access Information 4.5. Scoring fecal matter and mud on animals.

For all species, a simple 1, 2, 3, 4 scoring system can be used. This score is used for fecal matter, fecal matter mixed with earth, and fecal matter mixed with bedding that has adhered to the animal or bird. This scoring system can be used for dairy cows, feedlot cattle, pigs, sheep, and poultry.

1. Completely clean legs, belly/breast, and body. Birds must be completely clean.
2. Legs are soiled but belly/breast and body are clean (Fig. 4.2).
3. Legs and belly/breast are soiled.
4. Legs, belly/breast, and the sides of the body are soiled (Fig. 4.3).

Soil on animals that is caused by an animal's attempts to cool itself should not be confused with fecal matter on animals that is due to a failure to provide a dry place for the animals to lie down. Some systems will label clean as 0 and renumber degrees of dirtiness as 1, 2, 3.

T. Grandin

Condition of coat or feathers

When coats and feathers are being evaluated, it is important to determine the cause of the damage. The three major causes are: abrasion against a cage or feeder; damage inflicted by another bird; or animal and external parasites.

There are published scoring systems for feather condition in poultry. The best systems have separate scores for feather wear and damage caused by other birds. The LayWel website has good pictures for scoring feather condition in poultry (Fig. 4.4; www.laywel.eu). The type of housing system will have an effect on the pattern of feather damage. Hens on deep litter have greater damage on the head and neck, and caged birds have more wing and breast damage (Bilcík and Keeling, 1999; Mollenhorst *et al.*, 2005). Damage on the head is most likely due to pecking from other chickens, and wing damage is more likely to be caused by abrasion against equipment.

For livestock, poor coat condition is an indicator of problems such as untreated external parasites or mineral deficiencies. These can occur in organic systems where drugs for treating parasites are not allowed. In cattle, bald spots are an indicator of untreated external parasites. This would be a severe welfare problem.

Fig. 4.2. These dairy cows have been grazed on lush green pastures. Their bellies, udder, and upper legs are clean. They would score a 2 for cleanliness.

Sores and injuries

Pictures and charts for scoring will need to be provided for each species. In pigs, pressure sores on the shoulder are an indicator of poor management in

Fig. 4.3. This dirty cow has soil on her legs, udder, belly, and the sides of her body. She would score a 4 for being a dirty cow.

gestation stalls, and bites and wounds would be scored in group housing. For dairy cows living in freestall (cubicle) housing, lesions and swellings on the legs are an important measure (see Chapter 2, this volume). Cows in cubicles that are frequently bedded have fewer lesions (Fulwider *et al.*, 2007). Freestalls that are too small also increase lesions. Leg lesions on dairy cows are an excellent outcome measure of both poor management and poor stall design. On birds, injuries caused by pecking, cannibalism, or aggressive roosters can be quantified easily. In pigs, injuries such as bitten tails and ears can be quantified. In broiler chickens, lesions and burns on the feet can be scored with a chart showing different levels of severity (Fig. 4.5). Wet litter is one major cause of footpad lesions.

(a)

(b)

Fig. 4.4. LayWel feather condition scoring for laying hens. It is a four-point scale, where a 1 is almost bald and a 4 is normal. This figure shows (a) score 2 and (b) score 4. The LayWel website also has pictures for scoring breasts, backs, and necks. (Photographs courtesy of www.laywel.eu).

Obvious neglected health problems

Some examples of neglected health problems would be advanced necrotic cancer eye in cattle, bald spots due to untreated parasites, large infected untreated injuries, large untreated abscesses, necrotic prolapses, or other obvious neglected conditions.

Handling scoring

The following acts of abuse that can occur during handling should be grounds for an automatic audit failure. These obvious abuses are universally banned in all well-managed livestock and poultry operations.

If possible, 100 animals should be scored as they are handled for veterinary procedures during truck

---

Quick Access Information 4.6. Other important critical control points that should be scored with animal-based outcome measures.

- Temperature stress – open-mouth breathing in cattle indicates severe heat stress, or huddling or piling in pigs indicates cold stress.
- Condition of haircoat on mammals.
- Condition of feathers on poultry.
- Condition of feet, all species.
- Sores and injuries on the body.
- Swellings and injuries on the legs.
- Neglected health problems.
- Eye damage or lung lesions due to high ammonia or dust level.
- Size of flight zone. Good stockmanship: animals approach people. Poor stockmanship: animals run away.

Quick Access Information 4.7. Acts of abuse that would result in a failed animal welfare audit.

1. Beating, kicking, or throwing animals or birds.
2. Poking out eyes or cutting tendons.
3. Dragging live animals.
4. Deliberately driving animals over the top of other animals.
5. Breaking tails.
6. Poking animals in sensitive areas such as the rectum, eyes, nose, ears, or mouth to move them.
7. Deliberately slamming gates on animals.
8. Jumping or throwing animals off a truck that has no drop-down tailgate or unloading ramp.

T. Grandin

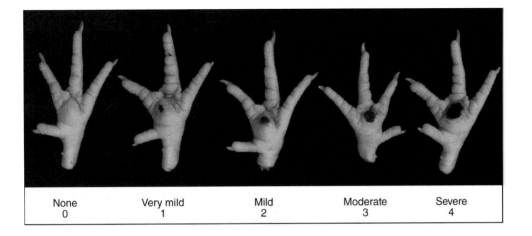

None	Very mild	Mild	Moderate	Severe
0	1	2	3	4

Fig. 4.5. Broiler chicken foot burn chart. Footpad lesions are caused by poor litter conditions. Lesions that are brown or otherwise discolored usually occur during a later stage of growth. Lesions that are white and not discolored at slaughter are more likely to have occurred early in the growing period.

Quick Access Information 4.8. Handling scoring – score when animals are being moved by a stockperson.

- Percentage falling.
- Percentage moved with an electric goad.
- Percentage vocalizing (moo, bellow, squeal) (do not use vocalization scoring for sheep).
- Percentage moved faster than a walk or trot.
- Percentage that run.
- Percentage that strike fences or gates.

(See Chapter 5, this volume, for more detail.)

---

loading/unloading or moving to slaughter at an abattoir. Each animal is scored on a per animal yes or no basis. For vocalization and electric prod use, each animal is scored as either silent or vocalizing, or moved with an electric prod or not moved with one. The main continuous variables that are scored are the percentage of animals. Further information on handling scoring is in Welfare Quality Network (2009), Grandin (2012, 2014), the OIE (2014a) beef cattle standards and OIE (2014d).

All five of these handling measures are outcomes of either poorly trained people or deficiencies in the facilities. For example, falling can be caused by either rough handling methods, slick flooring (if the floor is too slick, the animals will be more likely to slip and fall), or the animal not being fit for handling and transport. Vocalization scoring is a sensitive indicator of both equipment design problems and poorly trained people. Vocalizations during the handling and restraint of cattle and pigs are associated with the following bad practices (Grandin, 1998b, 2001):

- excessive use of electric goads
- slipping on the floor
- sharp edge in a restrainer
- malfunctioning stunning equipment
- slamming gates on animals
- excessive pressure applied by a restraint device
- being left alone in a race or stun box.

Vocalization scoring should not be used for sheep because severely abused sheep often remain silent. Many large slaughter plants have been able to achieve easily vocalization percentages of 3% or less of the cattle (Grandin, 2005, 2012). Industry data also indicate that in many plants, only 5% or less of the pigs squeal in the restrainer.

Heat and cold stress

Many animals have died or have had their welfare severely compromised by either heat or cold stress. The thermal neutral zone of an animal is very variable and it depends on genetics, length of hair/coat, access to shade, and many other factors. If animals or birds are panting, they are heat stressed, and shivering or huddling animals are cold stressed. Open-mouth breathing in cattle is a sign of severe

---

heat stress (Gaughan *et al.*, 2008) and either shade or water sprays should be provided to relieve it. Auditors and inspectors can count the number of panting animals easily. For use under commercial conditions, the scoring system should be simplified to open-mouth breathing, severe heat stress, or closed-mouth breathing. It is beyond the scope of this book to discuss all aspects of thermal comfort, but conditions that have caused death from thermal stress must be corrected. For heat stress, some of the corrective actions would be fans, sprinklers, shades, extra water, or changing animal genetics (Fig. 4.6). For cold stress, some possible corrective actions would be shelter, heat, additional bedding, or changing animal genetics. Some of the worst problems with thermal stress occur when animals are moved between different climatic regions. It takes several weeks for an animal's body to acclimatize to either hot or cold conditions. When animals have to move from a cold area to a hotter area, it is best to transport them to the hotter region during its cool season. Transporting cold acclimatized cattle to a hot desert region may result in higher death losses (see Chapter 1, this volume).

Abnormal Behavior Associated with Poor Welfare and Assessment of Behaviors Associated with Positive Emotions

Feather and coat condition scoring can be used to detect damage that is caused by other animals. Cannibalism in hens and wool pulling in sheep are two forms of abnormal behavior that can be detected by examination for injuries. In pigs, abnormal behavior such as biting tails and ears can be quantified by counting the percentage of pigs with injuries. The author has observed that some genetic lines of rapidly growing lean pigs have a higher percentage of tail biting compared to others. In bovines, tongue rolling and urine sucking are abnormal behavior that can be prevented by feeding roughage (Montoro *et al.*, 2013; Webb *et al.*, 2013). Animals housed in barren intensive environments may be more likely to engage in stereotypies such as bar biting or pacing. There are further discussions of abnormal behavior in Chapters 2, 8, and 15, this volume. Most animal welfare specialists agree that lameness, disease, or injury is bad for welfare, but there is likely to be

Fig. 4.6. Well-designed shades and sprinklers keep cattle cool in this feedlot located in the arid, dry south-western USA. The cattle stay clean and dry in the arid climate. Feedlot shades must be installed north and south so that the shadow will move and there is sufficient space for all animals to lie under the shade. This prevents mud build-up under the shade. Shades should be 3.5 m (12 ft) high. Places with rainfall under 50 cm/year (20 in/year) are the best locations for feedlots. Pens should have 2–3% slope away from the feed trough to provide drainage and prevent mud build-up.

T. Grandin

more disagreement on how heavily behavioral measures should be weighted. Bracke *et al.* (2007) have developed a technique where a number of different animal welfare scientists are polled about behavioral needs. Statistics were used to determine the behavioral needs that the specialists ranked as most important. The highest ranking was given to materials for pigs to root and explore, and prevention of tail biting. This group consensus approach will be important for writing guidelines in areas where the science may be less clear and decisions are based partially on ethical concepts. In 2001, the EU passed a directive that pigs must have access to fibrous materials such as straw to root and chew (see Chapter 8, this volume, on behavior).

Quick Access Information 4.9. Abnormal animal behaviors that are not acceptable.

Abnormal animal behavior should be prevented either with environmental enrichment, feeding more roughages, or changing animal genetics. Count the percentage of animals performing these abnormal behaviors. Use scoring to determine if interventions are effective for reducing abnormal behaviors.

- Tongue rolling in beef and dairy cattle.
- Tail biting in pigs.
- Urine sucking.
- Pacing – all species.
- Repetitive behaviors – all species.
- Feather pecking in poultry.
- Pulling out hair or wool.
- Bar-biting sows.

See Chapters 2, 8, and 15, this volume, for further information.

Positive behaviors and qualitative behavioral assessments (QBA)

To achieve a minimum level of animal welfare will require the prevention of injuries, health problems, and other problems discussed in this chapter. More and more welfare scientists see the importance of using indicators for truly positive animal welfare (Boissy *et al.*, 2007; Edgar *et al.*, 2013). Having a barely acceptable level of welfare may not be adequate. Chapter 8, this volume, contains a detailed discussion of emotional systems in an animal's brain and has a list of behaviors that are indicators of positive emotions (Quick Access Information 8.2). Some examples of behaviors that indicate a positive emotional state are: animals that have a small flight zone and approach people, play behavior, dust bathing in chickens, and normal grooming behaviors.

Qualitative Behavioral Assessment (QBA) is a method for assessing animal emotionality by having people assign words to animal behavior such as calm, curious, nervous, agitated, or angry. There is a definite statistical relationship between words assigned by observers and the physiological state of the animal (Rutherford *et al.*, 2012; Stockman *et al.*, 2012). This method is not sufficiently developed for routine use on commercial farms. In the future, powerful computer programs may make it easy to use.

How to Set Critical Limits for Animal-based Standards (Also Called Benchmarking)

To avoid vague guidelines, numerical limits have to be put on animal-based standards to determine passing and failing scores. The limits have to be set high enough to make the industry improve, but not so high that people will either say it is impossible or fight conforming to the standard. When the author worked on auditing slaughter plants, three out of four plants failed their first audit (Fig. 4.7). They were then given time to improve their practices, with no penalties. When a numerical scoring system is first introduced, baseline data will need to be taken and the limit for an acceptable rating should be set so the best 25–30% of the plants or farms will pass. The others have to be given a period of time to attain the standard. For example, slaughter plants should be required to correct minor problems within 30–60 days and major problems within 6 months. On farms, producers should be given up to 2 years to correct problems such as a high percentage of lame animals. This will make it easier to implement the system because the places with excellent scores can be used as examples to show the bad operators that they need to improve. In parts of the world where practices are really poor, the standards may need to be raised again after the system is implemented and the farms have improved.

Continuous Improvement Approach

Another approach is to not set strict numerical limits but to use the ISO 9001 method of continually working to improve (Main *et al.*, 2014). It is the author's opinion that a combination of thresholds

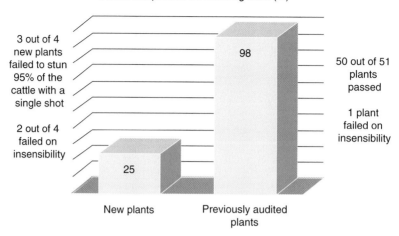

Plants that passed the stunning audit (%)

3 out of 4 new plants failed to stun 95% of the cattle with a single shot

2 out of 4 failed on insensibility

50 out of 51 plants passed

1 plant failed on insensibility

98

25

New plants

Previously audited plants

Fig. 4.7. Comparison of beef plants audited for 4 years versus new plants audited for the first time. Only 25% of the slaughter plants passed their initial animal welfare and humane slaughter audit. Plant managers did not know what to expect, and they often engaged in bad practices because they were their normal practices. An initial assessment should be considered as a training session to educate both the management and the employees.

(critical limits) for minimum numerical standards and continuous improvement should be used. Setting a minimum acceptable critical limit for a continuous measure such as lameness or body condition score prevents poor welfare. This approach is recommended for legislative enforcement or retailer auditing programs where a farm or slaughter plant is removed from an approved supplier list. Continuous improvement approaches can be used to improve welfare further. A list ranking farms on different measures can also be used to motivate continuous improvement.

Implementation of Successful Auditing and Assessment Programs

When the author first started doing audits of slaughter plants, many people did really bad things in front of her because they did not know any better. In the early years, there was no difference between announced and unannounced audits. After people learned that certain practices were wrong, some managers became very skilled at acting "good" during an audit and then bringing back the electric goads when the audit was over. Audits and assessments of handling and transport practices are more likely to have different results between those obtained on an announced visit and those gained on an unannounced audit, whereas measures of lameness, body condition, dirty animals, or maintenance of facilities

are more likely to be consistent between the two types of audit. Some slaughter plants in the USA have installed video auditing that can be viewed unannounced over a secure Internet link. Video auditing is most effective when it is monitored by people outside the plant, who look at it every day at different random times. This solves the problem of acting "good" during an audit and then reverting back to old, bad practices after the auditor leaves.

Combining animal welfare audits with other audits and inspections

Most government and industry auditing and inspection programs have people that conduct several types of inspections. In many countries, government veterinary inspectors carry out both meat hygiene and animal welfare enforcement. For most of the audits done by McDonald's, Tesco, and other large retailers, the same auditor has to audit both food safety and animal welfare. When farm assessments are being done, the same person may have to audit animal welfare, environmental compliance, and animal drug use. Bundling auditor and inspection duties is often required to cut down on many duplicate trips. Because people often have many duties, this is another reason why clear, easy-to-implement standards and guidelines are essential. To comply with EU standards, a person has to be designated as that animal welfare specialist.

T. Grandin

Structure of effective audit and inspection systems

The most effective audit and inspection systems have a combination of internal and external audits. The best systems that have been implemented by private industry have three parts:

- Internal audits (first party) done on a daily or weekly basis either by employees at the plant or by a farm's local veterinarian.
- Audits by people employed by the corporate office of a meat-buying customer (second party). They should visit a certain percentage of their suppliers each year.
- Third-party audits done by an independent auditing company. They are conducted once or twice a year on EVERY plant and farm.

One advantage of using a third-party independent auditing company is to prevent a conflict of interest. For example, a meat buyer in the company corporate office might favor a certain plant and be less strict because they want their cheap prices. However, it is the author's opinion that a corporation should not delegate all its auditing responsibilities to a third-party auditor. Corporate personnel need to make regular visits to show suppliers that they are serious about animal welfare.

On a farm, the farm's regular veterinarian or other professional can do the "internal" audits. Due to problems with conflict of interest, the entire responsibility for auditing should not be done by the farmer's local veterinarian. The veterinarian may be reluctant to fail a client farm because he or she would risk getting fired by the farmer. To avoid this potential conflict of interest problem, the auditor should not be the farmer's regular veterinarian. The person doing the audit should be on someone else's payroll. They should be paid by either the government, a third-party auditing company, a meat-buying corporation such as McDonald's, a livestock association, or work as field staff for a large, vertically integrated company. In many systems, farms and slaughter plants have to pay for audits. To avoid a conflict of interest, the farm or plant pays the auditing company and the auditing company pays the auditor's salary. A similar system is used for government inspection. The inspector's salary is paid by the government. In both of these systems, either the auditing company or the government makes inspector and auditor assignments. A farm or plant should never be allowed to choose their favourite inspector. To help prevent personality conflicts, a third-party auditing firm should rotate auditors to avoid having the same person always visiting the same farm or plant.

Each plant or farm is a separate unit

Big multinational companies own many plants and farms. Each plant or farm site should be treated as an independent unit, regardless of ownership. It either passes or it fails an audit. A big company may have bad operations in another country. If your job is to implement a welfare audit program in your country, it is best to concentrate on working with each individual plant or farm in your country. A plant or a farm is either taken off the approved supplier list or put back on the approved list. Meat-buying companies that have made the greatest progress in improving animal welfare have enough plant or farm suppliers so that a few can be taken off the approved supplier list and they will still have enough product. Even if all the meat comes from two big corporations, a buyer still has sufficient economic power to bring about change if he or she buys from several different plants owned by each company. In government programs, there are different levels of penalties, ranging from fines to shutdown of the farm or plant.

Procedures for handling noncompliance

Most government and industry systems have a formal procedure that is followed when there is a noncompliance. For both minor and major noncompliance, a letter (e-mail) has to be written explaining the corrective action that will be taken to rectify the problem. After a specified period of time, the farm or plant will have to have a new audit or inspection. Throughout this chapter, the author has emphasized the need for clear guidelines and standards that can be applied consistently by different auditors or inspectors.

The penalty for being one or two points below the minimum passing score should be much less severe than the penalty for a serious violation such as dragging a nonambulatory animal down the unloading ramp. The author has worked with many large meat buyers on decisions to remove a supplier from the approved list. Usually, one or two points below the minimum standard requires a corrective action letter and a re-audit. Buying from the plant would continue. Dragging the nonambulatory animal is a serious act of abuse and the meat buyer would stop buying from the plant for a minimum of 30 days. If they had another serious violation, they would be suspended for a longer time.

In conclusion of this section, the auditing and inspection process should be very objective, but sometimes great wisdom is required to determine the appropriate punishment. From 15 years of experience implementing auditing systems, the author has learned that suppliers with a poor, noncooperative attitude need harsher punishment than more cooperative suppliers. When the original McDonald's audits were started in 1999 and 2000, out of 75 pork and beef plants, three plant managers had to be fired before improvements in animal welfare occurred. To get these three plants to conform to the standard, all purchases from them were suspended until new managers were hired. With new management, one of the plants went from being one of the worst plants in the system to one of the best. For both food safety and animal welfare, the best places have managers who want to do the right thing.

Use of Remotely Viewed Video Cameras to Audit Animal Welfare

Two major meat companies in the USA, and several independents, are successfully using video cameras that can be viewed from a remote location to audit animal welfare at the slaughter plant. The cameras solve the problem of people "acting good" when they know they are being watched. The cameras are used to assess stunning efficacy and animal handling. The cameras have brought about the biggest improvements in animal handling and reducing electric procedures. To be effective, the cameras must be monitored by auditors who are part of a structured program. Reliance on plant management is not effective because they stop looking at the cameras after the novelty wears off.

Conclusion

The implementation of an effective auditing and inspection program can improve animal welfare greatly. Clear guidelines will improve the consistency of judgement between different people. This will help prevent the problem of one inspector being super strict and another failing to improve the treatment of animals.

References

Andreasen, S.N., Sandoe, P. and Forkman, B. (2014) A comparison of the Welfare Quality® protocol for dairy cattle and the simpler and less time-consuming protocol developed by the Danish Cattle Federation. *Animal Welfare* 23, 81–94.

Barnett, J.L., Hemsworth, P.H., Cronin, G.M., Jongman, E.C. and Hutson, G.D. (2001) A review of the welfare issues for sows and piglets in relation to housing. *Australian Journal of Agricultural Research* 52, 1–28.

Beattie, V.E., O'Connell, N.E., Kilpatrick, D.J. and Moss, B.W. (2000) Influence of environmental enrichment on welfare related behavioral and physiological parameters in growing pigs. *Animal Science* 70, 443–450.

Bilcík, B. and Keeling, L.J. (1999) Changes in feather condition in relation to feather pecking and aggressive behaviour in laying hens. *British Poultry Science* 40, 444–451.

Boissy, A., Manteuffel, G., Jensen, M.B., Moe, R.O., Spruijt, B., *et al.* (2007) Assessment of positive emotions in animals to improve welfare. *Physiology and Behavior* 92, 375–397.

Bracke, M.B.M., Zonderland, J.J. and Bleumer, E.J. (2007) Expert consultation on weighing factors of criteria for assessing environmental enrichment materials for pigs. *Applied Animal Behaviour Science* 104, 14–23.

Dawkins, M.S., Donnelly, C.A. and Jones, T.A. (2004) Chicken welfare is influenced more by housing conditions than stocking density. *Nature* 427, 342–344.

D'Eath, R.B. (2012) Repeated locomotion scoring of a sow herd to measure lameness: Consistency over time. The effect of sow characteristics and interobserver reliability. *Animal Welfare* 21, 219–231.

DeRosa, G., Tripaldi, C., Napolitano, F., Saltalamacchia, F., Grasso, F., *et al.* (2003) Repeatability of some animal related variables in dairy cows and buffaloes. *Animal Welfare* 12, 625–629.

deVries, M., Bokkers, E.A.M., Van Schaik, G., Botreau, R., Engel, B., *et al.* (2013) Evaluating the results of the Welfare Quality multi-criteria evaluation model for classification of dairy cattle welfare at the herd level. *Journal of Dairy Science* 96, 6264–6273.

Edgar, J.L., Mullen, S.M., Pritchard, J.C., McFarlane, U.J.C. and Main, D.C. (2013) Towards a good life for farm animals: development of resource tier framework to achieve positive welfare for laying hens. *Animals* 3, 584–605.

Edge, M.K. and Barnett, J.L. (2008) Development and integration of animal welfare standards into company quality assurance programs in the Australian livestock (meat) processing industry. *Australian Journal of Experimental Agriculture* 48(7), 1009–1013.

Edmonson, A.J., Lean, I.J., Weaver, L.D., Farver, T. and Webster, G. (1989) A body condition scoring chart of Holstein dairy cows. *Journal of Dairy Science* 72, 68–76.

T. Grandin

EFSA (European Food Safety Authority) (2012) Statement on the use of animal based measures to assess the welfare of animals. *EFSA Journal* 10(6), 2767 (29 pp), doi:10.2903/j.efsa.2012.2767.

Engblom, L., Lundeheim, N., Dalin, A.M. and Anderson, K. (2007) Sow removal in Swedish commercial herds. *Livestock Science* 106, 76–86.

Ferguson, J.O., Galligan, D.T. and Thomsen, N. (1994) Principal descriptors of body condition score in Holstein cows. *Journal of Dairy Science* 77, 2695–2703.

Fulwider, W.K., Grandin, T., Garrick, D.J., Engle, T.E., Lamm, W.D., *et al.* (2007) Influence of freestall base on tarsal joint lesions and hygiene in dairy cows. *Journal of Dairy Science* 90, 3559–3566.

Gaughan, J.B., Mader, T.L., Holt, S.M. and Lisle, A. (2008) A new heat load index for feedlot cattle. *Journal of Animal Science* 86, 226–234.

Gibbons, J., Haley, D.B., Cutler, J.H., Nash, C., Heyerhoff, J.Z., *et al.* (2014) Technical note: A comparison of 2 methods of assessing lameness prevalence in tiestall herds. *Journal of Dairy Science* 97, 350–353.

Grandin, T. (1998a) Objective scoring of animal handling and stunning practices at slaughter plants. *Journal of the American Veterinary Medical Association* 212, 36–39.

Grandin, T. (1998b) The feasibility of using vocalization scoring as an indicator of poor welfare during slaughter. *Applied Animal Behaviour Science* 56, 121–128.

Grandin, T. (2001) Cattle vocalizations are associated with handling and equipment problems in beef slaughter plants. *Applied Animal Behaviour Science* 71, 191–201.

Grandin, T. (2005) Maintenance of good animal welfare standards in beef slaughter plants by use of auditing programs. *Journal of the American Veterinary Medical Association* 226, 370–373.

Grandin, T. (2010) Auditing animal welfare at slaughter plants. *Meat Science* 86, 56–65.

Grandin, T. (2013) *Recommended Animal Handling Guidelines and Audit Guide*, 2013 edn. American Meat Institute, Washington, DC. Available at: www.animalhandling.org (accessed 9 May 2015)

Grandin, T. (ed.) (2014) *Livestock Handling and Transport*, 4th edn. CAB International, Wallingford, UK.

Grandin, T., Curtis, S.E. and Taylor, I.A. (1987) Toys, mingling and driving reduce excitability in pigs. *Journal of Animal Science* 65 (Supplement 1), 230 (Abstract).

Heath, C., Lin, Y.-C., Mullan, S., Brown, W.J. and Main, D.C.J. (2014) Implementing Welfare Quality® in UK assurance schemes: Evaluating the challenges. *Animal Welfare* 23, 95–107.

Hemsworth, P.H., Barnett, J.L., Coleman, G.J. and Hansen, C. (1989) A study of the relationships between the attitudinal and behavioral profiles of stock persons and the level of fear of humans and reproductive performance of commercial pigs. *Applied Animal Behaviour Science* 23, 301–314.

Hewson, C.J. (2003) Can we access welfare? *Canadian Veterinary Journal* 44, 749–753.

Hoffman, A.C., Moore, D.A., Wenz, J.R. and Vanegas, J. (2013) Comparison of modeled sampling strategies for estimation of dairy herd lameness, prevalence and cow level variables associated with lameness. *Journal of Dairy Science* 96, 5746–5755.

Jones, E.K.M., Wathes, C.M. and Webster, A.J.F. (2005) Avoidance of atmospheric ammonia by domestic fowl and the effect of early experience. *Applied Animal Behaviour Science* 90, 293–308.

Knowles, T.G., Kestin, S.C., Haslam, S.M., Brown, S.N., Green, L.E., *et al.* (2008) Leg disorders in broiler chickens, prevalence, risk factors and prevention. *PLOS One* 3(2). DOI: 10.1371/journal.pone.0001545.

Kristensen, H.H. and Wathes, C.M. (2000) Ammonia and poultry: A review. *World Poultry Science Journal* 56, 235–243.

Kristensen, H.H., Burgess, L.R., Demmers, T.G.H. and Wathes, C.M. (2000) The preferences of laying hens for different concentrations of ammonia. *Applied Animal Behaviour Science* 68, 307–318.

LayWel (2006) Welfare implications of changes in production systems for laying hens. Available at: http://ec.europa.eu/food/animal/welfare/farm/laywel_final_report_en.pdf (accessed 22 April 2015).

Main, D.C.J., Mulan, S., Atkinson, C., Cooper, M., Wrathall, J.H.M., *et al.* (2014) Best practice framework for animal welfare certification schemes. *Trends in Food Science and Technology* 37, 127–136.

Maria, G.A., Villarroel, M., Chacon, G. and Gebresenbet, G. (2004) Scoring system for evaluating the stress to cattle of commercial loading and unloading. *Veterinary Record* 154, 818–821.

Meluzzi, A., Fabbri, C., Folegatti, E. and Sirri, F. (2008) Survey of chicken rearing conditions in Italy: Effects of litter quality and stocking density on productivity, foot dermatitis and carcass injuries. *British Poultry Science* 49, 257–264.

Mollenhorst, H., Rodenburg, T.B., Bokkers, E.A.M., Koene, P. and de Boer, I.J.M. (2005) On-farm assessment of laying hen welfare – a comparison of one environment based and two animal based methods. *Applied Animal Behaviour Science* 90, 277–291.

Montoro, C., Miller-Cushon, E.K., DeVries, T.J. and Bach, A. (2013) Effect of physical form of forage on performance, feeding behavior, and digestibility of Holstein calves. *Journal of Dairy Science* 96, 1117–1124.

Mulling, C., Dopfer, D., Edwards, T., Larson, C., Tomlinson, D., *et al.* (2014) *Cattle Lameness, Identification, Prevention, and Control of Claw Lesions*, 2nd edn. Zinpro Corporation, Eden Prairie, Minnesota. To obtain write to contact@zinpro.com.

Nicole, R.A. (2014) A comparison of three animal welfare assessment programs on Canadian Swine Farms. Thesis, University of Guelph, Canada.

OIE (2014a) Chapter 7.9 *Animal welfare and beef cattle production systems. Terrestrial Animal Health Code*. World Organisation for Animal Health, Paris.

OIE (2014b) Chapter 7.10 *Animal welfare and broiler chicken production systems. Terrestrial Animal Health Code*. World Organisation for Animal Health, Paris.

OIE (2014c) Chapter 7.5. *Slaughter of animals. Terrestrial Animal Health Code*. World Organisation for Animal Health, Paris.

OIE (2014d) Chapter 7.3. *Transport of animals by land. Terrestrial Animal Health Code*. World Organisation for Animal Health, Paris.

Phythian, C.J., Toft, N., Cripps, P.J., Michalopoulou, E., Winter, A.C., et al. (2013) Interobserver agreement diagnostic sensitivity and specificity of animal based indicators of young lamb welfare. *Animal* 7, 1182–1190.

Rutherford, K.M., Donald, R.D., Lawrence, A.B. and Wemelsfelder, F. (2012) Qualitative behavioural assessment of emotionality in pigs. *Applied Animal Behaviour Science* 139, 218–224.

Smulders, D., Verbeke, G., Mormede, P. and Geers, R. (2006) Validation of a behavioral observation tool to assess pig welfare. *Physiology and Behavior* 89, 438–447.

Stockman, C.A., McGilchrist, P., Collins, T., Barnes, A.L., Miller, D., et al. (2012) Qualitative behavioural assessment of Angus steers during pre-slaughter handling and relationships with temperament and physiological responses. *Applied Animal Behaviour Science* 142, 125–133.

Stull, C.L., Reed, B.A. and Berry, S.L. (2005) A comparison of three animal welfare assessment programs in California dairies. *Journal of Dairy Science* 88, 1595–1600.

Thomsen, P.T., Munksgaard, L. and Togersen, F.A. (2008) Evaluation of lameness scoring of dairy cows. *Journal of Dairy Science* 91, 119–126.

Tuyttens, F.A.M., de Graaf, S., Heerkens, J.L.T., Jacobs, L., Nalon, E., et al. (2014) Observer bias in animal behavior research: Can we believe what we score, if we score what we believe? *Animal Behaviour* 90, 273–280.

University of Wisconsin (2005) Body Condition Score – What is Body Condition Score? What Does it Help Us Manage? University of Wisconsin.

Vasseur, E., Gibbons, J., Rushen, J. and dePassille, A. (2013) Development and implementation of a training program to ensure high repeatability of body condition scoring of dairy cows. *Journal of Dairy Science* 96, 4725–4737.

Velarde, A. and Dalmau, A. (2012) Animal welfare assessment at slaughter in Europe: Moving from inputs to outputs. *Meat Science* 92, 244–251.

von Borell, E., Bockisch, F.J., Büscher, W., Hoy, S., Krieter, J., et al. (2001) Critical control points for on-farm assessment of pig housing. *Livestock Production Science* 72, 177–184.

Webb, L.E., Bokkers, E.A.M., Heutinck, L.F.M., Engel, B., Buist, W.G., et al. (2013) Effects of roughage source, source amount and particle size on the behavior and gastrointestinal health of veal calves. *Journal of Dairy Science* 96, 7765–7776.

Webster, A.B., Fairchild, B.D., Cummings, T.S. and Stayer, P.A. (2008) Validation of three point gait scoring system for field assessment of walking ability of commercial broilers. *Journal of Applied Poultry Research* 17, 529–539.

Webster, J. (2005) The assessment and implementation of animal welfare: theory into practice. *Review of Science and Technology Office International des Epizooties* 24, 723–734.

Welfare Quality Network (2009) www.welfarequalitynetwork.net (accessed 9 May 2015).

Whay, H.R., Main, D.C.J., Green, L.E. and Webster, A.J.E. (2003) Assessment of the welfare of dairy cattle using animal based measurements: direct observations and investigation of farm records. *Veterinary Record* 153, 197–202.

Whay, H.R., Leeb, C., Main, D.C.J., Green, L.E. and Webster, A.J.F. (2007) Preliminary assessment of finishing pig welfare using animal based measurements. *Animal Welfare* 16, 209–211.

Wildman, E.E., Jones, G.M., Wagner, P.E., Boman, R.L., Troutt, H.F., et al. (1982) A dairy cow body condition scoring system and its relationship to selected production characteristics. *Journal of Dairy Science* 65, 495–501.

Winckler, C. and Willen, S. (2001) The reliability and repeatability of a lameness scoring system for use as an indicator of welfare in dairy cattle. *Acta Agriculturae Scandinavica: Section A, Animal Science* 30 (Supplement 1), 103–107.

5 How to Improve Livestock Handling and Reduce Stress

TEMPLE GRANDIN

Department of Animal Science, Colorado State University, Fort Collins, Colorado, USA

Abstract

Calm, low-stress methods for handling cattle, sheep, pigs, and other farm animals will result in better productivity and increased human safety. People who are handling animals should be trained to understand and use basic behavioral principles such as the flight zone, point of balance, and natural following behavior. Animals will move more easily through a race or alley if distractions such as shadows or reflections on shiny metal, or if people up ahead are removed from the animal's line of sight. Stress will be reduced if animals are acclimated both to handling procedures and to having people walk through their pens or pastures. Restraint of livestock will be easier and there will be less vocalization or struggling if the following principles are used. The most important features are nonslip flooring and supporting the body to avoid triggering the instinctual fear of slipping and falling. A common mistake is to apply excessive pressure with a restraint device.

Learning Objectives

- Understand the basic behavioral principles of moving animals, such as flight zone and point of balance
- Troubleshoot and remove visual distractions from handling facilities that stop animal movement
- Understand bull behavior and how raising bulls in social groups may help reduce dangerous aggression
- Principles of design and operation of restraint equipment
- Learn the importance of acclimating animals both to handling and to having people walking in their pens or pastures

Introduction: Benefits of Low-stress Farm Animal Handling

Calm animals are easier to handle, sort, and restrain than frightened, excited animals. If cattle become severely agitated, it takes 20–30 min for them to calm back down and for their heart rate return to normal (Stermer *et al.*, 1981). When they are brought into the corrals from pasture, it is often best to let them calm down and settle before working them in the race for veterinary procedures. If a horse becomes severely agitated during a veterinary procedure, it may be easier to handle if it is given 30 min to calm down before the procedure is attempted again.

Production benefits of good handling

Livestock that remain calm while they are restrained and walk calmly during handling will have improved weight gain, better reproduction and better meat quality compared to animals that become agitated during handling. Cattle that struggle when they are restrained or run wildly out of a race after handling had lower weight gains, higher cortisol, lower artificial insemination (AI) pregnancy rates, more dark cutters and tougher meat (Voisinet *et al.*, 1997a,b; Petherick *et al.*, 2002; Curley *et al.*, 2006; King *et al.*, 2006; Kasimanickam *et al.*, 2014). Dark cutting beef is a serious quality defect that makes meat darker and drier than normal. Cattle that were prodded six times with a battery-operated electric goad within 15 min of slaughter produced beef that was rated as tougher by a taste panel (Warner et al., 2007; Ferguson and Warner, 2008). Cattle that become agitated in the slaughter line or were the last animals in the single-file race had higher lactate and higher (tougher) Warner Bratzer shear force measurements (Gruber et al., 2010; Stockman et al., 2012). Work by Australian researcher, Paul Hemsworth, with pigs and dairy cows showed that negative handling reduced milk production in dairy cows and reduced weight gain in pigs (Barnett et al., 1992; Hemsworth et al., 2000) (see Chapter 7, this volume). Handling animals quietly will also help to reduce sickness and improve reproductive performance. Stress lowers immune function (Mertshing and Kelly, 1983). Stressful handling practices shortly after breeding will also lower conception rates in sheep, cattle, and pigs (Doney et al., 1976; Hixon et al., 1981; Fulkerson and Jamieson, 1982). Sows that were afraid of people and backed away from a person had fewer piglets (Hemsworth et al., 1981). Stress from poor handling practices can also make animals more susceptible to disease. Immune function, which helps animals fight disease, was reduced. Restraint stress reduced immune responses in pigs (Mertshing and Kelly, 1983). Multiple shocks with an electric goad are very detrimental to pigs. Pigs that were shocked multiple times were more likely to become nonambulatory and not able to walk (Benjamin et al., 2001). McGlone (2005, personal communication) reported that the occurrence of nonambulatory pigs was four times higher when an electric goad was used on over 50% of the pigs compared to 10%. The observations were made in a large commercial slaughter plant that processed large, 115-kg pigs. Electric goads also cause large increases in lactic acid and glucose in the pig's blood (Benjamin et al., 2001; Ritter et al., 2009). Edwards et al. (2010a,b) found that the use of electric goads in the stunning race, or getting jammed in the race, raised lactate levels and lowered the pH of the meat. Low pH meat is more likely to be pale soft exudative (PSE) meat. This is a severe quality defect.

Downed animals that cannot walk should never be dragged. Pigs can become nonambulatory due to either the porcine stress syndrome (PSS) (a genetic condition) or because they have been fed too high a dose of beta-agonists such as ractopamine (see Chapters 3 and 14, this volume). Careful, quiet handling will reduce the number of nonambulatory animals.

Safety benefits of good animal handling

Another advantage of low-stress, quiet cattle handling is improved safety for both people and animals. Both animals and people will have fewer injuries. A 10-year analysis of workman compensation claims indicated that claims for injuries associated with cattle handling represented the highest percentage of costly severe injuries (Douphrate et al., 2009). Cattle that are handled quietly during loading into a truck will also have half as many bruises (Grandin, 1981). Improvements in cattle handling in Brazil reduced bruising from 20% of the cattle to 1.3% (Paranhos da Costa, 2006, personal communication; Paranhos da Costa et al., 2014). Good handling will also help prevent pig death losses. A Spanish survey indicated that death losses increased when the time taken to load the pigs decreased (Averos et al., 2008). When people hurry, handling usually gets rougher.

Behavioral Principles for Handling Extensively Raised Animals That Are Not Completely Tame

Flight zone principles

Handlers need to understand the behavioral principles of the flight zone and point of balance for moving cattle, pigs, sheep, and other animals. To keep animals calm and move them easily, the handler should work on the edge of the flight zone (Grandin, 1980a,b, 2014a; Grandin and Deesing, 2008) (Fig. 5.1). If cattle rear up or become agitated in a single-file race, the handler should back up and get out of the animal's flight zone. Most animals will stop rearing and calm down when the handler backs out of the animal's flight zone. The flight zone is the animal's personal space, and the size of the flight can vary from 0 m from a person to over 50 m.

Hediger (1968) defines the process of taming an animal as removal of the flight zone to the point that the animal allows people to touch it. When a person penetrates the flight zone, the animals will turn and move away. When the animals turn and face the handler, the person is outside their flight zone. The size of the flight zone will vary depending on how wild or tame the animal is. Completely tame animals that have been trained to lead have no flight zone and they should be led instead of being driven. Extensively raised animals that seldom see people will have a larger flight zone than animals that see people every day. When cattle are handled quietly, their flight zone will become smaller.

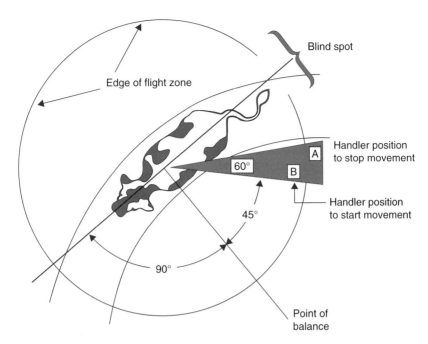

Fig. 5.1. Flight zone diagram for moving an animal through a single-file race or out of small pens. The handler should stand behind the point of balance at the shoulder to make an animal go forward (positions A and B) and in front of the point of balance at the shoulder to make an animal back up. When the handler is further away, the point of balance may move forward and be slightly behind the eye. A handler must never stand at the head and poke the rear. This gives the animal confusing signals.

The size of the flight zone is determined by three factors. They are: (i) the amount of contact with people; (ii) the quality of the contact with people, calm and quiet versus shouting and hitting; and (iii) animal genetics. Animals that see people every day when they are fed will usually have a smaller flight distance than animals on extensive pasture that see people only a few times a year.

To move extensively raised animals in a calm, controlled manner, the handler should work on the edge of the flight zone. The handler penetrates the flight zone to make the animals move and he backs up if he wants them to stop moving. The handler should avoid the blind spot behind the animal's rear. Deep penetration of the flight zone should be avoided because continuous deep penetration of the flight zone will cause the animals to run. The handler should also practice the principle of pressure and release. When the animals are moving in the desired direction, the handler should back out of the flight zone. When the animals slow down or stop moving, the flight zone should be re-entered.

Animals may become agitated when a person is inside their personal space and the animal is unable to move away due to being in a race or small pen. If animals turn back and run past the handler while they are being driven down a drive alley in the corrals, overly deep penetration of the flight zone is a likely cause. The handler should back up and retreat from inside the flight zone when the animals give the slightest indication that they will turn back.

The flight zone diagram (Fig. 5.1) shows the correct positions for the handler to move cattle through a single-file race or out of a small pen. The handler should stand behind the point of balance at the shoulder to make an animal go forward and in front of the point of balance at the shoulder to make an animal back up (Kilgour and Dalton, 1984; Smith, 1998; Grandin, 2014a,b). The most common mistake that people make is to stand in front of the point of balance and poke the animal on the hindquarters to make it go forward. This gives the animal conflicting signals and confuses it. The handler must be behind the shoulder when he wants to urge the animal forward. For pigs in small pens, the point of balance is usually just behind the eye. Handlers should take advantage of the natural behavior of pigs to circle around them (National Pork Board, 2014).

Genetic effects on behavior

Genetic differences in the animal's temperament will also affect the flight distance (Grandin and Deesing, 2013). During handling in a race, Holstein steers and heifers balked more than the beef breeds (Thomas, 2014). There are individual animals within a breed in every herd that will have a more flighty, excitable temperament and a larger flight zone. When raised under extensive conditions, British breeds such as Shorthorns were less likely to become agitated during handling compared to Brahman crosses (Fordyce *et al.*, 1988). Within a breed of livestock, temperament and the tendency to startle is definitely heritable (Hearnshaw and Morris, 1984). Holstein cattle balk more and require greater use of electric prods to move them through a single-file chute (race) compared to beef breeders (Thomas, 2014). There are also genetic differences in the tendency of cattle, pigs, or sheep to bunch and flock together (Jorgensen *et al.*, 2011; Dodd *et al.*, 2012). Merino and Rambouillet sheep will flock more tightly together than Suffolks, Hampshires, and Cheviots.

Point of balance principles to move a group

The second diagram (Fig. 5.2) illustrates how to use the point of balance at the animal's shoulder to move extensively raised livestock forward in a race. When a person walks quickly back past the point of balance in the OPPOSITE direction of desired movement, the cattle, pigs, and sheep will move forward. The principle is to walk quickly in the opposite direction of desired movement inside the flight zone and in the same direction outside the flight zone. When animals are out on pasture, a basic principle is that walking in the opposite direction of cattle movement speeds up herd movement, and walking alongside the herd in the same direction tends to slow the herd down. This principle applies to the movement of a person who is alongside a herd and not behind it. When groups of animals are handled and the handler is further away, the point of balance may move forward to just behind the eye. Figure 5.3 shows how, using the same pattern shown in Fig. 5.2, to move cattle out of a large pen.

For cattle and pigs – fill the crowd pen half full and use following behavior

One of the most common mistakes when handling cattle and pigs is overloading the crowd pen that leads to a single-file race or loading ramp. The crowd pen should be filled HALF full so that animals will have room to turn. All livestock will follow the leader, and handlers can take advantage of this natural behavior to move animals easily. Cattle and pigs will enter a single-file race more easily if it is allowed to become partially empty before attempting to fill it. A partially empty race takes advantage of following behavior because it provides enough space for four to ten animals to enter while following the leader. Small bunches of cattle and pigs should be moved. Low-stress handling in races and crowd pens will require more walking to put smaller groups of cattle or pigs in the crowd pen. If your crowd pen holds 20 animals when it is stuffed full, you should bring in only ten animals so it will be half full. During truck loading, market-weight pigs should be moved in groups of four or five. Sheep have such intense following behavior that they can be moved in one continuous large group. The principle of moving sheep is never to have a break in the flow. It is analogous to siphoning water. Once the flow is going, it should be kept going because restarting the flow may be difficult.

Principles of leading animals

Extensively raised sheep with a large flight zone will follow a trained lead animal. Both goats and sheep work well as lead animals. Figure 5.4 shows a goat that is leading sheep through a stockyard. Trained sheep work very well for leading sheep on and off trucks and through the lairage at slaughter plants. Cattle can also be trained easily to follow a lead person or vehicle when they are being rotated to another pasture. When animals become very tame, the principles of using the flight zone and the point of balance will no longer work. Tame animals MUST be led.

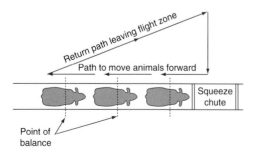

Fig. 5.2. Point of balance diagram. Walking quickly past the point of balance in the opposite direction of desired movement will make the animals move forward.

T. Grandin

Fig. 5.3. Movement pattern to move livestock out a large pen and into an alley. The triangle pattern controls animal movement to provide controlled movement out of the pen. The principle is to move inside the flight zone in the opposite direction of desired movement to speed up the animals and in the same direction outside the flight zone to slow down movement. This principle also works on large pastures. This movement pattern may not work in small lairage or auction market pens. Handlers need to learn to stop pushing from behind the group and move animals by working in the front of the group.

Acclimate animals on the farm to people walking through them

Pigs and cattle will be easier to load on to trucks and move if they have become acclimated to having people walk through their pens BEFORE it is time to handle them for transport, vaccinations, or at an abattoir. Pigs that have never been acclimated to people walking through them are more likely to pile up during truck loading. Cattle that have not been acclimated to people walking through their pens may have large flight zones and be difficult or dangerous to handle during truck loading movement through auctions or at the abattoir. Producers need to train animals to move quietly away from a person walking through them.

Moving cattle from large pens or pastures

When cattle or pigs are moved from a large pen through a gate, the handler should move to the front of the pen to control movement out through the gate. A triangular movement pattern is used to speed up or slow down flow (see Fig. 5.3).

Fig. 5.4. Trained goats or sheep will lead sheep through stockyards or lairages. Lead animals also work really well for moving sheep on and off trucks. A specialist lead sheep or goat should be used for different handling procedures. Training the lead animals will be easier if one leader is trained to unload trucks, another is trained to load trucks, and a third animal is trained to lead sheep out of pens and through the stockyards. In confined spaces such as trucks and small lairage pens at the slaughter plant, lead animals are recommended instead of dogs. Lead sheep have been working very successfully for many years at the largest sheep slaughter plant in the USA.

Livestock become fearful when alone

A lone cow that becomes agitated in an attempt to join its herdmates has caused many accidents that have injured people severely. Never get in a small, confined pen with a single, agitated bovine. Isolation away from herdmates is highly stressful for all livestock (Bates *et al.*, 2014). It may jump a fence or run into people. If a lone animal becomes agitated, put some other animals in with it.

Animal Sensory Perception

Hearing in farm animals

Farm animals have sensitive hearing and are sensitive to high-pitched sounds. They can hear high-frequency sounds that people cannot hear. The human ear is most sensitive to sounds in the 1000–3000 hz range, and cattle and horses are most sensitive to frequencies at 8000 hz and above (Heffner and Heffner, 1983). Sheep can hear up to 10,000 hz (Wollack, 1963). High-pitched, intermittent sounds are also more likely to cause cattle or pigs to react (Talling *et al.*, 1998; Lanier *et al.*, 2000). Handlers should watch where animals point their ears.

Horses, cattle, sheep and many other animals will point their ears towards things that attract their attention. Yelling and screaming at livestock is highly stressful (Waynert *et al.*, 1999; Pajor *et al.*, 2003; Macedo *et al.*, 2011).

Vision in cattle, sheep, and horses

Grazing animals are very sensitive to rapid movement. Sudden rapid movement may frighten cattle (Lanier *et al.*, 2000). Cattle, sheep, and horses have wide-angle vision and they can see all around themselves without turning their heads (Prince, 1970; Hutson, 1980; Kilgour and Dalton, 1984). Grazing animals such as horses and sheep have a horizontal band of sensitive retina (Shinozaki *et al.*, 2010). This enables them to scan their surroundings easily for danger during grazing. Grazing animals have depth perception (Lemmon and Patterson, 1964). It is likely that their depth perception is poor because animals will stop and put their heads down when they see a shadow on the floor.

Grazing animals are dichromats. The retina of cattle, sheep, and goats is most sensitive to yellowish-green (552–555 nm) and bluish-purple light (444–455 nm) (Jacobs *et al.*, 1998). The dichromatic vision of the horse is most sensitive at 428 nm and 539 nm (Carroll *et al.*, 2001). Dichromatic vision and the absence of a retina receptor for red may explain why livestock are so sensitive to sharp contrasts of light and dark, such as shadows or shiny reflections on handling equipment.

Vision in poultry

Poultry appear to have excellent vision. Chickens and turkeys possess four cone cell types in the retina, giving them tetrachromatic colour vision, as compared to the human trichromatic vision based on three cone cell types (Lewis and Morris, 2000). The spectral sensitivity of chickens is greater than that of humans, from 320–480 nm to 580–700 nm. Their maximum sensitivity is in a similar range (545–575 nm) to humans (Prescott and Wathes, 1999). The broader spectral sensitivity of poultry may make them perceive many light sources as being brighter than what a human would see. Poultry may be more docile during handling in blue light spectra (Lewis and Morris, 2000). Lighting conditions have a large effect on chicken behavior when they are shackled for slaughter (Jones *et al.*, 1998). During handling of poultry, the occurrence of flapping should be minimized. Changes in lighting may be used as one tool to keep birds calmer during handling. Quick Access Information 5.1 provides a summary of basic handling principles.

T. Grandin

Remove Visual Distractions from Handling Facilities that Cause Animals to Balk

Extensively raised livestock and animals that are trained to lead may often stop and refuse to move past little distractions that people do not notice. Animals are very sensitive to both high contrasts of light and dark and rapid movement. They notice visual detail in their environment that people often fail to notice. A calm animal will look right at a distraction such as a shirt hung on a corral fence. A frightened animal that is being forced to move towards the shirt will often turn back and try to run back past the handler. People need to be aware of the distractions that frighten animals. Handlers should walk through their races and corrals and look for distractions that can stop cattle movement. Distractions cause the most problems with animals that are not familiar with the handling facilities. Below is a list of common distractions that can cause animals to stop and refuse to move through a race or other handling facility. An experienced old dairy cow will walk over a distraction such as a shadow on a floor drain she sees every day. A new heifer that has never been in the facility is much more likely to stop at the drain or shadow.

- Shadows or bright sunbeams of light are often a problem on sunny days. Livestock may refuse to walk over them (Fig. 5.5). In Fig. 5.6, the pigs are avoiding walking on the shadows and they are balking at the metal strip. Allow the leader time to put its head down to investigate a shadow. After it has determined that the shadow is safe, it will lead the others over the shadow. If the handlers rush the leader's investigation of a shadow, the animals may turn back on the handlers. When an animal is being led, allow it time to put its head down to investigate the shadow before urging it to walk over it.
- Objects on the fence or in the race, such as loose chains or shirts, should be removed. In Fig. 5.7, the animal is refusing to move past a dangling rope. Cattle may refuse to move past a shirt that is hung on a fence. A little, loose chain that swings may also stop cattle movement. Bright yellow clothes and objects are especially bad. Objects with high contrasts of light and dark cause the most problems.
- Vehicles parked outside the corrals may also cause problems. A shiny reflection on the bumper of a car may stop animal movement. The vehicles should be moved.
- Livestock will often refuse to approach visible people who are standing up ahead. The people should either move or a shield can be installed so that the cattle do not see the people. Installing solid fences to block the animal's vision of distractions outside the facility often improves animal movement (Grandin, 1996, 2014a,b; Grandin and Deesing, 2008). The materials must be stiff and not flap. Flapping materials will scare the animals.

Fig. 5.5. Calf balking at a bright spot of sunlight as it exits the squeeze chute (crush). Note how the animal is pointing both its eyes and ears towards the sunbeam. A calm animal will look directly at the distraction. This facility has an excellent nonslip floor made from woven tyre treads.

Fig. 5.6. The pig is balking and refusing to walk over a metal strip. Animals will often avoid walking on shadows or beams of light.

Fig. 5.7. This animal is refusing to walk past the dangling rope. Remove ropes and chains that hang down in front of animals. Note that the truck driver left his electric goad on the top of the truck. If the dangling rope had been removed, the electric goad would not be needed. Cattle will unload easily from a truck and electric goads can be eliminated if all the distractions are removed.

- Driving livestock directly into either the rising or the setting sun is difficult because the animals are blinded. The best way to fix this problem is to change the time of day that the livestock are handled. When new facilities are built, avoid pointing races or truck loading ramps towards the sun.
- Animals may refuse to enter a dark building where a race is located (Fig. 5.8). They will enter more easily if they can see daylight through the other side of the building (Fig. 5.9). Removing a side wall may help. In new facilities, installing white, translucent skylights to admit lots of shadow-free daylight will usually improve animal movement. Cattle and pigs have a natural

tendency to move from a dark place to a brighter place unless they are approaching a blinding sun (van Putten and Elshof, 1978; Grandin, 1982; Tanida *et al.*, 1996). At night, a lamp can be used to attract animals into buildings or trucks. The light MUST NOT shine directly in their eyes. It should provide even, bright, indirect light.
- Remove paper cups, old feed bags, plastic bags, or other rubbish that is lying on the ground because animals may refuse to walk over it.
- Reflections in a puddle of water in a corral or race can stop animal movements. Either fill in the puddles with dirt or change the time of day that the animals are handled. In indoor facilities, changing the position of lamps will often eliminate a reflection. One person should get down

Fig. 5.8. Animals may refuse to enter this dark building. Opening up door or windows to admit light into the building will attract the animals in. This is especially a problem on bright, sunny days. At night, lamps providing indirect lighting can be used to attract the animals into the building.

Fig. 5.9. Animals will enter this veterinary building more easily because they can see through it and see daylight on the other side.

in the race to see an animal's eye view while the other changes the position of lamps.

- Abrupt changes in flooring will also stop animal movement. Examples are moving from a dirt floor to a concrete floor or walking up a wooden loading ramp into a vehicle with a metal floor. Spreading some dirt, hay, or other material on the junction between the two flooring types will remove the contrast. Often, a very small amount will work. In indoor facilities, a drain grating or metal drain cover may stop animal movement. When facilities are being designed, drains should be outside the areas where animals will walk. When portable handling facilities are used, metal pipes on the ground that hold the facility together should be covered with dirt. Even animals that are completely tame and trained to lead can panic and become agitated if they are forced to approach a strange thing. This is especially a problem if it is something they have never seen before. Horses may fear llamas, and pigs are scared of bison.

Nonslip Flooring is Essential

A nonslip surface for animals to stand on is essential (Grandin, 1983, 2014a). Animals panic and can become agitated when they slip. Some of the most important places to have non-slip flooring are in the single-file race, truck floors, veterinary facilities, stun boxes, and the crush. Small, repeated slips where one hoof moves back and forth rapidly are really scary for animals. A floor can be made nonslip by covering it with gravel, sand, or dirt. In facilities with metal or concrete floors, a grating made from steel bars can be made (Fig. 5.10). This works really well in high traffic areas such as stun boxes, loading ramps, and scales. It is best to use heavy rods with a minimum diameter of 2 cm for cattle and other large animals. The rods can be welded to make a grid in a 30 cm × 30 cm square pattern. The mat of rods MUST LIE FLAT.

Fig. 5.10. A grating constructed from steel rods will provide a nonslip surface on unloading docks, scales, stun boxes, and other high animal traffic areas. The rods must not be criss-crossed on top of each other. The rod mat should lay completely flat. For large animals, the rods are laid out in a 30 cm × 30 cm square pattern.

Do NOT criss-cross the rods on top of each other. The gaps can injure hooves. Nonslip flooring is essential for good animal welfare. Both handling facilities and trucks should have nonslip floors. Grates made from steel rods work well for providing nonslip flooring in vehicles. A rough broom finish on concrete will quickly wear smooth and animals will slip and fall. For pigs and sheep, a good finish for concrete is stamping the wet concrete with expanded metal mesh. This provides a rougher surface that is still reasonably easy to clean. Deeper grooves will be required for larger animals. In many countries, rubber mats that provide a good nonslip surface are available. They are especially useful in high traffic areas such as loading ramps, scales, squeeze chutes, and race exits.

Driving Aids for Moving Livestock

An electric goad should never be used as a person's primary driving tool, and the World Organisation for Animal Health (OIE, 2014a,b) code recommends that only battery-operated goads should be used. The OIE (2014a,b) code also states that electric goads should not be used on sheep, horses, or very young animals such as piglets. The goad should only be picked up when other nonelectrified methods fail to move an animal. Research has shown that the attitude of the handlers towards the animal is better when shocks are not being used (Coleman *et al.*, 2003). A brief shock on the hindquarters is preferable to hard tail twisting or beating an animal. After the stubborn animal is moved with the electric goad, the goad should be put away. A plastic bag taped to the end of a stick or a flag is an excellent tool for moving cattle (Fig. 5.11). Other commonly used tools are plastic paddles. For pigs, panels work really well in narrow alleys (McGlone *et al.*, 2004). Some other innovative tools for moving pigs are long strips of opaque plastic and the "witch's cape". The pigs respect the solid barriers and they will not run through them (Fig. 5.12). Driving aids must never be applied to sensitive parts of the animal such as the eyes, ears, nose, rectum, genitals, or udder. The OIE code states:

> Painful procedures (including whipping, tail twisting, use of nose twitches, pressure on eyes, ears or external genitalia) or the use of goads or other aids which cause pain and suffering (including large sticks, sticks with sharp ends, lengths of metal piping, fencing wire or heavy leather belts) should not be used.
>
> (OIE, 2014a,b)

T. Grandin

Fig. 5.11. Handlers in Uruguay use flags to move cattle. To promote the concept of animal welfare, the flags were provided by a local meat company and had the words "animal welfare" printed on them in Spanish.

Fig. 5.12. Pigs can be moved easily with a plastic cape. The top of the cape is stiffened with a rod. This alternative driving tool makes it possible to greatly reduce electric-prod use.

There are many situations where a driving aid is not required. People who become highly skilled at low-stress livestock handling can often direct animal movements by slight movements of their body position.

Measure Animal Handling with Animal-based Outcome Measures

People are in control of the things that they measure. Throughout her long career, the author has worked with many ranchers and feedlots on cattle handling. While the author was present, handling practices were excellent, but when the author returned a year later, the yelling and screaming and use of the electric prod had often increased. In many cases, people did not realize this had happened, because reverting back to old, rough practices often happens slowly. To prevent the return of these old, rough practices, handling should be measured with numerical scoring. This makes it possible to determine if practices are improving or getting worse.

To measure handling, keep score on the number of animals that have the following faults and determine the percentages:

- The percentage of animals that fall during handling – score a fall if the body touches the ground. The OIE (2014a,b) codes state that numerical scoring should be used to measure falling. Corrective action should be taken to improve either the handling practices or the facility if falling exceeds 1% of the animals (Grandin, 2014b).
- The percentage of animals that are moved at faster than a trot – animals should NOT be running and jumping. They should not be forced to move faster than a normal walking speed.
- The percentage of animals that run into a gate or fence.
- The percentage of animals moved with an electric goad – the OIE (2014a,b) codes state that numerical scoring should be used to measure electric goad use. On a farm or ranch, an electric goad should be used on 5% or less of the animals.
- The percentage of animals that vocalize (moo, bellow, or squeal) when they are caught by the head or body restraint – when the quality of handling is assessed, do not score after ear tagging or other procedures start.

Monitor continuous improvement in handling

If the handlers are doing a good job of low-stress handling, the percentage of animals that have any one of these faults will be very low. By keeping score every time you handle livestock, you can determine if handling is getting better or becoming worse.

Falling should be extremely infrequent. If more than 1% of the animals fall, there is a serious problem with either a slick floor or rough handling. Falling is a very good measure, because it is the outcome of either a slippery floor or poor handling methods.

Research the author conducted on vocalization showed that 99% of the time that cattle vocalized during active handling at a slaughter plant was due to a frightening or painful event (Grandin, 1998). Some of the events that caused vocalization were slipping on the floor, electric goads, (Munoz *et al.*, 2012), sharp edges on a restraint device, or being squeezed too hard by a restrainer or head stanchion.

When a head restraint that applied excessive pressures was loosened, the percentage of cattle that vocalized was reduced from 23% to 0% (Grandin, 2001). The goal of 5% or less of cattle vocalizing due to handling and restraint problems is achievable (Barbalho, 2007; Grandin, 2012). On farms, vocalization to assess the quality of handling should only be scored when animals are actively being moved in the single-file race and during restraint. If the bovine vocalizes the moment it is caught by either a head or body restraint, it is being hurt by the device. Woiwode *et al.* (2014) carried out a survey of cattle handling and restraint in squeeze chutes for veterinary practices in 28 feedlots in Kansas, Colorado, and Nebraska. The average vocalization score was 1.3% of the cattle and the average percentage of cattle moved with an electric prod was 3.8% (Table 5.1). Of the feedyards surveyed, 20% had a 0% electric prod use. Do not score vocalization from weaning animals or procedures such as branding when you are assessing handling. However, when you are measuring stress from a procedure such as weaning or branding, animal vocalizations associated with these procedures should be scored. Vocalizations caused by handling and veterinary procedures should be scored separately from vocalizations caused by weaning stress. Vocalization is an outcome measure of an event that either frightens the animal or causes pain. Cattle and sheep left alone will both vocalize because they are separated from herd mates. If a head restraint has a sharp edge, that could also cause vocalization. Vocalization scoring is a useful measure of animal distress. Vocalization in both cattle and pigs is associated with physiological measures of stress (Dunn, 1990; Warriss *et al.*, 1994; White *et al.*, 1995; Hemsworth *et al.*, 2011). Procedures that cause either pain or fear will usually increase vocalization

Table 5.1. Survey of cattle handling in squeeze chute in 28 large Nebraska, Kansas, and Colorado feedlots. (From Woiwode and Grandin, 2014.)

Percentage of cattle	Average	Range
Electric prod (%)	3.8	0–45
Vocalization per cent[a] when caught by the head restraint	1.3	0–6
Exit falling (%)	0.6	0–4.5
Exit stumble (%)	5.7	0–28
Miscaught (%)	4.2	0–16
Exit run (%)	30.7	0–25

[a]Note: No branding, castration, or dehorning.

T. Grandin

in cattle and pigs (Lay *et al.*, 1992a,b; White *et al.*, 1995; Watts and Stookey, 1999). Vocalization scoring cannot be used for assessing stressful or painful procedures in sheep, because sheep will not vocalize when they are hurt. Sheep are defenceless prey animals. In the wild, they do not want to advertise to predators that they are hurt. Sheep vocalize when they become isolated because they are seeking the safety of the flock, but they seldom vocalize when they are being handled or restrained.

Bull Behavior and Safety

Dairy bulls have a bad reputation for attacking humans, possibly due to the differences in the way beef bulls and dairy bulls are raised in some countries. Bulls are responsible for about half of the fatal accidents with livestock (Drudi, 2000). In India, bulls injure many people. Wasadikar *et al.* (1997) reported that a rural medical college treated 50 cases of bull-horn injuries over a 5-year period. Dairy bull calves are often removed from the cow shortly after birth and raised in individual pens, whereas beef bull calves are reared by the cow and kept in a larger social group.

Price and Wallach (1990) found that 75% of Hereford bulls reared in individual pens from 1 to 3 days of age threatened or attacked the handlers, whereas only 11% threatened handlers when they were hand reared in groups. These authors also report that they have handled over 1000 dam-reared bulls and have experienced only one attack. Bull calves that are hand reared in individual pens may fail to develop normal social relations with other animals, and they possibly view humans as a sexual rival (Reinken, 1988). It is important for young bulls to learn that they are cattle. Problems with bull attacks are a social-identity issue, not a tameness issue. If a young bull is going to start being aggressive towards people, the first attack is likely to occur at 18–24 months of age.

Both dairy and beef breed bulls will be safer if bull calves are either raised on a cow or kept in groups with other cattle. This provides socialization with their own species, and they will be less likely to direct attacks towards people. Similar aggression problems have also been reported in hand-reared male llamas (Tillman, 1981) and orphan buck deer. In one sad case, an orphan buck attacked and killed the person who raised him. Fortunately, hand rearing usually does not cause aggression problems in females or castrated animals. It will make these animals easier to handle. More information on bulls can be found in Smith (1998).

Before a bull attacks, he will turn sideways and do a broadside threat to show how big he is. Bulls that display a broadside threat towards humans are really dangerous and should either be slaughtered or sent to a secure bull stud. If a bull threatens a person, the person should slowly back away. Handlers must be instructed to NEVER turn their back on a bull. He is more likely to attack when the person is not looking at him. To help prevent dangerous attacks by intact males, orphan bull calves should either be castrated at an early age, raised on a nurse cow or penned with other calves.

Stress of Handling and Veterinary Procedures

Often, fear stress caused by restraining an extensively raised animal may be almost as stressful as hot-iron branding, ear tagging, or injections. Extensively raised cattle that were restrained in a squeeze chute had similar cortisol levels as cattle branded with a hot iron (Lay *et al.*, 1992a). When the same experiment was repeated with tame dairy animals, the hot-iron-branded animal had significantly higher cortisol (Lay *et al.*, 1992b). When more severe procedures such as dehorning are done, the procedure causes an elevation of cortisol for a longer period compared to handling (see Chapter 6, this volume). This occurs in both extensively and intensively raised animals. Therefore, stress from handling may often exceed the stress caused by minor procedures such as ear tagging and injections (Grandin, 1997a). For more severe procedures such as dehorning or castration, pain-relieving medication will reduce stress (Grandin, 1997a) (see Chapter 6, this volume). Isolating sheep and leaving them tied up for long periods is very stressful and will cause very high cortisol levels (Apple *et al.*, 1993). In another study, cortisol levels were extremely high after sheep were unloaded from a truck by picking each animal up.

Preventing fear memories

It is very important that an animal's first experience with something new should be a good experience. When a new corral or other animal-handling facility is built, the animal's first experience with it should be a positive one, such as feeding them in it. If the first experience is really bad, it may be difficult to get the animal to enter the corral in the future. Sheep have remembered a single stressful handling experience a year later (Hutson, 1985). Painful procedures should NEVER be associated with an animal's first experience with a new person or a new piece of equipment. Miller (1960) reported that if a rat received a severe shock the first time it entered a new arm of a maze, it would never go in that arm again. However, if its first experience in the new arm resulted in delicious food treats, the rat would tolerate gradually increasing shocks and keep re-entering the same arm to get the food reward. This same procedure may also apply to livestock. Hutson (2014) recommends that the first few times animals are handled in the facility, low-stress procedures such as sorting or weighing should be done first. Doing this may help make the animals more willing to re-enter the facility when a painful procedure has to be done. Frightening or painful first experiences with new things can create a permanent fear memory.

A question that often gets asked is: will doing a painful procedure on a 2-day-old calf or young lamb have an effect on handling in the future? Practical experience has shown that very young infant animals probably do not remember the experience and make an association with the person. When the calves and lambs get older, they have excellent memories for associating painful or frightening experiences to specific people, places, sounds, or sights.

Fear memories of experiences that have been really frightening or painful can never be erased (LeDoux, 1996; Rogan and LeDoux, 1996). Even if an animal is trained to tolerate something it had previously been afraid of, the fear memory can sometimes suddenly reappear. This is most likely to occur in animals that startle easily, such as Arabian horses or Saler cattle. This is the reason why rough, harsh training methods are especially harmful to breeds such as Arabians. If an Arabian horse is abused, it may no longer be a safe riding horse because it cannot stop being frightened. Calmer breeds of livestock such as draft horses or Hereford cattle have a greater ability to suppress past memories of scary or painful events. The fear circuits in an animal's brain have been fully mapped. The brain has a fear center called the amygdala. Scientists have proven that a fear center exists in the brain, because destroying the amygdala completely stops fear (Kemble et al., 1984; Davis, 1992). A review of this research can be found in Morris et al. (2011).

Sensory-based memory

Animals store memories as pictures, sounds, or other sensory impressions (Grandin and Johnson, 2005). Since they think in sensory-based memories, their memories are very specific. Research with horses showed that a horse sometimes failed to recognize a child's large toy when the toy was rotated to a different position (Hanggi, 2005). When the toy was rotated, it created a different picture, which the horse failed to recognize. Visual picture memory is a basic function of the brain. Even ants have the ability to remember visual images (Judd and Collett, 1998). For example, a cow or horse may fear men with beards or people with black hats (Grandin and Johnson, 2005). This would happen if the animal had been looking at a hat or beard when a really painful or frightening event occurred. Animals tend to associate a bad experience with an obvious feature on a person, such as a white coat, a beard, or blond hair. Since animal learning is sensory based, it is very specific. Training a horse to tolerate a blue and white umbrella does not prevent fear of a flapping canvas (Leiner and Fendt, 2011). The fact that cattle have learned to be calm when people feed them does not transfer to create calm behavior during handling in a novel new squeeze chute (Cooke et al., 2009). They also can learn to fear certain sounds associated with a painful or scary past experience. Animals can learn the voices of a trusted stockperson and the voice of a person who has hurt them (see Chapter 7, this volume). To help prevent fear of the regular stockpeople, painful procedures should be done: (i) by different people; or (ii) by people wearing distinctive clothes that are reserved for painful procedures; or (iii) in a specific place that is reserved for painful procedures.

Habituate animals to novel new experiences

A common complaint that the author often hears is "my cattle/my horse was gentle and calm at home, but my animal went crazy and wild at a showground or auction". This problem occurs because the animals are suddenly confronted with many new, scary things that they have never seen on the home farm. To prevent this problem, the animals must be acclimated to the sights and sounds of a fairground or auction BEFORE they go there. Flags, bikes, or balloons can be really frightening, even though an animal is completely tame and trained to lead. The author has observed tame show animals that have panicked and run through the fairground, knocking people over. Animals should be habituated carefully to items such as flags and bikes on the home farm.

The best way to teach an animal to tolerate scary, moving objects is to allow the animal to approach voluntarily flags or balloons that are tied to a pasture fence. An important principle is that new novel things are scary when an animal is forced to approach them, but they are attractive if the animal can explore them voluntarily. Animals with a high-strung, nervous temperament are more likely to get scared when novelty is suddenly introduced. Objects that move rapidly can cause severe problems. Most animals have probably learned that large, moving vehicles are safe, but a flag is something new. New moving objects that the animals have never seen before can be really frightening.

A large panel such as a sheet of plywood that is moved rapidly can also cause panic. The author observed a tame heifer on a lead rope that attempted to break away from her handler when a large 1.2 m × 3 m white panel was suddenly moved. When the panel was still, she ignored it. Training livestock to tolerate a large object that suddenly moves is highly recommended. Grazing animals have an instinctual fear of things that move rapidly. In the wild, predatory animals such as lions move rapidly. Animals can also get really scared and react violently when a novel object rapidly moves near their face while they are eating. Extensively raised cattle that had a calm temperament reacted violently when a ball was dropped near their head while they were licking salt in a single-animal salt feeder (Sebastian, 2007). In this type of feeder, the animal puts its head under a plastic hood that keeps rain off the salt. When the ball fell from the ceiling of the feeder, some of the animals may have instinctually reacted to get a "predator" off their head.

In developing countries, cattle and water buffalo often graze along the highways. They see many new things from a very early age, and they will be less likely to be frightened by an object such as a large piece of plywood. These animals have seen every kind of object passing by them on the road. The mother animal teaches her calf not to be afraid of rapid movement. The calf sees his mother grazing calmly by the highway, so he will continue to graze beside her.

Reid and Mills (1962) were two of the first researchers to introduce the idea that sheep could be acclimated to changes in routine. It is important to teach cattle gently that new people and new vehicles are safe. It is recommended to have different people and vehicles involved in feeding cattle. If the cattle are fed by only one person, they are more likely to panic if they have to be handled or fed by a new person. Animals also perceive a person on a horse and a person on the ground as two different things. It is important for cattle to learn how to move in and out of pens in the corral BOTH by a person on foot and by a person on a horse. The author has seen cattle that were gentle and easy to handle by a person on a horse but which became agitated and dangerous the first time a person on foot tried to move them out of a pen. Learning in animals is very specific. A person on a horse and a person walking on the ground are two different visual images.

Signs of Fear and Agitation During Handling in Grazing Animals

People working with farm animals need to be able to recognize the behavioral signs that an animal is becoming more and more agitated and fearful. Learning to recognize these signs will help prevent both people and animals from getting injured.

Tail switching

Horses and cattle will switch their tails when no flies are present. The speed of tail switching will increase as the animal gets more and more agitated. Failure to heed this warning signal may result in people getting kicked. Bison hold their tails straight up.

Defecation

Scared animals with high anxiety will defecate more. Loose diarrhea in a healthy animal may be a sign of emotional stress, unless it has been grazing on lush green fields. When animals are handled quietly, there is often less manure left in the facility to clean up. Defecation is one of the first signs of stress.

Whites of the eyes show

Bulging eyes where the whites show is a sign of distress. Research shows that visible eye white is a sign of emotional distress because the response can be blocked in cattle with the anti-anxiety drug, diazepam (Valium®) (Sandern et al., 2006). The percentage of eye white showing is highly correlated with scores on standard cattle temperament tests (Core et al., 2009).

Sweating with minimal physical exertion

This occurs mainly in horses.

Quivering skin

Animals that quiver all over when they are touched and have no flies on them may be becoming fearful.

Nostrils flaring

This is more visible in horses compared to cattle.

Both ears laid back indicates fear

When an animal lays both its ears back, it is a sign of fear (Boissy *et al.*, 2011). Calm, relaxed animals will hold their ears in a horizontal position. Cattle, sheep, and other grazing animals will also "watch" things with their ears. A cow or ewe may point one ear towards her baby and the other ear towards a person. When dairy cows received pleasant stroking, their ears were more likely to be in either a hanging or horizontal position (Proctor and Carder, 2014).

Head held up high

Both cattle and horses will raise their heads up high when they are fearful. This is an instinctual behavior that is used to look for predators.

Tonic immobility

This occurs mainly in poultry, *Bos indicus* cattle, and bison. The bird stays still and does not move. The length of time that a bird will stay still after it has been held by a person is used as a measure of fearfulness in poultry. Different genetic lines can be assessed for fearfulness by measuring the length of time they will remain in tonic immobility (Faure and Mills, 2013). The author and South American researchers have both observed that some *Bos indicus* cattle and American bison may also lie down and become immobile if they are shocked multiple times while they are held in a race from which they cannot escape. An animal that has become immobile will often calm down and get back up if people leave it alone for 15–30 min.

Principles of Training Animals to Cooperate with Handling and Restraint

Animals can be trained to cooperate completely with veterinary procedures such as blood sampling.

When an animal cooperates, the animal's cortisol levels will remain very low. Sheep, pigs, cattle, and wild ungulates such as antelopes can be trained easily to enter a restraint device voluntarily (Panepinto *et al.*, 1983; Grandin, 1989a,b). Almost baseline levels of cortisol (stress hormone) and glucose have been obtained from trained antelopes (Phillips *et al.*, 1998). Cortisol levels from wild antelopes captured in the bush were three to four times higher. The fear stress of being trapped and restrained can kill a wild animal by causing capture myopathy. Sometimes the animal dies right away and other times it may take up to 2 weeks to die. Chalmers and Barrett (1977) and Lewis *et al.* (1977) both contain excellent photographs and descriptions for diagnosing capture myopathy in elk and pronghorn antelopes. A basic principle is: when an animal cooperates voluntarily, fear stress will be very low. In all these cases, the animals were lining up eagerly to get the delicious treats.

Habituate animals with flighty genetics in small steps

Animals with a flighty, excitable temperament, such as antelopes, will take longer to train and must be trained more slowly compared to a calmer species such as cattle or sheep. Great care had to be taken to habituate antelopes carefully to the sights and sounds of new equipment. If the animal is frightened early in the habituation phases, it may become so afraid of the equipment that training will be impossible. With flighty antelopes, it took 10 days to habituate them to the sound of a sliding door opening. In the early stages of habituation (Grandin *et al.*, 1995), the animal was never pushed past the orienting stage. The orienting stage occurs when the animal points its eyes and ears towards a sight or sound. When it oriented towards the door, the trainer stopped moving it. Each day the door was moved a little more. After the 10-day habituation period, the door could be opened quickly. After careful habituation, food rewards were used to train the animal to stand still.

Acclimating livestock to being moved by people

Obviously, it is not practical to train large numbers of cattle and pigs to the point where there is total voluntary cooperation. However, livestock will be easier to handle if people work with them carefully. Sheep will move through a race more quickly if they get a feed reward (Hutson, 1985). Binstead (1977), Fordyce (1987), and Becker and Lobato (1997) all found that moving extensively raised young calves quietly through the handling races, or walking among them, produced calmer adult animals. These handling sessions were done every day for 10 days. Moving young heifers through the squeeze chute four times produced calmer adult animals that moved more slowly when released from the squeeze chute (Cooke *et al.*, 2012). Pigs that have been walked in the aisles of their barn or had people quietly walking through their pens were easier to drive and move in the future (Abbott *et al.*, 1997; Geverink *et al.*, 1998; Lewis *et al.*, 2008). Poultry that had socialized and been habituated to people at a young age were easier to handle, gained more weight, and had better immune function (Gross and Siegel, 1982). In all species, numerous positive experiences with handling at a young age will produce calmer, easier-to-handle adult animals.

Principles of Restraint for Holding Livestock for Veterinary Procedures or Slaughter

When restraining animals, either with a mechanical apparatus such as a squeeze chute (crush) or holding a small animal in a person's hand, the same behavioral principles of restraint apply. Below is a list of restraint principles to keep animals calmer and minimize struggling or vocalization. Animals should be restrained in a comfortable, upright position.

Nonslip flooring

This is essential because the fear of falling is a primal fear. Repeated small slips where the animal repeatedly slips and then pulls its foot back into position to maintain its balance will cause an animal to become frightened. Repeated slipping is likely to occur if an animal will not settle down while it is standing in a race or in a head stanchion.

Avoid sudden jerky motion

Avoid sudden jerky motion of either people or equipment. Smooth movements by people, ropes, or equipment will keep animals calmer (Grandin, 1992).

Support the body

When an animal is restrained with its feet off the ground, it will remain calmer if its body is fully supported. This applies to a small animal held in a person's hand or a larger animal held in a mechanical restraint device. When the body is fully supported, the animal will be comfortable and the fear of falling will be eliminated. Restraint devices such as the Panepinto sling (Panepinto *et al.*, 1983) and the double-rail restrainer (Westervelt *et al.*, 1976; Grandin, 1988) employ the principle of full body support in a balanced position (see Chapter 16, this volume). Upright restraint was less stressful and caused less vocalization compared to inverted restraint (Dunn, 1990; Velarde *et al.*, 2014). If inverted restraint is used, the device must fully support the body of the animal to prevent slipping and shifting during rotation (OIE, 2014a,b). Animals are more likely to struggle or vocalize if they are not held snugly.

Even pressure

Animals are calmed by even pressure over a wide area of the body (Ewbank, 1968). It is important that there are no concentrated pressure points or pinch points. The author fixed a restrainer for pigs that was causing them to squeal by replacing a narrow bar that hurt their backs with a wide board that spread the pressure evenly along the pig's back. This stopped the pigs from squealing.

Optimum pressure

Pressure should not be too loose or too tight. An animal needs to be held tight enough to give it the feeling of being held, but not so tight that it feels pain. Excessive pressure will cause struggling and vocalization (see Chapter 9, this volume, for further information; Grandin, 2001; Bourguet *et al.*, 2011).

Blocking vision

This is not required for completely tame animals with no flight zone. A blindfold made from completely opaque material will keep both extensively raised cattle and bison calmer (Mitchell *et al.*, 2004). It is essential that the material is completely opaque. If the animal can see moving shadows through it, it will not work. Solid sides installed on a race or a fully enclosed dark box will have a calming effect on cattle with a large flight zone (Grandin, 1980a,b, 1992; Hale *et al.*, 1987; Pollard and Littlejohn, 1994; Muller *et al.*, 2008). Species such as bison, antelopes, and wild horses will often rear up in a race. A solid top on the race will stop rearing.

Avoid painful restraint methods

Animals remember painful experiences. Nose tongs hurt and it is strongly recommended to use a halter (head collar) when the head has to be restrained for veterinary procedures (Sheldon *et al.*, 2006). When more comfortable restraint methods are used, the animals will be more willing to enter the veterinary race in the future.

Electrical immobilization is highly aversive

Devices that restrain animals by paralyzing and freezing the muscles are highly detrimental to animal welfare. These devices should not be confused with electrical stunning, which produces instant insensibility. OIE (2014b) codes state that these devices should not be used. Numerous scientific studies have shown that restraining a conscious animal with these devices is very stressful. Jephcott *et al.* (1986) found that immobilization significantly increased beta-endorphin levels, which is an indicator of pain. Another study showed that animals were more averse to electrical immobilization than to restraint in a tilting restraint table. When sheep were given a choice between the two methods of restraint, the tilting restraint table was preferred (Grandin *et al.*, 1986).

A series of studies carried out by Rushen (1986a,b) and Rushen and Congdon (1986a,b) all showed that electro-immobilization was highly aversive and should not be used. Other researchers have also come to the same conclusion, that electrical immobilization is a method that should not be used (Lambooy, 1985; Pascoe, 1986). The author immobilized her own arm with three different brands of electrical immobilizers. The author reported that "it felt like sticking my arm into an electrical socket". Four different research groups independently came to the same conclusion that electrical immobilization was very stressful.

The Use of Dogs for Moving Livestock

In confined areas such as trucks, lairages at slaughter plants, and stockyards, the use of dogs is not recommended. The use of trained lead sheep or goats is strongly recommended for moving sheep in confined areas. Kilgour and de Langen (1970)

found that dogs biting sheep was highly stressful. Dogs used in confined areas such as handling races at an abattoir were also highly stressful for sheep (Hemsworth *et al.*, 2011). The author has observed that cattle that have been bitten by dogs in confined places, such as races, where they are not able to move away, are more likely to kick at people. The author recommends that dogs should be limited to pastures, large pens, and other open areas where the animals have room to move away.

Appropriate Handling Equipment for Tame or High Flight Zone (Wild) Livestock

Simple facilities that will work for tame animals that are trained to lead are not appropriate for handling untamed, extensively raised animals with a large flight zone. One example is slaughter facilities that are used in the Middle East for extensively raised Australian sheep. Because the facilities were originally constructed for use with local sheep that were accustomed to close contact with people, they were not appropriate for extensively raised animals. When wild Australian sheep were brought to these facilities, animal welfare was poor because no races were available for handling the wild sheep. This resulted in pile-ups, sheep flipping over, and people grabbing them roughly. The welfare of extensively raised wild cattle in a facility designed for tame animals may be even worse. Since there is no single-file race or stun box, handlers may use cruel methods to restrain animals, such as poking out eyes or cutting tendons. Good animal welfare in these facilities will be impossible until equipment that is suitable for handling wild, extensively raised animals is installed. Information on the design of suitable equipment for extensively raised animals can be found in Grandin (1997b, 2014a,b) and Grandin and Deesing (2008) (see Chapters 9 and 16, this volume). If the animals are completely tame and trained to lead, no restraining or handling equipment may be required. The one essential component for good welfare with all types of animals is to have a nonslip surface for the animal to stand on.

Handling facilities for extensively raised cattle

Race and corral layouts that include curves often work more efficiently than layouts with straight races (Grandin, 1997b, 2014a,b) (Figs 5.13 and 5.14). Kasimanickam *et al.* (2014) reported that AI pregnancy rates were better in a semi-circular system than

Fig. 5.13. Plan for a basic round crowd pen and curved race system for handling cattle and loading trucks. To facilitate cattle movement through this system, the junction between the round forcing pen, the loading ramp and the single-file chute must be laid out exactly as shown. There is a drawing of a cow at the entrance of the single-file race. This layout makes it possible for her to see two body lengths up the race. The cow must be able to see a clear pathway up ahead before the race bends.

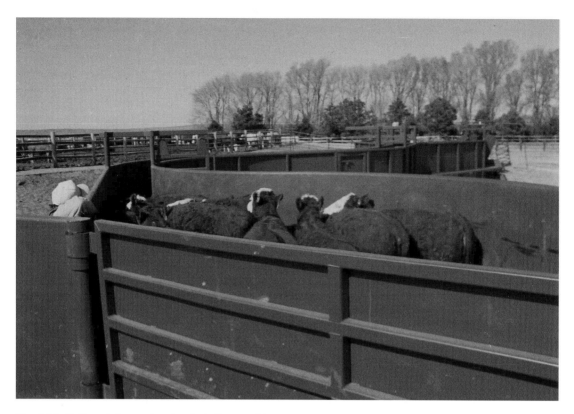

Fig. 5.14. Cattle quietly entering the single-file race. The correct number of cattle have been placed in the crowd pen. A solid crowd gate prevents cattle from turning back. The handler's elbow and head can be seen at the center pivot of the crowd gate. Sometimes the round crowd pen shown in Fig. 5.14 will work more efficiently if the handler stands at the pivot point of the crowd gate. It would work better if the handler had a small platform to stand on so he could easily reach over the fence and use a flag to move the cattle.

a system with lots of acute bends and angles. A curved, single-file race is efficient for two reasons:

1. Animals entering the race from the crowd pen do not see the people standing at the stun box or head stanchion.
2. Animals have a behavioral tendency to go back to where they came from. A curved, single-file and a full half-circle crowd pen race takes advantage of this tendency.

For these systems to work efficiently, they must be laid out correctly (Figs 5.13 and 5.15). The most common mistake is bending the single-file race too sharply where it joins on to the crowd pen. Figure 5.16 shows a well-designed sliding gate. Well-designed facilities will make handling safer for both people and animals.

Design of Pig and Sheep Handling Facilities

Cattle and sheep will move easily through a funnel-shaped crowd pen to enter a single-file race. Figure 15.17 shows a system for handling large numbers of sheep.

When pigs have to be moved into a single-file race, NEVER use a funnel-shaped crowd pen. Pigs will jam in a funnel. This is a species difference between cattle and pigs. The single-file race MUST have an abrupt entrance (Fig. 5.18). In developing countries, the use of a ramp is strongly recommended for loading trucks. Manually picking up sheep is highly stressful and results in high cortisol levels compared to using a ramp (Yardimci *et al.*, 2013).

T. Grandin

Fig. 5.15. Simple economical handling system with a round crowd pen that takes advantage of the natural tendency of animals to go back to where they came from. The handler stands on a small walkway located at the pivot point of the crowd gate. The cattle will circle around the handler easily and enter the single-file race.

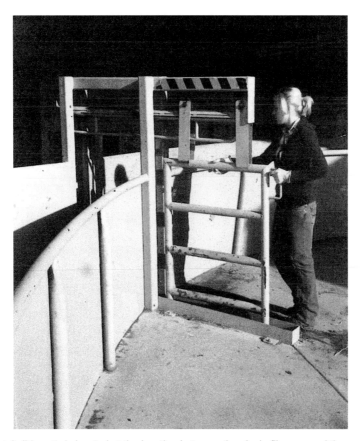

Fig. 5.16. A horizontal slide gate is located at the junction between the single-file race and the crowd pen. In many facilities, slide gates have replaced one-way backstop gates. Backstop gates that cattle can walk through prevent animals from backing up, but they also may cause animals to balk and refuse to enter. Animals will enter a single-file race more easily if backstops are either removed or tied open and replaced with gates that can be opened.

Fig. 5.17. Curved system for handling large numbers of sheep.

Fig. 5.18. The diagram shows a simple crowd pen design that has an offset step at the single-file race entrance. This helps prevent jamming by allowing one pig to step aside. The handler is shown in the correct position to move the pigs. The handler should work opposite to the straight side.

References

Abbott, T.A., Hunter, E.J., Guise, J.H. and Penny, R.H.C. (1997) The effect of experience of handling on pig's willingness to move. *Applied Animal Behaviour Science* 54, 371–375.

Apple, J.K., Minton, J.E., Parsons, K.M. and Unruh, J.A. (1993) Influence of repeated restraint and isolation

T. Grandin

stress an electrolyte administration on pituitary secretions, electrolytes, and blood constituents of sheep. *Journal of Animal Science* 71, 71–77.

Averos, X., Knowles, T.G., Brown, S.N., Warriss, P.D. and Gosalvez, L.F. (2008) Factors affecting the mortality of pigs being transported to slaughter. *The Veterinary Record* 163, 386–390.

Barbalho, P.C. (2007) Availacao de programas de trienamento em manejo racional de bovinos em frigorificos para melhoria do bem-estar animal. Dissertacao de mestrado. Programa de Pos-Graduacao em Zootecnia, Faculdade de Ciencias Agarias e Veterinarias, Universidade Estadual Paulista (UNESP), Jaboticabal-SP, Brazil.

Barnett, J.L., Hemsworth, P.H. and Newman, E.A. (1992) Fear of humans and its relationships with productivity in laying hens at commercial farms. *British Poultry Science* 33, 699–710.

Bates, L.S.W., Ford, E.A., Brown, S.N., Richards, G.J., Hadley, P.J., *et al.* (2014) A comparison of handling methods relevant to religious slaughter of sheep. *Animal Welfare* 23, 251–258.

Becker, B.G. and Lobato, J.F.P. (1997) Effect of gentler handling on the reactivity of Zebu crossed calves to humans. *Applied Animal Behaviour Science* 53, 219–224.

Benjamin, M.E., Gonyou, H.W., Ivers, D.L., Richardson, L.F., Jones, D.J., *et al.* (2001) Effect of handling method on the incidence of stress response in market swine in a model system. *Journal of Animal Science* 79 (Supplement 1), 279 (Abstract).

Binstead, M. (1977) Handling cattle. *Queensland Agriculture Journal* 103, 293–295.

Boissy, A., Aubert, A., Desire, L., Greiveldinger, L., Delval, E., *et al.* (2011) Cognitive sciences relate ear postures to emotions in sheep. *Animal Welfare* 20, 47–56.

Bourguet, C., Deiss, V., Tannugi, C.C. and Terlouw, E.M. (2011) Behavioural and physiological reactions of cattle in a commercial abattoir: Relationships with organisational aspects of the abattoir and animal characteristics. *Meat Science* 88, 158–168.

Carroll, J., Murphy, C.J., Neitz, M., Ver Hoeve, J.N. and Neitz, J. (2001) Photopigment basis for dichromatic color vision in the horse. *Journal of Vision* 1, 80–87.

Chalmers, G.A. and Barrett, M.W. (1977) Capture myopathy in pronghorns in Alberta, Canada. *Journal of the American Veterinary Medical Association* 171, 918–923.

Coleman, G.J., McGregor, M., Hemsworth, P.H., Boyce, J. and Dowling, S. (2003) The relationship between beliefs, attitudes, and observed behaviors in abattoir personnel in the pig industry. *Applied Animal Behaviour Science* 82, 189–200.

Cooke, R.F., Arthington, J.D., Araujo, D.B. and Lamb, G.C. (2009) Effects of acclimation to human interaction on performance, temperament, physiological responses,

and pregnancy rates in Brahman crossbred cows. *Journal of Animal Science* 87, 4125–4132.

Cooke, R.F., Bohnert, D.W., Cappellozza, B.I., Mueller, C.J. and Del Curto, T. (2012) Effects of temperament and acclimation to handling on reproductive performance of *Bos taurus* beef females. *Journal of Animal Science* 90, 3547–3555.

Core, S., Widowski, T., Mason, G. and Miller, S. (2009) Eye white as a predictor of temperament in beef cattle. *Journal of Animal Science* 87, 2168–2174.

Curley, K.O., Paschal, J.C., Welsh, T.H. and Randel, R.D. (2006) Technical note: Exit velocity as a measure of cattle temperament is repeatable and associated with serum concentration of cortisol in Brahman bulls. *Journal of Animal Science* 84, 3100–3103.

Davis, M. (1992) The role of the amygdala in fear and anxiety. *Annual Review of Neuroscience* 15, 353–375.

Dodd, C.L., Pitchford, W.S., Hocking-Edwards, J.E. and Hazel, S.L. (2012) Measure of behavioural reactivity and their relationships with production traits: A review. *Applied Animal Behaviour Science* 140, 1–15.

Doney, J.M., Smith, W.F. and Gunn, R.G. (1976) Effect of post mating environmental stress and administration of ACTH on early embryonic loss in sheep. *Journal of Agricultural Science* 87, 133–136.

Douphrate, D.I., Rosecrance, J.C., Stallones, L., Reynolds, S.J. and Gilkey, D.P. (2009) Livestock-handling injuries in agriculture: An analysis of Colorado workers' compensation data. *American Journal of Industrial Medicine*, 52, 391–407. Available at: http://www.ncbi.nlm.nih.gov/pubmed/19197949/ordinalpos=4&citool=EntrezSystem2.PEntr (accessed 22 April 2015).

Drudi, D. (2000) *Are Animals Occupational Hazards? Compensation and Working Conditions.* US Department of Labor, Washington, DC, pp. 15–22.

Dunn, C.S. (1990) Stress reactions of cattle undergoing ritual slaughter using two methods of restraint. *Veterinary Record* 126, 522–525.

Edwards, L.N., Engle, T.E., Correa, J.A., Paradis, M.A., Grandin, T., *et al.* (2010a) The relationship between exsanguination blood lactate concentration and carcass quality in slaughter pigs. *Meat Science* 85, 435–440.

Edwards, L.N., Grandin, T., Engle, T.E., Porter, S.P., Ritter, M.J., *et al.* (2010b) Use of exsanguination blood lactate to assess the quality of pre-slaughter pig handling. *Meat Science* 86, 384–390.

Ewbank, R. (1968) The behavior of animals in restraint. In: Fox, M.W. (ed.) *Abnormal Behavior in Animals.* W.B. Saunders, Philadelphia, Pennsylvania, pp. 159–178.

Faure, J.M. and Mills, A.D. (2013) Improving the adaptability of animals by selection. In: Grandin, T. (ed.) *Genetics and the Behavior of Domestic Animals*, 2nd edn. Academic Press (Elsevier), San Diego, California, pp. 233–265.

Ferguson, D.M. and Warner, R.D. (2008) Have we under-estimated the impact of pre-slaughter stress on meat quality in ruminants? *Meat Science* 80, 12–19.

Fordyce, G. (1987) Weaner training. *Queensland Agricultural Journal* 113, 323–324.

Fordyce, G., Dodt, R.M. and Wythes, J.R. (1988) Cattle temperaments in extensive herds in northern Queensland. *Australian Journal of Experimental Agriculture* 28, 683–688.

Fulkerson, W.J. and Jamieson, P.A. (1982) Pattern of cortisol release in sheep following administration of synthetic ACTH or imposition of various stress agents. *Australian Journal of Biological Science* 35, 215–222.

Geverink, N.A., Kappers, A., van de Burgwal, E., Lambooij, E., Blokhuis, J.H., *et al.* (1998) Effects of regular moving and handling on the behavioral and physiological responses of pigs to pre-slaughter treatment and consequences for meat quality. *Journal of Animal Science* 76, 2080–2085.

Grandin, T. (1980a) Livestock behavior as related to handling facilities design. *International Journal for the Study of Animal Problems* 1, 33–52.

Grandin, T. (1980b) Observations of cattle behavior applied to the design of cattle handling facilities. *Applied Animal Ethology* 6, 19–31.

Grandin, T. (1981) Bruises on southwestern feedlot cattle. *Journal of Animal Science* 53 (Supplement 1), 213.

Grandin, T. (1982) Pig behavior studies applied to slaughter-plant design. *Applied Animal Ethology* 9, 141–151.

Grandin, T. (1983) Welfare requirements of handling facilities. In: Baxter, S.H., Baxter, M.R. and MacCormack, J.A.D. (eds) *Farm Animal Housing and Welfare*. Martinus Nihoff, Boston, Massachusetts, pp. 137–149.

Grandin, T. (1988) Double rail restrainer for livestock handling. *Journal of Agricultural Engineering* 41, 327–338.

Grandin, T. (1989a) Behavioral principles of livestock handling. *Professional Animal Scientist* 5(4), 1–11.

Grandin, T. (1989b) Voluntary acceptance of restraint by sheep. *Applied Animal Behaviour Science* 23, 257–261.

Grandin, T. (1992) Observation of cattle restraint devices for stunning and slaughtering. *Animal Welfare* 1, 85–91.

Grandin, T. (1996) Factors that impede animal movement at slaughter plants. *Journal of the American Veterinary Medical Association* 209, 757–759.

Grandin, T. (1997a) Assessment of stress during handling and transport. *Journal of Animal Science* 75, 249–257.

Grandin, T. (1997b) The design and construction of handling facilities for cattle. *Livestock Production Science* 49, 103–119.

Grandin, T. (1998) The feasibility of using vocalization scoring as an indicator of poor welfare during slaughter. *Applied Animal Behaviour Science* 56, 121–128.

Grandin, T. (2001) Cattle vocalizations are associated with handling at equipment problems in beef slaughter plants. *Applied Animal Behaviour Science* 71, 191–201.

Grandin, T. (2012) Developing measures to audit welfare of cattle and pigs at slaughter, *Animal Welfare* 21, 351–356.

Grandin, T. (ed.) (2014a) Behavioural principles of handling cattle and other grazing animals under extensive conditions. In: Grandin, T. (ed.) *Livestock Handling and Transport*. CAB International, Wallingford, UK, pp. 39–64.

Grandin, T. (2014b) Improving welfare and reducing stress on animals at slaughter plants. In: Grandin, T. (ed.) *Livestock Handling and Transport*. CAB International, Wallingford, UK, pp. 421–450.

Grandin, T. and Deesing, M. (2008) *Humane Livestock Handling*. Storey Publishing, North Adams, Massachusetts.

Grandin, T. and Deesing, M. (2013) Genetics and behavior during handling, restraint and herding. In: Grandin, T. and Deesing, M. (eds) *Genetics and the Behavior of Domestic Animals*, 2nd edn. Academic Press, Elsevier, San Diego, California, pp. 115–158.

Grandin, T. and Johnson, C. (2005) *Animals in Translation*. Scribner (Simon and Schuster), New York.

Grandin, T., Curtis, S.E., Widowski, T. and Thurmon, J.C. (1986) Electro-immobilization versus mechanical restraint in an avoid-avoid choice test. *Journal of Animal Science* 62, 1469–1480.

Grandin, T., Rooney, M.B., Phillips, M., Cambre, R.C., Irlbeck, N.A. and Graffam, W. (1995) Conditioning of nyala (*Tragelaphus angasi*) to blood sampling in a crate using positive reinforcement. *Zoo Biology* 14, 261–273.

Gross, W.B. and Siegel, P.B. (1982) Influence or sequences or environmental factors on the response of chickens to fasting and to *Staphylococcus aureus* infection. *American Journal of Veterinary Research* 43, 137–139.

Gruber, S.L., Tatum, J.D., Engle, T.E., Chapman, P.L., Belk, K.E., *et al.* (2010) Relationships of behavioral and physiological symptoms of preslaughter stress on beef longissimus muscle tenderness. *Journal of Animal Science* 88, 1148–1159.

Hale, R.H., Friend, T.H. and Macaulay, A.S. (1987) Effect of method of restraint of cattle on heart rate, cortisol, and thyroid hormones. *Journal of Animal Sciences* 65 (Supplement 1), 217 (Abstract).

Hanggi, E.B. (2005) The thinking horse: cognition and perception reviewed. In: Broken, T.D. (ed.) *Proceedings of the 51st Annual Convention of the American Association of Equine Practitioners*, Seattle, Washington, 3–7 December 2005. American Association of Equine Practitioners, Lexington, Kentucky, pp. 246–255.

Hearnshaw, H. and Morris, C.A. (1984) Genetic and environmental effect on temperament score in beef cattle. *Australian Journal of Agricultural Research* 35, 723–733.

Hediger, H. (1968) *The Psychology and Behavior of Animals in Zoos and Circuses*. Dover Publications, New York.

Heffner, R.S. and Heffner, H.E. (1983) Hearing in large mammals: Horse (*Equus caballus*) and cattle (*Bos taurus*). *Behavioral Neuroscience* 97(2), 299–309.

Hemsworth, P.H., Barnett, J.L. and Hansen, C. (1981) The influence of handling by humans on the behavior, growth and corticosteroids in the juvenile female pig. *Hormones and Behaviour* 15, 396–403.

Hemsworth, P.H., Coleman, G.J., Barnett, J.L. and Borg, S. (2000) Relationships between human–animal interactions and productivity of commercial dairy cows. *Journal of Animal Science* 78, 2821–2831.

Hemsworth, P.H., Rice, M., Karlen, M.G., Calleja, L., Barnett, J.L., *et al.* (2011) Human–animal interactions at abattoirs: Relationships between handling and animal stress in sheep and cattle. *Applied Animal Behaviour Science* 135, 24–33.

Hixon, D.L., Kesler, D.J., Troxel, T.R., Vincent, D.L. and Wiseman, B.S. (1981) Reproductive hormone secretion and first service conception rate subsequent to ovulation control with Synchromate B. *Theriogenology* 16, 219–229.

Hutson, G.D. (1980) Visual field restricted vision and sheep movement in laneways. *Applied Animal Behaviour Science* 6, 175–187.

Hutson, G.D. (1985) The influence of barley feed rewards on sheep movement through a handling system. *Applied Animal Behaviour Science* 14, 263–273.

Hutson, G.D. (2014) Behavioural principles of sheep handling. In: Grandin, T. (ed.) *Livestock Handling and Transport*. CAB International, Wallingford, UK, pp. 193–217.

Jacobs, G.H., Deegan, J.F. and Neitz, J. (1998) Photo pigment basis for dichromatic colour vision in cows, goats, and sheep. *Visual Neuroscience* 15, 581–584.

Jephcott, E.H., McMillen, I.C., Rushen, J., Hargreaves, A. and Thorburn, G.C. (1986) Effect of electro-immobilization on ovine plasma concentrations of β-endorphin, β-lipotrophin, cortisol and prolactin. *Research in Veterinary Science* 41, 371–377.

Jones, R.B., Satterlee, D.G. and Cadd, G.G. (1998) Struggling responses of broiler chickens shackled in groups on a moving line: effects of light intensity hoods and 'curtains'. *Applied Animal Behaviour Science* 58, 341–352.

Jorgensen, G.H.M., Andersen, I.L., Holand, O. and Boe, K.E. (2011) Differences in the spacing behaviour of two breeds of domestic sheep (*Ovis aries*) – Influence of artificial selection. *Ethology* 117, 597–605.

Judd, S.P.D. and Collett, T.S. (1998) Multiple stored views and landmark guidance in ants. *Nature* 392, 710–714.

Kasimanickam, R., Schroeder, S., Assay, M., Kasimanickam, V., Moore, D.A., *et al.* (2014) Influence of temperament score and handling facility on stress, reproductive hormone concentrations and fixed time AI pregnancy rates in beef heifers. *Reproduction of Domestic Animals* 49, 775–782. doi:10.1111/rda.12368.

Kemble, E.D., Blanchard, D.C., Blanchard, R.J. and Takushi, R. (1984) Taming in wild rats following medical amygdaloid lesions. *Physiology and Behaviour* 32, 131–134.

Kilgour, R. and Dalton, C. (1984) *Livestock Behavior: A Practical Guide*. Westview Press, Boulder, Colorado.

Kilgour, R. and de Langen, H. (1970) Stress in sheep from management practices. *Proceedings, New Zealand Society of Animal Production* 30, 65–76.

King, D.A., Schuehle-Pfeiffer, C.E., Randel, R., Welsh, T.H., Oliphint, R.A., *et al.* (2006) Influence of animal temperament and stress responsiveness on the carcass quality and beef tenderness of feedlot cattle. *Meat Science* 74, 546–556.

Lambooy, E. (1985) Electro-anesthesia or electro-immobilization of calves, sheep, and pigs by Feenix Stockstill. *Veterinary Quarterly* 7, 120–126.

Lanier, J.L., Grandin, T., Green, R., Avery, D. and McGee, K. (2000) The relationship between reaction to sudden intermittent movements and sounds to temperament. *Journal of Animal Science* 78, 1467–1474.

Lay, D.C., Friend, T.H., Bowers, C.L., Grissom, K.K. and Jenkins, O.C. (1992a) A comparative physiological and behavioral study of freeze and hot-iron branding using dairy cows. *Journal of Animal Science* 70, 1121–1125.

Lay, D.C., Friend, T.H., Randel, R.D., Bowers, C.L., Grissom, K.K., *et al.* (1992b) Behavioral and physiological effects of freeze branding and hot-iron branding on crossbred cattle. *Journal of Animal Science* 70, 330–336.

LeDoux, J.E. (1996) *The Emotional Brain*. Simon and Schuster, New York.

Leiner, L. and Fendt, M. (2011) Behavioral fear and heartrate responses in horses after exposure to novel objects: Effects of habituation. *Applied Animal Behaviour Science* 131, 104–109.

Lemmon, W.B. and Patterson, G.H. (1964) Depth perception in sheep: Effects of interrupting the mother–neonate bond. *Science* 145, 835–836.

Lewis, C.R.G., Hulbert, L.E. and McGlone, J.J. (2008) Novelty causes elevated heart rate and immune changes in pigs exposed to handling alleys and ramps. *Livestock Science* 116, 338–341.

Lewis, P.D. and Morris, T.R. (2000) Poultry and coloured light. *World's Poultry Science Journal* 56, 189–207.

Lewis, R.J., Chalmers, G.A. and Barrett, M.W. (1977) Capture myopathy in elk in Alberta, Canada. *Journal American Veterinary Medical Association* 171, 927–932.

Macedo, G.G., Zuccari, C.E.S.N., Pinto de Abreu, U.G., Negrao, J.A. and da Costa e Silva, E.V. (2011) Human animal interaction, stress and embryo production in *Bos indicus* embryo donors under tropical conditions. *Tropical Animal Health and Production* 43, 1175–1182.

McGlone, J.J., McPherson, R. and Anderson, D.L. (2004) Case study: Moving devices for finishing pigs: efficacy of electric prod, board, paddle or flag. *Professional Animal Scientist* 20, 518–523.

Mertsching, H.J. and Kelly, A.W. (1983) Restraint reduces size of the thymus gland and PHA swelling in pigs. *Journal of Animal Science* 57 (Supplement 1), 175.

Miller, N.E. (1960) Learning resistance to pain and fear effects over learning, exposure, and rewarded exposure in context. *Journal of Experimental Psychology* 60, 137–145.

Mitchell, K., Stookey, J.M., Laturnas, D.K., Watts, J.M., Haley, D.B., *et al.* (2004) The effects of blindfolding on behaviour and heart rate in beef cattle during restraint. *Applied Animal Behaviour Science* 85, 233–245.

Morris, C.L., Grandin, T. and Irlbeck, N.A. (2011) Companion Animal Symposium, Environmental enrichment for companion, exotic and laboratory animals. *Journal of Animal Science* 12, 4227–4238.

Muller, R., von Keyserlinkg, M.A.G., Shah, M.A. and Schwartzkopf-Genswein, K.S.G. (2008) Effect of neck injection and handler visibility on behavioral reactivity of beef steers. *Journal of Animal Science* 86, 1215–1222.

Munoz, D., Strappini, A. and Gallo, C. (2012) Indicadores de bienestar animal para detectar problemas en al cajon de insensibilizacion de bovinos (Animal welfare indicators to detect problems in the cattle stunning box). *Archivos de Medicina Veterinaria* 44, 297–302.

National Pork Board (2014) www.pork.org, Des Moines, Iowa.

OIE (2014a) Chapter 7.3. *Transport of Animals by Land. Terrestrial Animal Health Code.* World Organisation for Animal Health, Paris.

OIE (2014b) Chapter 7.5. *Slaughter of Animals. Terrestrial Animal Health Code.* World Organisation for Animal Health, Paris.

Pajor, E.A., Rushen, J. and de Passille, A.M.B. (2003) Dairy cattle choice of handling treatments in a Y maze. *Applied Animal Behaviour Science* 80, 93–107.

Panepinto, L.M., Phillips, R.W., Norden, S., Pryor, P.C. and Cox, R. (1983) A comfortable minimum stress method of restraint for Yucatan miniature swine. *Laboratory Animal Science* 33, 95–97.

Paranhos da Costa, M.J.R., Huertas, S.M., Strappini, A.C. and Gallo, C. (2014) Handling and transport of cattle and pigs in South America. In: Grandin, T. (ed.) *Livestock Handling and Transport*, 4th edn. CAB International, Wallingford, UK, pp. 174–192.

Pascoe, P.J. (1986) Humaneness of an electrical immobilization unit for cattle. *American Journal of Veterinary Research* 10, 2252–2256.

Petherick, J.C., Holyroyd, R.G., Doogan, V.J. and Venus, B.K. (2002) Productivity, carcass and meat quality of feedlot *Bos indicus* cross steers grouped according to temperament. *Australian Journal of Experimental Agriculture* 42, 389–398.

Phillips, M., Grandin, T., Graffam, W., Irlbeck, N.A. and Cambre, R.C. (1998) Crate conditioning of bongo (*Tragelaphus eurycerus*) for veterinary and husbandry procedures at Denver Zoological Gardens. *Zoo Biology* 17, 25–32.

Pollard, J.C. and Littlejohn, R.P. (1994) Behavioral effects of light conditions on red deer in holding pens. *Applied Animal Behaviour Science* 41, 127–134.

Prescott, N.B. and Wathes, C.M. (1999) Spectral sensitivity of domestic fowl (*Galus g. domesticus*). *British Poultry Science* 40, 332–339.

Price, E.O. and Wallach, S.J.R. (1990) Physical isolation of hand reared Hereford bulls increases their aggressiveness towards humans. *Applied Animal Behaviour Science* 27, 263–267.

Prince, J.H. (1970) The eye and vision. In: Swenson, M.J. (ed.) *Dukes Physiological of Domestic Animals.* Cornell University Press, New York, pp. 696–712.

Proctor, H.S. and Carder, G. (2014) Can ear postures reliably measure positive emotional state of cows. *Applied Animal Behaviour Science* 161, 20–27.

Reid, R.L. and Mills, S.C. (1962) Studies in carbohydrate metabolism in sheep, XVI. The adrenal response of sheep to physiological stress. *Australian Journal of Agricultural Research* 13, 282–294.

Reinken, G. (1988) General and economic aspects of deer farming. In: Reid, H.W. (ed.) *The Management and Health of Farmed Deer.* Springer, London, pp. 53–59.

Ritter, M.J., Ellis, M., Anderson, D.B., Curtis, S.E., Keffaber, K.K., *et al.* (2009) Effects of multiple concurrent stressors on rectal temperature, blood acid base status and longissimus muscle glycolytic potential in market weight pigs. *Journal of Animal Science* 87, 351–362.

Rogan, M.T. and LeDoux, J.E. (1996) Emotion, systems cells, and synaptic plasticity. *Cell* 85, 469–475.

Rushen, J. (1986a) Aversion of sheep to electro-immobilization and physical restraint. *Applied Animal Behaviour Science* 15, 315–324.

Rushen, J. (1986b) Aversion of sheep for handling treatments: paired-choice studies. *Applied Animal Behaviour Science* 16, 363–370.

Rushen, J. and Congdon, P. (1986a) Relative aversion of sheep to simulated shearing with and without electro-immobilization. *Australian Journal of Experimental Agriculture* 26, 535–537.

Rushen, J. and Congdon, P. (1986b) Sheep may be more averse to electro-immobilization than to shearing. *Australian Veterinary Journal* 63, 373–374.

Sandern, A.I., Janczak, A.M., Salte, R. and Braastad, B.O. (2006) The use of diazepam as a pharmacological validation of eye white as an indicator of emotional state in dairy cows. *Applied Animal Behaviour Science* 96, 177–183.

T. Grandin

Sebastian, T. (2007) Temperament in beef cattle: Methods of measurement, consistency and relationship to production. University of Saskatchewan Library, Canada.

Sheldon, C.C., Sonesthagan, T. and Topel, J.A. (2006) *Animal Restraint for Veterinary Professionals*. Mosby/Elsevier, St Louis, Missouri.

Shinozaki, A., Hosaka, Y., Imagawa, T. and Uehara, M. (2010) Topography of ganglion cells and photo receptors in the sheep retina. *Journal of Comparative Neurology* 518, 2305–2315.

Smith, B. (1998) *Moving 'Em: A Guide to Low Stress Animal Handling*. Graziers Hui, Kamuela, Hawaii.

Stermer, R., Camp, T.H. and Stevens, D.C. (1981) Feeder cattle stress during transportation. American Society of Agricultural Engineers Paper No. 81–6001. American Society of Agricultural Engineers, St Joseph, Michigan.

Stockman, C.A., McGilchrist, P., Collins, T., Barnes, A.L., Miller, D., *et al.* (2012) Qualitative behavioural assessment of Angus steers during pre-slaughter handling and relationship with temperament and physiological responses. *Applied Animal Behaviour Science* 142, 125–133.

Talling, J.C., Waran, N.K., Wathes, C.M. and Lines, J.A. (1998) Sound avoidance in domestic pigs depends on characteristics of the signal. *Applied Animal Behaviour Science* 58, 255–266.

Tanida, H., Miura, A., Tanaka, T. and Yoshimoto, T. (1996) Behavioral responses of piglets to darkness and shadows. *Applied Animal Behaviour Science* 49, 173–183.

Thomas, M.L. (2014) Balking incidence and behavioral responses during handling of fed cattle: Genetic and carcass implications. Dissertation, University of Arkansas.

Tillman, A. (1981) *Speechless Brothers: The History and Care of Llamas*. Early Winters Press, Seattle, Washington.

van Putten, G. and Elshof, W.J. (1978) Observations of the effects of transport on the well being and lean quality of slaughter pigs. *Animal Regulation Studies* 1, 247–271.

Velarde, A., Rodriguez, P., Dalmau, A., Fuentes, C., Llonch, P., *et al.* (2014) Religious slaughter: Evaluation of current practices in selected countries. *Meat Science* 96, 278–287.

Voisinet, B.D., Grandin, T., Tatum, J.D., O'Connor, S.F. and Struthers, J.J. (1997a) Feedlot cattle with calm temperaments have higher average daily gains than cattle with excitable temperaments. *Journal of Animal Science* 75, 892–896.

Voisinet, B.D., Grandin, T., Tatum, J., O'Connor, S.F. and Struthers, J.J. (1997b) *Bos indicus*-cross feedlot cattle with excitable temperaments have tougher meat and a higher incidence of borderline dark cutters. *Meat Science* 46, 367–377.

Warner, R.D., Ferguson, D.M., Cottrell, J.J. and Knee, B.W. (2007) Acute stress induced by preslaughter use of electric prodders causes tougher meat. *Australian Journal of Experimental Agriculture* 47, 782–788.

Warriss, P.D., Brown, S.N., Adams, S.J.M. and Corlett, I.K. (1994) Relationship between subjective and objective assessment of stress at slaughter and meat quality in pigs. *Meat Science* 38, 329–340.

Wasadikar, P.P., Paunikar, R.G. and Deshmukh, S.B. (1997) Bull horn injuries in rural India. *Journal of the Indian Medical Association* 95(1), 3–4, 16.

Watts, J.M. and Stookey, J.M. (1999) Effects of restraint and branding on rates and acoustic parameters of vocalization in beef cattle. *Applied Animal Behaviour Science* 62, 125–135.

Waynert, D.F., Stookey, J.M., Schwartzkopf-Genswein, K.S.. Watts, J.M. and Waltz, C.S. (1999) Response of beef cattle to noise during handling. *Applied Animal Behaviour Science* 62, 27–42.

Westervelt, R.G., Kinsman, D., Prince, R.P. and Giger, W. (1976) Physiological stress measurement during slaughter of calves and lambs. *Journal of Animal Science* 42, 831–837.

White, R.G., DeShazer, J.A., Tressler, C.L., Borcher, G.M., Davey, S., *et al.* (1995) Vocalization and physiological response of pigs during castration with and without a local anesthetic. *Journal of Animal Science* 73, 381–386.

Woiwode, R., Grandin, T. and Kirch, B. (2014) Survey of BQA Cattle handling practices that occurred during processing feedlot cattle. Paper presented at: 2014 ADSA-ASAS-CSAS Joint Annual Meeting. 2014 July 20–24. Kansas City, MO.

Wollack, C.H. (1963) The auditory acuity of the sheep (*Ovis aries*). *Journal of Auditory Research* 3, 121–132.

Yardimci, M., Sahin, E.H., Cetingul, I.S., Bayram, I., Aslan, R., *et al.* (2013) Stress responses to comparative handling procedures in sheep. *Animal* 7, 143–150.

6 Painful Husbandry Procedures in Livestock and Poultry

KEVIN J. STAFFORD AND DAVID J. MELLOR

Massey University, Palmerston North, New Zealand

Summary

This chapter views common painful husbandry practices in cattle, sheep, pigs, poultry, and other livestock. It explains why the procedures are done, how they are done, and methods to alleviate pain both during and after the procedure. The procedures covered are dehorning, disbudding, castration, marking, beak trimming, and many other practices. Methods of alleviating pain that can be done easily on the farm are reviewed, and clear diagrams show the correct methods for doing procedures. Lignocaine, a local anesthetic, is effective for dehorning. Meloxicam, a NSAID that can be given orally either before or after a procedure, reduces pain in calves and lambs. Topical gel anesthetics containing two local anesthetics may reduce pain when applied carefully. It is possible to reduce pain, but not eliminate it. In the future, some of the procedures may be replaced with immunological castration, genetic selection in polled (hornless) cattle, and genetic selection for less aggressive pigs and poultry.

Learning Objectives

- Explain common painful husbandry procedures and why they are done
- Show the correct methods for performing the procedures
- Understand the three types of pain – acute, chronic, and pathological
- Learn about methods of pain alleviation that can be done under practical farm conditions

Introduction – Reasons for Doing Painful Procedures

The management of farm livestock, poultry, and other domesticated animals is based on the ability of people to control them and many aspects of their environments. Animal husbandry involves the provision of food and shelter, the control of reproduction, and the use of handling and restraint methods that are safe for both the stock handlers and the animals. It also includes measures designed to promote animal health by the prevention and treatment of disease. Some of these husbandry objectives are met by practices that are painful for the animals concerned. The purpose of this chapter is to provide both practical advice and guidance on methods to reduce pain during painful procedures. Discussing the ethics of these procedures is beyond the scope of the chapter. For each species, several such practices may be used (Table 6.1) for a range of reasons, which include the following:

- to minimize the risk of injury to animals and people (e.g. dehorning)
- to reduce aggressive behavior and make male animals easier to handle (e.g. castrating oxen)
- to prevent carcass damage such as bruising (e.g. dehorning)
- to enhance carcass quality (e.g. castration)
- to reduce the risk of flystrike (e.g. mulesing in sheep)
- to allow other husbandry practices (e.g. shearing) to be undertaken more quickly and efficiently (e.g. tail docking)
- to prevent damage to the environment (e.g. nose ringing in pigs)

Table 6.1. Routine husbandry practices carried out on farm livestock.

Procedure	Cattle	Sheep	Goats	Horses	Pigs
Castration	+++[a]	+++	++	+++	++
Docking	+	+++		+	+++
Disbudding	+++		+		
Dehorning	+++	+			
Spaying	+				
Ear notching	+++	+++	++		+
Ear tagging	+++	+++	++		+
Branding – hot	++			+	
Branding – freeze	++			++	
Mulesing		+			
Teeth grinding		+			
Teeth clipping					+
Tusk grinding					+
Nose ringing	+				+
Tail nicking				+	

Notes: [a]+ = in some management systems; ++ = in many management systems; +++ = in virtually all management systems.

- to aid in identification (e.g. ear marking or notching, branding)
- to enable the harvesting of products (e.g. velvet antler removal in deer).

Likewise, poultry are subjected to painful management practices such as beak trimming to minimize feather pecking and cannibalism, and trimming their wattles and comb (dubbing, desnooding) to minimize pecking injury by other birds.

The type of farming system determines when these procedures are carried out. For instance, animals managed indoors throughout life can be handled and treated at any time, whereas animals held on extensive pastoral properties may be several months of age before they are brought into yards for the first time, and may thereafter be handled infrequently throughout life. The value of individual animals is also a factor, as it influences the time that will be allotted to each one undergoing such procedures, and it also affects assessments of the cost-effectiveness of using anesthetic and analgesic drugs, veterinarians or veterinary technicians. In addition, the availability of analgesics and veterinary expertise may be important. For instance, in some developed countries, analgesic drugs may not be directly available to farmers, due to legal restrictions. A veterinarian would be required to administer them. In other countries, a farmer could buy the drugs if he had a prescription from a veterinarian. The situation is more difficult in some developing countries where pain medicines may not even be available to people in rural communities.

Most of the husbandry procedures used today were developed many years ago. They were usually selected to be carried out quickly, easily, cheaply, and with inexpensive equipment, and to be generally safe for the animals and people involved (Stafford and Mellor, 1993). At the time of their development, little significance was attributed to any associated animal pain, apart from completing the procedure quickly, and scientific understanding of pain and its management in animals was poor. This situation has now changed.

General Aspects of Animal Pain

In general terms, there are three main categories of pain associated with husbandry procedures that cause tissue damage – they are acute, chronic, and pathological pain (Flecknell and Waterman-Pearson, 2000; Gregory, 2004). Acute pain is generated by the initial act of damaging the tissues, and by subsequent chemical changes within those tissues. Acute pain is linked to characteristic behavioral and physiological responses that are used to indicate its presence and demonstrate the efficacy of alleviation methods (Lester et al., 1996; Mellor et al., 2000). Chronic pain, which is usually less intense than acute pain, persists for the days or weeks it takes for the tissue damage to resolve and healing to be complete (Molony et al., 1995; Sutherland et al., 2000). The assessment of low-level chronic pain and differentiating it from irritation are problematical, and knowledge about the alleviation of chronic pain is poor. Pathological pain occurs when barrages of nerve impulses in pain pathways during the acute pain phase modify the operation of those pathways by effects at the injury site itself or in the spinal cord or brain (Mellor et al., 2008). Animals affected by pathological pain are more sensitive to both painful and otherwise non-painful stimulation in the area of the injury or elsewhere. Pathological pain may last for weeks, months, or years. It is not clear whether tissue-damaging husbandry procedures may cause pathological pain in farm livestock, but there is some evidence that castration of lambs soon after birth might have such an effect. Pain in farm animals is often undertreated (Walker *et al.*, 2011).

Behavioral and physiological indicators of pain

Pain-inducing stimuli elicit a range of physiological and behavioral responses. They allow the presence or absence of significant pain to be assessed, and strategies for alleviating the pain to be developed (Mellor *et al.*, 2000). However, as pain is a subjective experience, which by definition is not quantifiable in absolute terms: all such variables are indirect indices of pain. Cautious interpretation is therefore required. Nevertheless, both physiological and behavioral indices have been used successfully in experimental settings to study animal pain responses. Physiological indices include changes in heart rate, blood pressure, rectal and skin temperatures, plasma levels of stress hormones and related metabolites, and brain electrical activity. Useful behaviors, which may be species specific and/or injury specific, enable acute, chronic, and pathological pain to be distinguished, and they indicate when pain is present and, importantly, when it is at very low levels or absent. Some behavior is overt and obvious, such as the initial marked restlessness of lambs after the application of rubber castration rings (Lester *et al.*, 1996). Vocalization (bellows, moos, or squeals) in cattle and pigs are associated with pain and distress during a procedure such as castration or branding. Cattle that were hot-iron branded had more vocalizations compared to controls that were restrained and not branded (Lay *et al.*, 1992a,b; Watts and Stookey, 1999). Vocalization (squeals) in piglets was reduced during castration when anesthetics were used (White *et al.*, 1995). Less overt behavior, such as the repetitive tail swishing in calves, persists for about 8 h after dehorning (Sylvester *et al.*, 2004). Ear flicking after a procedure is another indicator of pain. Calves that had their tails docked had more ear twitching (Eicher and Dailey, 2002). After dehorning, some behavioral indicators of pain are head shaking, rubbing the head, tail or ear flicking, and lying still (Stafford and Mellor, 2011). In clinical settings, behavior, demeanour, and appearance are used more commonly because they are more immediately observable than are some key physiological indices (Mellor *et al.*, 2008). Changes in stress hormone concentrations

or pain-related features of brain electrical activity require subsequent biochemical or mathematical analysis.

Pain or fear stress

Both pain and fear stress can increase cortisol. Extensively raised cattle had greater increases in cortisol after handling and restraint compared to dairy cows (Lay *et al.*, 1992a,b). For example, injecting a local anesthetic will reduce pain-related behavior both during and after dehorning. In extensively raised Brahman cattle, cattle injected with the anesthetic had higher cortisol (Carol Petherick, 2014, personal communication). This is probably due to increased fear stress because the wild cattle had to be restrained longer to receive the anesthetic. A combination of an excitable temperament and little contact with people may increase fear stress greatly during handling (Grandin, 1997; Grandin and Deesing, 2014).

Grazing animals may hide pain

Older animals that are not completely tame, such as 4- to 6-month-old extensively raised calves and sheep, may not exhibit obvious symptoms of pain after castration or dehorning. A common comment by ranchers is: "They must have little pain because my cattle ate and drank immediately after the operation". Grazing animals are prey species that often hide obvious pain behaviors to avoid being eaten by predators. This is most likely to occur in animals with a large flight zone that have some fear of people (T. Grandin, 2008, personal communication). Observation of 8-month-old bulls castrated with high-tension rubber bands illustrates this problem (Grandin and Johnson, 2005). Some bulls acted normally, whereas others stamped their feet repeatedly, and a few of them lay on the ground in strange, contorted postures. These pain-related behaviors disappeared when a person walked up to their pen, and animals that were rolling on the ground jumped up and ran over to their herdmates when they saw a person. In order to see the pain-related behaviors, the observer had to be hidden so the animals did not know they were being observed. For all species, the most accurate way to study pain-related behavior after a procedure is to hide the observer or use a video camera. Sheep are the most defenceless prey species animal. They may be more likely to cover up and hide pain-related behavior, especially when they see people watching.

Alleviation of Animal Pain During and After Painful Procedures in Cattle and Sheep

The behavioral and physiological responses, and by inference the pain, caused by particular husbandry procedures may be reduced by choosing methods that cause less pain, by carrying out the procedures in young rather than older animals, and by using pain-relieving drugs. There is evidence that performing painful procedures such as castration at an early age produces lower neurological, behavioral, and physiological responses compared to older calves (Dockweiler *et al.*, 2013). A fuller analysis of matters that should be considered when deciding whether or not, and how, to undertake particular painful husbandry procedures is presented in Quick Access Information 6.1 and discussed in the final section of this chapter.

Quick Access Information 6.2 lists words associated with pain relief.

Quick Access Information 6.1. Primary and secondary questions relevant to deciding whether or not, and how, to undertake a particular painful husbandry procedure. (From Mellor *et al.*, 2008.)

- *First*, is it necessary to perform the procedure?
- What are the anticipated benefits of the anatomical modification?
- Does the modification achieve the desired benefits?
- Does the modification benefit a significant proportion of the treated animals?
- How significant are the benefits (i.e. how pressing is the need to make the modification)?
- Can the same benefits be achieved in other, less invasive ways?
- *Second*, what harms are caused by the procedure?
- Does the procedure cause transient, short-term harm, such as acute pain and distress, when it is carried out?
- Does the procedure cause longer-lasting harm, such as chronic pain and distress, during recovery from it?
- Does the procedure itself cause harms that persist beyond recovery from the procedure, such as persistent adverse behavioral or functional changes?
- What are the magnitudes of such transient, longer-lasting and/or persistent harms – how bad are they?
- In what proportion of animals do such harms occur to a significant extent?
- Are there effective ways of reducing any such significant harms?
- *Third*, do the benefits of the procedure outweigh the harms?
- Do the act and the anatomical modification cause greater harms to the animals (individuals and groups) than they prevent?
- Alternatively, are there sufficient direct benefits to the animal?
- Are there sufficient indirect wider benefits, for instance, commercial, educational, recreational, scientific, or social benefits, to offset the harms caused by the procedure itself and the associated anatomical modification to individuals or groups?

Quick Access Information 6.2. Glossary of words associated with pain relief.

Alpha-2 adrenoreceptor agonists – a class of drug with sedative and analgesic effects (e.g. xylazine).
Analgesia – a state in which pain is absent or at very low levels.
Analgesic – a drug that provides pain relief, usually given by injection or orally.
Epidural anesthesia – local anesthetic is injected into the spinal canal.
Local anesthetic – drug used to prevent nerve transmission from a painful location such as lignocaine or bupivacaine.

NSAID (non-steroidal anti-inflammatory drug) – a drug that may have anti-inflammatory effects at the site of injury only, or that may, in addition, have analgesic effects on the spinal cord or within the brain (e.g. ketoprofen and meloxicam).
Systemic analgesia – analgesic drugs administered by various routes carried around the body in the bloodstream.

Local anesthetic nerve blockade

Local anesthetics produce their effects by disrupting the function of nerve membranes and thereby preventing electrical impulses from passing along nerves (Flecknell and Waterman-Pearson, 2000; Mellor et al., 2008). They are usually injected close to nerves in order to block sensations in particular parts of the body. The size of the area affected determines the name given to the type of nerve blockade, such that a "local" blocks a relatively small area (e.g. the scrotum and testicles), a "regional" blocks a larger area (e.g. parts of the flank and abdominal contents), and a "spinal" or "epidural" blocks a much larger region of the body (e.g. hind legs, perineal region, flank, uterus, and some abdominal contents). The effectiveness of local, regional, or epidural anesthetics can be demonstrated easily by the absence of escape behavior during tissue-damaging procedures (i.e. surgery).

Lignocaine is the most common local anesthetic drug in worldwide veterinary use. It is used to alleviate the acute pain experienced by animals during and for 1–2 h after a number of painful procedures, including some livestock husbandry practices (Mellor and Stafford, 2000; Stafford and Mellor, 2005a,b). It is a short-acting local anesthetic that is usually cleared from the site of injection quickly enough for its effects to last for about 60–120 min (Table 6.2). Lignocaine blockade of the cornual nerve prior to the dehorning of calves virtually eliminates the plasma cortisol response indicative of pain for about 2 h (Stafford and Mellor, 2005a). There are concerns about a carcinogenic metabolite of lignocaine, called 2,6 xylidine, because in a 2-year study of rats, about 50% of those receiving a high dose of the metabolite developed papillomata or carcinomata of the nasal cavity. In the UK, lignocaine has effectively been withdrawn from use in food animals, as the maximum, safe, residual levels for it have not yet been established. There are other local anesthetics (Table 6.2) such as bupivacaine and mepivacaine, with different characteristics, that may be useful in livestock. For instance, bupivacaine blockade of the cornual nerve of calves virtually eliminates the cortisol response to dehorning for about 4 h (Stafford and Mellor, 2005a). However, most such local anesthetics are not licensed for livestock.

Epidural nerve blockade is produced by injecting local anesthetic into the epidural space of the spinal cord (Flecknell and Waterman-Pearson, 2000; Mellor et al., 2008). Effective epidural analgesia can also be achieved by injecting alpha-2 agonists such as xylazine epidurally, and this is used when castrating adult cattle because it has the added advantage of being accompanied by sedation. A lignocaine–xylazine mixture produces effective epidural analgesia in cattle castrated using a castration clamp (Burdizzo®) and extends the duration of analgesia (Ting et al., 2003). Although epidural injections were once considered to have practical difficulties and side effects that made them unattractive when castrating large numbers of bulls, this technique is now used in cattle veterinary practice.

Gel topical anesthetics

Topical anesthetics have the advantage of being extremely easy to use, but are they effective? They can be applied easily at the time the painful procedure is conducted. Giffard et al. (2007) describes a 70% sorbital liquid gel that contains two local anesthetics, lignocaine and bupivacaine. The gel is effective for reducing pain from surface procedures such as mulesing (Espinoza et al., 2013) and branding (Carol Petherick, 2014, personal communication). When 9 ml of the gel was applied carefully to the exposed spermatic cords and the wound incision, pain behaviors were reduced (Lomax and Windsor, 2013). The gel has little ability to penetrate tissues, and for it to have a beneficial effect, sufficient amounts must be applied. The gel may also help reduce post-dehorning pain because it decreases wound sensitivity to touch for 24 h (Espinoza et al., 2013).

Systemic analgesia general principles and NSAID analgesic

Systemic analgesics are carried in the bloodstream throughout the body and thus have the capacity to

Table 6.2. Local anesthetics.

	Time to onset of action (min)	Duration of activity (h)
Lignocaine[a]	1–2	1–2
Bupivacaine	5–10	4–12
Mepivacaine	1–5	1–2
Procaine	5	1

Note: [a]Lignocaine is generally licensed for use in animals and is cheap and short acting.

act generally or on specific target tissues, depending on the biological actions of the particular drug (Flecknell and Waterman-Pearson, 2000; Stafford *et al.*, 2006; Mellor *et al.*, 2008). They may be administered orally, topically, per rectum or vagina, or by intravenous, intramuscular, or intraperitoneal injection. They are appropriate for the treatment of pain caused by disease, trauma, or surgery. Systemic analgesics are used infrequently in farm livestock at present. Their use is limited by concern about residues and cost.

In the past, when pain alleviation was applied, it was focused on the surgical act itself, making surgery easier and safer by using local or regional anesthesia together with sedation. Post-operative pain was given little or no attention, because of a dearth of effective analgesics. Three major groups of systemic analgesics are available. They are the opioids, alpha-2 adrenoreceptor agonists, and non-steroidal anti-inflammatory drugs (NSAIDs). Most of the research on the effect of these analgesics on ruminants has involved sheep rather than cattle.

The opioids have long been used in animals, but are not very effective in ruminants. Clinically, alpha-2 agonists are effective analgesics in sheep. Xylazine given intramuscularly produces significant analgesic effects in adult sheep, but it appears to be less effective in cattle. Xylazine given alone to calves reduces the plasma cortisol response to dehorning, but not to the degree achieved with a lignocaine blockade of the cornual nerve.

NSAID analgesia practical for on-farm use

The development of NSAIDs with analgesic effects has been a major breakthrough in pain relief in cattle and sheep, as these were the first really useful analgesics for these species. Giving calves 0.5 mg/kg to 1 mg/kg of oral meloxicam at the time of dehorning lowered cortisol levels, and increased time at the feedbunk (Theurer *et al.*, 2012). The analgesic effects of these drugs were considered to be due to their peripheral anti-inflammatory actions, but some of them have central nervous system effects as well. The effectiveness of any particular NSAID as an analgesic will depend on its rate of clearance from the body, on the cause of pain, and on poorly understood characteristics of the drug itself. Some NSAIDs, notably phenylbutazone, are more anti-inflammatory than analgesic; others such as carprofen are more analgesic than anti-inflammatory. The longer period of activity of some NSAIDs may

make them more useful than opioids for treating non-ruminants. Some of the NSAIDs, such as meloxicam, are very effective analgesics in cattle. The NSAID anti-inflammatory drug, meloxicam, given in conjunction with a local anesthetic, provides additional reduction in the physiological stress response after hot-iron disbudding of calves (Heinrich *et al.*, 2009; Stewart *et al.*, 2009).

Most NSAIDs are strongly protein bound and tend to have long meat-withholding times. However, ketoprofen has a very short biological half-life, and there are no significant residues in the meat and milk of cows given ketoprofen intravenously. Used with local anesthesia, it virtually eliminated the cortisol response, and by inference markedly reduced pain, for at least 12 h following the dehorning and surgical castration of calves. Also, when given alone, ketoprofen reduces the plasma cortisol response significantly, and by inference the pain, in the hours following dehorning, surgical castration and Burdizzo® (or clamp) castration of cattle.

Some NSAIDs are effective analgesics in horses, and have been used in the treatment of colic for years. Their use for post-operative pain has become popular recently. Opioids are also used in horses, and pethidine provides analgesia for a few hours if given intramuscularly. Alpha-2 adrenoreceptor agonists are also useful in horses, but they are used mainly as sedatives in this species. The use of analgesics in pigs is poorly studied, but some NSAIDs are effective for post-surgical pain control, while opioids are generally short acting in pigs and not very useful clinically. Our knowledge of analgesia in poultry is limited, but NSAIDs apparently reduce articular pain in chickens.

The future of pain relief

Since the mid-1990s, many advances have been made in the field of pain research, but much remains to be done. There are two major problem areas. One is scientific and involves determining what pain animals experience, especially in the longer term. The other is practical and relates to what pain-relieving drugs livestock farmers can afford and will be permitted to use within existing regulatory controls. The now well-established physiological and behavioral responses of farm livestock to procedures such as dehorning and castration will enable these techniques to be used as models for the routine testing of the efficacy of currently available and new analgesics. These models could also be used to screen the efficacy of analgesics for treating the pain caused by different types of injuries.

The painful husbandry procedures discussed below are covered on a species basis, with the more important procedures receiving the greatest attention. The *alleviation* of pain may be easy to bring about, but the *elimination* of pain is a much more difficult proposition. Some procedures may be eliminated in the future by using immunological castration and breeding polled animals without horns. In poultry, genetic selection may be used to reduce the need to beak trim poultry.

Cattle – Painful Husbandry Procedures

Cattle are subjected to five common painful husbandry procedures (disbudding, dehorning, castration, branding, and ear notching or tagging; Table 6.1). In addition, some heifers may be spayed, and dairy cattle may have their tails docked. Cattle are also subjected to procedures that may be stressful, but are not painful, so that these procedures will not be discussed here. They include pregnancy diagnosis, vaccination, different processes around artificial breeding, and treatment with anthelmintics (orally, by injection, or transcutaneous) and ixodicides (by shower or dipping).

Cattle – disbudding and dehorning

Although the terms "disbudding" and "dehorning" are considered to be synonyms by some people, here disbudding is used to mean the prevention of horn growth before it has become advanced, and dehorning means the amputation of horns at any stage after their growth has progressed beyond the early budding stage. Disbudding of calves at a very young age is recommended. Disbudding calves by cauterizing the horn bud causes a significantly lower cortisol response compared to horn amputation (Stafford and Mellor, 2005a).

A summary of the types of procedures used for disbudding and dehorning is provided in Quick Access Information 6.3.

Reasons for removing horns on cattle

In general, cows with horns are more dangerous and difficult to manage, so many cattle routinely have their horns removed (Fig. 6.1). On farms where predation is a problem, cows may be left with horns to protect their calves and themselves. Likewise on organic dairy farms, horns may not be removed, and this can result in injuries during aggressive interactions between cows. Horned cattle are also more dangerous to people. In some countries, it is forbidden to transport cattle with horns because they may injure other animals during the journey. Alternatively, before transport, some cattle may have the tips cut off their horns, a process called "tipping". Cattle with large horns do not fit easily into head bails (stanchions or headgates), and this makes restraint for oral drenching, ear-tag placement, and other procedures difficult. Cattle without horns require less trough space. In some cows, horn growth can be circular, which necessitates

---

Quick Access Information 6.3. Disbudding and amputation dehorning.

Cautery disbudding – this is carried out on calves from birth to until about 6 weeks of age. A bar, often concave in shape, is heated and placed against the horn bud and rotated for sufficient time to destroy the horn bud germinal tissue.

Caustic chemical disbudding – the use of caustic chemicals such as potassium hydroxide in a stick form to destroy the horn buds. The stick is rubbed on to the horn bud on very young calves before 2 days of age. Older calves will rub their head and the caustic method is not recommended (Villarroel, 2011).

Surgical disbudding – a knife is used to cut off the horn bud, or a scoop dehorner is placed around the base of the horn bud and closed, removing the bud and a 5 mm strip of skin around the horn bud.

Amputation dehorning – the horn and a 1 cm strip of tissue around the horn are removed by saw, embryotomy wire, guillotine shears, or scoop. Do not remove horns if horn base is larger than 2.5 cm (AABP, 2014).

Hemostasis – may be by cautery, a tourniquet, hemostatic powder, or the major blood vessels may be identified, picked up by forceps and twisted. A figure-of-eight tourniquet may be tied around the horns to prevent bleeding.

Local anesthesia – may be by cornual nerve block using 5–10 ml of lignocaine injected midway along the lateral edge of the crest of the frontal bone. A ring block may also be used. A ring block may be most effective (Fierheller *et al.*, 2012).

K.J. Stafford and D.J. Mellor

Fig. 6.1. Horns on cattle.

tipping every few years to prevent horns from growing back into the animal's head. Tipping should not be so extensive as to affect the pain-sensitive core of the horn (Fig. 6.2).

Although horn removal is a common practice, many breeds, particularly beef breeds, are now bred not to produce horns, that is they are polled.

The genetics for polled characteristics in *Bos taurus* cattle are well understood, but those for *Bos indicus* cattle are more complex. Cross-bred cattle bred from polled and horned cattle, or mixtures of them, may or may not produce horns. The best way to reduce the pain and distress caused by horn injuries or horn removal would be to select for polled cattle. At present, the common dairy breeds, Friesian and Jersey, are horned. It may be possible for such breeds to be bred polled if an effort were made to do so, and this would completely remove the necessity for disbudding or dehorning. Breeding polled *B. taurus* cattle will be easier than breeding polled *B. indicus* (Zebu, Brahman, or Nelore cattle) (Carol Petherick, 2014, personal communication).

Methods for disbudding of calves

In developed dairy industries, the horn buds of calves are routinely destroyed by application of a caustic paste or stick (sodium hydroxide, potassium hydroxide, or colloidion) or by cautery (Stafford and Mellor, 2005a). Horn buds and the skin immediately surrounding them may also be removed surgically using a scoop or knife (Fig. 6.3). The ideal age for destroying the horn buds is during the first few weeks after birth. When caustic chemicals (paste or stick) are used, caution is required because housed calves may lick the caustic material from the horn buds of other calves, and outdoors rain may wash

Fig. 6.2. Tipping horns.

the caustic material into the eyes of treated calves. When a cautery iron (heated electrically or by gas) is used, it must be hot enough and applied for long enough to destroy the generative tissue in the skin around the base of the horn bud. If all of this tissue is not destroyed, then horns may grow in a lopsided or distorted manner. Successful surgical disbudding is achieved by ensuring that each horn bud is removed together with the skin around its margin, which contains the generative tissue. Disbudding is carried out on dairy calves worldwide.

The use of potassium hydroxide caustic stick to destroy the horn buds in 4-week-old calves causes intense pain for about 4 h. As judged by cortisol stress responses and calf behavior, both cautery disbudding and surgical (scoop) disbudding cause pain immediately and for some hours afterwards, but the amount of acute pain caused is apparently less with cautery than with surgical disbudding or

Fig. 6.3. Scoop dehorner.

destruction by caustic chemicals (Morisse *et al.*, 1995; Petrie *et al.*, 1996a). Thus, cautery disbudding is preferable, but extreme escape behavior of the calf shows that application of the hot iron is painful. This behavior, and the often less obvious response that occurs with surgical disbudding, can be eliminated both by using a local anesthetic blockade of the cornual nerve or a ring block around the base of each horn bud.

Methods for dehorning of calves

Removal of horns from mature cattle without anesthetics and post-operative analgesia is very painful. Cutting horns on mature animals may be more painful than most other procedures. Crude methods of dehorning mature cattle, such as chopping off with a machete, should never be used.

Horns are removed from calves, weaners, or adult cattle by amputation using a variety of tools (Stafford and Mellor, 2005a), including saws (manual and electric), embryotomy wire, guillotine shears, and scoops (Fig. 6.4). During such dehorning, the horn and a small area of skin surrounding the horn base are removed. In larger cattle, the wound may open up the frontal sinus. It is a painful procedure and cattle attempt to escape during it. Effective restraint is required if the amputation is done without local anesthesia, and dehorning is

Fig. 6.4. Dehorning shears.

often carried out with the animal held in a head bail or with its legs tied.

Hemorrhage will occur after dehorning. This may be minimized by using an embryotomy wire for dehorning rather than a saw, guillotine shears, or scoop. Dehorning by embryotomy wire is harder work and takes longer than with a saw, but allows better control of the wound size. Guillotine shears are easy to use but if the head shifts during the procedure, the wound may be larger than required and the frontal bone may fracture. The size of scoops makes them useful on smaller cattle but not

K.J. Stafford and D.J. Mellor

large cattle. Hemorrhage may be controlled by identifying the bleeding arteries and twisting or cauterizing them, or by ligating them. Hemostatic pads may be used, as may hemostatic powder. A tourniquet may be applied before amputation using a figure-of-eight pattern around the horns.

Wounds may become infected after dehorning, and in summer they may become flystruck or be attacked by screwworm. Wounds need to be treated to prevent these problems from developing. In larger cattle the wound may open into the frontal sinus. If the cattle are housed, dirt may get into the sinuses and cause sinusitis. This may result in a purulent discharge from the wound, and the sinus may have to be drained by trephination. With a large wound, or when cattle are held indoors, it is wise to cover the wounds with pads containing antibiotic powder. Wound healing after amputation dehorning takes several weeks.

Pain alleviation for disbudding and dehorning

The behavioral response to the pain caused by the act of amputating the horn can be eliminated using local anesthetic blockade of the cornual nerve, or by a ring block around the horn base. Cornual nerve blockade is not always successful, and it is wise to test the skin around the base of the horn by needle prick before cutting begins to check whether more local anesthetic may be required. The plasma cortisol response follows a defined pattern after amputation dehorning (Fig. 6.5). The concentration initially rises rapidly to a peak, then declines to a plateau, and finally returns to pretreatment levels after about 8 h (McMeekan et al., 1998a). In one trial, the size of the wound did not affect the magnitude of the cortisol response, which suggests that within the small range studied, the wound size did not affect the acute pain caused by the procedure (McMeekan et al., 1997).

Effective cornual nerve blockade using local anesthetic eliminates the initial peak of the plasma cortisol response, and cortisol concentrations remain at pretreatment levels for about 2 h. Thereafter, the concentrations increase for about 6 h, before returning to pretreatment levels (McMeekan et al., 1998a). This is interpreted to indicate that effective cornual nerve blockade eliminates pain for about 2 h, after which some pain is experienced. This delayed pain is probably dull and not as acutely sore as the initial pain of horn amputation. To eliminate the acute cortisol response for at least 12 h, the dehorned animal needs to be given a systemic analgesic along with the local anesthetic; a combination of an NSAID with a local anesthetic is effective in this regard (McMeekan et al., 1998b; Stafford et al., 2003; Milligan et al., 2004; Stewart et al., 2009). Meloxicam given orally is effective (Allen et al.,

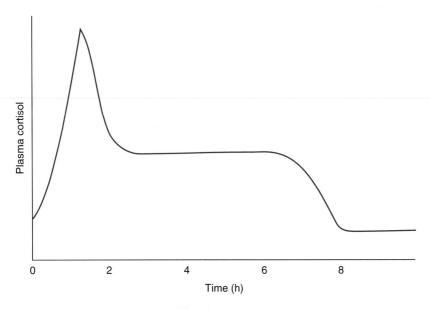

Fig. 6.5. Plasma cortisol response to amputation dehorning.

2013) and had effects on behavior for 5 days (Theurer *et al.*, 2012). Long-lasting systemic analgesics have the potential to alleviate the pain for up to 3 days. Alternatively, when local anesthetic is given before dehorning, cautery of the wound to control hemorrhage also reduces the plasma cortisol response, and by inference the pain experienced by cattle, for at least 24 h (Sutherland *et al.*, 2002). Dehorning alters feeding and rumination behavior for 2 or 3 days, which suggests that the animals experience pain during this period. There is limited evidence that dehorning may not cause protracted pathological pain. High-tension elastic bands used for castration should not be used for dehorning due to prolonged pain-related behavior and slow wound healing (Neely, 2013).

Castration

Methods and reasons for cattle castration

Castration is a standard husbandry practice in most cattle production systems. It makes male animals safer and easier to handle, and reduces the likelihood of unwanted breeding by poorer-quality bulls. Cattle are also castrated to become oxen or bullocks for ploughing and traction. Castration modifies the fat content of the beef carcass. Castrated cattle are sometimes given growth hormones to increase growth rates and accelerate finishing. In some countries, however, bulls are reared for beef. In New Zealand, for example, Friesian bull calves are reared intact for beef production and are usually slaughtered before they reach 2 years of age.

Castration is carried out using a variety of techniques (Stafford and Mellor, 2005b), including the application of tight rubber rings or larger bands, surgery, or a bloodless castration clamp (Burdizzo®, Fig. 6.6) (Quick Access Information 6.4). Small rubber rings are used on calves and larger rubber bands on large cattle (Fig. 6.7). Surgical castration entails removing the distal scrotal wall and then either pulling the testicles and spermatic cords until the latter break or cutting through the spermatic cords. Ligatures or clamps may be placed on the spermatic cords before they are severed, or an emasculator may be used to sever the cords (Fig. 6.8). Another method twists the cords (Warnock *et al.*, 2012). Bloodless castration clamps (Burdizzo®), which are applied across the spermatic cords, are available in different sizes.

Fig. 6.6. Bloodless castration clamp – Burdizzo®.

When analgesia is not available, it is probably best for animal welfare to castrate young calves with a rubber ring. This method is easy, cheap, and safe for the calves and the stock handlers using it. However, wounds formed by the rings are present for several weeks before the scrotum becomes detached. In intensive systems, removal of the rubber ring at day 9 is a possible alternative in 4- to 6-week-old calves, to promote healing (Becker *et al.*, 2012). There are few side effects of using rings on small calves, but tetanus may occur if the scrotum is dirty when the rings are put in place.

High-tension band castration of bulls

Band castration of older post-pubertal bulls is carried out because it is believed that these animals have growth advantages over animals castrated early in life. With grass-fed beef cattle, however, there is little evidence to support this idea, and it is not recommended (Knight *et al.*, 2000). Bands may cause large wounds in the abdominal wall if placed too close to the abdomen. Banding 6-month-old

K.J. Stafford and D.J. Mellor

Fig. 6.7. Rubber rings used for castration.

bulls may cause pain and reduce growth at the time the testicles fall off (L.A. Gonzales, 2009, personal communication). Mature bulls should not be castrated by banding because there is prolonged pain (Petherick *et al.*, 2014). There is an inflammatory response and higher cortisol that may last up to 4 weeks. For big bulls, castration should be surgical.

Clamp castration of cattle

Clamp castration results in more failures than either rubber-ring or surgical castration (Kent *et al.*, 1996; Stafford *et al.*, 2002). This is important when steers are run with heifers, as the unintended presence of fertile males may result in poorly developed heifers being bred accidentally, with all the

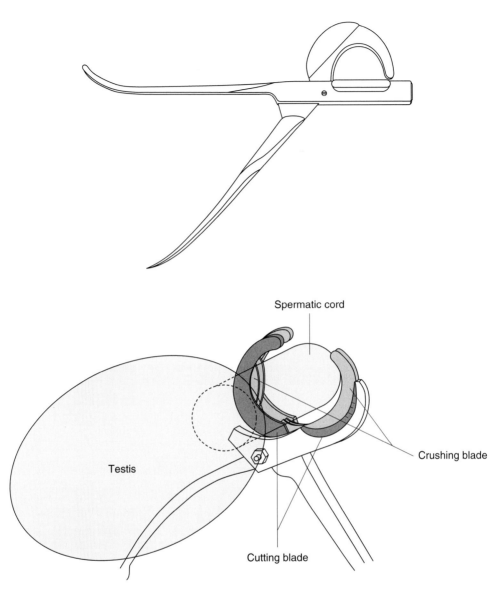

Spermatic cord

Crushing blade

Testis

Cutting blade

Fig. 6.8. Emasculator used to sever spermatic cord and prevent hemorrhage.

potential welfare problems that entails. Failure occurs when one or both spermatic cords are not crushed sufficiently. Misapplication of the clamp may also cause crushing of the penis and damage to the urethra, leading to urine leakage.

Surgical castration may be followed by hemorrhage, but this is seldom serious. The surgical wound may become dirty and infected. However, if the hole in the scrotum is large enough, drainage will be sufficient to keep the wound clean. Tetanus may also occur if drainage is inadequate.

Pain alleviation for castration of calves and bulls

Castration is painful (Stafford and Mellor, 2005b; Stafford *et al.*, 2006). When no form of analgesia is used, cattle respond behaviorally in a manner that suggests pain. In addition, plasma cortisol concentrations increase over the first hour or so after castration and then return to pretreatment levels after about 3–4 h (Fig. 6.9). The cortisol response to clamp castration is less than the response to rubber-ring or

K.J. Stafford and D.J. Mellor

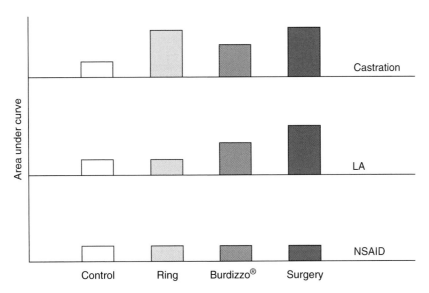

Fig. 6.9. Plasma cortisol response to different methods of castration. The area under the curve for 8 h after castration in cattle castrated by ring, Burdizzo®, or surgically is shown under three treatments: without local anesthetic (Castration); with local anesthetic (LA); and with local anesthetic and a non-steroidal anti-inflammatory drug (NSAID). The control animals were handled as for castration but were not castrated.

surgical castration, which suggests that the clamp is less painful (Stafford *et al.*, 2002). Local anesthesia placed in the distal scrotum and testicles eliminates or reduces the pain-related behaviors seen at castration by these methods. It eliminates the plasma cortisol response to castration by rubber ring or band. This is because the tight ring or band stops blood and lymph perfusion of the testicles and scrotum so that the local anesthetic remains in those tissues beyond the time required for anoxic death of the pain receptors and associated nerves. In contrast, the local anesthetic reduces the cortisol response to the clamp castration and has little overall effect on the cortisol response to surgical castration. The latter method, particularly when the spermatic cords are stretched or torn, results in damage to tissues in the inguinal canal and abdomen, areas that are not affected by local anesthetic placed in the testicles and scrotum. To eliminate the plasma cortisol response and related pain due to clamp and surgical castration, it is necessary to combine administration of local anesthetic with a systemic analgesic such as an NSAID (Stafford *et al.*, 2002, 2006; Ting *et al.*, 2003). Under extensive conditions where large bulls can be handled only once, ketoprofen at the time of castration is recommended (Petherick *et al.*, 2014). In 309-kg bulls, meloxicam was not effective in animals castrated with a high-tension band (Repenning

et al., 2013). NSAIDS may be more effective in surgically castrated cattle. When cattle from multiple auctions were received at a feedlot and surgically castrated, oral meloxicam reduced sickness (Allen *et al.*, 2013).

There is no obvious alternative to castration in the production of draft oxen, but for meat production, bulls can be reared and finished intact, as seen in management systems in New Zealand and Europe. Another possibility is the use of immunocastration. The vaccines used for this are now available. Studies show that immunocastration improves meat quality compared to intact bulls (Amatayakul-Chantler *et al.*, 2012, 2013; Miguel *et al.*, 2014).

Spaying cattle

Heifers are spayed to prevent mismating and to prevent mating of cull cows in extensive farming systems. The ovaries are removed either through the vagina in larger cows and heifers following epidural anesthesia or through an incision in the flank in smaller heifers (see Ohme and Prier, 1974). Special surgical instruments have been developed for cutting through the flank and for removing the ovaries. A flank incision is certainly painful and warrants local anesthesia, and probably post-surgical analgesia. Electro-immobilization should not be used to restrain cattle (Petherick *et al.*, 2012). The vaginal dropped ovary method is preferable to flank spaying (McCosker *et al.*, 2010). Rendering heifers infertile is now possible immunologically, and this will become popular if it is shown to be practicable and acceptable to consumers and producers.

Tail docking of dairy cows

Tail docking is carried out on dairy heifers, either as calves or after calving to make access to the cow's udder easier during milking in a herringbone or rotary dairy shed. Docking has no beneficial effect on milk quality, udder cleanliness, or on the incidence of mastitis or leptospirosis (see Stull *et al.*, 2002). Animals that had docked tails had more flies on them, and more fly avoidance behaviors such as foot stamping (Eicher *et al.*, 2001; Eicher and Dailey, 2002). Docking young calves by rubber ring or cautery is not especially painful, and local anesthetic given either as a tail ring block or an epidural can be used to alleviate the pain (Petrie *et al.*, 1996b). Rubber-ring docking is preferable as hemorrhage may be significant after cautery docking. It is a serious and more painful procedure in heifers and is not recommended for them. Tails may be docked short at the level of the vulva, or the tip of the tail together with the hair switch may be removed. The latter approach retains most of the tail but reduces the likelihood of stock handlers being hit by the tail hairs, with or without dung contamination, while accessing the udder. A non-surgical option is to trim the tail hairs to reduce the likelihood of the milker being hit by a tail covered with feces. Tail docking of dairy cows is prohibited in some countries, and in many natural and organic programs. It is a procedure that should be phased out, because there is no scientific justification for it (Sutherland and Tucker, 2011).

Ear tagging or ear notching of calves and cattle

Ear tagging is a common way of identifying individual animals, and a range of tags have been developed. Farmers often use large plastic ear tags to aid recognition at a distance. Individual identification facilitates management by allowing the easy recognition of superior animals for use in breeding programs. The emphasis on trace-back of meat products and on disease control has led many countries to make the identification of individual cattle mandatory.

Such identification requires an easily recognized ear tag, made of plastic or metal, to be placed in one or both ears. Ear tags or ear notches are used also to indicate the ownership of animals grazed on common ground or those likely to stray on to other properties, and also help to safeguard against theft. Ear notches of different shapes and/or placed in different positions are also used to mark the year of birth, vaccine status, and selection status for breeding.

The ear is a sensitive organ and ear tagging and notching are likely to be painful. Ear tagging side effects include infection, tears, and loss of tags that then have to be replaced. Tag placement involves either punching a hole in the ear with a specific tool and then placing the tag, or the use of a tagging device that both punches the hole and places the tag. The tag contains a stud that pierces the ear. This type of tag usually makes a smaller hole and may be less painful because each tag contains a new sharp ear-piercing stud. Failure to sharpen the tool used to install the other type of tag may increase pain. Little work has been done on reducing the pain caused by these procedures, and as it is not easy to anesthetize the ears using local anesthetic, systemic analgesia would be preferable.

Ear tagging is definitely preferable from a welfare standpoint compared to wattling or ear splitting. In some countries, cattle ears are slit and long strips of skin are left dangling. When wattling is done, strings of hide and skin are cut and left dangling from the neck. Both of these procedures would be banned in many countries, and they violate the voluntary industry welfare guidelines of many countries. These extreme cuts should not be confused with ear notching, where small notches are made in the ear.

Branding of cattle

Branding by hot iron or freeze branding are common means of identifying cattle, both for ownership and for recognizing individual animals (Lay *et al.*, 1992a,b). It is a common practice in countries with extensively raised cattle that are pastured on huge areas of land. It is used in Australia, the western USA, Central America, South America, and many other countries. Hot branding causes definite escape behavior in cattle, and its cauterizing action is obviously painful. Freeze branding, using irons first immersed in liquid nitrogen, causes the destruction of pigment-producing melanocytes in the skin. The brands are placed on the skin and held there for an appropriate time (seconds rather than minutes).

K.J. Stafford and D.J. Mellor

The skin freezes and when the brand is removed, the skin warms up and some time later a white mark is left. The animals probably experience pain during thawing of the skin and reperfusion of it with blood, but hot branding is probably more painful than freeze branding. Little work has been done on reducing the pain caused by these procedures. As multiple injections of local anesthetic in the area to be branded are impractical, systemic analgesia would probably reduce, but not eliminate, the pain experienced following branding. A sorbital gel containing local anesthetics may reduce pain from branding (Carol Petherick, 2014, personal communication). Animals should never be branded on sensitive parts of the body, such as the face.

Microchips to identify cattle are becoming a feasible option to replace branding for identification, and their placement will be less painful than branding. However, they are not really suited to everyday identification and it will take some time before handheld identity chip readers become a widely used tool on cattle farms. Another alternative for animal identification is DNA testing. The cost of DNA testing is falling steadily, and it may become economically feasible in the future.

Nose rings or ropes in cattle and water buffalo

Nose rings or nose ropes are placed in the nasal septum of cattle and water buffalo to help control fractious animals, especially bulls. They are usually placed while the animals are restrained securely or sedated, with or without local anesthetic. Placement without local anesthetic is likely to cause pain and related behavioral responses.

Sheep – Painful Husbandry Practices

The three most common painful husbandry procedures used in sheep are castration, tail docking, and ear tagging or notching. Merino sheep in Australia may be subject to mulesing if flystrike is a problem. These procedures are generally carried out on lambs. Shearing of the whole body or to remove fecal pads (dagging) or wool from the perianal region (crutching), as well as pregnancy diagnosis, bathing, or showering (backlining) with insecticides, oral drenching with anthelmintics, and vaccination are stressful procedures but are not painful, unless animals are injured by the procedure (e.g. cuts during shearing).

Castration of lambs – general considerations

Most male lambs in many countries are castrated. One reason is to prevent unwanted breeding of ewe lambs that are to be reared and finished for slaughter after puberty; the other is to prevent meat taint. Ram lambs that will be slaughtered before 5 or 6 months of age may not need to be castrated for these reasons, but may still be castrated to improve hygiene at slaughter, because in some breeds the intact scrotum may be large, woolly, and contaminated with fecal matter.

Where ewes are managed indoors during lambing, castration and tailing can be carried out at any age from the second day after birth, and certainly before lambs and ewes are let out on pasture. Avoiding the first day after birth promotes successful ewe–lamb bonding and a good colostrum intake. Maintaining clean deep litter to reduce floor contamination is important for indoor systems, to minimize neonatal and post-castration infections. When lambing is managed outdoors over a 3-week period, it may be possible to muster the ewes and lambs into yards and castrate the lambs at 4 or 5 weeks of age without problems of mismothering or abandonment of lambs. In contrast, when ewes lamb over 6–8 weeks in flocks that have not been divided according to lambing dates, they should not be yarded for castration and related procedures until the danger of mismothering of the youngest lambs has passed. In addition, it is important that when castration is carried out outdoors, it should be in dry yards or in temporary grass yards on fine days, without rain, sleet, or snow, as muddy conditions may predispose lambs to tetanus or other infections.

In New Zealand and Australia, the scrotum is removed from many lambs using a rubber ring applied below the testicles, leaving them intact. The lambs treated thus are called either cryptorchid lambs or short-scrotum lambs (Fig. 6.10). In New Zealand, 40% of male lambs are short-scrotum lambs, 20% are castrated, and 40% are left intact. The intact animals are slaughtered as prime lamb early in life, but short-scrotum lambs may be kept as stores and slaughtered after puberty. A small proportion of short-scrotum lambs may be fertile and behave sexually like a ram, but they are not usually considered a nuisance. In addition, short-scrotum lambs grow better than wethers.

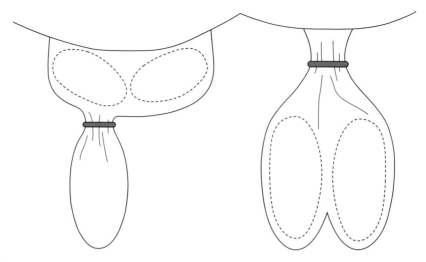

Fig. 6.10. Short scrotum and castration in lambs using a rubber ring.

Methods of castration for lambs

Castration is carried out using a variety of techniques, including the application of tight rubber rings, surgery, or a bloodless castration clamp (Burdizzo®) (Mellor and Stafford, 2000). These techniques are essentially the same as those described above for cattle. For full castration, the ring is placed around the neck of the scrotum above both testicles, although it is sometimes difficult to find both testicles in really young lambs as one may still be in the inguinal canal. With the short-scrotum technique, the ring is applied to the scrotum below the testicles, to push them against the entrance to the inguinal canals. This makes the majority of the lambs infertile by elevating intratesticular temperature, but testosterone secretion continues, thereby explaining why such lambs retain the growth and behavioral characteristics of uncastrated males. Rubber-ring placement in young lambs is simple and straightforward. The rings used on young lambs may not be appropriate for older, larger lambs.

Surgical castration involves removal or incision of the distal scrotal sac so that the testicles and spermatic cords may be drawn out of the scrotum until the cords break. A number of knives have been designed to carry out this surgery. In older lambs, the cords may be severed by knife or an emasculator rather than by traction, and in mature rams, the cords may be ligated to prevent hemorrhage.

With clamp castration, the clamp is applied either individually across each spermatic cord from outside the scrotum, or across the full width of the scrotal neck, including both spermatic cords at once. The latter approach is not recommended unless a ring is applied to the scrotal neck proximal to the clamp cut. In any case, if both rings and a clamp are available, the method of choice, overall, is rings. Clamp castration is easier on larger lambs with longer spermatic cords but, as described for calves, it is technically more difficult than either rubber-ring or surgical castration.

Pain alleviation for castration of lambs

All physical forms of castration cause pain (Mellor and Stafford, 2000). Ring castration causes a smaller, acute cortisol response than surgical castration, but lamb behavior is more active following ring castration than after surgical castration. This has led to disagreement as to which of these techniques causes more acute pain, but as the behavioral responses are specific to each technique, it is not possible to use behavior in a comparative way. However, the acute cortisol responses, and by inference the pain, are markedly lower and shorter lived with ring than with surgical castration, so that ring castration is preferable in these terms (see Mellor and Stafford, 2000). A novel smaller, tighter ring provides no advantage compared to conventional rings (Molony *et al.*, 2012). Yet, wound healing takes longer following ring castration than surgical castration. Lambs castrated by rubber ring still turn to look at their scrotal

K.J. Stafford and D.J. Mellor

area about 42 days after castration, unlike surgically castrated lambs (Molony et al., 1995). This suggests that they may still be experiencing some irritation or low-level chronic pain associated with ring-induced skin lesions before the scrotum finally becomes detached. There is evidence that ring castration of lambs within 3 days of birth causes less acute pain than subsequent castration, and, paradoxically, that it may increase the sensitivity of these lambs to other painful procedures conducted about 1 month later. These findings need to be elucidated.

Local anesthetic injected into the scrotum and testicles eliminates the plasma cortisol response, and by inference, the acute pain caused by rubber-ring castration (Dinniss et al., 1997; Molony et al., 2012). Kent et al. (1998) found that lignocaine injected into the neck of the scrotum was more effective than injection into the testicles, but Dinniss et al. (1997) showed that intratesticular injections were just as effective but practically much easier to execute than scrotal neck injections. Local anesthetics have less impact on responses to surgical or clamp castration, and with these techniques a systemic analgesic such as an NSAID is required, together with local anesthesia, to eliminate the pain caused by castration. Squirting local anesthetic or anesthetic gel into the scrotal neck and on to the spermatic cords before they are severed towards the inguinal area may reduce the pain experienced by lambs following surgical castration where the spermatic cords are drawn out and broken by traction (Lomax et al., 2010). Following surgical castration, the wound area must be treated to prevent screwworm infestation in areas where this parasite occurs.

Whenever possible, ram lambs should be left intact, but if they must be castrated, then it is important to determine whether the short-scrotum procedure, which causes a lower physiological stress response than ring or surgical castration (Lester et al., 1991), is acceptable to farmers, processors, and consumers. If acceptable, this should be carried out in preference to full castration. Immunocastration vaccines for lambs are likely to become available in the near future, and they could be used to eliminate the pain caused by castration if their use is acceptable throughout the production and consumption chain, and provided that they are cheap enough.

Tail docking of lambs

Tail docking is carried out on woolly sheep and not on hair sheep. It is one method of reducing the incidence of flystrike (external myiasis) by decreasing the likelihood of fecal pads (dags) adhering to wool in the perineal region. It is also easier to shear and to crutch (removal of wool from the perineal area) sheep that have been docked. Tails may be cut short on lambs that are reared for meat. A longer tail, perhaps to the tip of the vulva, should be left on breeding animals as this reduces the likelihood of dags developing by maintaining a channel along which urine and feces flow. Extremely short tail docking, flush with the body, must be avoided because it increases the incidence of rectal prolapse (Thomas et al., 2003).

Tail docking may be carried out using a rubber ring, a cautery iron or bar, or a sharp knife (Mellor and Stafford, 2000). Rubber-ring and cautery tail docking elicit similar acute plasma cortisol responses, suggesting that they cause similar amounts of acute pain. In contrast, the use of a cold, sharp knife produces a substantially greater acute plasma cortisol response, and probably significantly greater pain, than the other two methods. Local anesthesia may be injected as a ring block around the tail (subcutaneously), but this is difficult and time-consuming. The use of local anesthetics greatly reduced behavioral indicators of pain (Kent et al., 1998). Application of an anesthetic gel after either surgical or hot-iron docking is recommended (Lomax et al., 2010). Unshorn lambs living in environments where flystrike or screwworm occurs must be treated with anti-flystrike and anti-screwworm medication following docking.

It is possible to breed sheep with reduced wool in the perineal area and for shorter and less woolly tails (Scobie et al., 2007). This would reduce the necessity of docking to reduce flystrike, but shearing would still be impeded by the presence of the tail.

Ear tagging or ear notching of lambs

These procedures are similar to those carried out in cattle. Lambs' ears are smaller, however, and relatively more damage is done during tagging or marking (Fig. 6.11). Identity microchips may reduce the need for ear marking and tagging in the future, but they are economically and practically difficult to use on large commercial sheep farms at present.

Mulesing in sheep

Mulesing is a surgical procedure used to remove wrinkled skin from the breech of some Merino sheep

Fig. 6.11. Ear tagging and marking in sheep.

by cutting away skin from the perianal region. Care is taken to avoid the underlying muscle tissue. In addition, the tail will be docked. The procedure is typically performed by trained personnel with wool-trimming shears, or modified versions of them, while the animal is restrained in a cradle. Carried out in Australia, it is generally recommended as essential to prevent flystrike (myiasis) in Merino sheep. It is usually done at marking time, that is when lambs are castrated, tail docked, and ear tagged at 1–2 months of age. The procedure in its radical form helps to prevent flystrike because wool-free skin around the anus (and vulva in females) is pulled tight and the cuts heal by forming scar tissue that does not get fouled with feces and urine.

It is difficult to eliminate the pain caused by mulesing because an injectable local anesthetic would have to be given epidurally, and this is not feasible in young lambs under farm conditions. Applying local anesthetic to the wounds topically after the operation does reduce the pain (Lomax *et al.*, 2008, 2013). The use of both a topical local anesthetic and a long-acting NSAID would improve welfare. Carprofen plus a topical anesthetic abolished the

cortisol response (Paull *et al.*, 2008). Carprofen alone was not effective for reducing the cortisol response. Other, less painful ways of removing the breech folds or wool are currently being researched in Australia. These include an injectable protein product that, when injected subcutaneously in the perineal region, may have the desired effects, as may the placement of clips specially designed to remove excess skin. Breeding Merino sheep with fewer breech folds or less wool is also being attempted. The length of tail left after mulesing/docking is important: a short tail is more likely to result in fecal pads (dags) and flystrike, whereas medium-length docking appears to reduce the likelihood of dags forming, possibly by leaving a channel for urine and feces to travel through.

The Australian wool industry aims to phase out mulesing. This will be facilitated by breeding Merino sheep with fewer wrinkles in the perineal area and the use of anti-fly vaccination or long-lasting insecticides (Elkington and Mahony, 2007; Rothwell *et al.*, 2007; Scobie *et al.*, 2007).

Teeth grinding in ewes

It was thought that shortening the incisor teeth of older ewes could extend their grazing efficiency and prolong their productive life. The teeth were ground down towards the gum using an electrical grinder. Although the procedure was shown not to be particularly painful, its use was discontinued in Australia, where it was shown not to confer any significant production benefits.

Marking and processing of lambs

"Marking and processing" is the term used in some countries to describe the combination of procedures carried out on lambs when they are first brought into yards; the procedures may include castration, docking, ear tagging/notching, vaccination, and, in Australia at present, mulesing. Except for vaccination, these procedures cause pain and it is difficult to provide analgesia sufficient to eliminate it. It may be worthwhile considering eliminating the pain caused by some procedures, for example, by using local anesthetic for ring castration, and reducing the pain caused by other procedures, for example, by using a systemic analgesic for ear tagging/notching.

Local anesthetic and systemic analgesic might be given together some time before the painful procedures are carried out. This would be possible with

K.J. Stafford and D.J. Mellor

small flocks or when small numbers of lambs are marked at any one time, such as with flocks lambed indoors. However, when hundreds or thousands of lambs are being marked outdoors, it is not so feasible to handle them twice; that is, to catch lambs to administer a local anesthetic and systemic analgesic and then release them, and then catch again to complete the painful procedures. One option in such flocks may be to give systemic analgesia when the procedures are being carried out. This will alleviate some of the pain, but not eliminate it. More research is required to assess the beneficial effects of such approaches. Where flystrike occurs, many farmers spray lambs with an anti-fly ectoparasiticide after docking. If this spray were mixed with local anesthetic, it may reduce, but probably would not eliminate, the pain caused by castration and tail docking.

Goat – Painful Husbandry Procedures

Disbudding of goats

Disbudding is carried out on young goats before the horn bud becomes too large. The horn-generating tissue is destroyed using either a heated iron, which is placed over the horn bud, or by a chemical agent that corrodes the tissue. The germinal tissue may also be removed surgically using a purpose-designed scoop or knife, similar to those used for cattle, or a knife or scissors. Disbudding goats by cautery may cause injury to the brain by heat transference through the skull, so the heated iron should not be applied for more than 5 s, allowing the same interval between successive applications, if they are necessary. Local anesthetic is often used when disbudding goats.

Castration and ear tagging or ear notching

These painful procedures and the use of analgesia are conducted in goats in a manner similar to those used in cattle and sheep.

Pigs – Painful Husbandry Procedures and Pain Alleviation

The management of pigs ranges from intensive indoor to extensive outdoor approaches. Piglets are subject to a number of painful procedures, including castration, teeth clipping, tail docking, and if they are going to live outdoors, rings may be inserted into their noses. Generally, it is advised that these procedures be conducted as close to birth as practical, usually within 3–8 days, but not during the first 6 h.

Castration of piglets and pain alleviation

Traditionally, pigs have been castrated surgically to prevent boar taint. In some production systems, pigs are slaughtered at lighter weights, before boar taint becomes a problem, and so castration is not required. The physiological and behavioral (Taylor and Weary, 2000) responses to castration indicate that it is certainly painful, and pain relief is recommended. Local anesthetics prevented pain-related behaviors in 2-week-old piglets, but not 7-week-old piglets (McGlone and Heilman, 1988; Kluivers-Poodt et al., 2012). In Europe, general anesthetic is now being recommended for pigs being castrated, and some retailers intend accepting for sale only meat from pigs that have not been castrated. Gaseous anesthesia technology suited to pig farms is being developed, and it may soon become normal to give piglets a general anesthetic and then to castrate them, clip teeth, and dock tails at one time. A simple inhaler device can be built for US$100 and the cost of isoflurane was US$0.02–0.03 per pig (Hodgson, 2006). In older pigs, general anesthetic before castration can be produced in a number of ways. A common cheap practice is to inject intra-testicular pentobarbitone sodium. The pig becomes unconscious, and as the source of the drug is removed when the testicles are excised, they regain consciousness soon afterwards. However, intratesticular injections of a large volume of barbiturate may be painful. Another problem with barbiturates is that they can be abused by people, and in many countries, they would not be practical for on-farm use. Immunocastration is a possible alternative to surgical castration (Dunshea et al., 2001; Thun et al., 2006) and may become a more widely accepted practice. In Brazil, immunocastration with a vaccination has become a common practice, but in Europe it is not very popular. Unfortunately, meloxicam and other NSAIDS given at the time of castration are not very effective in pigs compared to cattle (Sutherland et al., 2012). Giving meloxicam by injection 10–20 min before castration did reduce pain (Keita et al., 2010; Tenbergen et al., 2014). Another possible approach that needs further research is giving the nursing sow meloxicam, and the analgesic is administered via the sow's milk (Bates et al., 2014).

Teeth clipping of piglets

A common procedure in intensive pig systems is to clip or grind off the canine teeth of piglets to reduce facial damage to litter mates and damage to the sow's udder. The significance of the damage done by piglets to litter mates and the sow's udder varies between indoor and outdoor management systems. Some large pig farms have eliminated the practice. If clipped close to the gum, the tooth may split and infection may enter the gum and tooth root area. Teeth clipping may not be needed in small litters. It is a painful procedure and there is no easy way to eliminate the pain, unless it is carried out under general anesthesia when castration is done. Grinding resulted in greater increases of cortisol compared to clipping (Marchant-Forde *et al.*, 2009).

Tail docking of piglets

Docking is carried out to prevent tail biting. It is often carried out before 8 days of age, without pain relief, and involves removing one-third to one-half of the tail using clippers, a searing iron, knife, or some other instrument that severs the tail immediately. Neuromas may develop, and these may be painful (Simonsen *et al.*, 1991). The effect of tail docking on tail biting is questioned. Different breeds may be more prone to tail biting than others, and modifying the environment, for example by providing straw bedding and more space, may reduce the problem. Tail docking is painful, and local anesthetic could be used to alleviate the pain in piglets that are not being castrated under general anesthesia. When docking was done without anesthesia, hot clipping resulted in higher frequency of squealing and reduced growth compared to cold clipping (Marchant-Forde *et al.*, 2009), although hot clipping (cautery) caused a lower cortisol response, suggesting that pain was less 60 min after the procedure (Sutherland *et al.*, 2008).

Ear notching of piglets

Ear notching is carried out to identify piglets. Microchip identification technology will probably eliminate the need for ear notching in large pig farms.

Castration, teeth clipping, tail docking, and ear notching are carried out at one time on young male piglets. The development of general anesthesia for castration will reduce the pain experienced by male piglets during these procedures, but female piglets will probably have their teeth clipped, tails docked, and ears notched without any anesthesia. The three procedures probably cause moderate but transient pain (Noonan *et al.*, 1994; Prunier *et al.*, 2005). The distress caused by general anesthesia may be more unpleasant than the pain caused by these procedures, but this would be exceptionally difficult, if not impossible, to quantify. If general anesthesia is not used, then alternative protocols to alleviate pain should be developed and used.

Tusk grinding of mature boars

Tusk grinding is undertaken in mature animals to reduce damage caused during fighting between aggressive boars. The canine teeth (tusks) are shortened using rotary grindstones, saws, or embryotomy wire. It may need to be repeated every few years. Protruding tusks on boars can also cause very serious injuries to people.

Nose ringing of sows

Pigs farmed outdoors, especially sows, often have rings, clips, or pieces of wire placed in their noses to prevent digging and rooting (Fig. 6.12). Sows housed on pasture will destroy pastureland unless rings are used to reduce rooting. Pastures with ringed sows had significantly more grass cover compared to pastures with control sows (Eriksen *et al.*, 2006). The rings may be purpose-designed for pigs and placed in the nasal septum (as with a bull ring), or clips or pieces of wire placed through the cartilage at the top of the snout. Rings are placed when the animal is restrained securely, often using a nose loop. Local anesthesia and sedation could be used to reduce the pain caused by ring placement, or general anesthesia could be used. As rooting is probably independent of hunger, and is an activity pigs favor, the pain induced by movement of the rings during foraging is probably quite significant. Some welfare specialists prefer septum rings. A popular high-welfare program in the USA permits septum rings and forbids the use of rings in the outer rim of the nose. A ring in the septum would permit gentle rooting. The use of a single septum ring may be a reasonable trade-off between the welfare of the sow and protecting the pasture when sows are housed outdoors.

K.J. Stafford and D.J. Mellor

Fig. 6.12. Pig with a single nose ring in the septum (upper illustration); pig with multiple rings on the outer rim of the nose (lower illustration).

Deer – Velvet Antler Removal and Pain Alleviation

The growing antler of deer is called "velvet" because it is covered by soft hair that looks like the fabric of that name. Velvet is living tissue supplied with pain nerves and its removal is forbidden in some European countries. However, velvet is a widely used medicine in Korea and China. It is harvested from farmed and wild deer throughout Asia and from farmed deer in New Zealand, usually in early summer before the antler hardens and its nerve supply becomes non-functional. Antlers are quite

dangerous weapons, and some people recommend that they be removed from breeding stags for the safety of stock handlers and other animals. They may also get caught in fences and cause injury to the stag itself. However, many stags are reared and farmed only for velvet production, a farming practice that has been questioned for welfare reasons. In New Zealand, velvet harvesting is controlled by a national authority. This authority issues permits to veterinarians who supervise the farmers undertaking the procedure, and audits the veterinarians and the farmers. During velvet removal, the deer are restrained either in a crush or by sedation, usually with xylazine. Prior to velvet removal, local anesthesia must be applied either to specific nerves, or better still as a ring block, to prevent or alleviate the acute pain caused by the procedure (Wilson and Stafford, 2002). The use of a compression device to reduce pain is not recommended – lidocaine anesthesia is superior (Woodbury *et al.*, 2002; Johnson *et al.*, 2005). However, the presence of chronic pain has not been examined and it is not mandatory to use a systemic analgesic to alleviate this, if it exists, after the effects of the local anesthetic wear off.

Horses – Painful Husbandry Procedures and Pain Alleviation

Horses are subject to only a few painful procedures, including castration. In some countries, docking is allowed, as is tail nicking.

Castration of colts

Many horses are castrated to make them easier and safer to handle and to prevent unwanted mating. Castration is carried out in many countries under general anesthesia, or standing under sedation and with local anesthesia (see Oehme and Prier, 1974), but in some countries neither anesthesia nor analgesia is used and horses are tied down during the operation. In some countries, only registered veterinarians can castrate horses. The age at castration will vary depending on the breed and the use to which the animal will be put, but most horses are castrated at about 12 months of age. The side effects include hemorrhage, eventration (prolapse of a part of the small intestine into the inguinal canal), infection, and tetanus. Post-operative analgesia is being used increasingly, and if an NSAID were used, this might also reduce post-castration swelling.

Tail docking of horses

Draft horses were often docked in the Middle Ages, as were hunters. This made it easier to keep draft horses clean and prevent tangling with harnesses and reins. Docking was also carried out to improve the appearance of the hindquarters and it was fashionable to dock hackney horses. Traditionally, docking shears were used and these were sometimes heated in a forge to sear the tissues.

The decline in the use of working horses resulted in a decline in this practice, and in some countries it has been banned. Indeed, in the late 1930s, the practice was condemned as cruel unless carried out because of irreparable tail damage or a malignant tumor. Docking horses is still practised in the USA for cosmetic reasons, and draft horses are shown there with docked tails. The tail is usually docked at the level of the ventral commissure of the vulva, or just below the ischial arch in stallions and geldings.

There are two basic techniques used to dock horses. In very young foals, breeders place a rubber ring on the tails. The hair is clipped to increase the speed of constriction and the loss of the tail. The alternative is to remove the tail surgically, and this is described in many modern American texts on veterinary surgery and older British texts. The horse is sedated and an epidural anesthetic given. The proposed amputation site is clipped and a tourniquet is placed proximal to it. Two large flaps of skin and coccygeal muscle are made on the ventral and dorsal side of the tail. The tail is amputated at an intracoccygeal joint and the flaps sutured together. The horses are immunized against tetanus.

Tail nicking of horses

Tails are modified in hackney horses and American Saddlebreds to induce high tail carriage by a technique called nicking or tail setting. The procedure, apparently introduced into England in the early 18th century from the Low Countries of Europe, involves transecting the sacrococcygeal muscles under the tail that depress it, so that the horse thereafter has to hold his tail aloft. The procedure is not described in modern veterinary surgery texts, but it was called a brutal operation as long ago as the mid-1920s. It is still carried out as a cosmetic exercise in the USA, but is banned in many other countries. Horses that have had this operation have difficulty swishing their tails to remove flies.

Poultry – Painful Husbandry Procedures and Pain Alleviation

Poultry are subjected to a few procedures that are painful. Laying hens may have their beaks trimmed as young chicks. Breeding birds may have their wattles removed (dubbing) and their claws removed. Turkeys, especially breeding turkeys, may have their snoods removed.

Beak trimming of poultry

Beak trimming is the most important amputation carried out on chickens and turkeys (Fig. 6.13). Its purpose is to prevent cannibalism, vent pecking, and feather pecking in laying birds and broiler breeder flocks. Young broilers grown for meat are not beak trimmed. It is undertaken on birds managed under all conditions. For instance, in Switzerland in 2000, where cages for layer chickens had not been used since 1992, 59% of flocks (61% of hens) were beak trimmed (Hane *et al.*, 2000). These birds were kept in deep litter, grid-floor or aviary systems, many with access to outdoor runs. Thus, even in non-cage-layer systems, beak trimming is common.

Beak trimming is carried out using a cauterizing blade or infrared beam, and chickens are trimmed either during the first 10 days after hatching or later at 16–18 weeks of age when moved from the rearing house into the laying house. The operator should not remove more than one-half of the upper beak and one-third of the lower beak, being about 3 mm and 2.5 mm in 1-day-old chicks, respectively. In 10-day-old birds, not more than 4.5 mm and 4 mm should be removed from the upper and lower beaks, respectively. In large commercial hatcheries, beaks are trimmed on egg-layer chicks at 1 day of age. The infrared-beam treatment is a highly automated system that uses an infrared beam and is used in many countries (Goran and Johnson, 2005). Proper maintenance and adjustment of the machine are essential. Unpublished industry data initially indicated that chickens that received the infrared-beam treatment performed more poorly than chicks trimmed with a hot blade, until the machines were adjusted properly. When the infrared-beam treatment machine is working correctly, it is superior to the hot blade. One advantage of the machine is that its operational effectiveness remains consistent throughout its use. Birds trimmed with a hot blade had more deviations of the beak compared to birds subjected to the infrared beam (Marchant-Forde

K.J. Stafford and D.J. Mellor

Fig. 6.13. Beak trimming in chicks.

resulted in acute pain (Marchant-Forde *et al.*, 2008). Heart rates were higher immediately after the hot blade compared to the infrared-beam treatment (T. Grandin, 2008, personal communication). Beak trimming may have to be repeated in older birds to prevent cannibalism. There is a great need for research on the use of pain-relief medications for beak trimming.

The beak is a sensitive organ, with many sensory receptors. Beak trimming is acutely painful, and food-intake levels decrease following it (Glatz, 1987). Behavioral and physiological evidence suggests that pain and beak sensitivity persists for weeks or months after trimming (see Craig and Swanson, 1994). Neuromas can form, leading to pain (Gentle, 1986; Gentle *et al.*, 1990). To prevent the formation of neuromas, less than 50% of the beak should be removed (Kuenzel, 2007). However, it may have long-term beneficial effects in that the weight of the adrenal glands and the plasma corticosterone levels of trimmed birds suggest that they are less stressed than birds with intact beaks. A local analgesic (bupivacaine and diethyl sulfoxide) administered to the cut beak of 6-week-old chicks after trimming prevents at least some of the acute pain and the reduction in feed intake usually seen in the first 24 h after trimming.

It is suggested that the pain of trimming is outweighed by the likelihood of beak-related injuries and the suffering caused by cannibalism, vent pecking, and feather plucking. It is strongly recommend that beak trimming be carried out only on chicks less than 10 days of age. It is possible to breed against the tendency to feather peck (Craig and Muir, 1993). Some birds that are highly selected for egg production are more prone to peck each other. Specific environmental features, such as perches, reduce the likelihood of cannibalism (Gunnarsson *et al.*, 1999), and giving hens the opportunity to forage may reduce feather pecking (Blokhuis, 1986). Thus, a combination of using specific genetic lines of laying hens that have a low tendency to feather peck, and the use of particular environmental features, might reduce cannibalism to the degree that beak trimming is not necessary.

Dubbing of roosters

Dubbing is the surgical removal of the comb and wattles using scissors. Male breeding birds are usually dubbed to prevent serious injuries to the comb and wattles, which, when left intact, are targeted

et al., 2008). Both beak-trimming methods resulted in weight loss, with the hot blade birds having more weight loss (Gentle and McKeegan, 2007). Hot blade birds drank less than infrared birds (Dennis and Cheng, 2012). Both beak-trimming methods during fighting between birds. It should be carried out during the first few days after hatching. Local anesthetic can be used to reduce the acute pain caused by the procedure in older birds, and in these hemorrhage may be a problem.

Desnooding of turkeys

Amputating the snood may help to prevent head injuries from pecking and fighting. It is also thought to reduce the risk of frostbite and the spread of erysipelas. The snood is removed from 1-day-old birds using the thumbnail and finger, and in older birds it can be cut off using scissors. Desnooding is apparently a less common practice today than it once was.

Claw removal of roosters and Tom turkeys

The last joint of the inside toes of male breeding birds is removed to prevent injuries to the hens during mating. This should be done at an early age, probably within 72 h of hatching. For older birds, only the nail should be trimmed. However, spurs may be removed or trimmed from roosters to reduce injuries to handlers and to other birds during fighting. Toes may be clipped off each foot of 1-day-old turkeys to prevent scratched and torn backs, which are important causes of carcass downgrading. These amputations are carried out without local anesthetic, using sharp scissors or shears. Hemorrhage is not usually much of a problem. Genetic selection for lower aggression could be one method that could be used to eliminate the need for this procedure. There are genetic effects on aggression in roosters (Millman and Duncan, 2000).

Conclusion

The decision to use a painful husbandry procedure and whether or not to use pain relief may be facilitated by reference to a series of questions (Quick Access Information 6.1). In essence, these questions are asked: Is the procedure justified in terms of direct benefits to the animals and/or by other benefits such as to the farming enterprise? What harms are caused, how bad are they, and can they be avoided or reduced? One harm is pain, so that providing relief from it helps to address our responsibility to take all practical steps to minimize that harm. In this context, additional issues need to be considered, such as the effectiveness, availability and cost of pain-relieving drugs, their ease of administration, possible side effects, and residue concerns. In some countries, many drugs have to be administered by veterinarians, and this raises the matter of the availability of veterinarians and the cost. A further matter relevant to cost is whether it will be recovered in farm-gate prices, or whether, in the world of today, the cost of providing pain relief is becoming a farm overhead that must be absorbed in order to do business with trade partners in welfare-sensitive markets. In the future, it is likely that painful procedures such as castration will be replaced with a vaccine that blocks male hormone. Dehorning can be replaced with breeding polled animals, and genetic selection of poultry for less feather pecking may replace beak trimming in poultry.

References

AABP (2014) Castration and dehorning guidelines. American Association of Bovine Practitioners, Auburn Alabama. Available at: www.aabp.org (accessed 22 September 2014).

Allen, K.A., Coetzee, J.F., Edwards-Calloway, L.N., Glynn H., Dockweiller, J., Kukanich, B., *et al.* (2013) The effect of timing of oral meloxicam administration on physiological responses in calves after cautery dehorning with local anesthesia. *Journal of Dairy Science* 96, 5194–5205.

Amatayakul-Chantler, S., Jackson, J.A., Stegner, J., King, V., Rubio, L.M.S., Howard, R., Lopez, E., *et al.* (2012) Immunocastration of *Bos indicus* × Brown Swiss bulls in feedlot with gonadotropin releasing hormone vaccine, Bopriva, provides improved performance and meat quality. *Journal of Animal Science* 90, 3718–3728.

Amatayakul-Chantler, S., Jackson, J.A., Stegner, J., King, V., Rubio, L.M.S., Howard, R., *et al.* (2013) Effects on performance and carcass and meat quality attributes following immunocastration with gonadotropin releasing factor vaccine, Boprova or surgical castration of *Bos indicus* bulls raised on pasture in Brazil. *Meat Science* 95, 78–84.

Bates, J.L., Kamiker, L.A., Stock, M.L., Pertzborn, K.M., Baldwin, L.G., *et al.* (2014) Impact of transmammary-delivered meloxicam on biomarkers of pain and distress in piglets after castration and tail docking. *PLOS One* 9(12), e113678. doi:10.1371/journal.pone.0113678.

Becker, J., Doherr, M.G., Bruckmaier, R.M., Bodmer, M., Zanolari, P. and Steiner, A. (2012) Acute and chronic pain in calves after different methods of rubber ring castration. *The Veterinary Journal* 194, 380–385.

Blokhuis, H.J. (1986) Feather-pecking in poultry: Its relation with ground pecking. *Applied Animal Behaviour Science* 16, 63–67.

Craig, J.V. and Muir, W.M. (1993) Selection for the reduction of beak inflicted injuries among caged hens. *Poultry Science* 72, 411–420.

Craig, J.V. and Swanson, J.C. (1994) Welfare perspectives on hens kept for egg-production. *Poultry Science* 73, 921–938.

Dennis, R.L. and Cheng, H.W. (2012) Effects of different infrared treatment protocols on chicken welfare and physiology. *Poultry Science* 91, 1499–1505.

Dinniss, A.S., Mellor, D.J., Stafford, K.J., Bruce, R.A. and Ward, R.N. (1997) Acute cortisol responses to lambs to castration using rubber ring and/or a castration clamp with or without local anaesthetic. *New Zealand Veterinary Journal* 45, 114–121.

Dockweiler, J.C., Coetzee, J.F., Edwards-Callaway, L.N., Bello, N.M., Glynn, H.D., *et al.* (2013) Effect of castration method on neurohormonal and electroencephalographic stress indicators in Holstein calves of different ages. *Journal of Dairy Science* 96, 4340–4354.

Dunshea, F.R., Colantoni, C., Howard, K., McCauley, I., Jackson, P., *et al.* (2001) Vaccination of boars with GnRH vaccine (Improvac) eliminates boar taint and increases growth performance. *Journal of Animal Science* 79, 2524–2535.

Eicher, S.D. and Dailey, J.W. (2002) Indicators of acute pain and fly avoidance behaviors in Holstein calves following tail docking. *Journal of Dairy Science* 85, 2850–2858.

Eicher, S.D., Morrow-Tesch, J.L., Albright, J.L. and Williams, R.E. (2001) Tail docking alters numbers of fly avoidance behaviors and cleanliness, but not physiological measures. *Journal of Dairy Science* 84, 1822–1828.

Elkington, R.A. and Mahony, T.J. (2007) A blowfly strike vaccine requires an understanding of host–pathogen interactions. *Vaccine* 25(28), 5133–5145.

Eriksen, J., Studnitz, M., Strudsholm, K., Kongsted, A.G. and Hermansen, J.J.E. (2006) Effect of nose ringing and stocking rate of pregnant and lactating sows on exploratory behavior, grass cover and nutrient loss potential. *Livestock Science* 104, 91–102.

Espinoza, C., Lomax, S. and Windsor, P. (2013) The effect of a topical anesthetic on the sensitivity of calf dehorning wounds. *Journal of Dairy Science* 96, 2894–2902.

Fierheller, E.E., Caulkett, N.A., Haley, D.B., Florence, D. and Doepel, L. (2012) Onset duration and efficacy of four methods of local anesthesia of horn bud in calves. *Veterinary Anaesthesia and Analgesia* 39, 431–435.

Flecknell, P. and Waterman-Pearson, A. (2000) *Pain Management in Animals*. W.B. Saunders, London.

Gentle, M.J. (1986) Neuroma formation following partial beak amputation (beak trimming) in the chicken. *Research in Veterinary Science* 41, 383–385.

Gentle, M.J. and McKeegan, D. (2007) Evaluation of the effects of infrared beak trimming in broiler chicks. *Veterinary Record* 160, 145–148.

Gentle, M.J., Waddington, D., Hunter, L.N. and Jones, R.B. (1990) Behavioural evidence for persistent pain following partial beak amputation in chickens. *Applied Animal Behaviour Science* 27, 149–157.

Giffard, A., Olsson, C.R. and Shell, M. (2007) A topical anesthetic composition EP1863468 AI. Australian patent AU 2006/000336.

Glatz, P. (1987) Effects of beak trimming and restraint on heart rate, food intake, body weight and egg production in hens. *British Poultry Science* 28(4), 601–611.

Goran, M.S. and Johnson, S.C. (2005) Beak treatment with tongue protection. US Patent 7,363,881B2. Patent Office, Washington, DC.

Grandin, T. (1997) Assessment of stress during handling and transport. *Journal of Animal Science* 75, 249–257.

Grandin, T., and Deesing, M. (2014) *Genetics and the Behavior of Domestic Animals*. Academic Press, San Diego, California.

Grandin, T. and Johnson, C. (2005) *Animals in Translation*. Scribner (Simon and Schuster), New York.

Gregory, N.G. (2004) *Physiology and Behaviour of Animal Suffering*. Blackwell Publishing, Oxford, UK.

Gunnarsson, S., Keeling, L.J. and Svedberg, J. (1999) Effects of rearing factors on the prevalence of floor eggs, cloacal cannibalism and feather pecking in commercial flocks of loose housed laying hens. *British Poultry Science* 40, 12–18.

Hane, M., Huber-Eicher, B. and Frohlich, E. (2000) Survey of laying hen husbandry in Switzerland. *World's Poultry Science Journal* 56, 21–31.

Heinrich, A., Duffield, T.F., Lissemore, K.D., Squires, E.J. and Millman, S.T. (2009) The impact of meloxicam on postsurgical stress associated with cautery dehorning. *Journal of Dairy Science* 92, 540–547.

Hodgson, D.S. (2006) An inhaler device using liquid injection of isoflurane for short-term anesthesia of piglets. *Veterinary Anaesthesia and Analgesia* 33, 207–213.

Johnson, C.B., Wilson, P.R., Woodbury, M.R. and Caulkett, N.A. (2005) Comparison of analgesic techniques for antler removal in halothane anesthetics red deer (*Cervus elaphus*) electroencephalographic responses. *Veterinary Anaesthesia and Analgesia* 32, 61–71.

Keita, A., Pagot, E., Prunier, A. and Guidarini, C. (2010) Pre-emptive meloxicam for post operative analgesia in piglets undergoing surgical castration. *Veterinary Anesthesia and Analgesia* 37, 367–374.

Kent, J.E., Thrusfield, M.V., Robertson, I.S. and Molony, V. (1996) Castration of calves: A study of the methods used by farmers in the United Kingdom. *Veterinary Record* 138, 384–387.

Kent, J.E., Molony, V. and Graham, M.J. (1998) Comparison of methods for the reduction of acute pain produced by rubber ring castration or tail docking of week-old lambs. *Veterinary Journal* 155, 39–51.

Kluivers-Poodt, M., Houx, B.B., Robben, S.R., Koop, G., Lambooij, E., *et al.* (2012) Effects of a local anaesthetic and NSAID in castration of piglets on the acute pain responses, growth and mortality. *Animal* 6, 1469–1475.

Knight, T.W., Cosgrove, G.P., Death, A.F., Anderson, C.B. and Fisher, A.D. (2000) Effect of method of castrating bulls on their growth rate and liveweight. *New Zealand Journal of Agricultural Research* 43, 187–192.

Kuenzel, W.J. (2007) Neurological basis of sensory perception: Welfare implications of beak trimming. *Poultry Science* 86, 1273–1282.

Lay, D.C. Jr, Friend, T.H., Randel, R., Bowers, C.L., Grissom, K.K., *et al*. (1992a) Behavioral and physiological effects of freeze and hot-iron branding on crossbred cattle. *Journal of Animal Sciences* 70, 330–336.

Lay, D.C. Jr, Friend, T.H., Bowers, C.L., Grissom, K.K. and Jenkins, O.C. (1992b) A comparative physiological and behavioral study of freeze and hot-iron branding using dairy cows. *Journal of Animal Science* 70, 1121–1125.

Lester, S.J., Mellor, D.J., Ward, R.N. and Holmes, R.J. (1991) Cortisol responses of young lambs to castration and tailing using different methods. *New Zealand Veterinary Journal* 39, 134–138.

Lester, S.J., Mellor, D.J., Holmes, R.J., Ward, R.N. and Stafford, K.J. (1996) Behavioural and cortisol responses of lambs to castration and tailing using different methods. *New Zealand Veterinary Journal* 44, 45–54.

Lomax, S. and Windsor, P.A. (2013) Topical anesthesia mitigates pain of castration in calves. *Journal of Animal Science* 91, 4945–4952.

Lomax, S., Sheil, M. and Windsor, P.A. (2008) Impact of topical anesthesia on pain alleviation and wound healing in lambs after mulesing. *Australian Veterinary Journal* 86, 159–168.

Lomax, S., Dickson, H., Sheil, M. and Windsor, P.A. (2010) Topical anesthesia alleviates short-term pain of castration and tail docking in lambs. *Australian Veterinary Journal* 88, 67–74.

Lomax, S., Sheil, M. and Windsor, P.A. (2013) Duration of action of a topical anaesthetic formulation for pain management of mulesing in sheep. *Australian Veterinary Journal* 91, 160–167.

McCosker, K., Letchford, P., Petherick, J.C., Mever, D. and McGowan, M. (2010) Morbidity, mortality, and body weight gain in surgically spayed, yearling Brahman heifers. *Australian Veterinary Journal* 88, 497–503.

McGlone, J.J. and Hellman, J.M. (1988) Local and general anesthetic effects on the behaviour and performance of two and seven week old castrated and uncastrated piglets. *Journal of Animal Science* 66, 3049–3058.

McMeekan, C.M., Mellor, D.J., Stafford, K.J., Bruce, R.A., Ward, R.N., *et al*. (1997) Effect of shallow and deep scoop dehorning on plasma cortisol concentrations in calves. *New Zealand Veterinary Journal* 45, 69-71.

McMeekan, C.M., Mellor, D.J., Stafford, K.J., Bruce, R.A., Ward, R.N., *et al*. (1998a) Effects of local anaesthesia of 4 or 8 hours duration on the acute cortisol response to scoop dehorning in calves. *Australian Veterinary Journal* 76, 281–285.

McMeekan, C.M., Stafford, K.J., Mellor, D.J., Bruce, R.A., Ward, R.N., *et al*. (1998b) Effect of regional analgesia and/or a non-steroidal anti-inflammatory analgesic on the acute cortisol response to dehorning in calves. *Research in Veterinary Science* 64, 147–150.

Marchant-Forde, R.M., Fahey, A.G. and Cheng, H.W. (2008) Comparative effects of infrared and one-third hot blade trimming on beak topography behaviour and growth. *Poultry Science* 87, 1474–1483.

Marchant-Forde, J.N., Lay, D.C. Jr, McMunn, K.A., Cheng, H.W., Pajor, E.A., *et al*. (2009) Post-natal piglet husbandry practices and wellbeing: the effect of alternative techniques delivered separately. *Journal of Animal Science* 87(4), 1479–1492.

Mellor, D.J. and Stafford, K.J. (2000) Acute castration and/or tailing distress and its alleviation in lambs. *New Zealand Veterinary Journal* 48, 33–43.

Mellor, D.J., Cook, C.J. and Stafford, K.J. (2000) Quantifying some responses to pain as a stressor. In: Moberg, G.P. and Mench, J.A. (eds) *The Biology of Animal Stress: Basic Principles and Implications for Welfare*. CAB International, Wallingford, UK, pp. 171–198.

Mellor, D.J., Thornber, P.M., Bayvel, D.A.C. and Kahn, S. (eds) (2008) Scientific assessment and management of animal pain. *OIE* [World Organisation for Animal Health] *Technical Series* 10, 1–218.

Miguel, G.Z., Faria, M.H., Rocca, R.O., Santos, C.T., Suman, S.P., *et al*. (2014) Immunocastration improves carcass traits and beef color attributes in Nellore and Nellore–Aberdeen Angus crossbred animals finished in feedlot. *Meat Science* 96, 884–891.

Milligan, B.N., Duffield, T. and Lissemore, K. (2004) The utility of ketoprofen for alleviating pain following dehorning of young dairy calves. *Canadian Veterinary Journal* 45, 140–143.

Millman, S.T. and Duncan, I.J.H. (2000) Strain differences in aggressiveness of male domestic fowl in response to a male model. *Applied Animal Behaviour Science* 66, 217–233.

Molony, V., Kent, J.E. and Robertson, I.S. (1995) Assessment of acute and chronic pain after different methods of castration of calves. *Applied Animal Behaviour Science* 46, 33–48.

Molony, V., Kent, J.E., Viniuela-Fernandez, I. Anderson, C. and Dyer, C.M. (2012) Pain in lambs castrated at 2 days using novel smaller tighter rings with and without local anesthetic. *Veterinary Journal* 193, 81–86.

Morisse, J.P., Cotte, J.P. and Huonnic, D. (1995) Effect of dehorning on behaviour and plasma cortisol responses in young calves. *Applied Animal Behaviour Science* 43, 239–247.

Neely, C.D. (2013) Comparison of the effects of three different dehorning techniques on behavior and wound healing in feeder cattle in a western Kansas feedlot. Thesis, Kansas State University, Manhattan, Kansas.

Noonan, G.J., Rand, J.S., Priest, J., Ainscow, J. and Blackshaw, J.K. (1994) Behavioural observations of piglets undergoing tail docking, teeth clipping, and ear notching. *Applied Animal Behaviour Science* 39, 203–213.

Oehme, F.W. and Prier, J.E. (1974) *Large Animal Surgery*. Williams and Wilkins, Baltimore, Maryland.

Paull, D.R., Lee, C., Atkinson, S.J. and Fisher, A. (2008) Effects of meioxicam or tolfenamic acid administration on the pain and stress responses of Merino lambs to mulesing. *Australian Veterinary Journal* 86, 303–311.

Petherick, J.C., McCosker, K., Mayer, D.G., Letchford, P. and McGowan, M. (2012) Evaluation of impacts of spaying by either dropped ovary technique or ovariectomy via flank laparotomy on welfare of *Bos indicus* beef heifers and cows. *Journal of Animal Science* 91, 382–394.

Petherick, J.C., Small, A.H., Mayer, D.G., Colditz, I.G., Ferguson, D.M., *et al*. (2014) A comparison of welfare outcomes for weaner and mature *Bos indicus* bulls surgically castrated with and without analgesic. 1. Behavioral responses. *Applied Animal Behaviour Science* 157, 23–34.

Petrie, N., Mellor, D.J., Stafford, K.J., Bruce, R.A. and Ward, R.N. (1996a) Cortisol responses of calves to two methods of disbudding used with or without local anaesthetic. *New Zealand Veterinary Journal* 44, 9–14.

Petrie, N., Mellor, D.J., Stafford, K.J., Bruce, R.A. and Ward, R.N. (1996b) Cortisol responses of calves to two methods of tail docking used with or without local anaesthetic. *New Zealand Veterinary Journal* 44, 4–8.

Prunier, A., Mounier, A.M. and Hay, M. (2005) Effects of castration, tooth resection, or tail docking on plasma metabolites and stress hormones in young pigs. *Journal of Animal Science* 83(1), 216–222.

Repenning, P.E., Ahola, J.K., Callan, R.J., French, J.T., Giles, R.L., *et al*. (2013) Impact of oral meloxicom administration before and after band castration on feedlot performance and behavioral responses in weanling beef bulls. *Journal of Animal Science* 91, 4965–4974.

Rothwell, J., Hynd, P., Brownlee, A., Dolling, M. and Williams, S. (2007) Research into alternatives to mulesing. *Australian Veterinary Journal* 85, 94–97.

Scobie, D.R., O'Connell, D., Morris, C.A. and Hickey, S.M. (2007) A preliminary genetic analysis of breech and tail traits with the aim of improving the welfare of sheep. *Crop and Pasture Science* 58, 161–167.

Simonsen, H.B., Klinken, L. and Bindseil, E. (1991) Histopathology of intact and docked pigtails. *British Veterinary Journal* 147, 407–412.

Stafford, K.J. and Mellor, D.J. (1993) Castration, tail docking and dehorning – what are the constraints? *Proceedings of the New Zealand Society of Animal Production* 53, 189–195.

Stafford, K.J. and Mellor, D.J. (2005a) Dehorning and disbudding distress and its alleviation in calves. *Veterinary Journal* 169, 337–349.

Stafford, K.J. and Mellor, D.J. (2005b) The welfare significance of the castration of cattle: A review. *New Zealand Veterinary Journal* 53, 271–278.

Stafford, K.J. and Mellor, D.J. (2011) Addressing the pain associated with disbudding and dehorning in cattle. *Applied Animal Behaviour Science* 135, 226–231.

Stafford, K.J., Mellor, D.J., Todd, S.E., Bruce, R.A. and Ward, R.N. (2002) Effects of local anaesthesia or local anaesthesia plus a non-steroidal anti-inflammatory drug on the acute cortisol response of calves to five different methods of castration. *Research in Veterinary Sciences* 73, 61–70.

Stafford, K.J., Mellor, D.J., Todd, S.E., Ward, R.N. and McMeekan, C.M. (2003) Effects of different combinations of lignocaine, ketoprofen, xylazine and tolazoline on the acute cortisol response to dehorning in calves. *New Zealand Veterinary Journal* 51, 219–226.

Stafford, K.J., Chambers, J.P. and Mellor, D.J. (2006) The alleviation of pain in cattle: A review. *Perspectives in Agriculture, Veterinary Science, Nutrition and Natural Resources* 1, No. 032.

Stewart, M., Stookey, J.M., Stafford, K.J., Tucker, C.B., Rogers, A.R., *et al*. (2009) Effects of local anaesthetic and a nonsteroidal anti-inflammatory drug on pain responses of dairy calves to hot-iron dehorning. *Journal of Dairy Science* 92, 1512–1519.

Stull, C.L., Payne, M.A., Berry, S.L. and Hullinger, P.J. (2002) Evaluation of the scientific justification for tail docking in dairy cattle. *Journal of the American Veterinary Medical Association* 220, 1298–1303.

Sutherland, M.A. and Tucker, C.B. (2011) The long and the short of it: A review of tail docking in farm animals. *Applied Animal Behaviour Science* 135, 179–191.

Sutherland, M.A., Stafford, K.J., Mellor, D.J., Gregory, N.G., Bruce, R.A., *et al*. (2000) Acute cortisol responses and wound healing in lambs after ring castration plus docking with or without application of a castration clamp to the scrotum. *Australian Veterinary Journal* 78, 402–405.

Sutherland, M.A., Mellor, D.J., Stafford, K.J., Gregory, N.G., Bruce, R.A., *et al*. (2002) Effects of local anaesthetic combined with wound cauterisation on the cortisol response to dehorning in calves. *Australian Veterinary Journal* 80, 165–167.

Sutherland, M.A., Bryer, P.J., Krebs, N. and McGlone, J.J. (2008) Tail docking of pigs: Acute physiological and behavioural responses. *Animal* 2, 292–297.

Sutherland, M.A., Davis, B.L., Brooks, T.A. and Coetzee, J.F. (2012) The physiological and behavioral response of pigs castrated with and without anesthesia or analgesia. *Journal of Animal Science* 90, 2211–2221.

Sylvester, S.P., Stafford, K.J., Mellor, D.J., Bruce, R.A. and Ward, R.N. (2004) Behavioural responses of calves to amputation dehorning with and without local anaesthesia. *Australian Veterinary Journal* 82, 697–700.

Taylor, A.A. and Weary, D.M. (2000) Vocal responses of piglets to castration: Identifying procedural sources of pain. *Applied Animal Behaviour Science* 70, 17–26.

Tenbergen, R., Friendship, R., Cassar, G., Amezcua, M.R. and Haley, D. (2014) Investigation of the use of meloxicam for reducing pain associated with castration and tail docking and improving performance in piglets. *Journal of Swine Health and Production* 22, 64–70.

Theurer, M.E., White, B.J., Coetzee, J.F., Edwards, L.N., Mosher, R.A., *et al*. (2012) Assessment of behavioral changes associated with oral meloxicam administration at time of dehorning in calves using remote triangulation device accelerometers. *BMC Veterinary Research* 8, 48.

Thomas, D.L., Waldron, D.F., Lowe, G.D., Morrical, D.G., Meyer, H.H., *et al*. (2003) Length of docked tail and the incidence of prolapse in lambs. *Journal of Animal Science* 81, 2725–2732.

Thun, R., Gajewski, Z. and Janett, F. (2006) Castration of male pigs: Techniques and animal welfare issues. *Journal of Physiological Pharmacology* 57 (Supplement 8), 189–194.

Ting, S.T., Earley, B., Hughes, J.M. and Crowe, M.A. (2003) Effect of ketoprofen, lidocaine local anesthesia and combined xylazine and lidocaine caudal epidural anesthesia during castration of beef cattle on stress responses, immunity, growth, and behavior. *Journal of Animal Science* 81, 1281–1293.

Villarroel, A. (2011) Dehorning calves with paste. Oregon State University, PNW 626. https://catalog.extension.oregonstate.edu/pnw626 (accessed 24 April, 2015).

Walker, K.A., Duffield, T.F. and Weary, D.M. (2011) Identifying and preventing pain during and after surgery in farm animals. *Applied Animal Behavioural Science* 135, 259–265.

Warnock, T.M., Thrift, T.A., Irsik, M., Hersom, M.J.H., Yelich, J.V., *et al*. (2012) Effect of castration technique on beef calf performance, feed efficiency and inflammatory responses, *Journal of Animal Science* 90, 2345–2352.

Watts, J.M. and Stookey, J.M. (1999) Effects of restraint and branding on rates and acoustic parameters of vocalization in beef cattle. *Applied Animal Behaviour Science* 62, 125–135.

White, R.G., DeShazer, J.A., Tressler, C.J., Borcher, G.M., Davey, S., *et al*. (1995) Vocalization and physiological response of pigs during castration with and without anesthetic. *Journal of Animal Science* 73, 381–386.

Wilson, P.R. and Stafford, K.J. (2002) Welfare of farmed deer in New Zealand. 2: Velvet antler removal. *New Zealand Veterinary Journal* 50, 221–227.

Woodbury, M.R., Caulkett, N.A. and Wilson, P.R. (2002) Comparison of lidocaine and compression for velvet antler removal in Wapiti. *Canadian Veterinary Journal* 43, 869–875.

Further Reading

American Veterinary Medical Association (2012) *Backgrounder, Welfare Implications of Castration of Cattle*. American Veterinary Medical Association, Schaumberg, Illinois.

Anderson, D.E. and Edmondson, M.A. (2013) Prevention and management of surgical pain in cattle. *Veterinary Clinics of North America, Food Animal Practices* 29, 157–184.

Anon. (2005) *Animal Welfare (Painful Husbandry Procedures) Code of Welfare 2005*. National Animal Welfare Advisory Committee (NAWAC), c/o Ministry of Agriculture and Forestry, Wellington, New Zealand, ISBN 0-478-29800-5, pp. 1–36.

Appleby, M.C., Hughes, B.O. and Elson, H.A. (1992) *Poultry Production Systems: Behaviour Management and Welfare*. CAB International, Wallingford, UK.

Coetzee, J.F. (2013a) Assessment and management of pain associated with castration in cattle. *Veterinary Clinics of North America, Food Animal Practice* 29, 75–101.

Coetzee, J.P. (2013b) A review of analgesic compounds used in food animals in the United States. *Veterinary Clinics of North America, Food Animal Practice* 29, 11–28.

Faucitano, L. and Schaefer, A.L. (2008) *Welfare of Pigs*. Wageningen Academic Publishers, Wageningen, the Netherlands.

Hall, L.W. (1971) Wright's *Veterinary Anaesthesia and Analgesia*. Bailliere Tindall, London.

Knight, T.W., Cosgrove, G.P., Death, A.F., Anderson, C.B. and Fisher, A.D. (2000) Effect of method of castrating bulls on their growth rate and liveweight. *New Zealand Journal of Agricultural Research* 43,187–192.

Mellor, D.J. and Stafford, K.J. (1999) Assessing and minimizing the distress caused by painful husbandry procedures. *In Practice* 21, 436–446.

Stafford, K.J. and Mellor, D.J. (2006) The assessment of pain in cattle: A review. *Perspectives in Agriculture, Veterinary Science, Nutrition and Natural Resources* 1, No. 013.

Stafford, K.J., Mellor, D.J. and McMeekan, C.M. (2000) A survey of methods used by farmers to castrate calves in New Zealand. *New Zealand Veterinary Journal* 48, 16–19.

Weaver, A.D., St Jean, G. and Steiner, A (2005) *Bovine Surgery and Lameness 2nd Edition*. Wiley-Blackwell, London.

K.J. Stafford and D.J. Mellor

7 The Importance of Good Stockmanship and Its Benefits to Animals

Jeffrey Rushen and Anne Marie de Passillé

University of British Columbia, British Columbia, Canada

Summary

Good stockmanship will improve both animal welfare and productivity. Dairy cows, pigs, and other animals that are fearful of people will have lower weight gain, lower milk production, and poorer reproductive productivity. Farms where animals are willing to approach people may be more productive. This chapter reviews many studies that clearly show the relationship between aversive (bad) treatment and lower production. Animals that have been hit or shocked may become fearful of all people. Stockpeople who have a positive attitude towards animals often have animals with increased productivity. Studies also show how training can be used to improve the attitudes of stockpeople. The animal's relationship with the stockperson is not the only factor that determines productivity. Farm cleanliness and a stockperson's attention to good management practices is also extremely important. To help stockpeople maintain a positive attitude, they must not be worked to the point of becoming exhausted.

Learning Objectives

- Understand that good stockmanship will improve animal welfare
- Learn about the detrimental effects of fear on productivity
- Positive attitudes towards animals are important for improving both welfare and productivity
- Train stockpeople to have more positive attitudes

Introduction

With all of the attention that is given to intensive housing of farm animals, such as the use of battery cages for laying hens, stalls for sows, etc., we might assume that the welfare of farm animals is affected most by the way that they are housed. However, a very important component of farming that affects both animal welfare and animal productivity is the people who care for the animals. The decisions about how animals are housed, fed, and managed are made by people, and it is people who actually perform procedures like breeding, farrowing, milking, debeaking, dehorning, etc. There are many ways in which the stockpeople or caretakers can affect the welfare of the animals in their care. The knowledge or technical competence of the stockperson can play a major role if it leads to improper choice of housing or poor feeding methods. The quality and diligence with which routine tasks, such as cleaning the barn, are done can also be important. In addition, research has now shown that the way that animals are handled by people can have a major effect on their welfare and their productivity. In this chapter, we review some of the research that shows the importance of good stockmanship for animal welfare, animal productivity, and worker satisfaction.

Effect of Good Stockmanship on Animal Welfare and Productivity

The stockperson can most obviously affect the welfare of animals through the way that routine animal care tasks, such as feeding, cleaning, etc., are done. The care with which routine animal care tasks are performed is responsible for some of the differences between farms and farmers in the level of animal welfare. For example, dairy farms in which calves are cared for by females tend to have lower calf mortality than farms where men are responsible for care of the calves (Losinger and Heinrichs, 1997). A survey of 108 dairies in Minnesota, USA, showed that stockmanship training of employees was associated with better milk production (Sorge *et al.*, 2014).

Lensink *et al.* (2001a) examined the role of stockmanship in affecting the health and productivity of veal calves on 50 veal farms. The farms were operated by a single company, were located in the same region, and used similar management techniques, animal feed, etc. Farmers were interviewed to assess their attitudes to the animals (e.g. whether or not they believed that calves were sensitive to human contact) and their attitudes to the work (e.g. how important cleaning procedures were). The farms were also scored for cleanliness, and the performance of various management routines was noted. High-producing farms (i.e. those with high daily weight gains, good feed conversion efficiencies, and low mortality) were cleaner, had pens disinfected by an outside company, had Sunday-evening feedings of the calves, and were run by farmers whose own parents had managed a veal farm. This latter point was important because it resulted in the farmer having a greater experience in raising calves. The size of the effects on the productivity of the farms was large. The cleanliness of the barns accounted for 19% of the differences between farms in daily weight gain and 22% of the differences between farms in feed efficiency. The health of the calves was correlated with the attitudes of the farmers; for example, the more the farmer believed that calves were sensitive to human contact, and the more the farmer felt that cleaning was important, the better the health of the calves. The results show the importance of general stockmanship for the welfare and productivity of calves.

Handling and Animals' Fearfulness of People

Research has now shown quite clearly that when farm animals are frightened of the people who care for or handle them, this can have a marked effect on their welfare. This research has been reviewed a number of times previously (Rushen *et al.*, 1999b; Waiblinger *et al.*, 2004, 2006; Hemsworth and Coleman, 2010). Fear of people can be a major source of stress and a cause of lost production in most species of farm animals. Research with Nelore cattle in South America indicated that cows used in embryo production had 19% lower viable embryos when they were handled roughly by people hitting them compared to gentle handling (Macedo *et al.*, 2011). There are marked differences between animals and between farms in the degree to which the animals are frightened of people, and the degree of fear of the animals on a farm is associated strongly with the productivity of the animals on the farm.

Detrimental effects of fear on pigs and poultry

Researchers have found large differences between farms in the level of productivity achieved by farm animals and the extent that the animals are frightened of people. In one of the first studies, Hemsworth *et al.* (1981) compared one-man pig farms and found that the degree of fear that the pigs showed to people, as measured by their readiness to approach people (Fig. 7.1), accounted for a substantial amount of the difference between farms in farrowing rate and number of piglets born. In a later study, Hemsworth *et al.* (1999) found correlations between the behavioral response of lactating

Fig. 7.1. A common way of measuring how fearful animals are of people is to measure how closely they approach people, particularly when they are feeding. The degree of fear is thought to be apparent in the extent that they withdraw from the feed when people are near. However, animals may avoid people for many reasons. For example, animals that are fed by hand are more likely to approach people than animals that are not fed by hand. Yet this does not necessarily mean that their welfare is better or that they have been better handled. Caution is required in assessing the fearfulness of animals from their reactions to people when different farms are being compared (de Passillé and Rushen, 2005).

sows to people and the percentage of stillborn piglets. Farms on which sows were quicker to withdraw from an approaching person (a sign of fearfulness) had a higher stillbirth rate than those who allowed the person to approach closely. The difference between gilts in their response to the person accounted for about 18% of the difference between farms in percentage of stillborn piglets. The results indicate that high levels of fear of people by sows may affect the survival of their piglets strongly. Similarly, in poultry, feed conversion efficiency (that is, the amount of feed needed to produce a certain number of eggs) has been found to be lower on poultry farms where the birds are kept a larger distance from people (Hemsworth *et al.*, 1994a).

Sometimes, the effect of stockmanship may not be due to the use of rough handling but through more subtle effects. For example, Cransberg *et al.* (2000) found a higher mortality among broiler chicks on farms where the stockpeople moved quickly (presumably frightening the birds), while mortality was lower the more time the stockperson stayed still in the shed. Again, the size of the effect was large: the speed of movement of the stockperson accounted for 15% of the differences between farms in broiler mortality.

Fearful cows give less milk

Perhaps because dairy cows are handled often, much research on how handling methods affect the fearfulness of the animals has involved dairy cattle (Fig. 7.2). Seabrook (1984) showed that the way that dairy cows were handled by the stockperson and the degree of fear the animals showed towards people could be a major factor underlying the differing productivity of different stockpeople. He observed the behavior of cows that were being handled by "high-producing" stockpeople (that is, stockpeople whose cows produced a large amount of milk) and compared this to cows handled by "low-producing" stockpeople. He found that the cows with the high-producing stockperson were spoken to and touched more often, appeared less frightened and were moved more easily, and were more likely to approach the stockperson.

Fig. 7.2. Of all farm animals, dairy animals have the closest contact with people because of regular milking. Good handling practices, then, are particularly important for dairy animals, and much research has shown that the way dairy cows are handled can have a large impact on their productivity. There are large differences between farms and between milkers in how much milk can be obtained at milking, and low milk yields have been associated with rough handling and cows that are frightened of people.

Breuer et al. (2000) found substantial relationships between the levels of milk production on dairy farms, the way that the animals were handled, and the degree of fear the cows showed towards people. Thirty-one commercial dairy farms in Australia were visited and the stockpeople were observed while they moved and handled the animals during normal milkings. The degree to which the milkers used rough or aversive handling during milking varied widely between farms. Furthermore, the use of highly aversive handling techniques (e.g. forceful slaps, hits, and tail twists) could account for almost 16% of the differences between farms in annual milk yield.

The cows' fearfulness towards people was measured by observing the time that the cows spent close to a person. Again, large differences between farms were noted in the cows' degree of fearfulness: on the farm with the least fearful cows, the cows spent six times as long close to the person compared to the farm with the most fearful cows. Milk yield was significantly lower as the fearfulness of the cow increased, especially during milking. A multiple regression analysis suggested that differences between farms in the degree of fearfulness of the cows could account for 30% of the differences between farms in annual milk production: a sizeable figure! A subsequent report from the same research group (Hemsworth et al., 2000) found that farms with more fearful cows had a significantly smaller proportion of cows conceiving at the first insemination. Differences between farms in the level of fearfulness accounted for 14% of the variance between farms in conception rates. Many of these results were replicated by Waiblinger et al. (2002), who observed the behavior of cows and stockpeople during milking on 30 dairy farms in Austria, showing that the effects were not specific to the extensive dairy production system of Australia. Dairy cows' fear of people may also have effects on their health: Fulwider et al. (2008) found that dairies with cows that were most likely to approach or touch people had lower somatic cell counts.

Rough handling on the farm increases transport stress

Undesirable effects of animals' fear of people can be particularly important during pre-slaughter handling. A history of aversive interactions between stockpeople and veal calves has been found to affect ease of moving and transporting the animals and meat quality after slaughter. Lensink et al. (2001b) compared veal calves from farms where the farmer acted predominantly gently towards the animals with calves from farms where the farmer acted more roughly. Calves from farms with the rough farmers needed more effort to be loaded on to the trucks, had higher heart rates during loading and unloading (signs of stress), showed more fearful behavior and had more traumatic accidents (e.g. falling down or hitting structures) at lairage, and had poorer meat quality. A similar relationship between farmers' behavior and the meat quality of older beef bulls (Mounier et al., 2008) and of pigs (Hemsworth et al., 2002a) has also been reported.

Together, these studies present convincing evidence, for the major farm animal species, of a strong relationship between the methods used to handle the animals, the degree of fear shown by the animals towards people, and the level of productivity of the farm.

Evidence-based Research on the Effects of Aversive Handling on Production

The studies discussed above show that low levels of productivity (e.g. poor egg production, low growth rates) tend to occur on farms where the animals are handled roughly and are frightened of people. However, they do not show that rough handling is actually the cause of the fear or the reduced productivity. To establish this, researchers have experimentally altered the way that farm animals are handled and have looked at the changes in productivity that have occurred. In one of the first such studies, Hemsworth et al. (1981) subjected young female pigs (gilts) to either a pleasant (gentle stroking) or an unpleasant (briefly shocked with an electric prod) handling treatment for a relatively short period, namely for 2 min duration three times per week from 11 to 22 weeks of age. Gilts in the unpleasant handling treatment spent less time close to the handler, showing that they were more fearful. In addition, they had a lower growth rate. A second study (Hemsworth et al., 1986) showed that gilts in the unpleasant treatment had a lower pregnancy rate at the second estrus when mated than gilts in the pleasant treatment (33.3 and 87.5%, respectively), and that boars handled in the same unpleasant way had smaller testicles at 23 weeks of age and attained a coordinated mating response at a later age than boars in the pleasant treatment (192 and 161 days, respectively). The effect of aversive handling on the growth rates of pigs was confirmed by Gonyou *et al.* (1986), who noted that aversively handled pigs (i.e. those given electric shocks if they did not avoid the handler) grew during the first 3 weeks of life at only 85% of the rate of pigs that had little contact with people.

These studies show convincingly that aversive or rough handling of pigs can have effects on growth rates, reproduction, and sexual development, and that the magnitude of the effects can be substantial. They support the suggestion that the relationship between animals' fear of people and low productivity, which was noted in the previous section, may be due to the way that the animals are handled by the stockpeople (see Chapter 5, this volume, for good handling methods).

Poor stockmanship increases an animal's fear of people

How could poor stockmanship lead to such large reductions in the productivity of farm animals? To explain these effects, Hemsworth and Coleman (2010) proposed a simple model of the relationship between the stockperson's beliefs and attitudes, their behavior when handling animals, and the impact this has on the animals, which is illustrated in Fig. 7.3. The stockperson's specific beliefs about animals have a direct influence on how the animals are handled. For example, a belief that pigs are not sensitive and are difficult to move can lead the stockperson to use rough or aversive handling. The importance of the stockperson's beliefs about

animals will be discussed later. As a result of rough handling, the animals learn to associate the aversive handling with the stockperson and become frightened of him or her. This, in turn, results in the physiological changes that typically occur when an animal is stressed, which have deleterious effects on the animal's welfare and productivity. This model is a useful way of conceptualizing the sequence of events leading from the stockperson's beliefs and attitudes.

Direct evidence that aversive handling causes animals to become more fearful of people, and that this is responsible for the effects on production, comes from experimental studies on dairy cows (Munksgaard *et al.*, 1997; Rushen *et al.*, 1999a). These experiments examined whether the same cows could be made fearful of one person but not of another. In these studies, dairy cows were handled repeatedly by two people, one of whom always handled the cows gently (talking softly, patting and stroking the cow, and occasionally giving feed rewards), while the other handled the cows aversively (hitting, shouting, and occasional use of an electric cattle prod). The cow's fear of each person was tested by each person standing in front of the cow's stall and measuring how closely the cow approached. The results showed that after repeated handling the cows stood further from the aversive handler than from the gentle handler (Munksgaard *et al.*, 1997). Rushen *et al.* (1999a) also tested whether the degree of fearfulness elicited by the handling was sufficient to reduce milk yield, by milking cows in the presence of the handlers.

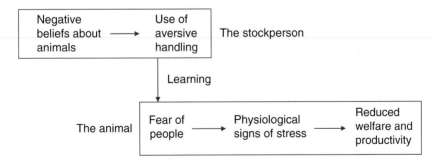

Fig. 7.3. A simple model describing the causal link between a stockperson's beliefs about animals and the effect on their welfare and productivity. According to this model, a stockperson's beliefs about animals (e.g. how easy they are to move) will influence how he or she handles the animal. Stockpeople who believe that animals are stupid, for example, are probably more likely to use rougher handling. The animal will learn to associate this rough handling with the person and thus become frightened of the person (or people in general). This fear of people will produce physiological changes associated with stress, which will reduce the animal's productivity and welfare. (From Hemsworth and Coleman, 2010.)

The gentle handler stood close to the cows for one milking and the aversive handler stood close to the cows for the other milking. It is important to note that the handler did not touch or interact with the cow during the milking. Just the presence of the aversive handler during milking was sufficient to increase residual milk by 70% (which is a sign of a stress-induced block of milk ejection) and tended to reduce milk yield. There were also some physiological signs of fear: the presence of the aversive handler increased heart rate during milking (Table 7.1). These small-scale studies provide support for the model presented in Fig. 7.3 by showing that aversive handling will make dairy cows frightened of people, and that this fear may at times be sufficient to reduce milk yield.

Rough treatment by one person may cause generalized fear of all people

An obvious question is whether an animal that is handled roughly by one person will become fearful of all people, or whether they will become fearful only of that particular individual. It seems that both results are possible, depending on how well the animals can recognize individual people. A number of experiments showed that farm animals responded the same way to different people. For example, Hemsworth et al. (1994b) found that pigs did not differentiate between different people, and that they did not behave differently with familiar and unfamiliar people. An experiment with sheep showed that rough handling by one person might make the animals frightened of all people (Destrez et al., 2013).

However, in some situations, farm animals can learn to recognize individual people and become frightened of particular people (reviewed in Rushen et al., 2001). The studies mentioned above with dairy cows provide an example. Other examples of animals recognizing individual people come from

pigs (Tanida and Nagano, 1998; Koba and Tanida, 2001) and sheep (Boivin et al., 1997; Destrez et al., 2013). Sheep were able to recognize the people who had treated them gently. What cues might the animals be using to recognize individual people? In pigs, sheep, and cattle, visual cues, especially those associated with the clothing worn, seem important in recognizing people (Rushen et al., 2001). Rybarczyk et al. (2003) showed that even very young dairy calves could distinguish between two people wearing different colors of clothing, learning to choose to approach the person who gave them milk. However, Taylor and Davis (1998) showed that cattle, in fact, could learn to distinguish people who wore the same color clothes. Rybarczyk et al. (2001) presented evidence that at least some adult cattle could recognize people by their faces (Fig. 7.4).

Clearly, farm animals show a good ability to recognize individual people using quite subtle recognition cues. If different people treat the animals in different ways, then the animals will respond differently to each person.

Identifying Which Types of Handling are Aversive or Positive

Because it is clear that the type of handling used by the stockperson has major effects on the fearfulness of the animals, a necessary first step in improving the relationship between animals and the stockperson is to identify the particular behaviors that the animals find aversive or rewarding (Quick Access Information 7.1).

Researchers have tried to discover exactly which handling practices cattle find aversive and which are most likely to lead them to become frightened of people. Pajor et al. (2000) compared treatments that are often used when moving cows, such as hitting with the hand, shouting, tail twisting (but not strongly enough to break the tail!), and use of

Table 7.1. Effects of the presence of different types of handler on milk yield and behavior during milking. (From Rushen et al., 1999a.)

	Aversive handler (hitting and shouting)	Gentle handler (talks softly, stroking cows)
Milk yield (kg)	18.48	19.2
Residual milk (kg)[a]	3.6	2.1
Kicks by cow/min during udder preparation	0	0.93
Heart rate change during milking (bpm)[b]	5.94	3.42

Notes: [a]Residual milk is the amount of milk a cow holds back; [b]an increase in heart rate is a sign of stress; bpm = beats per minute.

J. Rushen and A.M. de Passillé

Fig. 7.4. Research has now shown that most species of farm animals can recognize individual people, using predominantly visual cues. Some farm animals can use quite subtle cues to recognize people. Adult cows can learn quickly to approach one person, rather than another, if they receive a food reward only when they make the correct choice. Once the cows have learned to recognize the people, the cues that they use can be examined. For example, Rybarczyk et al. (2001) showed that when the people's faces were covered up and they were of the same height, the cows could no longer recognize them. This shows that some cows can recognize people by their faces.

Quick Access Information 7.1. Stockperson behavior towards livestock and poultry.

Aversive (bad) behaviors that may lower both productivity and welfare	Positive behaviors that may improve both productivity and welfare
Shouting	Calm, soft talking
Hitting	Stroking the animals
Shocking with an electric goad	Calm movement
Sudden movements made by stockpeople	
Kicking animals	

---

an electric prod. Cattle were placed in a runway and at the end of this they were restrained and handled. The experimenters measured the speed that the animals moved down the runway, as well as the effort required by the handler to move them. Based on these measures, all treatments appeared to be aversive to some extent. However, hitting with the bare hand and twisting the tail gently did not differ significantly from no handling, suggesting that the cattle perceived these treatments as relatively mild, although this conclusion probably depends on the force used. Unsurprisingly, the use of the cattle prod was aversive, but interestingly, shouting appeared to be as aversive as the cattle prod. A subsequent study (Pajor et al., 2003), which

allowed cattle to choose between pairs of treatments, confirmed these results: cattle showed no preference for shouting over the electric prod, and no preference for no treatment over gentle tail twisting. To teach the cow to move, the tail must be released the instant she moves forward. Such methods will help us get a better understanding of which handling techniques are the most aversive to animals and will make them most afraid of people.

There is interest in identifying positive behaviors that might improve human–animal relationships. For example, Schmied et al. (2008) found that stroking the neck of cows reduced the degree that the cows avoided people and resulted in the cows being more likely to approach people.

Improving Stockpeople's Attitudes and Beliefs

Stockpeople who had positive attitudes about the humane treatment of animals had significantly more piglets per litter in first parity gilts (Kauppinen *et al.*, 2012).

Grandin (2003) has noted that changing animal handling practices on the farm can be surprisingly difficult. She has noted that people are often more willing to purchase new, expensive equipment than they are to make low-cost changes in their handling methods. This may reflect the fact that the way stockpeople handle animals is a result of long-held, deep-seated beliefs about how animals need to be handled, and personality attributes in general (Hemsworth, 2003). A survey of 500 Dutch dairy farmers indicated that 25% did not believe that cows felt pain (Bruijnis *et al.*, 2013). Simply making recommendations about better ways to handle animals may not be sufficient to overcome the influence of these beliefs. A considerable amount of research has now shown that the way that stockpeople handle animals is a reflection of the specific beliefs of the stockperson (Fig. 7.3), and that altering these beliefs may be an effective means of improving the way animals are handled (Hemsworth, 2003; Hemsworth and Coleman, 2010).

Coleman *et al.* (2003) examined the attitudes towards pigs of handlers at an abattoir and observed how these handlers moved the pigs. Handlers that believed that pigs were greedy, aggressive, and gluttons were more likely to use an electric prod when moving the animals. The use of the prod was also associated with the belief that the way the pigs were handled had no effect on their behavior.

Can such beliefs be changed? Hemsworth *et al.* (2002b) examined the effect of a "cognitive-behavioral intervention" on dairy farmers' attitudes towards cows. The intervention involved multimedia presentations emphasizing the research results, which showed the negative effects of poor handling on the fearfulness and productivity of cattle, along with some clear examples of good and poor handling techniques. The intervention clearly improved the attitude of the farmers towards dairy cattle, specifically reducing the belief that considerable force was needed to move dairy cows. Visits to the farms showed that these changes in beliefs resulted in a reduced use of aversive handling techniques, reduced fearfulness, and tended to improve milk yield (Table 7.2). This study clearly shows the potential for such interventions to improve at least one component of stockmanship, and to improve both the welfare and the productivity of the cattle.

A similar study on pig farms found similar improvements in the way pigs were handled when stockmen followed a training course that challenged their attitudes and behavior towards pigs (Coleman *et al.*, 2000). Interestingly, an unsuspected additional positive effect was found: after the completion of the study, the retention rate for employees who had participated in the training program was higher (61%) compared to those who had not participated (47%). The results suggest that stockpeople who have been trained to handle animals better may be more likely to remain in the job, perhaps because of increased job satisfaction. The results of Maller *et al.* (2005), who found that dairy stockpeople who had positive beliefs about the ease of moving cows also tended to be more

Table 7.2. Effect of handling on cow behavior and milk yield. (From Table 2 and Table 3 in Hemsworth *et al.*, 2002b.)[a]

	Control	Intervention to change stockpeople's attitudes and beliefs about cows
Stockperson's behavior		
Frequency of mildly aversive handling techniques/ cow/milking	0.43	0.24
Frequency of strongly aversive handling techniques/ cow/milking	0.02	0.005
Frequency of gentle handling techniques/cow/milking	0.045	0.11
Cows' responses		
Flight distance (m)	4.49	4.16
Flinch, step, kick responses during milking	0.1	0.13
Milk yield (l/cow/month)	509	529

Note: [a]The tests were carried out twice on 99 different farms.

J. Rushen and A.M. de Passillé

positive about working in a dairy, reinforce the link between positive attitudes towards animals and the quality of life of stockpeople. Improving stockmanship in this way can thus improve the welfare of both animals and farmers.

Effects of the Stockperson's Personality

As well as depending on specific attitudes and beliefs about the importance of routine care, the quality of stockmanship appears to be related to general personality attributes. Seabrook (1984) used a questionnaire to examine correlations between aspects of the personality of stockpeople on a dairy farm and the level of milk production on the farms where they were working. Stockpeople who worked on dairy farms with a high level of milk production were found to be: "not easy going, considerate, not meek, patient, unsociable, not modest, independently minded, persevering, not talkative, confident, uncooperative, and suspicious of change"! These results suggest that the personality of the stockperson can have an effect on milk production. More recently, Seabrook (2005) has investigated the important personality characteristics of good pig stockpeople as rated by other stockpeople.

Most considered that a good stockperson should be "fussy", "confident", and "inward-looking" and not "harsh", "carefree", "unsympathetic" or "aggressive". Coleman *et al.* (2003) found that handlers at an abattoir who were rated as "tough-minded" were more likely to use an electric prod when moving pigs. Such research may have practical consequences in improving employee selection.

However, we should not focus exclusively on the stockperson's beliefs about the animals. As mentioned in the section, 'Effect of Good Stockmanship on Animal Welfare and Productivity', stockpeople's attitudes to routine tasks, such as cleanliness, can also have a major impact on animal welfare independently of how the animals are handled. Figure 7.5 shows a more complete model of the relationship between attitudes, beliefs, and stockmanship.

Identifying Why Stockpeople Mishandle Animals

To improve the ways animals are handled, it helps to know what circumstances lead to animals being handled roughly. Seabrook (2000) reported on a study he undertook to try to identify why dairy farmers were aggressive towards animals. Stockpeople

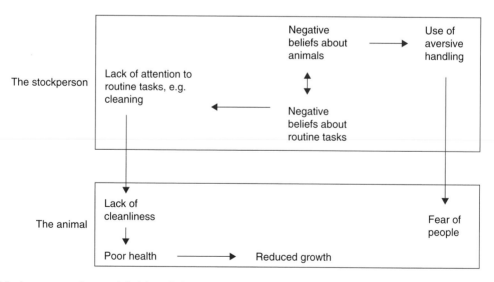

Fig. 7.5. A more complete model of the relationship between the stockperson's attitudes and beliefs and the impact on farm animals. This model is based on studies by Lensink *et al.* (2002) on veal calves. In this case, the largest effect on the calf's growth occurs through a lack of attention to cleaning routines rather than through the animal becoming frightened of people. The calves' fear of people is a correlated response that could arise because stockpeople who have negative beliefs about the importance of routine care may also have negative beliefs about the animals, and also use aversive handling techniques (adapted from Rushen *et al.*, 2008).

were asked to reflect on the times they had acted aversively towards the animals and to try to explain why they had acted that way. A number of reasons repeatedly came up. These included time pressures and the complexity of the work, frustration with the animals being "stubborn" (e.g. kicking off the milk clusters) or with equipment not working, and family and home problems. The study further highlighted the importance of general job satisfaction for good stockmanship. Managers must be careful not to understaff farms and overwork stockpeople. It is difficult for a stockperson to have a positive attitude if they are worked until they are exhausted. Unpublished industry data show that death losses increase after 6 h of work loading trucks with pigs or poultry. The people become too tired and fatigued to care.

Difficulties in moving animals can be a cause of frustration for stockpeople, and this can be responsible for much of the rough handling that occurs. The physical design of the work environment plays an important role. Maller *et al.* (2005) noted that several aspects of the design of milking sheds (e.g. number of turns the animals had to make to exit, stall gates, etc.) were reported to be associated with difficulties in cow movement, which in turn might have led to a greater need for the dairymen to intervene (often roughly) to help move cows. Similarly, Rushen and de Passillé (2006) reported that slippery flooring, as well as slowing cows' walking speed and increasing the risk of slipping, resulted in more need for handlers to try to encourage the cows to move forward, which increased the risk of rough handling. Difficulties with moving animals can be a particular problem in slaughterhouses, and this can be a major cause of improper stunning. Grandin (2006) has outlined some of the improvements in installations in slaughterhouses than can improve animal movement. These include installing nonslip flooring and eliminating distractions to animals that may cause them to balk. Good handling facilities are also particularly important for handling extensively managed animals such as beef cattle, which are more likely to be fearful of people. Grandin (1997) has produced useful guidelines for such facilities (see Chapter 5, this volume). Clearly, it is essential to have well-designed areas for handling and moving animals in order to reduce both the frustration of the workers and the tendency for such workers to take out their frustrations on the animals. Finding better ways to move animals is likely to lead to significant reductions in the use of aversive handling. This could be achieved either through improved raceway design or by finding other means of improving animal movement; Ceballos and Weary (2002) found that providing small food rewards when dairy cows entered the milking parlor reduced the time the cows took to enter the parlor, and reduced the need for the stockperson to push the cows or use other aversive handling techniques.

Stockperson Job Satisfaction

From other professions, it is recognized that the diligence with which a job is done depends very much on the level of job satisfaction; low job satisfaction often results in careless work. Recognizing the importance of this for stockmanship, Seabrook and Wilkinson (2000) interviewed dairy stockpeople in the UK to determine what factors affected their level of job satisfaction, focusing particularly on the routine tasks stockpeople enjoyed and those they found most unpleasant. Importantly, the stockpeople clearly valued and enjoyed their interactions with their animals, and the nature of their interactions with animals was largely responsible for the difference between a "good day" and a "bad day". Milking was widely rated the most important routine task in dairying, and was rated the most enjoyable. This may, however, not be true for all countries: Maller *et al.* (2005) reported that only about one-quarter of Australian dairy farmers preferred milking to other farm jobs. Possibly, this difference may relate to the differences in the design of milking sheds in the two countries.

Unfortunately, Seabrook and Wilkinson (2000) found that maintaining herd health and welfare was considered of relatively low importance, especially by younger stockpeople. Stockpeople from higher-producing herds, however, rated herd health and welfare more important than those from lower-producing herds. Interestingly, the rated importance of herd health and welfare increased when stockpeople were re-interviewed after quality assurance schemes had been introduced. The authors felt that these may have led the stockpeople to alter their priorities. Despite their importance for maintaining good animal health and welfare, cleaning buildings/parlors and foot trimming were rated the most unpleasant routine tasks. Foot trimming (Fig. 4.6) was disliked mainly because of inadequate equipment and holding facilities, and because it was generally considered a dangerous task. The authors

J. Rushen and A.M. de Passillé

concluded that the obvious dislike of foot trimming and cleaning routines showed the need to find improved means of cleaning cows and barns, for better designed parlors that facilitated cleaning, and for improved equipment and facilities for foot trimming. This type of research is clearly useful in showing what types of improvements are needed in order to improve job satisfaction, and hence job quality. Another issue that affects job satisfaction is pay. Being a good stockperson is a highly skilled job, and unfortunately, the salaries of stockpeople are often very low. Stockmanship needs to receive better recognition for the skills required.

Assessing Stockmanship on Farm Audits

Hemsworth (2007) has noted the need to establish appropriate standards for stockmanship, given the increasing importance of animal welfare for the public image of agriculture. Given the importance of stockmanship to the welfare of farm animals, there has been a lot of interest in including some measure of stockmanship in on-farm animal welfare assessment or audits. Usually, this involves taking some measure of how fearful the animals appear in response to people (Welfare Quality, 2009). For example, Rousing and Waiblinger (2004) visited commercial dairy farms and conducted two tests: a voluntary animal-approach test based on cows' latencies to approach and touch a stationary person; and an avoidance test, based on cows' avoidance of an approaching person. The inter-observer repeatability of both tests was found to be high, and the tests were highly correlated. The avoidance test was less affected by the familiarity of the people, and the authors concluded that the avoidance test was valid and useful in on-farm assessment of the human–animal relationship. Other studies have shown that the distance cattle keep from people can be measured reliably on the farm (Windschnurer et al., 2008), and similar tests are being developed for poultry (e.g. Graml et al., 2007). However, we have been critical of the use of such tests to assess stockmanship in on-farm audits (de Passillé and Rushen, 2005), not because the tests are unreliable, but because they have not been validated adequately as measures of stockmanship on farms, which is a more serious problem. First, the extent to which animals approach or avoid people can be affected by many factors other than the use of aversive handling (Boissy and Bouissou, 1988). For example, feeding animals by hand is one certain way of getting them to approach people. Animals that are not fed by hand may not approach people as readily, but that does not mean that their welfare is necessarily poorer. Second, even if there are differences between farms in the extent that animals approach people, it is difficult to establish a clear cut-off point where we can say that the handling of the animals has been unacceptably rough. This is partly because many differences between farms will affect the precise distance that animals approach people, so it will be difficult to make precise comparisons between different farms. Third, and most important, as we have discussed in this chapter, there is a lot more to good stockmanship than the way that the animals are handled (Fig. 7.5). The degree to which stockpeople perform important routine tasks, or their attitudes to animals and to such tasks, may be more important to measure in on-farm audits. In other

Fig. 7.6. A poor facility for hoof trimming of dairy cows. Performance of routine tasks, such as routine hoof trimming of dairy cows, is an essential aspect of good stockmanship. The degree of diligence and care with which such tasks are performed can have a major effect on the welfare of the animal. For example, inadequate hoof trimming of dairy cows is known to increase the chance of cows becoming lame. However, many dairy farmers report that hoof trimming is one of the least enjoyable tasks on a dairy farm (Seabrook and Wilkinson, 2000). As a consequence, there is a danger that it will not be done as often as necessary. This situation may be improved by ensuring that farmers have the right equipment and training to do such tasks. There are many well-designed units for holding the cow that are commercially available. Replacing the dangerous set-up with good equipment will make hoof trimming easier and safer.

situations besides on the farm, other aspects of animal handling may be more important (and more easily measured) in an animal welfare audit than the degree of fear that the animals show towards people. For example, noting the use of electric prods to move animals and incorrect stunning procedures should be an essential part of animal welfare audits of slaughterhouses (Grandin, 1998). Thus, while assessment of the quality of stockmanship is likely to be an important component of on-farm animal welfare audits, measures of the reaction of animals to people may not be the most appropriate measure as a welfare measure *between* different farms, but it would be appropriate to use it to assess improvements in stockmanshsip *within* the same farm.

Conclusion

From the research reviewed in this chapter, it is clear that the people who care for the animals can have a major, and possibly a decisive, role to play in affecting their welfare and productivity. Differences between stockpeople may be responsible for many of the differences between otherwise similar farms in the level of animal welfare and productivity. Research provides concrete examples of how poor stockmanship can lead to poor animal welfare, and to show some of the ways that stockmanship can be improved. Notably, these improvements, as well as improving animal welfare, often lead to substantial benefits to the farmers themselves, either through increased health and productivity of the animals, through increased efficiency and safety of operations that involve handling the animals, or through increased job satisfaction, and probably self-esteem. Research to date has focused on the more obvious aspects of stockmanship: how the animals are handled and how they become fearful of people. However, stockmanship involves much more, and the relationships that can develop between people and animals can be quite subtle. Research needs to consider a broader range of qualities associated with stockmanship. Seabrook concludes:

The welfare of farm animals is crucially dependent upon the actions of the stockpersons who day in day out handle, observe and monitor the animals in their charge. The animals' well-being and performance depend as much o[n] the stockperson as the building and the space allocation. Aversive action can arise through ignorance, but usually staff know what to do to handle animals correctly. They know what to do

for good welfare but sometimes they do not act correctly and in their words "let ourselves and our animals down".

(Seabrook, 2000, p. 30)

Acknowledgements

We thank all of our colleagues and collaborators for useful discussions on this topic, in particular: Paul Hemsworth, Lene Munksgaard, and Hajime Tanida. Our research in this area was supported by grants from the Natural Sciences and Engineering Research Council of Canada, Dairy Farmers of Canada, Novalait Inc., and Agriculture and Agri-Food Canada.

References

Boissy, A. and Bouissou, M.F. (1988) Effects of early handling on heifers' subsequent reactivity to humans and to unfamiliar situations. *Applied Animal Behaviour Science* 20, 259–273.

Boivin, X., Nowak, R., Desprès, G., Tournadre, H. and Le Neindre, P. (1997) Discrimination between shepherds by lambs reared under artificial conditions. *Journal of Animal Science* 75, 2892–2898.

Breuer, K., Hemsworth, P.H., Barnett, J.L., Matthews, L.R. and Coleman, G.J. (2000) Behavioural response to humans and the productivity of commercial dairy cows. *Applied Animal Behaviour Science* 66, 273–288.

Bruijnis, M., Hogeveen, H., Garforth, C. and Stassen, E. (2013) Dairy farmer's attitudes and intentions towards improving dairy cattle foot health. *Livestock Science* 155, 103–113.

Ceballos, A. and Weary, D.M. (2002) Feeding small quantities of grain in the parlour facilitates pre-milking handling of dairy cows: a note. *Applied Animal Behaviour Science* 77, 249–254.

Coleman, G.J., Hemsworth, P.H., Hay, M. and Cox, M. (2000) Modifying stockperson attitudes and behaviour towards pigs at a large commercial farm. *Applied Animal Behaviour Science* 66, 11–20.

Coleman, G.J., McGregor, M., Hemsworth, P.H., Boyce, J. and Dowling, S. (2003) The relationship between beliefs, attitudes and observed behaviours of abattoir personnel in the pig industry. *Applied Animal Behaviour Science* 82, 189–200.

Cransberg, P.H., Hemsworth, P.H. and Coleman, G.J. (2000) Human factors affecting the behaviour and productivity of commercial broiler chickens. *British Poultry Science* 41, 272–279.

de Passillé, A.M. and Rushen, J. (2005) Can we measure human–animal interactions in on-farm animal welfare assessment? Some unresolved issues. *Applied Animal Behaviour Science* 92, 193–209.

J. Rushen and A.M. de Passillé

Destrez, A., Coulon, M., Deiss, V., Delval, E., Boissy, A., et al. (2013) The valence of the long lasting emotional experiences with various handlers modulates discrimination and generalization of individual humans in sheep. *Journal of Animal Science* 91, 5418–5426.

Fulwider, W.K., Grandin, T., Rollin, B.E., Engle, T.E., Dalsted, N.L., et al. (2008) Survey of dairy management practices on one hundred thirteen north central and northeastern United States dairies. *Journal of Dairy Science* 91, 1686–1692.

Gonyou, H.W., Hemsworth, P.H. and Barnett, J.L. (1986) Effects of frequent interactions with humans on growing pigs. *Applied Animal Behaviour Science* 16, 269–278.

Graml, C., Niebuhr, K. and Waiblinger, S. (2007) Reaction of laying hens to humans in the home or novel environment. *Applied Animal Behaviour Science* 113, 98–109.

Grandin, T. (1997) The design and construction of facilities for handling cattle. *Livestock Production Science* 49, 103–119.

Grandin, T. (1998) Objective scoring of animal handling and stunning practices at slaughter plants. *Journal of the American Veterinary Medical Association* 212, 36–39.

Grandin, T. (2003) Transferring results of behavioral research to industry to improve animal welfare on the farm, ranch and the slaughter plant. *Applied Animal Behaviour Science* 81, 215–228.

Grandin, T. (2006) Progress and challenges in animal handling and slaughter in the US. *Applied Animal Behaviour Science* 100, 129–139.

Hemsworth, P.H. (2003) Human-animal interactions in livestock production. *Applied Animal Behaviour Science* 81, 185–198.

Hemsworth, P.H. (2007) Ethical stockmanship. *Australian Veterinary Journal* 85, 194–200.

Hemsworth, P.H. and Coleman, G.J. (2010) *Human–Livestock Interactions*, 2nd edn. CAB International, Wallingford, UK.

Hemsworth, P.H., Brand, A. and Willems, P. (1981) The behavioural response of sows to the presence of human beings and its relation to productivity. *Livestock Production Science* 8, 67–74.

Hemsworth, P.H., Barnett, J.L. and Hansen, C. (1986) The influence of handling by humans on the behaviour, reproduction and corticosteroids of male and female pigs. *Applied Animal Behaviour Science* 15, 303–314.

Hemsworth, P.H., Coleman, G.J., Barnett, J.L. and Jones, R.B. (1994a) Behavioural responses to humans and the productivity of commercial broiler chickens. *Applied Animal Behaviour Science* 41, 101–114.

Hemsworth, P.H., Coleman, G.J., Cox, M. and Barnett, J.L. (1994b) Stimulus generalization: The inability of pigs to discriminate between humans on the basis of their previous handling experience. *Applied Animal Behaviour Science* 40, 129–142.

Hemsworth, P.H., Pedersen, V., Cox, M., Cronin, G.M. and Coleman, G.J. (1999) A note on the relationship between the behavioural response of lactating sows to humans and the survival of their piglets. *Applied Animal Behaviour Science* 65, 43–52.

Hemsworth, P.H., Coleman, G.J., Barnett, J.L. and Borg, S. (2000) Relationships between human–animal interactions and productivity of commercial dairy cows. *Journal of Animal Science* 78, 2821–2831.

Hemsworth, P.H., Barnett, J.L., Hofmeyr, C., Coleman, G.J., Dowling, S., et al. (2002a) The effects of fear of humans and pre-slaughter handling on the meat quality of pigs. *Australian Journal of Agricultural Research* 53, 493–501.

Hemsworth, P.H., Coleman, G.J., Barnett, J.L., Borg, S. and Dowling, S. (2002b) The effects of cognitive behavioral intervention on the attitude and behavior of stockpersons and the behavior and productivity of commercial dairy cows. *Journal of Animal Science* 80, 68–78.

Kauppinen, T., Vesala, K.M. and Valros, A. (2012) Farmer attitudes towards improvement of animal welfare is correlated with piglet production. *Livestock Science* 143, 142–150.

Koba, Y. and Tanida, H. (2001) How do miniature pigs discriminate between people? Discrimination between people wearing coveralls of the same colour. *Applied Animal Behaviour Science* 73, 45–58.

Lensink, B.J., Veissier, I. and Florand, L. (2001a) The farmers' influence on calves' behaviour, health and production of a veal unit. *Animal Science* 72, 105–116.

Lensink, B.J., Fernandez, X., Cozzi, G., Florand, L. and Veissier, I. (2001b) The influence of farmers' behavior on calves' reactions to transport and quality of veal meat. *Journal of Animal Science* 79, 642–652.

Lensink, B.J., Boivin, X., Pradel, P., LeNeindre, P. and Vessier, L. (2002) Reducing veal calves reactivity to people by providing additional human contact. *Journal of Animal Science* 78, 1213–1218.

Losinger, W.C. and Heinrichs, A.J. (1997) Management practices associated with high mortality among preweaned dairy heifers. *Journal of Dairy Research* 64, 1–11.

Macedo, G.G., Zuccari, C.E., de Abreu, U.G., Negrao, J.A. and da Costa e Silva, E.V. (2011) Human animal interaction, stress, and embryo production in *Bos indicus* donors under tropical conditions. *Tropical Animal Health* 43, 1175–1182.

Maller, C.J., Hemsworth, P.H., Ng, K.T., Jongman, E.J., Coleman, G.J., et al. (2005) The relationships between characteristics of milking sheds and the attitudes to dairy cows, working conditions, and quality of life of dairy farmers. *Australian Journal of Agricultural Research* 56, 363–372.

Mounier, L., Colson, S., Roux, M., Dubroeucq, H., Boissy, A., et al. (2008) Positive attitudes of farmers and pen-group conservation reduce adverse reactions of bulls during transfer for slaughter. *Animal* 2, 894–901.

Munksgaard, L., de Passillé, A.M.B., Rushen, J., Thodberg, K. and Jensen, M.B. (1997) Discrimination of people by dairy cows based on handling. *Journal of Dairy Science* 80, 1106–1112.

Pajor, E.A., Rushen, J. and de Passillé, A.M.B. (2000) Aversion learning techniques to evaluate dairy cattle handling practices. *Applied Animal Behaviour Science* 69, 89–102.

Pajor, E.A., Rushen, J. and de Passillé, A.M. (2003) Dairy cattle's choice of handling treatments in a Y-maze. *Applied Animal Behaviour Science* 80, 93–107.

Rousing, T. and Waiblinger, S. (2004) Evaluation of on-farm methods for testing the human–animal relationship in dairy herds with cubicle loose housing systems – test-retest and inter-observer reliability and consistency to familiarity of test person. *Applied Animal Behaviour Science* 85, 215–231.

Rushen, J. and de Passillé, A.M. (2006) Effects of roughness and compressibility of flooring on cow locomotion. *Journal of Dairy Science* 89, 2965–2972.

Rushen, J., de Passillé, A.M.B. and Munksgaard, L. (1999a) Fear of people by cows and effects on milk yield, behavior and heart rate at milking. *Journal of Dairy Science* 82, 720–727.

Rushen, J., Taylor, A.A. and de Passillé, A.M.B. (1999b) Domestic animals' fear of humans and its effect on their welfare. *Applied Animal Behaviour Science* 65, 285–303.

Rushen, J., de Passillé, A.M.B., Munksgaard, L. and Tanida, H. (2001) People as social actors in the world of farm animals. In: Gonyou, H. and Keeling, L. (eds) *Social Behaviour of Farm Animals*. CAB International, Wallingford, UK.

Rushen, J., de Passillé, A.M., von Keyserlingk, M. and Weary, D.M. (2008) *The Welfare of Cattle*. Springer, Dordrecht, the Netherlands, 303 pp.

Rybarczyk, P., Koba, Y., Rushen, J., Tanida, H. and de Passillé, A.M. (2001) Can cows discriminate people by their faces? *Applied Animal Behaviour Science* 74, 175–189.

Rybarczyk, P., Rushen, J. and de Passillé, A.M. (2003) Recognition of people by dairy calves using colour of clothing. *Applied Animal Behaviour Science* 81, 307–319.

Schmied, C., Boivin, X. and Waiblinger, S. (2008) Stroking different body regions of dairy cows: Effects on avoidance and approach behavior toward humans. *Journal of Dairy Science* 91, 596–605.

Seabrook, M.F. (1984) The psychological interaction between the stockman and his animals and its influence on performance of pigs and dairy cows. *Veterinary Record* 115, 84–87.

Seabrook, M.F. (2000) The effect of the operational environment and operating protocols on the attitudes and behaviour of employed stockpersons. In: Hovi, M. and Bouilhol, M. (eds) *Human–Animal Relationship: Stockmanship and Housing in Organic Livestock Systems*. Proceedings of the Third Network for Animal Health and Welfare in Organic Agriculture (NAHWOA) Workshop, Clermont-Ferrand, France, 21–24 October 2000. University of Reading, Reading, UK, pp. 21–30. Available at: http://www.veeru.reading.ac.uk/organic/ProceedingsFINAL.pdf (accessed 27 April 2015).

Seabrook, M.F. (2005) Stockpersonship in the 21st century. *Journal of the Royal Agricultural Society of England* 166, 1–12.

Seabrook, M.F. and Wilkinson, J.M. (2000) Stockpersons' attitudes to the husbandry of dairy cows. *Veterinary Record* 147, 157–160.

Sorge, U.S., Cherry, C. and Bender, J.B. (2014) Perception of the importance of human–animal interaction on cattle flow and worker safety on Minnesota dairy farms. *Journal of Dairy Science* 97, 4632–4638.

Tanida, H. and Nagano, Y. (1998) The ability of miniature pigs to discriminate between a stranger and their familiar handler. *Applied Animal Behaviour Science* 56, 149–159.

Taylor, A. and Davis, H. (1998) Individual humans as discriminative stimuli for cattle (*Bos taurus*). *Applied Animal Behaviour Science* 58, 13–21.

Waiblinger, S., Menke, C. and Coleman, G. (2002) The relationship between attitudes, personal characteristics and behaviour of stockpeople and subsequent behaviour and production of dairy cows. *Applied Animal Behaviour Science* 79, 195–219.

Waiblinger, S., Menke, C., Korff, J. and Bucher, A. (2004) Previous handling and gentle interactions affect behaviour and heart rate of dairy cows during a veterinary procedure. *Applied Animal Behaviour Science* 85, 31–42.

Waiblinger, S., Boivin, X., Pedersen, V., Tosi, M.V., Janczak, A.M., et al. (2006) Assessing the human–animal relationship in farmed species: A critical review. *Applied Animal Behaviour Science* 101, 185–242.

Welfare Quality (2009) http://www.welfarequalitynetwork.net/network (accessed 27 April 2015).

Windschnurer, I., Schmied, C., Boivin, X. and Waiblinger, S. (2008) Reliability and inter-test relationship of tests for on-farm assessment of dairy cows' relationship to humans. *Applied Animal Behaviour Science* 114, 37–53.

8 Animal Well-being and Behavioral Needs on the Farm

LILY N. EDWARDS-CALLAWAY

JBS, Greeley, Colorado, USA

Summary

Many scientists agree that health and productivity are not the only factors for determining whether or not an animal has good welfare. There is increasing emphasis on assessing the emotional or affective state of the animal. Jaak Panksepp, a neuroscientist, has identified the core emotional system, which is located in the subcortex of the brain in all mammals. These systems drive behavior. The core emotional systems are: FEAR, which makes animals avoid danger (negative emotion); PANIC, separation distress (negative emotion); SEEK, explore (positive emotion); RAGE (negative emotion); LUST–SEX (positive emotion); CARING, mother young nurturing (positive emotion); and PLAY (positive emotion). Examples of behavior initiated due to negative emotions would be attempts to escape, hiding, kicking, defecation during restraint, vocalization when separated from other animals, large flight zone, and biting. Behaviors initiated by positive emotional systems are exploration, play, normal grooming, dust bathing in chickens, and sex. Problems with abnormal behavior are also discussed.

Learning Objectives

- Understand different definitions of animal welfare
- Understand that animal emotions have a neurological basis
- Emotions in animals are required for survival
- Differentiate positive and negative emotion
- Identify and understand abnormal displacement behaviors and stereotypies

Introduction

Animal welfare is such an important concept in today's society, yet it is a term that is difficult for animal scientists, producers, veterinarians, and consumers to define adequately and decisively and to recognize collectively. Scientists have provided numerous statements defining good animal welfare and identifying significant components of animal well-being (Duncan and Dawkins, 1983; Moberg, 1985; Broom, 1986; Duncan and Petherick, 1991; Mendl, 1991; Mason and Mendl, 1993; Fraser, 1995; Ng, 1995; Sandoe, 1996; Dawkins, 2006). There is general consensus among stakeholders as to what parameters (e.g. health, behavior, productivity, environment) should be considered to assess an animal's welfare; the lack of consensus revolves around the precise measurements of the parameters and how important each one of those factors is in the overall determination of animal welfare status, i.e. what is the level of consideration for each component when making recommendations on animal husbandry practices. This variation in consideration of components has been extremely apparent in discussions of different housing types for both hens and sows – does the ability for an animal to express normal behaviors outweigh the improvements seen in animal health and injury between differing housing types?

Scientists have not been able to create a mathematical formula into which we plug all measures of welfare (biological measures, behavioral measures, and psychological components, etc.) to be given a quantitative, all-encompassing measure on a simple scale of 1–10. Unfortunately, we have not even been able to achieve this daunting feat with humans, perhaps even an easier task than with animals considering our first-hand insight into human nature.

Different Definitions of Welfare

Although each welfare specialist may have their own definition of what "good" welfare means, it is safe to assume that many of the included components are similar. Some specialists maintain that an animal's welfare is based primarily on health (Moberg, 1985). Can the animal reproduce? Is it free from disease? Looking deeper at specific physiological mechanisms, levels of various hormones (e.g. cortisol, lactate, epinephrine, etc.) are measured to assess welfare status (Moberg, 1985). Scientists use their knowledge of stress physiology and immune response to help identify an animal's welfare status. For example, calves respond to dehorning and disbudding with physiological, behavioral, and immunological responses that indicate pain and stress (Petrie et al., 1996; Sylvester et al., 1998; Stafford and Mellor, 2005; Heinrich et al., 2009; Stilwell et al., 2009, 2010), likewise with castration (Coetzee, 2011; see Chapter 6, this volume).

Many welfare specialists extend the concept of animal welfare beyond what may be considered the more objective physiological stress measurements, health status, and productivity; animal welfare must also include emotional (affective) states, feelings, and wants (Duncan and Dawkins, 1983; Dawkins, 1988; Duncan and Petherick, 1991; Sandoe, 1996). Are the cows "happy"? Are all their needs being met? Are they distressed? Some scientists state that the performance of natural behaviors is a necessary component for good welfare. Is the sow in the gestation crate able to perform nesting behavior? If she cannot nest build, how is this affecting her well-being?

The definition of good animal welfare will change depending on whom you ask. The plethora of parameters used to assess animal well-being not only sets the platform for extreme debate, but additionally, it calls attention to the enormous task at hand for all those involved with assessing, determining, ensuring, and simply just understanding animal welfare.

Welfare concepts in different countries

In the past decade, there has been a global initiative to consider and define animal welfare standards.

The World Organisation for Animal Health (OIE) first included animal welfare under their global umbrella of consideration in their 2001–2005 Strategic Plan. As an international reference organization for animal health, it seemed suitable to provide leadership in defining international standards for animal welfare. Within the organization's Terrestrial Animal Health Code, animal welfare:

> means how an animal is coping with the conditions in which it lives. An animal is in a good state of welfare if (as indicated by scientific evidence) it is healthy, comfortable, well nourished, safe, able to express innate behaviour, and if it is not suffering from unpleasant states such as pain, fear and distress.
>
> (OIE, 2012)

The OIE has used this definition of welfare to develop various sets of standards to improve animal welfare worldwide. In 2012, the OIE adopted the "10 General Principles for the Welfare of Animals in Livestock Production Systems" (OIE, 2012), which have been thoroughly reviewed by Fraser et al. (2013). Prior to OIE involvement, individual countries had also been working towards the development and growth of animal welfare culture. In the USA, the Animal Welfare Act (Office of the Federal Register, 1989) is the federal law, first initiated in 1960, that provides minimum protection to certain species of animals, primarily companion animals and those used in research, for handling, care, housing, and treatment. Unfortunately, the law's definition of an animal does not include livestock species raised for food or fiber (it also excludes birds, rats, and mice used for research purposes). Farm animals were exempt from this law due to political pressure from livestock groups, and remain exempt to this day. There is one amendment, officially made in 1985, that requires facilities housing nonhuman primates to provide an "environment adequate to promote the psychological well-being of the animals". Although clearly stated as a necessity, the amendment provides no suggestion as to how to satisfy this requirement. In this law, the mental component of animal well-being applies only to nonhuman primates.

Almost 50 years ago in the UK, the Brambell Commission (Brambell, 1965), a governmental committee, was created to propose the conditions that farm animals should experience to ensure their welfare. Since the revision by the UK Farm Animal Welfare Council (FAWC) in 1979, these animal conditions are now universally referred to as the "five freedoms" and are as follows: (i) freedom from thirst, hunger, and malnutrition; (ii) freedom from discomfort; (iii) freedom from pain, injury, and disease; (iv) freedom to express normal behavior; and (v) freedom from fear and distress. The "five freedoms" are more inclusive of all aspects of animal welfare and have been the basis for welfare assessment in the UK and many other countries.

Within the European community, The Welfare Quality® project, described in Botreau *et al.* (2009), has worked towards integrating animal welfare into the food-quality chain by attempting to create a standardized method to measure and assess animal welfare. The project tackled the very complex nature of animal welfare and established four guiding principles to ensure animal welfare: good housing, good feeding, good health, and appropriate behavior. As seen in many initiatives, there is a broad sentiment to have a unified voice towards accepted animal welfare standards.

Components of comprehensive animal welfare

Dividing the different components of welfare into categories will make it easier to implement effective welfare programs. Several decades after the five freedoms had been composed, Dr David Fraser (2008) identified the three main approaches to animal welfare assessment and/or basis for definition: (i) biological functioning – physical health, growth, reproduction; (ii) affective states – fear, anxiety, frustration, pain, hunger, and thirst; and (iii) natural living – a lifestyle that approximates natural living conditions and the fulfilment of behavioral needs (see Chapter 1, this volume, for more information). At the 2008 American Society for Animal Science meeting, Fraser presented his reorganization of the components of animal welfare into four categories. They are: (i) maintain basic health; (ii) reduce pain and distress; (iii) accommodate natural behaviors and affective states; and (iv) provide natural elements in the environment (see Chapter 1, this volume). These four categories include all concepts of animal welfare, ranging from health and lack of pathological conditions to emotional states and feelings. Animal welfare may be best defined as a combination of all of these categories; however, the exact weighting of each component is unclear (i.e. is there one component that is more important, if two are satisfied can the third be forgotten?, etc.). The Brambell Commission (Brambell, 1965) additionally stated that animal welfare "embraces both the physical and the mental well-being of the animal", similarly drawing on the mental status of the animal, as seen in the Animal Welfare Act amendment.

Animal Abuse Versus Animal Welfare

One thing to note in this discussion is the difference between poor animal welfare and animal abuse. Animal abuse is a deliberate act that inflicts obvious, unnecessary harm and suffering. Some examples are torturing, beating, poking out eyes, and depriving of survival needs such as water. In many countries, animal cruelty laws were created as regulations against these deliberate acts of abuse. In some countries, there are even special humane officers with police powers who enforce these statutes. Although to have good welfare an animal must not experience obvious acts of abuse, the concept of good animal welfare extends much further. Animal welfare also includes good health, pain relief, reduction of fear, and environments that provide for an animal's behavioral needs. As concerns with animal abuse have transformed into concerns for an animal's total welfare, many more areas in which we use animals (research, agriculture, etc.) have come under scrutiny.

The Animal's Mental State is an Important Part of Welfare

As shown, the emphasis on the emotional (affective) component of the animal's life in the consideration of animal well-being and welfare is consistent among stakeholders. Some researchers have clearly emphasized this mental component, indicating that an animal's welfare is solely dependent on the psychological needs of the animal, the physical needs being met once the mental ones are met (Duncan and Petherick, 1991). Although the mental (affective) states are an important component of welfare, basic health and prevention of injury should never be neglected. When considering the mental state of an animal, some difficult questions arise: how do you determine that an animal has good psychological well-being, and more importantly, how do we measure it? Some people have a hard enough time understanding if they themselves are happy in certain environments, let alone another human, and even another animal.

To reach a state of well-being, humans want to minimize the amount of suffering in their lives. Suffering includes a multitude of negative feelings and emotions such as pain, anxiety, frustration, fear, and any other unpleasant sensations associated with physical and/or mental harm to the individual. A human can suffer from chronic, intense pain from physical trauma. A human can also suffer from fear, fear of the economic consequences of losing his or her job. Or from frustration, the frustration of being stuck in a marriage he or she is unhappy with but feels he or she cannot escape from. (As a side note, the American population seems to be experiencing widespread mental suffering when looking at the number of people taking prescription antidepressants and anxiety medications. The National Center for Health Statistics (2011) reported that the use of antidepressants among all ages increased 400% from 1988–1994 to 2005–2008.) There are many ways in which a human can suffer, all of which we would agree have a negative impact on his or her overall psychological well-being. Humans try to avoid these situations, looking for situations that are more pleasing, situations that play to the expression of positive emotions. How does this apply to animals?

Animal emotions: different, not less

Many people feel more comfortable thinking about what goes on in the minds of other scientists and humans than what goes on in the minds of farm animals. Scientists work to discover the unknown by collecting data, numbers and information that are easily measurable (i.e. objective measurements). An animal's feelings about its surroundings are harder to define and do not come with an easily interpretable number. When it becomes necessary to use subjective measurements (i.e. feelings), there is often hesitation to evaluate animals in the same manner (Yoerg, 1991; Ng, 1995).

In the past, it was commonly taught that it was not wise to compare human and animal emotions (i.e. human and animal suffering). This attribution of human characteristics to nonhuman objects (e.g. animals) is called anthropomorphism. The use of anthropomorphism is seen by some as dangerous (although I do think this is a dwindling sentiment among animal scientists), because people make assumptions about things that are too foreign and subjective to ever truly be understood. A contrary view is that anthropomorphism may not be the sin that some scientists used to teach it was (Panksepp, 2005a), and actually it may be helpful for predicting an animal's psychological state. One example of how we have used anthropomorphism beneficially is in the analysis of opioid control of separation distress in animals. The human psychological responses to opioids led to precise predictions of the behavioral responses of animals (i.e. psychological response) to these pharmacological agents (Panksepp, 2005a). Quick Access Information 8.1 outlines the core emotional circuits in all mammals (Panksepp, 2011). Furthermore, while some scientists may be reluctant to attribute emotions to animals, extensive experimental work has been, and is being, conducted in animals to explore the effectiveness of drugs to relieve pain and anxiety (Rollin, 1989). There must be some congruities if we use information from animal pain studies to extrapolate to human effects.

Quick Access Information 8.1. Core emotional circuits that are in the subcortex of all mammals (Panksepp, 1998, 2011). These core emotions have separate brain circuits.

Basic emotion	Purpose	Type of emotion
SEEKING	Motivates animals to explore and approach new things	Positive
RAGE – Anger	Fight a predator	Negative
FEAR – Anxiety	Avoid danger and predators	Negative
LUST – Sex	Reproduction	Positive
CARE – Nurturing	Raise young and social cooperation	Positive
PANIC – Separation distress	Motivates animals to stay with herdmates and keep mother and young together	Negative
PLAY – Joy	Teach social skills	Positive

L.N. Edwards-Callaway

What some people forget is that animal welfare is no trickier an entity to define than human welfare (Dawkins, 1998; Broom, 2001). We make assumptions about the emotions of other humans on a daily basis. Although we think we understand the mental capacities of other humans, there is no way of knowing if we are correct in our assumptions. Different humans interpret and perceive the same situation in different ways. For example, I hate roller coasters and any amusement park ride that goes fast, upside down, or sideways; I get stressed out and do not enjoy it. My daughter, on the other hand, is an adrenalin junkie at the age of four, it seems. She has not been on a roller coaster yet, but loves the feeling in your stomach when driving over hills fast, getting pushed very high on the swing, and getting tossed in the air as high as my husband can. People process situations in different manners. If we can understand this caveat for humans yet still make assumptions about human mental states, why can we not do the same for animals and give them the benefit of the doubt?

It is important to note that anthropomorphism can be risky when hard-wired instinctual behaviors are misinterpreted. For example, in humans, eye contact elicits positive emotions and indicates that someone is engaged, but many animals perceive eye contact and staring as a threat. Tail wagging across species also can mean very different things: cats wagging their tails are agitated, dogs are excited, and if a recently castrated calf is flicking its tail it may be in pain, but a cow might just be swatting flies.

It is an interesting phenomenon that sometimes people are more willing to attribute feelings to particular animals and not to others. Fear of animal feelings faces many in the workplace, whether it is at the farm, in the laboratory, or at the clinic, but it disappears when observing family pets wagging their tails when we come home from work. The animals we use in research and agriculture are perceived differently from pets. The beagle kennelled at the research facility is no different from the family pet kennelled in the house so he will not chew the carpet to bits while the owners are at work. The difference is our relationship with the animals and how we perceive them. This does not change the fact that dogs, pigs, rats, cats, and cows are all animals. If we think our guinea pig feels safe and content when we give it a paper towel roll to nest in, why can we not think the hen prefers having a place to nest?

Processing feelings in humans and animals

Animals do have the capacity to experience emotions. As stated earlier, the anxiety surrounding something so spooky and nebulous as animal feelings has prevented many scientists from acknowledging the existence of animal emotions. Emotions are scary because their subjectivity makes them difficult to explain with reliable and repeatable scientific (objective) measures. This can sometimes be translated into a reason for denouncing their existence. Ironically, there is a lot of science that can be used to prove not only the existence of animal emotions but also the similarities that exist between humans and animals.

Beginning with simple brain structure and neural networks, it has been demonstrated that, although in different proportions, the basic structure and regional function in human and animal brains are highly similar (Jerison, 1997). MacLean was the first to discuss the concept of the triune brain (1990). This is a simplified conceptualization of the brain indicating the adaptive change between reptilian, mammalian, and human brain regions. Through this tri-layer concept, it can be seen that humans and animals share these brain regions, and in particular the limbic brain region. The limbic system is the home of human emotions and because this region is shared in both humans and animals, it is also the seat of animal emotions (Rinn, 1984; Heath, 1996; Panksepp, 1998, 2003; Damasio, 1999; Liotti and Panksepp, 2004). Various emotional circuits (i.e. seeking, rage, fear, panic) are found in this common area (Panksepp, 1998). Anger (Siegel, 2005), fearfulness (Panksepp, 1990), sexuality (Pfaff, 1999), and maternal nurturance (Numan and Insel, 2003) are modulated in this subcortical region; they are all identifiable human and mammalian emotions. It is likely that the main difference between human and animal emotions is the complexity of emotional expression. Humans have a larger cortex than animals to process emotions. It is a misconception that the neocortex in the human brain provides the capacity for emotional experience. In fact, emotional states cannot be generated with neocortical electric or chemical stimulation (Panksepp, 1998). Additionally, when there is injury to the neocortex, certain emotional states, manifested by particular behaviors, such as play, reward seeking and self-stimulation (pleasure), are still present in animals (Huston and Borbély, 1973; Kolb and Tees, 1990; Panksepp *et al.*, 1994).

Although animals and humans experience similar emotions, animals probably do not process these basic emotions in the same manner as humans. This does not mean that they feel less, and could perhaps even mean that they feel more. Humans have a greater ability to regulate and inhibit their emotions with the use of higher brain functions (i.e. a highly developed neocortex). As higher brain function developed, perhaps lower brain function (i.e. emotional systems) became more and more regulated and inhibited (Liotti and Panksepp, 2004). This hypothesis suggests that animals, with their lower cognitive abilities, may in fact experience more raw unprocessed emotions than humans. Panksepp (2005b) identifies this distinction in emotional experience as emotional awareness, experienced by humans, versus emotional affect, experienced by animals. If one compares animals and human children, this theory gains more strength. Children, who have not reached full cognitive and forebrain development, often are more emotional than their parents and experience things such as fear, play, and panic, with less inhibition and moderation (Burgdorf and Panksepp, 2006). Anyone who has interacted with young children has probably witnessed this inhibition even in simple conversations. Humans suffering from traumatic prefrontal brain damage also lack the ability to control their "gut feelings" the way normal adult humans can (Damasio, 1994). People with frontal cortex head injuries are more prone to going into a rage with a minor provocation (Mason, 2008). In these patients, the brain damage prevented the control and inhibition of basic emotions and the result was lowered inhibition, perhaps similar to children and animals, both of whom have limited forebrain development. It is not that animals do not have subjective states of minds; they just do not have the capacity to inhibit emotional reactions like humans do.

Let us discuss calf weaning and kindergartners to illustrate this concept of differences in emotional processing. The kindergartener is nervous and scared to go to school; he is going to be all alone, away from his mother, in a novel environment. The calf is similarly put into a novel situation with other young calves with whom he is unfamiliar. As both mothers are "taken away", there may be some crying, from both the mother and the son. The human mother says, "Remember, you will only be here for a couple of hours and then I will come to pick you up". Despite her sadness and concern, the human mother knows that her son will be fine and that she will see him in no time at all. The bovine mother cannot comfort her calf the same way, because, like her calf, she does not understand the reason for the separation, how long it will last, or any further reasoning other than that her offspring is being taken away. The kindergartener can remind himself that his mother will be back to get him soon. The calf will cry constantly for his mother and his thoughts will bring him no solace because he cannot rationalize the fear of separation that he is feeling. Although that example is not perfect, it begins to illustrate that an animal's inability to foresee the end of a painful and fearful experience may make the experience more distressing.

Looking at it from an alternative perspective, there are some instances in which an animal's limited ability to process information could make it suffer less than humans. Consider the case of castration. When producers castrate calves, the animal suffers from fear of restraint, handling, a novel environment, and the pain from the procedure. If someone were to castrate an unwilling young male human, there would be an additional aspect of suffering. The male human would not only suffer from the things listed, he would also suffer from knowing that he was about to lead the rest of his life without testicles, a concept likely to be unknowable by the young calf. I think many human men would probably argue that this situation would be a source of extreme suffering. Sometimes ignorance is bliss, or at least a little less psychologically distressing. Unlike an animal, a man knows the pleasures he will never experience.

In the documentary on Temple Grandin's life ("Temple Grandin"), her mother puts it best, "Different, not less." This concept applies precisely to animal emotions and feelings.

Similar Neurotransmitters in Humans and Animals, and Emotions Motivate Animals to Avoid Danger

More evidence of the similarities between human and animal brains is in the homology of the neurochemical networks. The chemicals, neurotransmitters, that transmit signals between neurons, are the same. Researchers have demonstrated that the neurochemistry of addiction, specifically of opiate and psychostimulants, is highly similar between all mammals (Knutson *et al.*, 2002; Panksepp *et al.*, 2002). The brain networks that respond to addictive drugs are found in the subneocortical brain regions, the ancestral regions of the brain that we share with animals (McBride *et al.*, 1999). One way to study the way drugs interact with the brain is through place preference studies (Bardo and Bevins, 2000). In these types of experiments, particular drugs are injected into subcortical areas of the brain when the animal is in particular locations. If the drug produces positive and euphoric feelings, the place that the drug was applied becomes a conditioned place preference. If the drug is aversive, the animal also develops an aversion (avoidance) to that location. Many studies have been done indicating that drugs we consider euphorogenic (or recreational) are also rewarding to animals, producing place preferences, and stimulating the same brain systems (Panksepp, 2004, 2005c). Rats respond with positive vocalizations to places where they have received "good" drugs and negative vocalizations where they have received "bad" drugs (Burgdorf *et al.*, 2001; Knutson *et al.*, 2002). The fact that animals exhibit learned place preferences indicates that they are most likely experiencing positive emotional states.

Humans search to make their lives more enjoyable and pleasant by seeking situations that promote positive feelings; animals do the same. Animals respond to anti-anxiety and antidepression drugs. Antidepressants and anti-anxiety medications have become commonplace in many small animal practices to treat pets (Overall, 1997; Seksel and Lindeman, 1998; King *et al.*, 2000; Romich, 2005); the family pet comes in for some Prozac® (fluoxetine) to be relieved of separation anxiety while the owners are at work. There are many scientific papers that detail the effects various psychoactive drugs have on animals. There is a review article that alone cites close to 100 studies that have examined the effects of antidepressants on animal (mostly rat and mice) behavior (Borsini *et al.*, 2002). Animals have also been shown to self-medicate and increase consumption of painkillers (in feed) when experiencing pain from an injured joint or lameness (Colpaert *et al.*, 1980; Danbury *et al.*, 2000). This same phenomenon has been observed in animals in the field using wild plants with medicinal properties (Huffman, 2003; Provenza and Villalba, 2006).

Emotions required for survival

Moving away from brain anatomy and neurochemistry, the presence of animal emotions can also be supported from an evolutionary standpoint. Animals,

like us, use emotions to determine how to act in their environments. If an animal is fearful of entering a certain place in the forest, it will use that perception of its environment to avoid a potentially threatening situation. Likewise, if my car had broken down on the side of the road while driving home at night and a man approached me offering to help in an insincere manner, my feeling of uneasiness would deter me from accepting his offer. Why would we, humans and animals, have emotions if we did not need them? If certain emotions were useless, they probably would have been eliminated by natural selection a long time ago (Baxter, 1983). It is these "gut" ancestral feelings that drive an animal's survival; emotions provide the drive to seek valuable resources and avoid harm (Panksepp, 1994). Animals seldom hesitate before determining the appropriate action to take in a life-threatening situation. It would be an anomaly to see a gazelle locking eyes with a lion, pacing back and forth deciding whether or not to run away or wait a couple of minutes and see if the lion leaves. That gazelle instantly runs away from the predator's reach, driven by an intense feeling of fear, an emotion that enables it to survive.

Another example of how emotions assist animal survival is distress caused by separation anxiety. We see this distress in both humans and animals. The strong social bond between an offspring and its parent is developed as a means of survival, and when that bond is broken (at an unnatural, earlier time in particular), emotional reactions, and in many species separation vocalizations, are triggered (Panksepp, 2005a). The motivation to be with herdmates is linked to animal survival – if you are a baby separated from your mother, you decrease your chance of survival, and likewise, if you are an individual separated from your herd, you lose the herd protection.

The Positive Core Emotions

We have all experienced both positive and negative emotions at one time or another. We often focus on the absence of negative emotions in the determination of welfare states (probably because the negative feelings are identified so easily and are detrimental), but the presence of positive emotions is also essential to overall well-being (Boissy *et al.*, 2007; Yeates and Main, 2008; Mellor, 2014a,b). There has been recent growth in the realm of positive psychology (Linley *et al.*, 2006), in both the human and animal research arena, which will be important for understanding the expression of emotions in farm animals. For example, studies exploring the impact of painful procedures on calves, such as castration, have traditionally included psychological measurements, gross behavior measurements, and the presence of negative behavioral responses (tail flicking, foot stamping). More recently, the presence of play behavior in young animals has become a measure to determine the impact of painful procedures and pain mitigation on welfare state (Mintline *et al.*, 2012). Quick Access Information 8.2 is an outline of some of the behaviors associated with positive emotions.

Both the positive and the negative core emotional circuits are located in the subcortical, more primitive parts of the brain, which are the same in all mammals. Emotional systems within the brain have been mapped, and it has been demonstrated that the neural activity of the limbic system is the emotional center; this includes brain structures like the amygdala and hippocampus. Specifically, the nucleus accumbens and the parts of the amygdala complex are the locations within the brain housing the circuits that control positive, goal-directed behaviors (Ikemoto and Panksepp, 1999; Faure *et al.*, 2008). Morris *et al.* (2001) provide additional discussion of the emotions of animals and the supporting neuroscience. Jaak Panksepp has written extensively about affective neuroscience, providing

significant insight into the emotional worlds of animals. Panksepp (1998, 2005a) has described the positive core emotions, the most important being seeking or goal-directed behavior. Positive core emotions include play, lust, and care (e.g. mother–young nurturing behavior) (Panksepp, 2005a). In the mid-1990s, Panksepp's laboratory began to explore specific vocalizations (chirping) in juvenile rats that occurred during play behavior and anticipation of play (Siviy and Panksepp, 1987; Knutson *et al.*, 1998). Since the initial experimentation, Panksepp and his colleagues have continued to study the chirping, which has been called rat "laughter" as it can be initiated by "tickling" the rats, i.e. mimicking the rough and tumble play behaviors (Panksepp, 2007). In pigs, behaviors are associated with positive emotions and play, barks, and tail movements. Negative emotions are associated with defecating, urinating, high-pitched screams, and attempting to escape (Reimert *et al.*, 2013). Foraging is a good example of seeking/goal-directed behavior in many different species, and is a highly motivated behavior. Manipulating and rooting straw or other fibrous material is a highly motivated behavior in pigs (Berlyne, 1960; Wood-Gush and Vestergaard, 1989; Day *et al.*, 1995; Studnitz *et al.*, 2007), and the importance of this type of exploratory behavior in pigs has been reviewed by Studnitz *et al.* (2007). Although not everyone has this as a hobby, shopping for antiques

---

Quick Access Information 8.2. Behaviors associated with positive and negative emotions.

Behaviors associated with positive emotions	Behaviors associated with negative emotions
Play behaviors – animals running around when let out of the barn	Pain-related behaviors outlined in Chapters 1 and 6, this volume
Mother licking her baby – normal grooming behavior. Must not be confused with abnormal behaviors where animals damage each other	Freezing
	Attempting to escape
	Urination or defecation during handling
Mother nursing her baby	High-pitched vocalizations during restraint or painful procedures
Sexual behavior	
Exploring a non-threatening novel object	Kicking at handlers
Pigs rooting in the ground	Ears pinned back
Chasing prey	Suddenly startled and jumps away from a sudden novel stimulus
Eating	
Animal presents its body so people or other animals can stroke or groom it	Hides when threatened by a perceived threat
	Biting people or other animals
Dust bathing in poultry	

L.N. Edwards-Callaway

is a good demonstration of a goal-directed behavior in humans – the hunt for the treasure is driven by the satisfaction when the reward is found. Several studies in animals have been able to distinguish differences in the seeking system associated with wanting and liking, both in the brain activity associated with each emotion and also in the specific behaviors demonstrated that can be observed (Smith and Berridge, 2007; Berridge *et al.*, 2009).

Negative Core Emotions

The emotions of rage, panic (separation anxiety), and fear are three of the negative core emotions (Panksepp, 1998). As discussed, the emotional circuits of the brain have been mapped and include areas within the limbic region. Electrical stimulation of the thalamus (also considered part of the limbic system) in cats and rats will make them go into a rage (Panksepp, 1971; Olds, 1977; Heath, 1996; Siegel and Shaikh, 1997; Siegel, 2005). The second negative core emotion is separation anxiety. Livestock species are prey animals and are gregarious to varying degrees (i.e. sheep are extremely driven to be with their herd, and pigs, although herd animals, are less gregarious). Being separated from herdmates, being put in new unfamiliar surroundings, or being separated from a mother has been demonstrated to cause separation distress (Boissy and Le Neindre, 1997; Carter, 1998; Rushen *et al.*, 1999; Panksepp, 2005a).

The third negative core emotion is fear, probably the most widely studied in various species. (Forkman *et al.* (2007) have published a critical review of the variety of methods to study fear.) The circuit in the fear emotional system motivates animals to avoid predators, and they have been researched and studied extensively (LeDoux, 1992, 2000; Rogan and LeDoux, 1995, 1996; Jones and Boissy, 2011). Fear comes from the animal's fear center, the amygdala, found in the limbic region of the brain with the other emotional circuitry (Davis, 1992; LeDoux, 1992). The amygdala has also been shown to affect fear conditioning in humans (Bechara *et al.*, 1995; Büchel *et al.*, 1998; LaBar *et al.*, 1998). In animals, damage to the amygdala blocks both learned and unlearned fear responses (Davis, 1992). Some examples of unlearned fear responses would be a horse rearing and becoming agitated when it sees a balloon, or a dog running away from a firecracker. Examples of learned fear responses would be an animal that

becomes afraid of a person who beat it, or cattle that learn to avoid an electric fence after they get shocked.

Research studies have indicated that the fear and panic (separation anxiety) systems are separate systems (Faure and Mills, 1998; McHugh *et al.*, 2004). The natural wild quail shows both high fear and high separation anxiety. French scientists were able to breed four separate genetic lines of quail. They were: high fear and high separation anxiety; high fear and low separation anxiety; low fear and high separation anxiety; low fear and low separation anxiety (Faure and Mills, 1998). In their experiment, fear was measured by the tonic immobility test (see Chapter 2, this volume). The time the bird remained immobile was used as an index of fearfulness. Separation anxiety (called social reinstatement by Faure) was measured by placing the bird on a moving conveyor, which moved it away from a cage full of flock mates. The strength of separation anxiety was measured by the length of time the bird would "treadmill" on the conveyor to stay near its flock mates.

People do not like feelings of fear or panic, and so they avoid situations that stimulate these negative emotional systems. When a child in a grocery store suddenly finds she is no longer with her mother, she panics and begins to search for her. When I am walking my dog in the park alone at night and hear footsteps behind me, I get scared and I walk a little bit faster to get home. People try to escape from the negative emotions and we embrace the positive ones (Grandin and Johnson, 2009).

Both humans and animals express negative core emotions, just in a different manner. (The discussion of suffering even extends to aquaculture (Chandroo *et al.*, 2004).) My dog experiences a mild form of rage when he is frustrated that he cannot find the little piece of meat I dropped under the kitchen cabinet. Instead of communicating his frustration with an annoyed tone of voice, as a mildly frustrated human would do, he whines a little bit, letting me know how he is feeling. Fear in animals is easy to identify. Some examples of fear-motivated behavior are: a dog that crouches down and cowers when a person approaches it with a stick because it is afraid of being beaten, or an animal that struggles when it is restrained for veterinary procedures. When cattle are scared of an object, they move away from it or balk and will not walk forward (Grandin, 1997). Later sections discuss how emotions such as fear drive animal behavior.

Behavior Provides Indicators of the Animal's Emotional State

The emotional systems are the drivers that motivate an animal to engage in different behaviors. Animal emotions may not be able to be seen, but they are expressed through an animal's behavior (among other responses) and can thus be studied. Humans not only communicate how we feel by how we act and our particular behaviors but also we have the ability to express our emotions verbally when asked (and have the questioner understand what we mean). Unfortunately, not all of us have the talents of Dr Doolittle and we cannot talk to animals. Therefore, we must rely heavily on an animal's behavior to determine how it possibly perceives its environment. Behavior is extremely significant in human–animal communication. A veterinarian, whose job it is to cure disease, cannot simply ask an animal how it is feeling. He or she must use clues such as alterations in behavior (e.g. decreased eating, listlessness, increased water consumption) to determine what is wrong with the animal. Behavior is our key to understanding an animal's experience in the environments in which we raise it (Darwin, 1872; Dantzer and Mormede, 1983). People use behavioral cues from animals on a daily basis, whether it is the feedlot manager moving cattle through a chute or the pet owner teaching her puppy to sit and stay. It is important to recognize that animal behavior plays a large part in how we interact and communicate with animals.

Both animal instincts and learned behavior are driven by animal emotions. A cow's instinct to move away from a predator entering its flight zone is driven by fear; a pig's instinct to root for food is driven by its seeking emotion; a dog's instinct to do a play bow to a fellow puppy is driven by its motivation to play.

Positive goal-directed behaviors

Behavior can be examined as patterns of actions performed to achieve a specific goal. Positive goal-directed behaviors can be dissected into three phases: (i) searching for the goal; (ii) behavior directed at the goal; and (iii) quiescence following achievement of the goal (Manning, 1979). The searching phase is called the appetitive phase and the goal-directed phase is called the consummatory phase. A basic principle in animal behavior is that the searching phase is more flexible and more dependent on learning compared to the consummatory phase. The searching phase has to be flexible to enable the animal to survive in different environments. When wolves and other predators hunt, they develop their hunting skills through learning. When the predator kills its prey, the killing bite is a hard-wired instinctual behavior, but what to hunt and eat is learned. In both cattle and sheep, the young animal learns from its mother which forages are good to eat and which are not (Provenza et al., 1993). This would be the searching phase. The consumption of the forage, chewing and rumination, is hard-wired. This would be the consummatory phase. Feeding behavior and food-seeking behaviors provide good examples of this staged breakdown. Picture a feral cat hunting in the field. She spends some time stalking her prey and searching for her meal, the appetitive behavior. At the right time, she attacks the field mouse, killing her target (or "goal"), the consummatory behavior. After she has eaten the mouse and licked her lips, she rests. Feeding behavior is not as good an example when discussing grazing animals since they are constantly eating, not having as distinct quiescence phase. These goal-directed behaviors are usually highly motivated as there is a reward (the goal) once the behaviors are performed.

Animals have different levels of motivation or drive to perform a variety of goal-directed behaviors. These types of behaviors are directed at achieving a "goal", the goal being the situation that terminates the behavior (Manning, 1979). The "goal" of a hunting leopard is catching prey. The "goal" of a dog's begging behavior is to get food. Once it gets the food, the begging behavior ceases. Some behaviors are more highly motivated than others; the more drive there is, the harder the animal will work to perform them. Animals are highly motivated to eat, drink, sleep, copulate, locomote, play, explore, and find sensory stimulation and interact socially (Harlow et al., 1950; Brownlee, 1954; Panksepp, 1980; Dellmeier et al., 1985; Dellmeier, 1989). These goal-directed behaviors are motivated by the positive core emotions of lust, seeking, care, and play (Panksepp, 1998).

Behavior motivated by negative emotions

It is especially important to provide environments that reduce stimulation of the core emotions of fear, rage and panic (separation anxiety). Most of our common farm animals are prey species such as cattle, sheep, and chickens. They may become more fearful than predators such as wolves, dogs, or lions, because fear motivates them to avoid being eaten. Hens are highly motivated to have a hidden secluded place to lay their eggs (Appleby and McRae, 1986; Duncan and Kite, 1987; Cooper and Appleby, 1995, 1996b). When the jungle fowl, the ancestor of the modern hen, lived in the wild, the hens that hid in the bushes to lay their eggs survived and the ones that laid their eggs out in an open clearing got eaten by predators. The fear emotion motivates hens to hide during egg laying.

Just like humans, animals are driven to perform by their feelings and emotions (Manning, 1979); that is, animals too are motivated by their core emotions. Temple Grandin emphasizes the concept of designing animal environments and putting animals in situations that reinforce the positive emotional circuits and avoid the negative emotional circuits, to improve an animal's well-being (Grandin and Johnson, 2009). Animals will stimulate these positive emotional circuits through their behavior because these behaviors are highly motivated. When an animal can no longer express these highly motivated behaviors, some of them seemingly basic behaviors that most of us would expect to be able to do on a daily basis, their environment is no longer ideal.

Hard-wired Instinctual Behaviors

The behaviors that we observe in animals are either learned or instinctual hard-wired behaviors. We observe many learned behaviors in animals. Dairy cows learn to line up at milking time. Many zoos have taught primates and various other species to present limbs or stand still for husbandry and veterinary procedures (Grandin, 2000; Savastano et al., 2003). Dogs learn to do a multitude of behaviors during obedience training. Animals learn without the help of humans as well, an example being that of cheetah cubs learning how to kill prey efficiently and effectively. Cheetahs can kill at a young age, but take some time and lessons from their mother to develop their skills (Caro, 1994).

Hard-wired behaviors are different. These are innate behaviors that the animal does not have to learn. There are many examples of these hard-wired behaviors. Display behaviors performed by birds during courtship and mating are instinctual hard-wired behaviors. The nesting behavior we see sows perform before farrowing is hard-wired (Stolba and Wood-Gush, 1984; Jensen, 1986). The prey-catching behavior we see when dogs chase squirrels in the park is an example of an instinctual behavior. We observe many hard-wired behaviors

in livestock: the point of balance and flight zone responses in cattle during handling (see Chapter 5, this volume), rooting behavior in pigs, dust-bathing behavior in chickens, and using a nest box in hens (Vestergaard, 1982; Appleby and McRae, 1986; Newberry and Wood-Gush, 1988; Stolba and Wood-Gush, 1989; Studnitz et al., 2007). When hens are housed outdoors, they make greater use of an outdoor pasture if trees or bushes are provided for cover (see Fig. 8.1). The birds have a hard-wired instinctual fear of aerial predators. These behaviors are referred to as fixed action patterns (FAPs) because they are behavioral sequences that are always performed in the same manner. A classic example of an FAP is egg-rolling behavior in Greylag geese (Lorenz and Tinbergen, 1938). When an egg rolls out of a goose's nest, the goose performs a highly predictable pattern of movement to bring the egg back towards the nest. These FAPs are triggered into action by a sign stimulus. Mating displays are triggered by potential mates, prey chasing in dogs is triggered by rapid movement, and egg pulling is triggered by an egg (or egg-like object) that is out of the nest. FAPs are hard-wired but the particular sign stimulus that triggers the behavioral response can be influenced by learning and emotional experiences. For example, a cow's instinctual response to a human moving into its flight zone is to turn and face the person but keep a safe distance. The size of the flight zone can be modified by learning and experience with people. The cow's flight zone may shrink after becoming habituated to, and therefore less fearful of, a human's presence (see Chapter 5, this volume, for more information). The instinct to turn and face a person is a hard-wired instinct motivated by fear. The turn-and-face behavior will disappear when the cow has become so tame that she no longer has any fear.

Measuring the Strength of an Animal's Motivation

Scientists can measure objectively an animal's motivation to perform specific behaviors, a useful tool to determine how valuable a resource (e.g. food) or the ability to perform a behavior (e.g. nest building) is to an animal (Fraser and Nicol, 2011). The classic example of measuring motivation is the Mason et al. (2001) study of farm-raised mink to determine how hard they will work, or how much they will "pay", to gain access to various environments created to stimulate natural (wild) behaviors.

Fig. 8.1. Trees have been planted to provide these free-range hens with cover. The birds will forage more widely if trees or bushes provide cover from aerial predators.

The environments that were created for the mink included novel play items, a water pool, a nest site, and a vertical platform. The entrance doors to the various environments were weighted with increasing amounts of weight to determine how much effort the mink were willing to expend to get into the alternative environments. Results indicated that the mink were willing to lift the heaviest door to gain access to the water pool, their motivation being very high to have access to water. Additionally, it was found that when access to these resources was blocked, the animals had higher levels of cortisol, indicative of a stress response.

Many studies have been done measuring motivation for various valuable resources (e.g. food, water, floor substrate, social contact, nest-building material, etc.) for a range of species, including mice (Sherwin, 2004), rabbits (Seaman et al., 2001), pigs (Pedersen et al., 2002), and hens (Olsson et al., 2002). Several studies have been conducted exploring a hen's motivation to gain access to litter for dust bathing, a behavior constrained in many commercial systems due to a lack of bedding material with which to dust bathe (Vestergaard, 1982; Dawkins and Beardsley, 1986; Petherick et al., 1990, 1991; Matthews et al., 1993; Widowski and Duncan, 2000). Dust bathing is a behavior that has been observed in wild and domestic fowl and is considered a motivated behavior. Some studies have found that hens will work, by either pecking

a key or pushing a weighted door, to gain access to dust-bathing material such as wood shavings and peat moss, indicating a motivation for dust bathing (Matthews et al., 1993; Widowski and Duncan, 2000). Other studies have found results that are not as straightforward. For example, one study found that hens would break a photo beam with their beak to open a door that provided access to litter, after many trials, but they would not learn to peck a key to gain access to the litter (Dawkins and Beardsley, 1986). Other studies have found that hens would peck a key to enlarge their cages, but the number of key pecks would not increase substantially when cage enlargement also meant access to litter (Lagadic and Faure, 1987; Faure, 1991). Due to differences in the results of these motivation studies, it has been suggested that hens may not have as great a need for litter as was once thought (i.e. dust bathing may not be a highly motivated behavior). Rather than dismissing dust bathing as a behavioral need, these types of results can actually provide valuable information as to what drives the motivation to dust bathe. For example, certain behaviors could follow a particular biological rhythm (i.e. only occur at certain times of day), making scheduling a necessary component of experimental design. Additionally, visual stimuli are factors affecting motivation for dust bathing in hens, particularly the ability for the hens to see the litter. In the studies indicating that dust bathing

L.N. Edwards-Callaway

was not important, the substrate was not visible (Dawkins and Beardsley, 1986), and in the studies indicating that dust bathing was important, the substrate was visible (Matthews et al., 1993). These examples indicate the complexity of behavioral motivation and thus the importance of thorough assessment of experimental results.

The motivation to dust bathe is probably driven by one of the positive core emotions. Research on motivation to obtain a secluded private nest box shows clearly the hen's need for a nest box. This fear-motivated behavior has a much stronger motivation compared to dust bathing. A hen values access to a private nest box prior to egg laying more than gaining access to food following 4 h of feed deprivation (Cooper and Appleby, 2003). Hens will lift weighted doors (Duncan and Kite, 1987; Smith et al., 1990; Cooper and Appleby, 2003) and squeeze through narrow spaces to gain access to a nest box (Cooper and Appleby, 1995, 1996a, 1997; Bubier, 1996). One can conclude from these studies that a nest box is essential for good welfare and access to dust-bathing materials would have a lower priority (see Chapter 12, this volume).

In herding animals, motivation to be with a companion, rather than being in isolation, can be measured. It has been demonstrated that calves are motivated to get access to another calf (Holm et al., 2002). The motivation to be with another calf was measured by counting the number of times the calf would press a switch to gain access to its herdmate. Researchers also determined that calves would pay (more switch presses) more for full social contact versus only head contact. Similar to the mink study, the information from this study could be used to determine best-management practices for housing, and individual versus group pen systems in the case of the calf. Motivation studies are useful for determining an animal's behavioral needs, which behaviors satisfy the core emotions, and thus which are important to the animal.

Factors that affect motivation strength

There are different clues that help us determine if a particular behavior is highly motivated in the species we are studying. Some have suggested that if an animal performs a behavior, even when the materials or stimulus to accomplish the ultimate goal are not present, that behavior is highly motivated (Black and Hughes, 1974; Van Putten and Dammers, 1976). These are called vacuum behaviors; vacuum behaviors are FAPs carried out even in the absence of the appropriate sign stimulus. An example of this is when sows perform nest-building behaviors (i.e. pawing and rooting for straw and grasses) prior to parturition, even when there are no nesting materials present (Vestergaard and Hansen, 1984). Another vacuum behavior we can observe is the non-nutritive sucking of calves (i.e. suckling on pen mates and other objects) (de Wilt, 1985). Calves can exhibit increased motivation to suckle due to an absence of opportunities to suckle on their mother (Sambraus, 1985). Vacuum behaviors originate from an animal's frustration with not being able to express highly motivated behaviors effectively (Lindsay, 2001). These types of behaviors are important for the animal to be able to perform.

Another indication that an animal has a need to perform a particular behavior is an increased tendency to perform the behavior after a period of deprivation. This has been called a "rebound effect"; when the behavior is being prevented, the drive to perform it grows (Vestergaard, 1980; Nicol, 1987). The behavioral responses of laying hens to space restriction indicated that when given more space after being restricted, hens performed certain behaviors (e.g. leg stretching, wing flapping, feather raising) at a higher rate (Nicol, 1987). The authors acknowledged that although there was potential that this "rebound" resulted from an increased motivation to perform these behaviors during restriction, it could also simply be a response to a novel environment. Oral deprivation and the result of subsequent feeding behavior has been studied in dogs (Lawson et al., 1993). Experimental dogs were fed intragastrically, receiving no oral stimulation for a set number of days. After the restriction period, dogs were observed to over-eat, or rebound, when finally given food. We can imagine this in our dogs at home, too. If we get home late from work and feeding time for our dog is later than usual, it may consume its food at a faster rate than if fed on time. One study exploring the amount of work hens will do (push a weighted door) to gain access to a perch for roosting indicated that the hens that became habituated to perch access exhibited frustration behavior when the resource was taken away (Olsson and Keeling, 2002).

The use of preference and motivation tests in research have become a useful tool to animal welfare scientists to help assess the importance of certain resources or the preference for certain environments in farm animal species. Kirkden and Pajor (2006) provide information regarding the types of research questions that preference and motivation tests are often designed to answer: (i) whether an animal is motivated to obtain or avoid a resource; (ii) whether it has a preference among alternative resources; (iii) how strong is its motivation or preference; (iv) whether its preference, or the strength of its motivation or preference, is altered by changes in its internal or external environment. Their paper details the strengths and weaknesses of various experimental designs to answer a multitude of questions surrounding animal preference and motivation.

Abnormal Displacement Behaviors and Stereotypies

When an animal's ability to perform these highly motivated behaviors, such as dust bathing, foraging, or nesting, is thwarted, they may begin to develop displacement behaviors, for example bar biting, pacing, rocking, self-narcotizing behavior, increased aggression, etc. (Quick Access Information 8.3). Some of these behaviors are stereotypies. A stereotypy is defined as a repetitive behavior that repeats itself in a pattern that seldom changes and which serves no obvious purpose (Mason *et al.*, 2007; Price, 2008). The development of these repetitive, non-goal-oriented behaviors is a mechanism for the

animal to cope with the frustration of not being able to perform certain innate behaviors. Continuing with the food-seeking example, consider a bear in a zoo that rocks back and forth in one area of its pen prior to feeding time. In a zoo, the bear does not have to search for its food; it gets fed at a particular time each day. This rocking behavior has taken the place of long hours of food searching in the wild. Sows in gestation crates sometimes exhibit bar biting (Fig. 8.2) shortly before being fed, perhaps a reaction to the inability to food search and root, innate motivated behaviors (Lawrence and Terlouw, 1993; Day *et al.*, 1995). Straw has been proposed by many as a substrate to provide to multiple livestock species to enrich their environments and provide them with the opportunity to exhibit species-specific behavior. Tuyttens (2005) has provided a comprehensive review of the use of straw to improve welfare in pigs and cattle. Rodenburg and Turner (2012) and Grandin and Deesing (2014) discuss genetic influences on abnormal behavior. Animals that are bred to be highly productive have to eat large amounts of food to support high production. This may explain why abnormal behavior often involves the mouth.

Some animals may be self-medicating, filling a behavioral need with a maladaptive behavior. It has been demonstrated that the performance of stereotypical

Quick Access Information 8.3. Common stereotypical repetitive abnormal behaviors and other abnormal behaviors. The presence of these abnormal behaviors is an indicator that the animal's needs are not being fulfilled. Steps should be taken to provide environmental enrichment to prevent these behaviors.

Behavior	Species	Description of the behavior
Bar biting	Sows	Animal bites or mouths a bar or other object rhythmically
Tongue rolling	Cattle	Tongue is extended and moved back and forth rapidly (Fig. 8.3)
Feather pecking	Layer hens	Pecking at another hen, which damages feathers or causes injury, may be abnormal foraging behavior (Rodenburg *et al.*, 2010)
Weaving	Horses	Animal sways back and forth
Cribbing (wind sucking)	Horses	Upper jaw is placed on a fence and the animal bites and sucks wind rhythmically
Pacing	Minks and foxes	Animal circles the cage in a pattern that seldom changes
Wool, hair eating	Sheep, antelope	Pulling wool or hair out of another animal
Belly nosing	Pigs	Animal roots and rubs its nose on other animals
Non-nutritive sucking	Calves	Sucking navels or urine
Tail and ear biting	Pigs	Injures the tail or ears of other animals

L.N. Edwards-Callaway

Fig. 8.2. Bar biting by a sow housed in a gestation stall. There are large individual differences between sows. Some will be bar biters and others will not.

behaviors is related to the release of endorphins, providing the animal with some relief from the stressful environment (Cronin *et al.*, 1986; Dantzer, 1986). Supporting this concept of stereotypical behavior as a medication is the finding that when anti-addictive pharmacological agents such as nalmefene are given to horses, pigs, and mice who perform stereotypies, the stereotypies stop (Cabib *et al.*, 1984; Cronin *et al.*, 1985; Dodman *et al.*, 1988). Animals develop these stereotypical behaviors as a means of coping with a barren environment, an indication that their environment needs to be improved. Animal behavior can be used as an indicator of an animal's state of well-being (Duncan, 1998).

Behavioral Indicators of Poor Welfare

When animals are observed in confinement captivity, some of these stereotypical and displacement behaviors are often seen: a chicken performing dust-bathing behaviors with no dust; a sow in a gestation crate bar biting; a chained-up dog licking its paw repetitively; or a zoo jaguar pacing the perimeter of its cage. An animal's need to express certain behaviors (e.g. dust bathing, nesting, foraging, locomotion, social interaction, seeking, etc.) is blocked in some confinement livestock production systems. This causes potential problems with animal welfare. The agricultural industry has created

some housing systems in which animals are not able to behave naturally. Their drive to perform natural behaviors is thwarted and as a result they can potentially develop a higher level of frustration in their housing, manifested in various ways but particularly in their expression of behavior. Producers can raise livestock and poultry economically in intensive confinement systems. The question is: should we raise them this way? (Bernard Rollin, 2008, personal communication). Chapter 3 by Bernard Rollin provides a further discussion of the ethics of intensive confinement systems. The animals are never going to be able to "tell" us how they feel, and therefore we must try to understand them in a different manner (i.e. through their behavior). Some people may still fear making assumptions about an animal's well-being through its actions (even though they most likely do this on a daily basis with fellow humans). But, as one researcher profoundly stated, it is important to be "roughly right on something important than to be accurate but wrong or irrelevant" (Ng, 1995). Animal well-being is that "something important".

Environmental Enrichment Helps Enhance Animal Well-being

One of the five freedoms for captive animals is the freedom to express normal behaviors. As many of us are well aware, it is often difficult to provide this opportunity to the animals we raise and live with. Whether it is the space restrictions in zoos, the leash laws in our towns, or the nature of the confinement of livestock operations, an animal's ability to express normal behaviors has become somewhat limited in today's society. Although we may not be able to allow animals to act as they do in the wild on a daily basis, we must look for ways to encourage animals to perform important behaviors with activities that accomplish the same goal, i.e. environmental enrichment. Hal Markowitz, a founder of enrichment for zoo animals, demonstrates this concept of substitution. One example of his endeavors was the development of acoustic "prey" as a form of environmental enrichment for captive African leopards (Markowitz *et al.*, 1995). He was able to develop computer-controlled hunting simulation and exercise opportunities for captive felines, ultimately altering their behavior and enhancing overall well-being. Markowitz was able to replace hunting with a similar experience that achieved the same goal.

Providing environmental enrichment can reduce stereotypic abnormal behavior (Mason *et al.*, 2007). The emphasis has to be on preventing abnormal behavior from starting. Once it has started, it is difficult to stop. (Just think of your friend who has a high-strung dog that starts to lick at certain "hot" spots on its paw while your friend is out – it is pretty hard to get the dog to stop once he starts.) The goal of providing environmental enrichment such as straw for pigs or nest boxes for hens is to prevent abnormal behavior and allow the animals to perform more species-typical behaviors. It should help the biological functioning of the animal (Newberry, 1995). Another advantage of providing environmental enrichment is to help prevent behaviors that may cause injuries or damage to the animals.

In cattle, tongue rolling (see Fig. 8.3) can be reduced by providing adequate roughage (Redbo, 1990; Redbo and Nordblad, 1997). Providing hay forages and other roughages will help prevent stereotypic behavior in horses (McGreevy *et al.*, 1995; Goodwin *et al.*, 2002; Thorne *et al.*, 2005). Horses fed lots of concentrates and pellets may be more prone to stereotypies. Providing litter, straw, or materials to forage in may reduce feather pecking and injuries in chickens and turkeys. Providing turkeys with straw and hanging chains has reduced injuries from aggression (Sherwin *et al.*, 1999; Martrenchar *et al.*, 2001). Providing ducks with additional foraging materials had no effect (Riber and Mench, 2008). In pigs, providing straw or other fibrous materials for rooting can help prevent

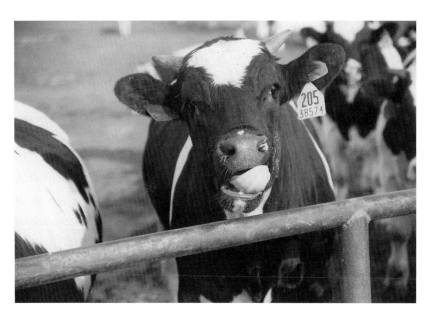

Fig. 8.3. Tongue rolling in cattle. Feeding cattle more roughage will help reduce tongue rolling. There are genetic influences on the tendency of different breeds to tongue roll. (From Grandin and Deesing, 1998.)

L.N. Edwards-Callaway

tail biting (Day *et al.*, 2002; Bolhuis *et al.*, 2006; Chaloupkova *et al.*, 2007). It is important to provide the straw or other foraging material for the entire life of the animal. If the straw is taken away, tail biting may increase (Day *et al.*, 2002; Bolhuis *et al.*, 2006). Figure 8.4 shows sows bedded on straw.

In systems with liquid manure and slatted floors, it may be difficult to provide straw or forage. Hanging objects such as ropes or chains can also help prevent pigs from directing biting, chewing, or nosing towards other pigs. Pigs prefer objects that they can chew up and destroy. Hanging cloth strips and rubber hoses were preferred compared to chains (Grandin, 1989). Jensen and Pedersen (2007) have reported similar results. If nondestructable objects such as chains or balls are used, they MUST be changed frequently to maintain the pig's interest. For caged laying hens, providing hanging strands of white string reduced feather damage (McAdie *et al.*, 2005). The same strings maintained their interest over a period of weeks. In one experiment, they pecked at the strings on the first day for about the same amount of time that they pecked them on the 47th day. Poultry may have less need for novelty compared to pigs. Layers in a cage system with access to strands of string had significantly better feather condition at 35 weeks of age. The birds in this experiment were not beak trimmed.

Genetic Effects on Abnormal Behavior

One must remember that there are strong genetic influences on the occurrence of behaviors that could impact an animal's well-being and productivity negatively, such as feather pecking in hens (Kjaer and Hocking, 2004; Muir and Cheng, 2014) and tail biting in piglets (Breuer *et al.*, 2005). In poultry, genetics play a substantial role (Kjaer and Hocking, 2004). Strains of chickens bred for high egg production often have higher rates of cannibalism, aggression, and feather pecking (Craig and Muir, 1998). More recent research indicates that both feather pecking and tail biting may be abnormal foraging behavior. Pigs and hens from genetics with high levels of feather pecking or tail biting spend more time manipulating things in their environment and eat more of a high fiber diet (Kalmendal and Bessei, 2012; Daigle *et al.*, 2014). In ducks, Muscovys are more likely to be cannibalistic compared to Peking ducks. In pigs, many producers and lairage managers have observed that certain genetic lines of lean, rapidly growing pigs had a higher incidence of tail biting. Farm managers have also reported higher levels of belly nosing in some lean lines of rapidly growing pigs. Although aggression is a normal component of an animal's behavioral repertoire, genetic selection to minimize

Fig. 8.4. Sows housed in a pen bedded with straw. Providing straw will help reduce abnormal behavior. These systems must be managed carefully to ensure that producers use sufficient straw to keep the animals clean. One of the biggest problems with these systems is failure to use enough straw. This will result in a dirty animal laying in muck.

aggression in sows has been a focus of several studies, as it has become an increasingly important component as many producers begin to move away from gestation stalls to open pen systems. Fear response and the heritability of fearfulness (or temperament) have been studied in a variety of species (Broadhurst, 1975; Mills and Faure, 1991; Boissy *et al.*, 2005). Taken from the OIE general principles for the welfare of animals in livestock production systems, "Genetic selection should always take into account the health and welfare of animals" (OIE, 2012). Reviews can be found in Rodenburg and Turner (2012) and Grandin and Deesing (2014).

The Dilemma of Maximizing Production

We as an agricultural industry are being tasked to do more with less. The livestock industry is faced with the difficult task of balancing animal performance and efficiency with animal welfare. With the ever-growing concern for animal welfare in society coupled with the increasing demand for food, particularly meat, producers and scientists are tasked daily with finding new, efficient and animal-friendly ways to raise animals. With the great strides in genetics, animals have been bred to produce large amounts of milk, meat, and/or eggs. As we approach heightened levels of efficiency and production, it seems that many domestic animal species may have been pushed to their biological limits to reach the desired levels of production. The biology of the animals will not sustain further manipulation of their biological limits. The agricultural industry was able to develop maize (corn) to reach unimaginable heights of production. Maize has been bred and genetically engineered so that it can withstand tight crowding, growing and thriving with little distance between itself and its identical neighbor. Unlike our livestock species, maize plants do not experience fear, pain, or frustration with their environment. We have tried to raise our livestock somewhat similarly, providing less space and maximizing genetic potential for production. By making environments limited, we have restricted the animal from truly acting like an animal, in the whole sense of the word. Societies around the world realized that housing wild animals in small, sterile environments in zoos was inhumane and detrimental for animal welfare. Long before the zoo animals received attention, suffering of companion animals was stopped with anti-cruelty laws and legislation. (Companion animals have a different role in society,

some pets are treated more like humans than animals because owners easily recognize their dog's capacity for pain, fear, and frustration.) The suffering of our beloved companion animals was thwarted at every chance, perhaps even to a fault. Society has also recognized the need to protect laboratory animals from unnecessary suffering. This can be seen in the creation of legislation protecting laboratory animals and the implementation of the animal care and use committees in research institutions around the world. What about livestock? The welfare of farm animals is increasingly becoming a concern in society, as seen by the changes in consumer demands and the increase in regulations.

As indicated through the previous discussion, limiting natural behavior and thus inducing potential frustration, fear, anxiety, and ultimately suffering, is going against the current trend towards an improved livestock animal welfare status. Ironically, as we limit certain behaviors, other behavioral cues, whether it be lack of behavior, appearance of abnormal behaviors, or exaggerated natural behaviors, begin to tell us that the current system is not working, at least in terms of keeping the animals physically and mentally content. It is necessary that in the next stage of animal agriculture we construct animal-centered outcomes (Hewson, 2003), focus on fitting the farms (now the confinement operations) to the animals, not the animals to the farms (Kilgour, 1978). Manipulating maize to meet our needs for high yields was easier than changing animals because plants do not have the capacity to suffer. We cannot completely breed natural behavior, and thus the capacity for psychological suffering, out of animals. Fear-motivated behaviors are needed for basic survival and therefore would be difficult to breed out of an animal. However, the intensity of some behavioral needs can be reduced with breeding. Mother-nurturing behavior has been reduced in some animals with breeding. The Holstein cow has less separation anxiety when her calf is removed compared to most beef breeds. Broodiness, the hen's behavior to sit and incubate her eggs, has been greatly reduced in some chickens (Hays and Sanborn, 1939; Hutt, 1949). A review of research in Chapters 1 and 2 indicates that certain behaviors should be accommodated. As animal scientists and producers, we need to find the balance between animal performance and efficiency, certainly needed to produce food, and an appropriate state of animal welfare for our livestock species.

L.N. Edwards-Callaway

References

Appleby, M.C. and McRae, H.E. (1986) The individual nest box as a super-stimulus for domestic hens. *Applied Animal Behaviour Science* 15, 169–176.

Bardo, M.T. and Bevins, R.A. (2000) Conditioned place preference: What does it add to our preclinical understanding of drug reward? *Psychopharmacology* 153, 31–43.

Baxter, M.R. (1983) Ethology in environmental design for animal production. *Applied Animal Ethology* 9, 207–220.

Bechara, A., Tranel, D., Damasio, H., Adolphs, R., Rockland, C., et al. (1995) Double dissociation of conditioning and declarative knowledge relative to the amygdala and hippocampus in humans. *Science* 269, 1115–1118.

Berlyne, D.E. (1960) *Conflict, Arousal and Curiosity*. McGraw-Hill Book Company, New York.

Berridge, K.C., Robinson, T.E. and Aldridge, J.W. (2009) Dissecting components of reward: 'Liking', 'wanting', and learning. *Current Opinion in Pharmacology* 9, 65–73.

Black, A.J. and Hughes, B.O. (1974) Patterns of comfort behaviour and activity in domestic fowls: a comparison between cages and pens. *British Veterinary Journal* 130, 23–33.

Boissy, A. and Le Neindre, P. (1997) Behavioral, cardiac and cortisol responses to brief peer separation and reunion in cattle. *Physiology and Behavior* 62, 693–699.

Boissy, A., Fisher, A.D., Bouix, J., Hinch, G.N. and Le Neindre, P. (2005) Genetics of fear in ruminant livestock. *Livestock Production Science* 93, 23–32.

Boissy, A., Manteuffel, G., Jensen, M.B., Moe, R.O., Spruijt, B., et al. (2007) Assessment of positive emotions in animals to improve their welfare. *Physiology and Behavior* 92, 375–397.

Bolhuis, J.E., Schouten, W.G.P., Schrama, J.W. and Wiegant, V.M. (2006) Effects of rearing and housing environment on behaviour and performance of pigs with different coping characteristics. *Applied Animal Behaviour Science* 101, 68–85.

Borsini, F., Podhorna, J. and Marazziti, D. (2002) Do animal models of anxiety predict anxiolytic-like effects of antidepressants? *Psychopharmacology* 163, 121–141.

Botreau, R., Veissier, I. and Perny, P. (2009) Overall assessment of animal welfare: Strategy adopted in Welfare Quality®. *Animal Welfare* 18, 363–370.

Brambell, R.W.R. (chairman) (1965) Report of the Technical Committee to Enquire into the Welfare of Animals Kept Under Intensive Livestock Husbandry Systems. Command Paper 2836. Her Majesty's Stationery Office, London.

Breuer, K., Sutcliffe, M.E.M., Mercer, J.T., Rance, K.A., O'Connell, N.E., et al. (2005) Heritability of clinical tail-biting and its relation to performance traits. *Livestock Production Science* 93, 87–94.

Broadhurst, P.L. (1975) The Maudsley reactive and non-reactive strains of rats: A survey. *Behavior Genetics* 5, 299–319

Broom, D.M. (1986) Indicators of poor welfare. *British Veterinary Journal* 142, 524–526.

Broom, D.M. (2001) *Coping with Challenge: Welfare in Animals including Humans*. Dahlem University Press, Berlin.

Brownlee, A. (1954) Play in domestic cattle in Britain: an analysis of its nature. *Nature* 173, 1081–1082.

Bubier, N.E. (1996) The behavioural priorities of laying hens: the effect of cost/no cost multi-choice tests on time budgets. *Behavioural Processes* 37, 225–238.

Büchel, C., Morris, J., Dolan, R.J. and Friston, K.J. (1998) Brain systems mediating aversive conditioning: an event-related fmri study. *Neuron* 20, 947–957.

Burgdorf, J. and Panksepp, J. (2006) The neurobiology of positive emotions. *Neuroscience and Biobehavioural Reviews* 30, 173–187.

Burgdorf, J., Knutson, B., Panksepp, J. and Ikemoto, S. (2001) Nucleus accumbens amphetamine microinjections unconditionally elicit 50-khz ultrasonic vocalizations in rats. *Behavioural Neuroscience* 115, 940–944.

Cabib, S., Puglisi-Allegra, S. and Oliverio, A. (1984) Chronic stress enhances apomorphine-induced stereotyped behavior in mice: involvement of endogenous opioids. *Brain Research* 298, 138–140.

Caro, T.M. (1994) *Cheetahs of the Serengeti Plains*. Chicago University Press, Chicago, Illinois.

Carter, C.S. (1998) Neuroendocrine perspectives on social attachment and love. *Psychoneuroendocrinology* 23, 779–818.

Chaloupkova, H.G., Illman, G., Bartoš, L. and Spinka, M. (2007) The effect of pre-weaning housing on the play and agonistic behaviour of domestic pigs. *Applied Animal Behaviour Science* 103, 25–34.

Chandroo, K.P., Duncan, I.J.H and Moccia, R.D. (2004) Can fish suffer? Perspectives on sentience, pain, fear and stress. *Applied Animal Behaviour Science* 86, 225–250.

Coetzee, J.F. (2011) A review of pain assessment techniques and pharmacological approaches to pain relief after bovine castration: Practical implications for cattle production within the United States. *Applied Animal Behaviour Science* 135, 192–213.

Colpaert, F.C., De Witte, P., Maroli, A.N., Awouters, F., Niemegeers, C.J., et al. (1980) Self-administration of the analgesic suprofen in arthritic rats: evidence of *Mycobacterium butyricum*-induced arthritis as an experimental model of chronic pain. *Life Science* 27, 921–928.

Cooper, J.J. and Appleby, M.C. (1995) Nesting behaviour of hens: Effects of experience on motivation. *Applied Animal Behaviour Science* 42, 283–295.

Cooper, J.J. and Appleby, M.C. (1996a) Demand for nest boxes in laying hens. *Behavioural Processes* 36, 171–182.

Cooper, J.J. and Appleby, M.C. (1996b) Individual variation in prelaying behaviour and the incidence of floor eggs. *British Poultry Science* 37, 245–253.

Cooper, J.J. and Appleby, M.C. (1997) Motivational aspects of individual variation in response to nestboxes by laying hens. *Animal Behaviour* 54, 1245–1253.

Cooper, J.J. and Appleby, M.C. (2003) The value of environmental resources to domestic hens: a comparison of the work-rate for food and for nests as a function of time. *Animal Welfare* 12, 39–52.

Craig, J. and Muir, W. (1998) Genetics and the behavior of chickens, welfare, and productivity. In: Grandin, T. (ed.) *Genetics and the Behavior of Domestic Animals*. Academic Press, San Diego, California, pp. 265–297.

Cronin, G.M., Wiepkema, P.R. and van Ree, J.M. (1985) Endogenous opioids are involved in stereotyped behaviour of tethered sows. *Neuropeptides* 6, 527–530.

Cronin, G.M., Wiepkema, P.R. and van Ree, J.M. (1986) Endorphins implicated in stereotypies of tethered sows. *Experientia* 42, 198–199.

Daigle, C.L., Rodenburg, T.B., Bolhuis, J.E., Swanson, J.C. and Siegford, J.M. (2014) Use of dynamic and rewarding environmental enrichment to alleviate feather pecking in non-cage laying hens. *Applied Animal Behaviour Science* 161, 75–85.

Damasio, A.R. (1994) *Descartes' Error: Emotion, Reason and the Human Brain*. Grosset/Putnam, New York.

Damasio, A.R. (1999) *The Feeling of What Happens*. Harcourt Brace, New York.

Danbury, T.C., Weeks, C.A., Chambers, J.P., Waterman-Pearson, A.E. and Kestin, S.C. (2000) Self-selection of the analgesic drug caprofen by lame broiler chickens. *Veterinary Record* 146, 307–311.

Dantzer, R. (1986) Symposium on 'indices to measure animal well-being': behavioral, physiological and functional aspects of stereotyped behavior: A review and a re-interpretation. *Journal of Animal Science* 62, 1776–1786.

Dantzer, R. and Mormede, P. (1983) Stress in farm animals: A need for re-evaluation. *Journal of Animal Science* 57, 6–18.

Darwin, C. (1872) *The Expression of the Emotions in Man and Animals*. University of Chicago Press, Chicago, Illinois.

Davis, M. (1992) The role of the amygdala in conditioned fear. In: Aggleton, J.P. (ed.) *The Amygdala: Neurobiological Aspects of Emotion, Memory and Mental Dysfunction*. Wiley-Liss, New York, pp. 255–306.

Dawkins, M.S. (1988) Behavioural deprivation: A central problem to animal welfare. *Applied Animal Behaviour Science* 20, 209–225.

Dawkins, M.S. (1998) Evolution and animal welfare. *Quarterly Review of Biology* 73, 305–328.

Dawkins, M.S. (2006) A user's guide to animal welfare science. *Trends in Ecology and Evolution* 21, 77–82.

Dawkins, M.S. and Beardsley, T. (1986) Reinforcing properties of access to litter in hens. *Applied Animal Behaviour Science* 15, 351–364.

Day, J.E.L., Kyriazakis, I. and Lawrence, A.B. (1995) The effect of food deprivation on the expression of foraging and exploratory behaviour in the growing pig. *Applied Animal Behaviour Science* 42, 193–206.

Day, J.E.L., Spoolder, H.A.M., Burfoot, A., Chamberlain, H.L. and Edwards, S.A. (2002) The separate and interactive effects of handling and environmental enrichment on the behavior and welfare of growing pigs. *Applied Animal Behaviour Science* 75, 177–192.

Dellmeier, G.R. (1989) Motivation in relation to the welfare of enclosed livestock. *Applied Animal Behaviour Science* 22, 129–138.

Dellmeier, G.R., Friend, T.H. and Gbur, E.E. (1985) Comparison of four methods of calf confinement. II. Behavior. *Journal of Animal Science* 60, 1102–1109.

de Wilt, J.G. (1985) *Behavior and Welfare of Veal Calves in Relation to Husbandry Systems*. PhD Thesis, Wageningen Agricultural University, Wageningen, the Netherlands.

Dodman, N.H., Shuster, L., Court, M.H. and Patel, J. (1988) Use of narcotic antagonist (nalmefene) to suppress self-mutilative behaviour in a stallion. *Journal of the American Veterinary Association* 192, 1585–1586.

Duncan, I.J. (1998) Behavior and behavioral needs. *Poultry Science* 77, 1766–1772.

Duncan, I.J.H. and Dawkins, M.S. (1983) The problem of assessing 'well-being' and 'suffering' in farm animals. In: Smidt, D. (ed.) *Indicators Relevant to Farm Animal Welfare*. Springer Science+Business Media B.V., the Netherlands, pp. 13–24.

Duncan, I.J.H. and Kite, V.G. (1987) Some investigations into motivation in the domestic fowl. *Applied Animal Behaviour Science* 18, 387–388.

Duncan, I.J.H. and Petherick, J.C. (1991) The implications of cognitive processes for animal welfare. *Journal of Animal Science* 69, 5017–5022.

Faure, J.M. (1991) Rearing conditions and needs for space and litter in laying hens. *Applied Animal Behaviour Science* 31, 111–117.

Faure, J.M. and Mills, A.D. (1998) Improving the adaptability of animals by selection. In: Grandin, T. (ed.) *Genetics and the Behavior of Domestic Animals*. Academic Press, San Diego, California, pp. 235–264.

Faure, A., Reynolds, S.M., Richard, J.M. and Berridge, K.C. (2008) Mesolimbic dopamine in desire and dread: Enabling motivation to be generated by localized glutamate disruptions in nucleus accumbens. *Journal of Neuroscience* 28, 7184–7192.

Forkman, B., Boissy, A., Meunier-Salaun, M.C., Canali, E. and Jones, R.B. (2007) A critical review of fear tests used on cattle, pigs, sheep, poultry and horses. *Physiology and Behavior* 92, 340–374.

Fraser, D. (1995) Science, values and animal welfare: exploring the 'inextricable connection'. *Animal Welfare* 4, 103–117.

Fraser, D. (2008) Toward a global perspective on farm animal welfare. *Applied Animal Behaviour Science* 113, 330–339.

Fraser, D. and Nicol, C.J (2011) Preference and motivation research. In: Appleby, M.C., Mench, J.A., Olsson, I.A.S. and Hughes, B.O. (eds) *Animal Welfare*, 2nd edn. CAB International, Wallingford, UK, pp. 183–199.

Fraser, D., Duncan, I.J.H., Edwards, S.A., Grandin, T., Gregory, N.G., *et al.* (2013) General principles for the welfare of animals in production systems: The underlying science and its application. *The Veterinary Journal* 198, 19–27.

Goodwin, D., Davidson, H.P.B. and Harris, P. (2002) Foraging environment for stabled horses: Effects on behaviour and selection. *Equine Veterinary Journal* 34, 686–691.

Grandin, T. (1989) Effects of rearing environment and environmental enrichment on the behavior and neural development of young pigs. Dissertation, University of Illinois, Urbana, Illinois.

Grandin, T. (1997) Assessment of stress during handling and transport. *Journal of Animal Science* 75, 249–257.

Grandin, T. (2000) Habituating antelope and bison to cooperate with veterinary procedures. *Journal of Applied Animal Welfare Science* 3, 253–261.

Grandin, T. and Deesing, M.J. (1998) Genetics and animal welfare. In: Grandin, T. (ed.) *Genetics and the Behavior of Domestic Animals*. Academic Press, San Diego, California, pp. 319–346.

Grandin, T. and Deesing, M.J. (2014) Genetics and animal welfare. In: Grandin, T. and Deesing. M.J. (eds) *Genetics and the Behavior of Domestic Animals*, 2nd edn. Academic Press (Elsevier), London, pp. 435–472.

Grandin, T. and Johnson, C. (2009) *Animals Make Us Human*. Houghton Mifflin Harcourt, Boston, Massachusetts.

Harlow, H.F., Harlow, M.K. and Meyer, D.R. (1950) Learning motivated by a manipulation drive. *Journal of Experimental Psychology* 40, 228–234.

Hays, F.A. and Sanborn, R. (1939) Breeding for egg production. *Massachusetts Agricultural Experimental Station Bulletin* 307. Agricultural Experimental Station, Amherst, Massachusetts.

Heath, R.G. (1996) *Exploring the Mind–Brain Relationship*. Moran Printing, Baton Rouge, Louisiana.

Heinrich, A., Duffield, T.F., Lissemore, K.D., Squires, E.J. and Millman, S.T. (2009) The impact of meloxicam on postsurgical stress associated with cautery dehorning. *Journal of Dairy Science* 92, 540–547.

Hewson, C.J. (2003) Can we assess welfare? *Canadian Veterinary Journal* 44, 749–753.

Holm, L., Jensen, M.B. and Jeppesen, L.L. (2002) Calves' motivation for access to two different types of social contact measured by operant conditioning. *Applied Animal Behaviour Science* 79, 175–194.

Huffman, M.A. (2003) Animal self-medication and ethnomedicine: Exploration and exploitation of the medicinal properties of plants. *Proceedings of the Nutrition Society* 62, 371–381.

Huston, J.P. and Borbély, A.A. (1973) Operant conditioning in forebrain ablated rats by use of rewarding hypothalamic stimulation. *Brain Research* 50, 467–472.

Hutt, F.B. (1949) *Genetics of the Fowl*. McGraw-Hill Book Company, New York.

Ikemoto, S. and Panksepp, J. (1999) The role of the nucleus accumbens dopamine in motivated behavior: A unifying interpretation with special reference to reward-seeking. *Brain Research Reviews* 31, 6–41.

Jensen, M.B. and Pedersen, L.J. (2007) The value assigned to six different rooting materials by growing pigs. *Applied Animal Behaviour Science* 108, 31–44.

Jensen, P. (1986) Observations on the maternal behaviour of free-ranging domestic pigs. *Applied Animal Behaviour Science* 16, 131–142.

Jerison, H.J. (1997) Evolution of the prefrontal cortex. In: Krasnegor, N.A., Lyon, G.R. and Goldman-Rakic, P. (eds) *Development of the Prefrontal Cortex: Evolution, Neurobiology and Behavior*. Paul H. Brooks Publishing Co., Baltimore, Maryland, pp. 9–27.

Jones, B. and Boissy, A. (2011) Fear and other negative emotions. In: Appleby, M.C., Mench, J.A., Olsson, I.A.S. and Hughes, B.O. (eds) *Animal Welfare*, 2nd edn. CAB International, Wallingford, UK, pp. 78–97.

Kalmendal, R. and Bessei, W. (2012) The preference for high fiber feed in laying hens selected divergently for feather pecking. *Poultry Science* 91, 1785–1789.

Kilgour, R. (1978) The application of animal behavior and the humane care of farm animals. *Journal of Animal Science* 46, 1478–1486.

King, J.N., Simpson, B.S., Overall, K.L., Appleby, D., Pageat, P., *et al.* (2000) Treatment of separation anxiety in dogs with clomipramine: Results from a prospective, randomized, double-blind, placebo-controlled, parallel-group, multicenter clinical trial. *Applied Animal Behaviour Science* 67, 255–275.

Kirkden, R.D. and Pajor, E.A. (2006) Using preference, motivation and aversion tests to ask scientific questions about animals' feelings. *Applied Animal Behaviour Science* 100, 29–47.

Kjaer, J.B. and Hocking, P.M. (2004) The genetics of feather pecking and cannibalism. In: Perry, G.C. (ed.) *Welfare of the Laying Hen*. CAB International, Wallingford, UK, pp. 109–122.

Knutson, B., Burgdorf, J. and Panksepp, J. (1998) Anticipation of play elicits high-frequency ultrasonic vocalizations in young rats. *Journal of Comparative Psychology* 112, 65–73.

Knutson, B., Burgdorf, J. and Panksepp, J. (2002) Ultrasonic vocalizations as indices of affective states in rats. *Psychological Bulletin* 128, 961–977.

Kolb, B. and Tees, R.C. (1990) *The Cerebral Cortex of the Rat*. MIT Press, Cambridge, Massachusetts.

LaBar, K.S., Gatenby, J.C., Gore, J.C., LeDoux, J.E. and Phelps, E.A. (1998) Human amygdala activation during conditioned fear acquisition and extinction: a mixed-trial fMRI study. *Neuron* 20, 937–945.

Lagadic, H. and Faure, J.M. (1987) Preferences of domestic hens for cage size and floor types as measured by operant conditioning. *Applied Animal Behaviour Science* 19, 147–155.

Lawrence, A.B. and Terlouw, E.M. (1993) A review of behavioral factors involved in the development and continued performance of stereotypic behaviors in pigs. *Journal of Animal Science* 71, 2815–2825.

Lawson, D.C., Schiffman, S.S. and Pappas, T.N. (1993) Short-term oral sensory deprivation: Possible cause of binge eating in sham-feeding dogs. *Physiology and Behavior* 53, 1231–1234.

LeDoux, J.E. (1992) Emotion and the amygdala. In: Aggleton, J.P. (ed.) *The Amygdala: Neurological Aspects of Emotion, Memory and Mental Dysfunction*. Wiley-Liss, New York, pp. 339–351.

LeDoux, J.E. (2000) Emotion circuits in the brain. *Annual Review of Neuroscience* 23, 155–184.

Lindsay, S.R. (2001) *Handbook of Applied Dog Behavior and Training: Etiology and Assessment of Behavior*, Vol 2. Iowa State University Press, Ames, Iowa.

Linley, P.A., Joseph, S., Harrington, S. and Wood, A.M. (2006) Positive psychology: Past, present, and (possible) future. *The Journal of Positive Psychology* 1, 3–16.

Liotti, M. and Panksepp, J. (2004) Imaging human emotions and affective feelings. Implications for biological psychiatry. In: Panksepp, J. (ed.) *Textbook of Biological Psychiatry*. Wiley, Hoboken, New Jersey, pp. 33–74.

Lorenz, K. and Tinbergen, N. (1938) Taxis and instinctive behavior pattern in egg-rolling by the greylag goose. In: Lorenz, K. (ed.) *Studies in Animal Behavior and Human Behavior No. 1*. Harvard University Press, Cambridge, Massachusetts, pp. 316–359.

McAdie, T.M., Keeling, L.J., Blokhuis, H.J. and Jones, R.B. (2005) Reduction in feather pecking and improvement in feather condition with presentation of a string device to chickens. *Applied Animal Behaviour Science* 93, 67–80.

McBride, W.J., Murphy, J.M. and Ikemoto, S. (1999) Localization of brain reinforcement mechanisms: Intracranial self-administration and intra-cranial place-conditioning studies. *Behavioural Brain Research* 101, 129–152.

McGreevy, P.D., Cripps, P.J., French, N.P., Green, L.E. and Nicol, C.J. (1995) Management factors associated with stereotypic and redirected behaviour in the thoroughbred horse. *Equine Veterinary Journal* 27, 86–91.

McHugh, S.B., Deacon, R.M.J., Rawlins, J.N.P and Bannerman, D.M. (2004) Amygdala and ventral hippocampus contribute differentially to mechanisms of fear and anxiety. *Behavioral Neuroscience* 118, 63–78.

MacLean, P.D. (1990) *The Triune Brain in Evolution*. Plenum Press, New York.

Manning, A. (1979) *An Introduction to Animal Behavior*. Addison-Wesley, Reading, Massachusetts.

Markowitz, H., Aday, C. and Gavazzi, A. (1995) Effectiveness of acoustic prey: Environmental enrichment for a captive African leopard (*Panthera pardus*). *Zoo Biology* 14, 371–379.

Martrenchar, A., Huonnig, D. and Cotte, J.P. (2001) Influence of environmental enrichment on injurious pecking and perching behaviour in young turkeys. *British Poultry Science* 42, 161–170.

Mason, G.J. and Mendl, M. (1993) Why is there no simple way of measuring animal welfare? *Animal Welfare* 2, 301–319.

Mason, G.J., Cooper, J. and Clarebrough, C. (2001) Frustrations of fur-farmed mink. *Nature* 410, 35–36.

Mason, G.R., Clubb, R., Latham, N. and Vickery, S. (2007) Why and how should we use environmental enrichment to tackle stereotyped behaviour? *Applied Animal Behavioural Science* 102, 163–188.

Mason, M.P. (2008) *Head Cases: Stories of Brain Injury and its Aftermath*. Farrar, Straus and Giroux, New York.

Matthews, L.R., Temple, W., Foster, T.M. and McAdie, T.M. (1993) Quantifying the environmental requirements of layer hens by behavioural demand functions. In: Nichelmann, M., Wierenga, H.K. and Braun, S. (eds) *Proceedings of the International Congress on Applied Ethology*, Berlin, 26–30 July. Kuratorium fur Technik und Bauwesen in der Landwirtschaft, Berlin, pp. 206–209.

Mellor, D. (2014a) Positive animal welfare states and reference standards for welfare assessment. *New Zealand Veterinary Journal* 60, 17–23.

Mellor, D. (2014b) Enhancing animal welfare by creating opportunities for positive affective engagement. *New Zealand Veterinary Journal* 63, 3–8.

Mendl, M. (1991) Some problems with the concept of a cut-off point for determining when an animal's welfare is at risk. *Applied Animal Behaviour Science* 31, 139–146.

Mills, A.D. and Faure, J.M. (1991) Divergent selections for duration of tonic immobility and social reinstatement behavior in Japanese quail (*Coturnix coturnix japonica*) chicks. *Journal of Comparative Psychology* 105, 25–38.

Mintline, E.M., Stewart, M., Rogers, A.R., Cox, N.R., Verkerk, G.A., *et al.* (2012) Play behavior as an indicator of animal welfare: Disbudding in dairy calves. *Applied Animal Behaviour Science* 144, 22–30.

Moberg, G.P. (1985) Biological response to stress: Key to assessment of animal well-being? In: Moberg, G.P. (ed.) *Animal Stress*. American Physiological Society, Betheseda, Maryland, pp. 27–49.

Morris, C.L., Grandin, T. and Irlbeck, N.A. (2001) Companion Animals Symposium: Environmental enrichment for companion, exotic, and laboratory animals. *Journal of Animal Science* 89, 4227–4238.

Muir, W.M. and Cheng, H.W. (2014) Genetic influences on the behavior of chickens associated with welfare and productivity. In: Grandin, T. and Deesing, M. (eds) *Genetics and the Behavior of Domestic Animals*, 2nd edn. Academic Press, Elsevier, San Diego, California, pp. 435–472.

National Center for Health Statistics (2011) *Health, United States 2010: With Special Feature on Death and Dying*. United States Department of Health and Human Services, Centers for Disease Control and Prevention, Hyattsville, Maryland.

Newberry, R.C. (1995) Environmental enrichment: Increasing the biological relevance of captive environments. *Applied Animal Behaviour* 44, 229–243.

Newberry, R.C. and Wood-Gush, D.G.M. (1988) Development of some behaviour patterns in piglets under semi-natural conditions. *Animal Production* 46, 103–109.

Ng, Y. (1995) Towards welfare biology: Evolutionary economics of animal consciousness and suffering. *Biology and Philosophy* 10, 255–285.

Nicol, C.J. (1987) Behavioural responses of laying hens following a period of spatial restriction. *Animal Behaviour* 35, 1709–1719.

Numan, M. and Insel, T.R. (2003) *The Neurobiology of Parental Behavior*. Springer, New York.

Office of the Federal Register (1989) Final Rules: Animal Welfare Act: 9 Code of Federal Regulations (CFR) Parts 1 and 2. Federal Register Vol. 54 No. 168, 31 August. Office of the Federal Register, Washington, DC, pp. 36112–36163.

OIE (2012) Article 7.1.4. Introduction to the recommendations for animal welfare. In: *Terrestrial Animal Health Code*, 21st edn. World Organisation for Animal Health (OIE), Paris, p. 306.

Olds, J. (1977) *Drives and Reinforcements: Behavioral Studies of Hypothalamic Function*. Raven Press, New York.

Olsson, I.A.S. and Keeling, L.J. (2002) The push-door for measuring motivation in hens: Laying hens are motivated to perch at night. *Animal Welfare* 11, 11–19.

Olsson, I.A., Keeling, L.J. and McAdie, T.M. (2002) The push-door for measuring motivation in hens: An adaptation and a critical discussion of the method. *Animal Welfare* 11, 1–10.

Overall, K. (1997) *Clinical Behavioral Medicine for Small Animals*. Mosby, St Louis, Missouri.

Panksepp, J. (1971) Aggression elicited by electrical stimulation of the hypothalamus in albino rats. *Physiology and Behavior* 6, 321–329.

Panksepp, J. (1980) Social deprivation and play in rats. *Behavioural and Neural Biology* 30, 197–206.

Panksepp, J. (1990) The psychoneurology of fear: Evolutionary perspectives and the role of animal models in understanding human anxiety. In: Burrows, G.D.,

Roth, M. and Noyes, R. Jr (eds) *Handbook of Anxiety No. 3, The Neurobiology of Anxiety*. Elsevier/North-Holland Biomedical Press, Amsterdam, pp. 3–58.

Panksepp, J. (1994) Evolution constructed the potential for subjective experience within the neurodynamics of the mammalian brain. In: Eckman P. and Davidson, R.J. (eds) *The Nature of Emotion: Fundamental Questions*. Oxford University Press, Oxford, UK, pp. 336–399.

Panksepp, J. (1998) *Affective Neuroscience: the Foundations of Human and Animal Emotions*. Oxford University Press, Oxford, UK.

Panksepp, J. (2003) At the interface of the affective, behavioral, and cognitive neurosciences: Decoding the emotional feelings of the brain. *Brain and Cognition* 52, 4–14.

Panksepp, J. (2004) Affective consciousness and the origins of human mind: A critical role of brain research on animal emotions. *Impuls* 57, 47–60.

Panksepp, J. (2005a) Affective consciousness: Core emotional feelings in animals and humans. *Consciousness and Cognition* 14, 30–80.

Panksepp, J. (2005b) On the embodied neural nature of core emotional affects. *Journal of Consciousness Studies* 12, 158–184.

Panksepp, J. (2005c) Affective-social neuroscience approaches to understanding core emotional feelings in animals. In: McMillan, F.D. (ed.) *Animal Mental Health and Well-being*. Blackwell Publishing Ltd, Oxford, UK, pp. 57–76.

Panksepp, J. (2007) Neuroevolutionary sources of laughter and social joy: Modeling primal human laughter in laboratory rats. *Behavioral Brain Research* 182, 231–244.

Panksepp, J. (2011) The basic emotional circuits of mammalian brains: Do animals have emotional lives? *Neuroscience and Biobehavioral Reviews* 35, 1791–1804.

Panksepp, J., Normansell, L., Cox, J.F. and Siviy, S.M. (1994) Effects of neonatal decortication on the social play of juvenile rats. *Physiology and Behavior* 56, 429–443.

Panksepp, J., Knutson, B. and Burgdorf, J. (2002) The role of brain emotional systems in addictions: A neuro-evolutionary perspective and new 'self-report' animal model. *Addiction* 97, 459–469.

Pedersen, L.J., Jensen, M.B., Hansen, S.W., Munksgaard, L., Ladewig, J., *et al.* (2002) Social isolation affects the motivation to work for food and straw in pigs as measured by operant conditioning techniques. *Applied Animal Behaviour Science* 77, 295–309.

Petherick, J.C., Duncan, I.J.H. and Waddington, D. (1990) Previous experience with different floors influences choice of peat in a *y*-maze by domestic fowl. *Applied Animal Behaviour Science* 27, 177–182.

Petherick, J.C., Waddington, D. and Duncan, I.J.H. (1991) Learning to gain access to a foraging and dustbathing substrate by domestic fowl: Is 'out of sight out of mind'? *Behavioural Processes* 22, 213–226.

Petrie, N.J., Mellor, D.J., Stafford, K.J., Bruce, R.A. and Ward, R.N. (1996) Cortisol responses of calves to two methods of disbudding used with or without local anaesthetic. *New Zealand Veterinary Journal* 44, 9–14.

Pfaff, D.W. (1999) *Drive: Neurobiological and Molecular Mechanisms of Sexual Motivation*. MIT Press, Cambridge, Massachusetts.

Price, E.O. (2008) *Principles and Applications of Domestic Animal Behavior*. CAB International, Wallingford, UK.

Provenza, F.D. and Villalba, J.J. (2006) Foraging in domestic vertebrates: Linking the internal and external milieux. In: Bels, V.L (ed) *Feeding in Domestic Vertebrates: From Structure to Behaviour*. CAB International, Wallingford, UK, pp. 210–240.

Provenza, F.D., Lynch, J.J. and Nolan, J.V. (1993) The relative importance of mother and toxicosis in the selection of foods by lambs. *Journal of Chemical Ecology* 19, 313–323.

Redbo, I. (1990) Changes in duration and frequency of stereotypies and their adjoining behaviours in heifers, before, during and after the grazing period. *Applied Animal Behaviour Science* 26, 57–67.

Redbo, I. and Nordblad, A. (1997) Stereotypies in heifers are affected by feeding regime. *Applied Animal Behaviour Science* 53, 193–202.

Reimert, I., Bolhuis, J.E., Kemp, B. and Rodenburg, T.B. (2013) Indicators of positive and negative emotions and emotional contagion in pigs. *Physiology and Behavior* 109, 42–50.

Riber, A.B. and Mench, J.A. (2008) Effects of feed and water based enrichment on the activity and cannibalism in Muscovy ducklings. *Applied Animal Behaviour Science* 114, 429–440.

Rinn, W.E. (1984) The neuropsychology of facial expression: A review of the neurological and psychological mechanisms for producing facial expressions. *Psychological Bulletin* 95, 52–77.

Rodenburg, T.B. and Turner, S.P. (2012) The role of breeding and genetics in the welfare of farm animals. *Animal Frontiers* 2, 16–21.

Rodenburg, T.B., De Haas, E.N., Nielsen, B.L. and Buitenhuis, A.J. (2010) Fearfulness and feather damage in laying hens divergently selected for high and low feather pecking. *Applied Animal Behaviour Science* 128, 91–96.

Rogan, M.T. and LeDoux, J.E. (1995) Ltp is accompanied by commensurate enhancement of auditory-evoked responses in a fear conditioning circuit. *Neuron* 15, 127–136.

Rogan, M.T. and LeDoux, J.E. (1996) Emotion: Systems, cells, synaptic plasticity. *Cell* 85, 469–475.

Rollin, B.E. (1989) *The Unheeded Cry: Animal Consciousness, Animal Pain and Science*. Iowa State University Press, Ames, Iowa. p. 330.

Romich, J.A. (2005) *Fundamentals of Pharmacology for Veterinary Technicians*. Thomson Delmar Learning, Clifton Park, New York.

Rushen, J., Boissy, A., Terlouw, E.M. and de Passille, A.M. (1999) Opioid peptides and behavioral and physiological responses of dairy cows to social isolation in unfamiliar surroundings. *Journal of Animal Science* 77, 2918–2924.

Sambraus, H.H. (1985) Mouth-based anomalous syndromes. In: Fraser, A.F. (ed.) *Ethology of Farm Animals*. Elsevier, Amsterdam, pp. 391–422.

Sandoe, P. (1996) Animal and human welfare: Are they the same kind of thing? *Acta Agriculturae Scandinavica Section A* 27, 11–15.

Savastano, G., Hanson, A. and McCann, C. (2003) The development of an operant conditioning training program for new world primates at the bronx zoo. *Journal of Applied Animal Welfare Science* 6, 247–261.

Seaman, S., Waran, N.K. and Appleby, M.C. (2001) Motivation of laboratory rabbits for social contact. In: Garner, J.P., Mench, J.A. and Heekin, S.P. (eds) *Proceedings of the 35th International Congress of the International Society of Applied Ethology*, 4–8 August. University of California Davis, The Center for Animal Welfare, Davis, California, p. 88.

Seksel, K. and Lindeman, M.J. (1998) Use of clomipramine in the treatment of anxiety-related and obsessive-compulsive disorders in cats. *Australian Veterinary Journal* 76, 317–321.

Sherwin, C.M. (2004) The motivation of group-housed laboratory mice, *Mus musculus*, for additional space. *Animal Behaviour* 67, 711–717.

Sherwin, C.M., Lewis, P.D. and Perry, G.C. (1999) The effects of environmental enrichment and intermittent lighting on the behaviour and welfare of male domestic turkeys. *Applied Animal Behaviour Science* 62, 319–333.

Siegel, A. (2005) *The Neurobiology of Aggression and Rage*. CRC Press, Boca Raton, Florida.

Siegel, A. and Shaikh, M.B. (1997) The neural bases of aggression and rage in the cat. *Aggression and Violent Behaviour* 2, 241–271.

Siviy, S. and Panksepp, J. (1987) Sensory modulation of juvenile play. *Developmental Psychobiology* 20, 39–55.

Smith, K.S. and Berridge, K.C. (2007) Opioid limbic circuit for reward interaction between hedonic hotspots of nucleus accumbens and ventral pallidum. *Journal of Neuroscience* 27, 1594–1605.

Smith, S.F., Appleby, M.C. and Hughes, B.O. (1990) Problem solving by domestic hens: Opening doors to reach nest sites. *Applied Animal Behaviour Science* 28, 287–292.

Stafford, K.J. and Mellor, D.J. (2005) Dehorning and disbudding distress and its alleviation in calves. *Veterinary Journal* 169, 337–349.

Stilwell, G., Campos de Carvalho, R., Lima, M.S. and Broom, D.M. (2009) Effect of caustic paste disbudding using local anesthesia with and without analgesia, on behavior and cortisol of calves. *Applied Animal Behaviour Science* 116, 35–44.

Stilwell, G., Carvalho, R.C., Carolino, N., Lima, M.S. and Broom, D.M. (2010) Effect of hot-iron disbudding on behaviour and plasma cortisol of calves sedated with xylazine. *Research in Veterinary Science* 88, 188–193.

Stolba, A. and Wood-Gush, D.G.M. (1984) The identification of behavioural key features and their incorporation into a housing design for pigs. *Annales de Recherches Veterinaries* 15, 287–299.

Stolba, A. and Wood-Gush, D.G.M. (1989) The behaviour of pigs in a semi-natural environment. *Animal Production* 48, 419–425.

Studnitz, M., Jensen, M.B. and Pedersen, L.J. (2007) Why do pigs root and in what will they root? A review on the exploratory behaviour of pigs in relation to environmental enrichment. *Applied Animal Behaviour Science* 107, 183–197.

Sylvester, S.P., Mellor, D.J., Stafford, K.J., Bruce, R.A. and Ward, R.N. (1998) Acute cortisol responses of calves to scoop dehorning using local anaesthesia and/or cautery of the wound. *Australian Veterinary Journal* 76, 118–122.

Thorne, J.B.D., Goodwin, D., Kennedy, M.J., Davidson, H.P.B. and Harris, P. (2005) Foraging enrichment for individually housed horses: practicality and effects on behaviour. *Applied Animal Behaviour Science* 94, 149–164.

Tuyttens, F.A.M. (2005) The importance of straw for pig and cattle welfare: A review. *Applied Animal Behaviour Science* 92, 261–282.

Van Putten, G. and Dammers, J. (1976) A comparative study of the well-being of piglets reared conventionally and in cages. *Applied Animal Ethology* 2, 339–356.

Vestergaard, K. (1980) The regulation of dustbathing and other patterns in the laying hen: A Lorentzian approach. In: Moss, R. (ed.) *The Laying Hen and its Environment*. Martinus Nijhoff, The Hague, the Netherlands, pp. 101–120.

Vestergaard, K. (1982) Dust-bathing in the domestic fowl – diurnal rhythm and dust deprivation. *Applied Animal Ethology* 8, 487–495.

Vestergaard, K. and Hansen, L.L. (1984) Tethered versus loose sows: Ethological observations and measures of productivity. I. Ethological observations during pregnancy and farrowing. *Annales de Recherches Veterinaries* 15, 245–256.

Widowski, T.M. and Duncan, I.J.H. (2000) Working for a dustbath: Are hens increasing pleasure rather than reducing suffering? *Applied Animal Behaviour Science* 68, 39–53.

Wood-Gush, D.G.M. and Vestergaard, K. (1989) Exploratory behaviour and the welfare of intensively kept animals. *Journal of Agricultural Ethics* 2, 161–169.

Yeates, J.W. and Main, D.C.J. (2008) Assessment of positive welfare: A review. *The Veterinary Journal* 175, 293–300.

Yoerg, S.I. (1991) Ecological frames of mind: the role of cognition in behavioral ecology. *Quarterly Review of Biology* 66, 287–301.

9 Improving Livestock, Poultry, and Fish Welfare in Slaughter Plants with Auditing Programs

TEMPLE GRANDIN

Department of Animal Science, Colorado State University, Fort Collins, Colorado, USA

Summary

Animal welfare during slaughter can often be improved by making simple changes in procedures and equipment, such as installing nonslip flooring in the stun box. Training employees to move small groups of animals and to stop yelling can reduce stress. Pigs, cattle, sheep, and other livestock are sensitive to visual distractions. Adding a light to illuminate a dark race entrance, or moving a light to eliminate a reflection, can often reduce balking and facilitates animal movement. People manage the things they measure. To determine if procedures are improving or becoming worse, numerical scoring should be used. The variables that should be measured are: percentage of animals stunned effectively with one application of the stunner; percentage vocalizing (moo, bellow, squeal); percentage falling; percentage moved without an electric prod; and percentages rendered insensible before hoisting. A major cause of poor captive bolt stunning is poor gun maintenance or damp cartridges. Dehydrated animals may be difficult to stun electrically. When controlled atmosphere stunning is being evaluated, behavior before loss of posture (ability to stand) should be monitored. Attempts to escape from the container are not acceptable.

Learning Objectives

- Assess slaughter procedures with numerical performance scoring
- Understand principles of all types of stunning methods
- Evaluation of gas and low atmospheric pressure stunning (LAPS)
- How to determine insensibility (unconsciousness)
- Understand welfare issues associated with slaughter without stunning
- Troubleshoot and solve animal handling and stunning problems

Introduction – Finding the Major Cause of Animal Welfare Problems

There are four basic types of problems that cause poor animal welfare in a slaughter plant, and to be effective in solving problems that cause animal suffering one must determine the cause of the problem correctly. The four basic types of problems are:

1. **Poorly maintained equipment or facilities.** Examples would be a broken stunner or a slippery floor that causes falling in the stun box or on the unloading ramp.

2. **Untrained employees or unsupervised employees** who mistreat animals. Examples would be poking a stick in an animal's rectum or slamming an animal down on to the floor with a heavy guillotine gate.

3. **Minor design faults** that can be fixed easily. Examples would be welding steel rods on a poorly designed slippery stun box floor to provide a nonslip surface or eliminating a reflection on shiny metal that makes animals balk and stop moving.

4. **Major design faults** where a plant will need to purchase a major piece of equipment or rebuild the facility. Examples would be replacing shackling and hoisting of live animals for religious slaughter with a restraining box that holds them in an upright position, or replacing old, undersized equipment with new, larger-capacity equipment

that can handle a plant's increased line speed. Plants handling wild, extensively raised animals that lack basic equipment such as races will need to build them (see Chapter 5, this volume, for information on design).

Some of the most serious animal welfare problems exist in parts of the world where modern equipment for handling and restraining large animals such as cattle or water buffalo is not available (Ahsan *et al.*, 2014). Welfare problems become worse if the animal is wild and untamed and not accustomed to being handled by people. Tame animals that are trained to lead are much easier to handle without modern equipment.

Practices that Should be Banned and Severe Welfare Problems that Need Immediate Corrective Action

In many countries, both government regulations and guidelines developed by industry have prohibitions against the worst practices. The World Organisation for Animal Health (OIE) has guidelines for handling and stunning of animals that include lists of practices that cause suffering and which should not be used (OIE, 2014). When a slaughter plant is being evaluated, the first step is to eliminate practices or bad employee behavior that should be prohibited. The OIE (2014) code specifically states that methods of restraint that cause avoidable suffering should not be used in conscious animals because they cause severe pain and stress, such as the following (See Quick Access Information 9.1 below).

The OIE (2014) and many countries have regulations that contain detailed specifications for stunning practices used on livestock and poultry. Animals must be rendered insensible and unconscious before dressing procedures begin. Cutting off limbs, skinning or scalding an animal that shows any sign of return to sensibility should NEVER be permitted. The animal must be re-stunned immediately before dressing procedures begin. This is a requirement of both the OIE and the regulations in many countries. If any of these cruel practices occurs, the plant has committed an egregious violation of animal welfare standards and humane slaughter regulations. The EU standards require that a plant has a designated animal welfare officer. United States Department of Agriculture (USDA) standards do not have this requirement, but many companies have hired animal welfare specialists.

Quick Access Information 9.1. Practices that cause suffering that should never be used (OIE, 2014).

- Suspending or hoisting animals (other than poultry) by the feet or legs.
- Indiscriminate and inappropriate use of stunning equipment (refers to practices such as using an electric stunner to move animals).
- Mechanical clamping of an animal's leg or feet (other than shackles used in poultry and ostriches) as the sole method of restraint.
- Breaking legs, cutting leg tendons, or blinding animals in order to immobilize them.
- Severing the spinal cord (e.g. using a puntilla or dagger) or using electric currents to immobilize animals, except for proper stunning (see Chapter 5, this volume, on the detrimental effects of immobilization). Electrical immobilization is highly stressful and aversive because it paralyses the animal but it remains sensible. Proper electrical stunning induces instantaneous insensibility. Research studies clearly show that immobilization is highly aversive and very detrimental to animal welfare

(Lambooij and Van Voorst, 1985; Grandin *et al.*, 1986; Pascoe, 1986; Rushen, 1986).
- Conscious animals should not be thrown, dragged, or dropped.
- Animals for slaughter should not be forced to walk over the top of other animals.
- The use of devices such as electric goads should be limited to battery-powered goads on the hindquarters of pigs and large ruminants, and never on sensitive areas such as the eyes, mouth, ears, anogenital region, or belly. Such instruments should not be used on horses, sheep, or goats of any age, or on calves or piglets.
- Painful procedures (including whipping, tail twisting, use of nose twitches, pressure on eyes, ears, or external genitalia) or the use of goads or other aids that cause pain and suffering (including large sticks, sticks with sharp ends, lengths of metal piping, fencing wire, or heavy leather belts) should not be used to move animals.

Assessing Welfare at the Slaughter Plant with Numerical Scoring of Performance

In Chapters 1, 2, and 4, this volume, the advantages of using numerical scoring are covered. Numerical scoring has improved greatly both animal handling and stunning (Grandin, 2003a, 2005). Measurement has also been used to show how the training of employees reduced bruises and carcass damage in both livestock and poultry (Paranhos da Costa *et al.*, 2014). A numerical scoring system developed by Grandin (1998a, 2013) to score handling and stunning is used extensively in the USA and many other countries. Others have also used numerical scoring successfully to evaluate handling (Maria *et al.*, 2004). The advantage of using numbers to evaluate practices makes it possible to determine if practices are improving or deteriorating. This is the same principle that is used in Hazard Analysis Critical Control Points (HACCP) programs for monitoring bacteria counts on meat.

The OIE code (2014) supports the use of numerically scored performance standards. They are much more effective than audits that are based largely on paperwork or subjective evaluations. The scoring system developed by Grandin (1998a) for handling and stunning is an animal-based outcome scoring system (i.e. all scores are per animal) where things that can be directly observed can be assessed. To pass the audit, an acceptable numerical score is required on all of the critical control points (CCPs). For large plants, 100 animals are scored for each CCP, while for small plants 1 h of production is monitored and scored. The five CCPs are listed (See Quick Access Information 9.2 below).

There are also acts of abuse that would result in an automatic audit failure. Some of the acts of abuse are listed in Quick Access Box 1.1 in Chapter 1 and Quick Access Box 9.1. There would also be an automatic failure if employees deliberately slammed gates on animals. In sensible animals, broken bones, bruises, or falls caused by powered devices such as head holders, stunbox gates, and other mechanical devices would also automatically result in a failed audit. Repeated use of obviously malfunctioning or broken stunning equipment would also be an act of abuse.

Objective performance scoring improves welfare

When performance is measured, plant management can determine if operations are getting better or becoming worse. In setting the critical limit, it had to

---

Quick Access Information 9.2. Numerical performance scoring of welfare at slaughter.

1. The percentage of animals rendered insensible with one application of the stunner (for all species of animals, poultry, and fish). When a captive bolt is used, an acceptable score is 95% and an excellent score is 99%. All animals, where the first shot failed, must be re-stunned immediately. For electrical stunning of both mammals and poultry, the stunner must be positioned correctly on 99% of the animals. In mammals, tong placement is scored, and in poultry, the percentage of effectively stunned birds is scored.
2. The percentage of animals rendered insensible before hoisting, skinning, scalding, or plugging of the esophagus, or other dressing procedures. The score must be 100% to pass an audit. For pigs and poultry, the audit is failed if a pig or bird enters the scalder showing any sign of sensibility. In poultry plants, the audit is failed if a red, discolored bird is found that does not have its throat cut. It is likely that the bird was scalded alive.
3. The percentage of cattle or pigs that vocalize (moo, bellow, or squeal) during handling and stunning. All vocalizing animals that occur in the stun box

or restrainer should be scored. In the lead-up race, vocalizations that occur during active handling when the animals are being moved into the stun box or restrainer are scored (see Chapter 5, this volume, for more information). Each animal is scored as either silent or vocalizing. The goal is 5% or less of the animals vocalizing. Vocalization scoring is not used for sheep at the slaughter plant. Do not score vocalizations in the lairage.
4. The percentage of animals that fall down (i.e. body touches the floor) during handling (use for all mammals). All parts of the facility should be scored, including the unloading ramp and stun box. If more than 1% of the animals fall, the plant has a problem that should be corrected.
5. The percentage of animals moved with no electric goads. Use for all mammals – 75% is acceptable and 95% is excellent.
6. For poultry only – the percentage of birds with broken or dislocated wings. The goal is 1% or less on a per bird basis.

---

be high enough to force poor plants to improve, but not so high that people think it is not attainable.

Some animal welfare advocates have complained that allowing 1% of the animals to fall or 5% to vocalize is not very good welfare. In slaughter plants that have really good internal and external auditing, the actual numbers are much lower. In 1996, before the restaurant audits started, the worst plant had 32% of the cattle vocalizing and the average was 7.7% (Grandin, 1997a, 1998b). Since the audits started, the average vocalization percentage is under 2% and the worst plant is 6% (Grandin, 2005). Reductions in falling were even more striking. In 2005, data were examined from the most heavily audited plants, which were audited by both McDonald's and Wendy's. Over 3000 cattle and pigs were scored in over 30 plants, and 0% fell down. Before the welfare audits were started, 100% of the cattle and pigs were moved with an electric prod in many plants. Today, the average electric goad use is under 20%. In most plants, the only place the electric goad is used is at the entrance of the stun box or restrainer. During the first 4 years of audits, 17.5% of cattle were moved with an electric goad (Grandin, 2005). Young fed cattle can be moved more easily than old cull cows. The average for young fed beef was 15.2% electric goaded and 29% for old cull Holstein cows (Grandin, 2005). Data collected from the 2005 audits of 72 beef and pork plants indicated that only one plant failed on electric goad use, and 23% of the beef plants used electric prods on 5% or less of the cattle. Cattle stunning scores averaged 97.2% on the first shot and 91% (60 out of 66 beef plants) passed the insensibility audit (Grandin, 2005). In 2005, 100% of the beef cattle were rendered insensible prior to hoisting in 42 plants. When a head holder was not used, a skilled operator shot 95% of the cattle accurately and a less experienced operator had a score of 81% (Atkinson et al., 2013).

Vocalization during the handling and restraint of cattle

Vocalization (moos or bellows) are scored during handling and in the stunning box or restrainer. Vocalization in cattle during handling and restraint is an indicator of stress (Dunn, 1990; Hemsworth et al., 2011). Grandin (1998b) found that 99% of cattle vocalizations occurred in response to an aversive event such as missed stuns, slipping on the floor, electric prod use, or excessive pressure from a restraint device. Vocalization scoring is especially useful for identifying severe welfare problems. A plant with a restraint device that applied excessive pressure had 25% and 35% of the cattle vocalizing (Grandin, 1998b; Bourguet et al., 2011). In another study, excessive pressure applied by a head-holding device caused 23% of the cattle to bellow, but when the pressure was reduced, vocalization dropped to 0% of the animals (Grandin, 2001a). Electric prod use on high percentages of animals is associated with high percentages of animals vocalizing. Two studies where over 60% of the cattle were moved with electric prods had high vocalization scores of 10% and 13% (Miranda de la Lama et al., 2012; Probst et al., 2014). Vocalizations are scored during both conventional slaughter with stunning and ritual slaughter without stunning. In plants with single-file races, vocalizing animals are scored in the race when they are being moved. All vocalizing animals in the stun box or religious slaughter restraint box or restrainer conveyor are scored. In plants that do not have races or stun boxes, vocalization is scored any time an animal is being moved by people. All vocalizations are also scored in the pen where animals are stunned or slaughtered. The critical limit is 3% of the cattle vocalizing. If the plant uses a head-holding device for either conventional or ritual slaughter, the standard is 5%. Yes or no scoring is used on a per animal basis – an animal is scored as either a quiet animal or a vocalizing animal. Plants can achieve this standard easily (Grandin, 1998a, 2005, 2012). Data collected by the author in ten beef plants that conducted ritual slaughter indicated that this standard could be achieved when the animal was held with a well-designed head-holding device. For both conventional and ritual slaughter, the audit is failed if any cattle vocalize after hoisting (Grandin, 2012).

Vocalization during the handling and restraint of pigs

Squealing in pigs is associated with stressful handling and painful procedures (Warriss et al., 1994; White et al., 1995). Unlike cattle, it is difficult to determine which pig is squealing in a single-file race full of pigs. A sound-level meter could be used by the quality assurance department to monitor sound levels. One innovative pork plant installed a traffic signal. When the volume of pig squealing during handling was too high, it turned red, and when pig squealing lessened, it turned green. Sound level scoring works well within a plant, but works poorly for comparing one plant with another plant, due to differences in plant design and the number of pigs in a room. For comparisons between plants, squealing of individual pigs can be scored in the following areas: in plants with races, the percentage of pigs squealing in the conveyor, restrainer or stun box are scored; in plants with gas stunning, each individual pig is scored when the gondola is loaded; in plants that stun pigs in groups on the floor, squealing should be scored on each individual pig as it is being handled for stunning. Yes or no scoring is used on a per animal basis – a pig is scored as either silent or vocalizing. The limit should be 5% of the pigs vocalizing. The audit is failed if any pig vocalizes after stunning.

Do not use vocalization scoring in sheep

Do not use vocalization (baa baa) scoring for sheep to assess the stressfulness of handling and restraint. As sheep are ultimately a defenceless prey animal, they often do not vocalize even when they are stressed and in pain. The inhibition of vocalization when a sheep is threatened by predators (Dwyer, 2004) may explain why sheep do not vocalize during aversive distressful handling procedures. Sheep may struggle when restrained, but they usually do not vocalize. A lone sheep will vocalize because it is separated from its herdmates (Boissy et al., 2005).

Vocalization scoring of goats

The vocalization standard for goats needs to be developed.

Electric prod (goad) use – critical limits

The critical limit for cattle, sheep, pigs, goats, and all other mammals is that 75% must be moved with no electric goad. An excellent score is 95% moved with no electric goad. Flags or other non-electrified items should be the primary driving aids. The electric goad should only be picked up when it is needed and then put away after use. If the electric goad is replaced with painful driving methods such as beating, dragging, breaking tails or other abusive practices the audit is failed. Yes or no scoring is used – the animal is either moved with an electric goad or not moved with one. Since it is very difficult to determine if the switch on an electric goad is pushed, electric goad use is scored if the animal

is touched with the electrical device. Plants can achieve the standard easily (Grandin, 2003a, 2005, 2014).

Falling during handling – critical limits

The critical limit for falling in cattle, sheep, goats, pigs, and all other mammals is that no more than 1% of the animals should fall anywhere in the facility, from the unloading trucks until the animal is stunned or ritually slaughtered. A fall is scored if the animal's body touches the ground. Stun boxes that are designed to make animals fall will result in a failed audit. Plants can achieve this standard easily (Grandin, 2012). The OIE (2014) codes also use numerical scoring for falling. Corrective action needs to be taken if more than 1% of the animals fall.

Poultry handling – critical limits

Handling of broiler chickens is assessed by counting the percentage of birds that have broken or dislocated wings. Broken wings should be scored with the feathers on to avoid confusing handling damage, which is a welfare issue, with damage caused by the feather-removal equipment. In dump module systems, the minimum acceptable score is 1% (of birds with broken wings) in lightweight birds and 3% (of birds with broken wings) in heavy birds over 3 kg. An excellent score for birds weighing over 3 kg is 1% or less. The industry can achieve this standard easily (Grandin, 2014). Data from restaurant audits of 22 poultry plants with dump module systems in the USA that process heavy chickens weighing 3 kg or more showed that all of the plants had broken wing scores of 3% or less. Six excellent plants processing heavy chickens had 1% or less broken wings. When the audits first started, most plants had 5–6% of birds with broken wings, but their scores became better when they improved their handling practices. In some countries with low labor costs, each bird is carefully picked up and held in an upright position and placed in the container. In countries where this method is used, broken wings may drop to less than 0.5%.

Principles of Captive Bolt Stunning Methods

See Chapter 10, this volume, and the American Veterinary Medical Association (AVMA, 2013) for descriptions and instructions for captive bolt stunning methods. The Humane Slaughter Association has an excellent paper on captive bolt stunning (HSA, 2014). The frontal shooting positions on the animal's forehead, shown on the diagrams in the chapter on euthanasia, are the preferred positions. They are usually more effective. There are situations where shooting behind the poll may be required. To be effective, the captive bolt must be aimed at the correct angle to hit the brain (Gregory *et al.*, 2009). Careful maintenance and cleaning of the captive bolt is essential.

Principles of Electric Stunning Methods for Mammals

When electric stunning is used, sufficient amperage (current) must be passed through the brain to induce instantaneous insensibility. Electricity flows like water. Amperage is the volume of electricity and voltage is analogous to water pressure. Modern electronically controlled stunners can be set to vary the voltage automatically to deliver the required amperage. On older stunning units, the voltage is set and the amperage will vary depending on the resistance of the animal. Wetting the animal will reduce electrical resistance and is strongly recommended. Quick Access Information 9.3 shows the minimum amperage (current) levels for electric stunning (OIE, 2014). To induce instantaneous insensibility, electric stunning must induce a grand mal epileptic seizure (Croft, 1952; Warrington, 1974; Lambooij, 1982; Lambooij and Spanjaard, 1982; HSA, 2000; AVMA, 2013). When an animal is stunned effectively, there will be a tonic (rigid)

phase where the animal is stiff, followed by a clonic (kicking) phase.

For pigs, sheep, cattle, or goats, there are two basic types of electrical stunning: head-only stunning and head-to-body cardiac arrest. Figure 9.1 shows correct positioning of a head-only electric stunner. Figures 9.2 and 9.3 show the correct positioning of the head electrode, which is critical. It must never be placed on the neck because this will cause the current to bypass the brain. The body electrode placement may be on the back, side of the body, brisket, or belly. Passing the current through the feet is not recommended. Head-only stunning is reversible and the lamb or pig must be bled within 15 s to prevent return to sensibility (Blackmore and Newhook, 1981; Lambooij, 1982). When cardiac arrest stunning is used, the stunning-to-bleed interval can be longer because the animal is killed by

Fig. 9.1. Head-only tong-type electric stunner placed in the correct position on the head. The current must pass through the brain to induce a grand mal epileptic seizure. This will make the animal instantly insensible.

Quick Access Information 9.3. **Minimum current levels for head-only stunning.** (From HSA 2000; AVMA, 2013; OIE, 2014).

Species	Minimum current levels (A)[a,b]
Cattle	1.5
Calves (bovines <6 months of age)	1.0
Pigs	1.25
Sheep and goats	1.0
Lambs	0.7
Ostriches	0.4

Notes: [a]In all cases, the correct current level should be attained within 1 s of the initiation of stun and maintained for at least 1–3 s, and in accordance with the manufacturer's instructions. [b]For head-to-body cardiac arrest stunning that kills the animal, higher amperages may be required because the current has to travel a greater distance through the body.

Fig. 9.2. Head-to-body cardiac arrest electric stunner. The head electrode must NEVER be placed on the neck. It must either be in the hollow behind the ears or on the forehead. Placing the body electrode on the side of the body helps reduce blood spotting in the meat. This photo shows the correct placement.

Fig. 9.3. Head-to-body cardiac arrest electric stunner for sheep. To ensure contact through the wool, water is injected through the electrodes. For electrical safety, the entire unit is constructed from plastic.

Fig. 9.4. In small plants, many pigs that are head-only stunned regain sensibility because the hoist is very slow. A simple solution to this problem is to apply the stunner to the head first and then apply it a second time to the chest to stop the heart. (Photograph courtesy of Erika Voogd.)

stopping the heart. Cardiac arrest will mask the tonic and clonic phase of the epileptic seizure. For more information on electrical stunning, refer to Wotton and Gregory (1986), Gregory (2007, 2008) and Weaver and Wotton (2008). Figure 9.4 shows a simple way to achieve cardiac arrest in a plant that has simple equipment. The stunner is applied to the head first to achieve insensibility. After the animal is rendered insensible, the wand or tong is applied to the chest (Vogel et al., 2010).

To reduce blood spotting (petechial hemorrhages in pork or poultry), some stunners use frequencies higher than the standard 50–60 cycle that comes from the electrical mains. Higher frequencies will reduce the length of time that pigs will remain insensible (Anil and McKinstry, 1992). Very high frequencies of 2000–3000 Hz fail to induce insensibility and should not be used (Croft, 1952; Warrington, 1974; Van der Wal, 1978). Frequencies of 1592 Hz sine wave or 1642 Hz square wave will induce insensibility in small pigs (Anil and McKinstry, 1992). Some effective commercial systems apply an 800 Hz current to the head, followed by a 50 Hz current applied to the body. This combination is effective (Lambooij et al., 1997; Berghaus and Troeger, 1998; Wenzlawowicz et al., 1999). High frequencies applied to induce cardiac arrest must never be used. Practical experience with a European tong-type stunner showed that 400 Hz applied to the chest failed to induce cardiac arrest in pigs.

Principles of Poultry Stunning Methods

Chickens and turkeys are stunned and rendered insensible in commercial poultry plants with either electric, gas, or low atmospheric pressure stunning (LAPS). To render poultry instantaneously insensible, the current (amperage) levels and electrical frequencies shown in Quick Access Information 9.4 must be used (OIE, 2014). The EU requirement for use of higher amperages has resulted in blood spotting and damage to breast meat. Wotton et al. (2014) report that 100 hz AC at 200 milliamps complies with both

T. Grandin

EU legislative requirements for an effective stun and halal requirements for a heartbeat after stunning. Use of these parameters may reduce meat damage. HSA (2001) has recommendations for small flocks.

Quick Access Information 9.5 shows a comparison of the advantages and disadvantages of electric stunning and gas stunning systems for poultry.

Mechanical design of gas systems for poultry

From an engineering standpoint, there are two basic types of gas stunning systems – the first type is an open system, where birds in the transport containers move down a tunnel or pit in a continuous flow on conveyors. The entrance and exit of the system are open to the atmosphere. The second type of system is a closed system (positive pressure). In this system, batches of birds in containers are placed in a chamber that has sealed doors. Carbon dioxide (CO_2) will work well in an open system because it is heavier than air and it will stay in the bottom of a chamber. Argon will also work well in an open system because it is also heavier than air. The main problem with argon is high cost. For best results, open systems should use gases that are heavier than air. This same principle is used for CO_2 stunning of pigs. Pigs in gondolas (elevator cars) are lowered into a deep pit filled with CO_2. In a closed system, the gas is introduced into a closed chamber under positive pressure. The gas moves through a ventilation system that recirculates it while the chamber is being filled to the specified gas concentration. One advantage of a closed chamber is that the gas mixture can be controlled more precisely compared to an open system. All kinds of gases will work well in a closed system. The disadvantage of a closed system is that it uses a greater volume of gas compared to a CO_2 chamber where CO_2 sits in the bottom of a pit or tunnel. In a closed system, all the gas has to be evacuated before the next batch of birds is put in. Closed systems would work really well with inert nitrogen gas because almost all the oxygen can be flushed out of the chamber. Inert gas systems work best and the bird's reactions to the gas will be reduced if the oxygen level is 2% or less. This is very difficult to achieve in an open chamber. Open systems that use nitrogen have been a commercial failure due to high numbers of broken wings. Further, nitrogen works poorly in an open system because it is not heavier than air. Nitrogen is cheap and can be used in a well-designed closed system.

Low atmospheric pressure stunning (LAPS) of poultry

Since the publication of the first edition of this book, a new stunning method for chickens has been developed and is used commercially (Vizzier-Thaxton et al., 2010; HSA, 2013). Containers full of chickens are

---

Quick Access Information 9.4. Current levels[a] and electrical frequencies for stunning poultry. (From OIE, 2014.)[b]

Species	Current (mA) per bird
Broilers	100
Layers (spent hens)	100
Turkeys	150
Ducks and geese	130

Frequency (Hz)	Chickens (mA)	Turkeys (mA)
<200	100	250
200–400	150	400
400–1500	200	400

Notes: [a]All currents are sine wave AC. Other types of currents may require higher amperages. [b]Additional information on electric stunning of poultry can be found in Raj (2006).

Quick Access Information 9.5. Welfare comparison of electric versus gas stunning and low-pressure systems for poultry.

	Stunning	
	Electric[a]	Gas or low atmospheric pressure (LAPS)
Advantages	• Instantaneous insensibility (Raj and O'Callaghan, 2004).	• Birds do not have to be removed from the transport containers for stunning. This greatly reduces handling stress. • All birds in the container will be stunned. • Less likely to have problems with employees abusing birds because they do not handle live birds.
Disadvantages	• Birds are hung live inverted on the shackle line. This is stressful for the birds and increases plasma corticosterone (Kannen *et al.*, 1997; Bedanova *et al.*, 2007). • Small birds mixed with large birds miss the water bath and may not be stunned. • Supervising and training employees is more difficult. Problems with people abusing birds are more likely because people handle each individual live bird. • In poorly designed systems, the birds may get small pre-shocks before they enter the water bath.	• Insensibility is not instantaneous. The distress and discomfort to the birds prior to the loss of insensibility will vary depending on the gas mixture or the speed of air removal. • Requires very careful continuous monitoring of the gas mixtures. Wind around the plant building, changes in plant ventilation, and open and closing doors may change gas mixtures in some systems. These problems do not occur in LAPS systems or closed gas systems. • Much higher cost to install equipment compared to electrical stunning.

Note: [a]Electric stunning takes place in a water bath.

placed in a chamber where the air pressure is gradually reduced. Chickens stunned by this method had no escape movements and flapping did not occur until after loss of posture (LOP), when the birds were unconscious. McKeegan *et al.* (2013) report that the lack of behavior responses before LOP indicate that LAPS is a humane method. The air must be withdrawn from the chamber at a slow rate and the total stun cycle kills the chickens in 280 s. The patent by Cattaruzzi and Cheek (2012) contains further specifications for the process. LAPS equipment is economical to operate and easier to install than gas stunning. Safeguards must be installed to ensure that the air removal cycle is never speeded up because this would be highly stressful for the birds. Vizzier Thaxton *et al.* (2010) recommends continuous video monitoring of the birds' reactions. Initial experiments with turkeys indicate that air pressure parameters used for chickens are not effective and turkeys will require different parameters (Kurt Vogel, 2014, personal communication). Low atmospheric pressure is not recommended for mammals due to differences in anatomy.

Backup systems and bleeding methods for LAPS and gas stunning of poultry

Both gas and LAPS systems for poultry should have a full electrical stunning backup system. Many plants have more than one chamber or line and backup electrical stunning removes the temptation for plant management to speed up the remaining systems after a breakdown. Speeding up the LAPS air removal cycle or raising CO_2 levels quickly would cause distress to poultry.

Bird and Pig Reactions to Different Gas Mixtures

There is great controversy in the scientific community on the correct gas mixtures for poultry, and it is beyond the scope of this book to review all the literature fully. CO_2 systems use either pure CO_2 or CO_2 mixed with oxygen, argon, or nitrogen. Other systems exclusively use inert gases such as nitrogen or argon. A mixture of 90% argon with 2% or less of oxygen is not aversive to pigs or chickens. The animals voluntarily entered a chamber filled with this mixture to obtain food (Raj and Gregory, 1990, 1995). In CO_2-only systems, chickens should never be introduced suddenly into concentrations of CO_2 higher than 30% unless oxygen is added, otherwise this will result in violent flapping and is highly detrimental to bird welfare. Observations by the author of commercial systems that use pure CO_2 indicate that, in order to reduce the bird's reaction to CO_2, the concentration of CO_2 must rise very gradually over a period of 6 min. The author observed a video where a camera was placed in a container with the birds. A method where CO_2 was raised gradually during a 6 min, ten-step procedure caused very little reaction before the birds lost the ability to stand. Further, practical experience in commercial facilities indicates that a smooth, gradual increase in CO_2 from 0% to more than 50–55% reduces bird reactions prior to falling over (LOP) (Gerritzen et al., 2013). Chickens require a more gradual slow increase of CO_2 concentration for a longer period of time than turkeys. A slow, gradual increase in CO_2 concentration will prevent wing flapping and panicked attempts to escape from the container. When chickens lose the ability to stand and have LOP, they are unconscious (Benson et al., 2012; AVMA, 2013).

Another gas system that is used commercially is the biphasic system. In this system, the chickens are initially exposed to an atmosphere of 40% CO_2, 30% oxygen, and 30% nitrogen for 60 s. The second stage is the euthanasia phase, which has an atmosphere of 80% CO_2 and air. The biphasic system with added oxygen during the anesthetic phase has advantages from both a welfare and a carcass-quality standpoint (McKeegan et al., 2007a,b; Coenen et al., 2009). The two most common commercial gas systems for chickens are CO_2 only with a slow induction and the biphasic system.

There is a species difference between birds and pigs. Pigs should be introduced suddenly into a high level of 90% CO_2 (Hartung et al., 2002; Becerril-Herrera et al., 2009). In 90% CO_2 when pigs lose the ability to stand (LOP), they are unconscious (Craig Johnson, Massey University, 2014, personal communication). Pigs stunned with 80% CO_2 have more pale soft exudative (PSE) meat, which is a severe quality defect that makes meat pale, soft, and watery (Gregory, 2008). Lower levels of CO_2 at 70% are aversive to pigs and are not recommended (Becerril-Herrera et al., 2009). The author has observed pigs in a chamber where they could be watched easily. Pigs free of the porcine stress syndrome (PSS) gene immersed in 90% CO_2 had very little behavioral reaction and remained silent until after they fell over (LOP) (Grandin, 2003a). Pigs with the PSS gene had more vigorous reactions (Troeger and Wolsterdorf, 1991). Observations in a commercial plant indicated that pig genetics might have an effect on a pig's reaction to CO_2. Some pigs remained calm and others violently attempted to escape from the container (Grandin, 1988). Studies on the pig's reaction to CO_2 have variable results. Some studies show that CO_2 is not aversive (Forslid, 1987; Jongman et al., 2000). Other studies show that it is aversive (Hoenderken, 1978, 1983). The Forslid (1987) study used purebred Yorkshire pigs, and in other studies the breed of the pigs was unknown. The author's observations indicate that genetic differences in the pigs may explain some of the variations in the results of different studies.

How to Determine Insensibility (Unconsciousness)

Vogel et al. (2010) explains that the corneal reflex is the involuntary eyelid-closure response to protect the eyes from foreign objects. It involves two cranial nerves, one sensory and one motor, which converge in the brainstem. When the cornea senses a stimulus, such as the touch of a finger or pen, an impulse is sent through the sensory cranial nerve to a center in the brainstem. A reflex impulse is then sent from the brainstem to the eyelids to trigger the closure, which is known as the corneal reflex. This test is commonly used to judge brainstem abnormalities in human medicine. However, the corneal reflex only indicates brainstem activity, which is not indicative of perception by the stunned animal. Electrically or gas-stunned animals with a weak corneal reflex triggered by the tip of a pen and no other signs of return to sensibility would be in a state of surgical anesthesia. If the animal has spontaneous, natural blinking that occurs when the eye is not touched, the animal is definitely sensible (conscious), and must be re-stunned (Verhoeven *et al.*, 2015). People who are assessing insensibility should look at live animals in the lairage so they will know what spontaneous blinking looks like. For animals stunned with either gunshot penetrating or non-penetrating captive bolt, the corneal reflex and all eye movements must be absent (Gregory, 2008; AVMA, 2013). The eyes should open into a wide, blank stare and not be rotated (Gregory, 2008). Do NOT use a finger or other thick, blunt object to poke the eyes of animals with small eyes, such as pigs and sheep, when testing small animals for eye reflexes. This causes confusing signs that are difficult to interpret (Grandin, 2001b). A finger may be used on animals with large eyes, such as cattle. Multiple indicators of return to consciousness must all be absent (Verhoeven *et al.*, 2015). Never rely on a single indicator.

The following signs of insensibility can be used when assessing stunning efficacy in cattle, pigs, sheep, and other mammals (Quick Access Information 9.6 and 9.7) (HSA, 2000, 2001, 2013, 2014; AVMA, 2013; Grandin, 2013). Additional information can be found in Gregory (2008) and Verhoeven *et al.* (2015).

For all types of stunning, ignore kicking and other leg movements. Kicking reflexes occur when the brain is disrupted because the neural circuits for back and forth walking are located in the spine (Grillner, 2011). The walking circuits become overactive when the brain is disrupted. Reflexes can still occur after the head has been removed. If the animal has stiff curled tongue, it has been poorly stunned. A limp flaccid tongue is an indicator of a properly stunned animal.

Quick Access Information 9.7 and 9.8 from Grandin (2013), which originally appeared in Gregory (2007).

Practical Methods to Improve Stunning in the Abattoir

In-plant evaluation of electrical stunning

A person evaluating animal welfare will not know the amperage of a stunning system if it has no

---

Quick Access Information 9.6. Signs of insensibility (unconsciousness) in animals shot with properly applied captive bolt or gunshot that cause physical brain concussion.

All four signs MUST BE ABSENT before invasive dressing procedures are started:

1. *Rhythmic breathing* – must be absent. Occurs if ribs move in and out at least twice.
2. *All eye movements* – must be absent – including corneal reflex, palperal reflex, natural blinking, and nystagmus. Eyes fully open in a blank stare, with a fixed dilated pupil.
3. *Vocalization* – must be absent – bellow, moo, or squeal. It is permissible to have a small grunt due to chest compression when the animal falls.
4. *Righting reflex* – must be absent – when hung on the rail (Fig. 9.5), this can be observed as an arching of the back and sustained backward lifting of the head. This should not be confused with a momentary flop of the head, which occurs when the back legs exhibit reflexive kicking. If the animal is on the floor, complete loss of posture (LOP) and loss of the ability to raise its head or stand. After the shot, a properly stunned animal will fall instantly and lose the ability to stand.

Fig. 9.5. Properly stunned, insensible beef animal. The head hangs straight down. The hind legs or forelegs may move and should be ignored. People assessing insensibility should be trained to look at the head. When the body moves, the head should be loose and floppy.

T. Grandin

Electrical stunner effectiveness may also be difficult to evaluate if a plant uses an immobilizing current to hold the carcass still after stunning. To observe the clonic and tonic spasm, the immobilizer current must be turned off. If the stunner is not capable of inducing an epileptic seizure, it should not be used. The use of low currents or very high frequencies will result in paralyzing a sensible animal. Electrical immobilization MUST NEVER be used as a substitute for effective electric stunning that induces a grand mal seizure (OIE, 2014). It is highly aversive and detrimental to welfare (Lambooij and VanVorst, 1985; Grandin *et al.*, 1986; Pascoe, 1986).

Correct tong placement for electric stunning

Correct placement of the stunning tongs (electrodes) is important to ensure that the current passes through the brain of 99% of the animals. This standard can be achieved easily (Grandin, 2001b, 2003a). Correct placement is essential to induce instantaneous insensibility (Anil and McKinstry, 1998). Figures 9.1, 9.2, and 9.3 show correct placement. The head electrodes must never be placed on the neck because the current will bypass the brain. The stunning wand or tongs also must not be applied to the eyes, ears, or rectum of the animal. Acceptable electrode positions on the head are:

- Forehead to body (cardiac arrest) – score position of head electrode only
- Hollow behind the ear to body (cardiac arrest) – score position of head electrode only

amperage gauge or equipment for measuring volts, amperage, and frequency. An easy way to evaluate a stunner is to test its ability to induce the tonic (rigid, still phase) followed by the clonic spasms (paddling, kicking phase) of an epileptic seizure. If the plant is using a one-piece head-to-body cardiac arrest stunning wand, the seizure may be masked by the cardiac arrest. The only way to test this system is to connect the electrical leads from the stunner power box to a tong-type stunner that can be applied to the head only. The stunner is acceptable if it induces a seizure. The presence of the tonic and clonic spasm is evaluated AFTER the electric stunner tong is removed from the head. When testing for tonic and clonic spasms, the tong should be held on for 1–3 s. Holding the tong on too long for over 5 s may depolarize the spine and mask the seizure.

- Between eye and ear on both sides of the head with a tong-type stunner (head only)
- Top of the head and under the jaw with a tong-type stunner (head only).

For a poultry water bath system, 99% of the birds must be positioned in the water bath so that the current will pass through the brain.

Hot wanding should occur in only 1% or fewer of the animals. Hot wanding occurs when the electrode is energized before the tong or other device is pressed firmly against the animal. Score a hot wand if the animal vocalizes (squeals, bellows) or pulls away in direct response to stunner application. Do not score hot wanding vocalizations in sheep. For poultry, score pre-shocks at the water bath entrance as hot wanding. Count each bird as pre-shocked or not pre-shocked.

In chickens, the behavior that indicates a seizure is different. Birds that are stunned properly while hanging on the rail will have vibrating tremors of their wings. Their wings will be held close to the body. When the birds regain sensibility, they may extend their wings fully and do full flapping motions.

Welfare evaluation of gas or low atmospheric pressure stunning

To evaluate gas stunning, a system should have either windows or video cameras so that the birds or pigs can be viewed during the induction phase before they fall over (LOP) and become insensible. It is the author's opinion that the animal-based outcome that should be measured for auditing and monitoring of commercial systems is the animal's reaction during the induction phase. The author has observed chickens and turkeys in several types of commercial systems. In the best poultry CO_2 systems where the gas is introduced gradually, very few birds flapped their wings, and some birds shook their heads and gasped before LOP. This is a sign of respiratory distress (Webster and Fletcher, 2001). None of the birds attempted to escape from the container or did continuous violent flapping before LOP. The author has observed similar responses in chickens stunned with LAPS. One study of CO_2 stunning with turkeys showed that 6.2% of the birds flapped their wings violently prior to falling over (Hansch et al., 2009). Head shaking occurred in 37% of birds and deep breaths occurred in 18% (Hansch et al., 2009). It is the author's opinion that violent flapping prior to loss of sensibility is not acceptable and the gas mixture or

rate of air removal must be adjusted to avoid this. When evaluating gas stunning or LAP systems, the emphasis should be on observing the animal's or bird's reaction before it loses posture and becomes insensible. The science for specifying the best gas mixture and methods of induction for poultry is still being researched.

Evaluating the entire poultry or pig handling and stunning system

When a poultry stunning system is being evaluated, you should look at BOTH the stunning method and the handling method as a complete system. Gas stunning or LAPS greatly improve the handling of the poultry and some discomfort during the induction phase may be a reasonable trade-off to eliminate live shackling and greater opportunities for employees to abuse the birds. It is the author's opinion that improving the pre-stun handling of the birds will result in overall better bird welfare, even if gas stunning or LAPS cause gasping and head shaking. Gerritzen et al. (2013) state that chickens lose consciousness below 40% CO_2 and that discomfort from multistage CO_2 stunning compares favorably with stresses from handling or shackling. If the birds do carry out full, vigorous wing flapping before LOP, or try to escape from the container, bird welfare would probably be better with electric stunning. Pigs must have sufficient room to all be able to stand in the gondola without being on top of each other. Poultry must have sufficient room to lie down in the transport container without being on top of each other.

Evaluation of captive bolt stunning and gunshot in the plant

For a minimum acceptable score, 95% of the animals must be rendered insensible with a single shot. This score is achieved easily in well-managed plants (Grandin, 2000, 2002, 2005). An excellent score is 99%. All animals requiring a second shot must be rendered insensible prior to hoisting, bleeding, or any other invasive procedure. The penetrating captive bolt is more effective than the nonpenetrating captive bolt (Zulkifli et al., 2013). Shots that are slightly off target are more likely to fail with a nonpenetrating captive bolt. Animals must be bled within 60 s after stunning with a nonpenetrating captive bolt. Cattle and pigs should be shot in the middle of the forehead (see Chapter 10, this volume). Due to a very thick skull, sheep should be shot on the top of the head (see Chapter 10, this volume). Brahman cattle, water buffalo, and other animals with a very thick skull may have to be shot behind the poll (hollow behind the horns) (Gregory et al., 2009). For further information on captive bolt stunning methods, refer to Gregory (2007, 2008) and Grandin (2002) (see Chapter 10, this volume).

All animals must be insensible and unconscious before invasive dressing procedures

All animals must show no signs of return to sensibility prior to hoisting to the bleed rail. All poultry that miss the stunner or automatic knife must be cut by the backup bleeder person. There is a zero tolerance for starting any dressing procedure such as scalding, skinning, or leg removal on an animal or bird showing signs of return to sensibility. This standard applies to both standard slaughter with stunning and ritual slaughter (kosher or halal) without pre-slaughter stunning. For CO_2 and other controlled-atmosphere stunning, there is a need to develop a scoring system for monitoring animal welfare during induction. Windows should be provided so that the induction phase (until the animal or bird loses posture and falls over) can be observed and scored (see the sections headed "Welfare evaluation of gas or low atmospheric pressure stunning" and "Evaluating the entire poultry or pig handling and stunning system"). When performance is measured, a plant can determine if it is getting better or worse. In setting these standards, the critical limit had to be made high enough to force poor plants to improve.

Economic Benefits of Improved Animal Handling

There are benefits to improved animal handling and stunning. Safety for employees is a sufficient reason for eliminating abusive handling practices of cattle and other large animals. Cruel practices such as shackling and hoisting live cattle prior to ritual slaughter are extremely dangerous. The installation of modern restraint equipment that holds the animal in a comfortable upright position greatly reduces accidents and injuries to employees (Grandin, 1988). Other benefits are reduced bruises and better meat quality. Stressful handling and overuse of electric goads shortly before stunning will increase greatly both blood lactate and meat

quality problems such as PSE pork (D'Souza et al., 1999; Hambrecht et al., 2005a,b; Edwards et al., 2010; Dokmanovic et al., 2014). Multiple shocks from an electric prod will increase greatly lactate levels in the blood and the percentage of downer (nonambulatory) pigs (Benjamin et al., 2001). The author has worked with plants that were able to export 10% more quality pork to Japan after they improved handling practices. A basic principle is that keeping cattle and pigs calm and reducing electric prod use shortly before stunning will reduce PSE in pigs and improve tenderness in beef (Hambrecht et al., 2005a,b; Warner et al., 2007). Grain-fed cattle that become agitated during truck loading at the feedlot had twice the risk of having their hides contaminated with Salmonella spp. (Dewell et al., 2008).

Economical methods to improve welfare

In the vast majority of plants in the USA, Europe, Australia, and South America, major capital costs such as building totally new handling systems were not required. In the USA and Canada, only three plants out of over 75 pork and beef plants had to rebuild the entire animal handling facility. The other plants had to make lots of small improvements in the facility and improve employee training and maintenance of equipment. Improvements in equipment usually cost under US$2000, and in many plants under US$500. The most common simple improvements were installation of nonslip flooring in the stunning box, changes in lighting to improve animal movement, and installation of solid panels to prevent approaching animals from seeing people and moving equipment (Grandin, 1982, 1996, 2005) (see Chapter 5, this volume). In poultry, the reduction of broken wings was accomplished with better management of catching employees and improved equipment maintenance.

The management of the plant is a big factor. In three plants, audits were failed until the plant manager was changed. Management is a much more important factor than line speed. Electric goad scores for cattle were very similar at different line speeds ranging from 50 to over 300 cattle/hour (Grandin, 2005) (Table 9.1). Good welfare can be maintained at high line speeds if the plant is designed and staffed for the speed. The worst problems occur when equipment is overloaded or the plant is understaffed. One of the worst cases of overloaded equipment observed by the author was

Table 9.1. Percentage of cattle moved with an electric goad at different line speeds. (From Grandin, 2005.)

Line speed (cattle/hour)	Number of plants	Moved with electric goads (%)
<50	16	20
51–100	13	27
101–200	10	12
201–300	21	24
>300	6	25

a small cattle plant that increased its line speed from 26 cattle/hour to 35 cattle/hour. This resulted in slamming the stun box door on the cattle. A plant that worked well at 26 cattle/hour was terrible at 35 cattle/hour. The main reasons that three plants had to be remodelled were either undersized equipment or the wrong type of equipment for their line speed.

Troubleshooting and Problem-solving Guides

The following are lists of commonly encountered problems at slaughter plants that can cause audit failure and some tried-and-tested solutions to these problems.

Captive bolt stunning (all species)

Problems and corrective actions:

1. Poor maintenance of the gun is a major cause of audit failure. Implement a daily maintenance and cleaning program (Grandin, 2002). The use of a test stand to measure bolt velocity is strongly recommended (see Chapter 10, this volume, for further information on bolt velocity).
2. Air pressure is too low in a pneumatic stunner for effective stunning. Most stunners will require a dedicated air compressor. Air accumulator tanks will only work in very small plants that process four or five cattle per hour.
3. Damp cartridges are a major cause of audit failure for cartridge-fired stunners (Grandin, 2002). Cartridges must be stored in a dry environment such as the office or in a sealed container.
4. Agitated animals make stunning difficult. Two solutions are: (i) improve handling practices so animals will enter the stunning area quietly (e.g. reduce electric goad use and stop employees from yelling

and whistling); and (ii) install a nonslip floor grating to prevent animals slipping on the floor (weld 2 cm diameter steel rods in a 30 × 30 cm square pattern to the floor of the stun box; the rods should be welded flat against the floor and not overlapped).
5. The stunning-to-bleed interval is too long when a nonpenetrating captive bolt is used. After using a nonpenetrating captive bolt, bleed within 60 s.
6. Long hair on cattle may reduce the effectiveness of a nonpenetrating captive bolt.
7. Problems with stunning mature bulls even when they are shot accurately. In one survey, three times as many bulls showed signs of being inadequately stunned (Atkinson *et al.*, 2013). Use a pneumatic stunner with adjustable air pressure and increase air pressure for bulls. Since publication of the first edition of this book, new heavier cartridge-fired captive bolt guns have become available. Another alternative is using a rifle (AVMA, 2013) with a bullet larger than 0.22 caliber.
8. Failure of nonpenetrating captive bolt stunning. Install a head holder because a nonpenetrating captive bolt requires more precise positioning than a penetrating captive bolt.

Electric stunning (all species)

Problems and corrective actions:

1. Insufficient amperage to induce insensibility. The minimum amperage settings are 1 A for sheep, 1.25 A for pigs, and 1.5 A for cattle (see Quick Access Information 9.3). Old or large animals may need higher settings.
2. Electrodes are in the wrong position to pass the current through the brain. To correct this, retrain the employees on correct electrode positioning. The electrodes (tongs) may have to be redesigned or adjusted to facilitate correct positioning in both manually operated and automated systems.
3. If the animal's body surface is too dry, stunning may not be effective. Wet either the animal or the electrodes to improve conductivity.
4. Dehydrated animals and birds are harder to stun. This is especially a problem in old animals or animals that have traveled a long distance. Providing water during long trips and in the stockyard (lairage) pens at the plant will help prevent dehydration.

Gas stunning (pigs and poultry)

Problems and corrective actions:

1. Gas concentration is either too low or the wrong gas mixture is used (see the section headed "Evaluating the entire poultry or pig handling and stunning system" for more information). For pigs, 90% CO_2 is recommended (Becerril-Herrera *et al.*, 2009); 70% is aversive for pigs. Another problem that can occur is that the gas is not distributed in the chamber. This may be due to a design fault in the chamber, or the ventilation system in the plant may be sucking the gas out of the chamber. Correcting this problem may require the expertise of an engineering professional skilled in ventilation systems. In open gas stunning systems such as the CO_2 systems used for pigs and poultry, the following factors can cause problems: wind blowing around the plant building; changes in the number of ventilation fans turned on in the plant; or opening and closing doors near the chamber. These factors can cause "stack pressure" that may cause the gas to be sucked out. Stack pressure is a common cause of a sudden appearance of conscious animals emerging from a chamber that has been operating effectively. This may occur when a certain specific sequence of either opening doors or turning on ventilator fans occurs. Doors that slam hard by themselves are moved by differences in air pressure between two rooms or between a room and the outdoors. Differences in pressure in rooms near the gas stunner may cause the stunner to fail. Stack pressure problems will have no effect on positive pressure (closed) systems where the gas is introduced into the chamber with a ventilation system.

2. An overloaded, undersized machine is one of the worst problems with gas stunning. As a plant increases production, the machine may become overloaded. Plant managers who are purchasing a gas stunning machine should purchase a large enough machine to handle future increases in production. An overloaded machine has to be replaced. Specific signs of an overloaded machine are: (i) animals are not rendered insensible because the exposure time has been decreased by speeding up the conveyor; (ii) gondolas or containers are overloaded and pigs or birds do not have enough room to stand or lie down without being on top of each other. Pigs should never be forced to jump on top of other pigs when a gondola is being loaded.

Problems with return to consciousness after stunning

These problems and corrective actions apply to animals that are stunned:

1. Stunning is not done correctly to induce insensibility – refer to the parts above on stunning in this section on problem-solving guides.
2. Stun-to-bleed interval is too long when head-only reversible electric stunning is used. The stun-to-bleed interval for head-only reversible electric stunning should be 15 s or less (Blackmore, 1984; Wotton and Gregory, 1986). For cardiac arrest electrical stunning, the stun-to-bleed interval can be 60 s.
3. Insufficient blood flow after bleeding. This was a major cause of audit failure in many pork plants. The employees should be trained in more effective bleeding methods. In pork, making a larger wound improved blood flow and corrected return-to-sensibility problems (Grandin, 2001b).

Methods to facilitate more rapid loss of consciousness after slaughter without stunning

In plants doing religious slaughter, insensibility will occur more rapidly when the cut is performed on a calm animal (Grandin, 1994). A rapid cut with a very sharp knife is usually more effective. Dull knives that lack a sharp edge should not be used. Excessive pressure applied by a restraint device may cause agitation and delay the onset of insensibility. High vocalization scores are a sign that the restrainer is hurting the animal. A plant with a restraint device that applied excessive pressure had 35% of the cattle vocalizing (Grandin, 1998b). Excessive pressure applied by a head-holding device caused 23% of the cattle to bellow. When the pressure was reduced, vocalization dropped to 0% of the animals (Grandin, 2001a). The knife should be long enough so that the end of the knife remains outside of the neck during the cut. The wound should not be allowed to close back over the knife during the cut. The animal should be restrained in a comfortable upright position and the cut should be performed within 10 s after the head is restrained. If more than 5% of the cattle vocalize, immediate corrective action should be taken. Great care must be taken to avoid excessive pressure from being applied by the rear pusher gate or any other part of the restraint device. Immediately after the cut, the head holder, belly lift, rear pusher gate, and other devices that press against the animal should be loosened immediately to facilitate bleed out. Cutting the throat close to the jaw in the C1 (cervical vertebrae 1) position is recommended (Gregory *et al.*, 2012; Kolesar *et al.*, 2014). The animal should not be removed from the box until it has collapsed and has become dead, with a fixed dilated pupil (Kolesar *et al.*, 2014).

Table 9.2 shows the difference between plants with good and bad procedures for the kosher slaughter of cattle (Erika Voogd, 2008, personal communication). The times in Table 9.2 are similar to results in the research literature (Blackmore and Newhook, 1981, 1983; Blackmore *et al.*, 1983; Gregory and Wotton, 1984a). Sheep lose sensibility more quickly after slaughter without stunning compared to cattle. This is due to differences in the anatomy of sheep and cattle (Baldwin and Bell, 1963a,b). In sheep, the average time to onset of insensibility is 2–14 s (Blackmore, 1984; Gregory and Wotton, 1984b). Time-to-collapse LOP scoring should be used to minimize the time that the animals remain sensible after the throat is cut. The animal is unconscious when it loses the ability to stand and does not respond to a hand waved near the eye (menace reflex) (EFSA, 2013). Poorly done slaughter without stunning can result in prolonged periods of sensibility (Newhook and Blackmore, 1982; Blackmore, 1984; Gregory and Wotton, 1984a,b,c). See Grandin (1986) for a review of these studies. Daly *et al.* (1988) reported that the time to loss of sensory responsiveness of the bovine brain varied from 19 to 126 s. To reduce this variability and shorten the time to loss of insensibility will require careful scoring and monitoring of collapse (LOP) times.

Table 9.2. Time to eye rollback and collapse (loss of posture) during kosher slaughter of cattle. (From Erika Voogd, 2008, personal communication).

	Good technique	Poor technique
Average time to collapse (s)	17	33
Longest time (s)	38	120
Cattle collapsed within 30 s (%)	94	68
Number of cattle	17	19

Correcting problems with high levels of cattle bellowing or pig squealing (vocalization)

Problems and corrective actions for both conventional and ritual slaughter:

1. Excessive electric goad use is a major cause of vocalization. Train employees to use a flag or other nonelectric aid as their primary method for moving animals (see Chapter 5, this volume). They should only pick up the electric goad when it is needed to move a stubborn animal and then put it back down.
2. Missed stuns cause vocalization – to correct this problem, see stunning recommendations under the heading "Captive bolt stunning (all species)".
3. Falling or many small rapid slips in the stun box, restrainer or lead-up race may cause vocalization. The remedy is the installation of nonslip flooring. Animals panic when they "jig" and make many repeated small slips in the stun box.
4. Excessive pressure from a head holder or other restraint device that causes pain results in vocalization (Bourquet *et al.*, 2011). Install pressure-limiting devices on hydraulic or pneumatic-powered equipment. A head-holding device will need to have its own separate pressure control. It must be set at a lighter pressure than the pressure control for heavy gates and other parts of the apparatus that would require more pressure. There is a problem with the restraint device if animals vocalize in direct response to application of the device.
5. Sharp edges stick into the animal and cause vocalization. Surfaces that contact animals should be smooth. Even a small, sharp edge may stick into the animal. It must be found and removed.

Prevention of falling during handling

Problems and corrective actions for both conventional and ritual slaughter:

1. Slick, slippery floors are the primary cause of animals falling. There are a number of ways to provide nonslip flooring. Diamond-plate or checker-plate floors wear out rapidly and become slippery. Stun boxes should have a flat, nonslip floor.
   (a) Install a nonslip floor grating made from 2 cm diameter metal rods welded in a 30 × 30 cm square pattern to form a flat mat (see Chapter 5). Do not overlap the rods. Gratings work best in high-traffic areas such as stun boxes, races, crowd pens, scales, and unloading areas.

(b) Make grooves in the floor with a concrete-grooving machine (can be done by a concrete contractor or the machine can be rented). This method is recommended for large areas of existing slick floors.

(c) Recommendations for the construction of new floors for cattle and other large animals include grooving the floor in pens and races in a 20 × 20 cm square or diamond pattern. The grooves should be a minimum of 2 cm deep by 2 cm wide. For pigs, sheep, and other small animals, smaller grooves can be used that are spaced closer together. Imprinting the pattern of expanded metal mesh into the wet concrete works well. A rough broom finish is not recommended as it wears out too quickly.

(d) Install rubber mats with a ribbed surface.

Reducing electric goad use

Problems and corrective actions:

1. Poorly trained employees tend to overuse electric goads when handling animals. Employees should be trained in the behavioral principles of handling animals (Grandin, 2014). Employees should not be allowed to yell or whistle at animals (see Chapter 5, this volume).

2. Overloading of the crowd pen that leads up to the single-file race makes handling more difficult. Cattle and pigs should be moved into the crowd pen in small, separate groups and the crowd pen should be filled half full. Sheep can be handled in large groups as one continuous flow (see Chapter 5, this volume).

3. Animals balking and refusing to move is a major cause of excessive use of electric goads. This is a facility problem that must be corrected. To improve the flow of animals, distractions that attract their attention must be removed from the facility. Refer to Quick Access Information 9.9 to troubleshoot distractions in the race, stun box, and restrainers that must be removed. To find the cause of balking, people should get down in the race and see what the animals are seeing. A calm animal can be used to help locate distractions, as a calm animal will point its eyes and ears towards distractions. A clearly lit entrance to the single-file race or stun box can improve animal movement (Fig. 9.6). Quick Access Information 9.10 provides assistance for solving problems such as vocalization or agitated animals in the stun box or restrainer.

Fig. 9.6. Placing a lamp at the entrance of the single-file race may improve animal movement by attracting the pigs into the race.

Quick Access Information 9.9. Troubleshooting guide for finding and removing distractions that cause animals to balk, turn back, or refuse to move.[a]

Distractions that cause balking	How to improve animal movement
• Race or stun box entrance too dark.	• Add indirect lighting that does not shine into the approaching animal's eyes (Fig. 9.6). If sunlight makes the entrance look dark, a shade may need to be installed to block the sunlight.
• Seeing moving people or equipment up ahead or seeing people in front of the stun box.	• Install solid sides on races and install shields for people to stand behind. Block the animal's view of moving equipment that causes balking and change locations where people stand. Experiment with large pieces of cardboard.
• Air blowing into the face of an approaching animal.	• Change ventilation so that air will not be moving at the stun box entrance.
• Excessive noise.	• Ensure that there is no yelling, silence hissing air exhausts, and install rubber pads on clanging equipment.
• Reflection on metal or wet floors.	• Adding or moving lamps will usually eliminate reflections. A person must get in the race and view it from the animal's eye level to determine if the reflection has been removed. Do many experiments with a portable lamp. Existing lamps may need to be moved.
• Small objects that cause balking.	• Remove loose plastic, clothes hung on fences, and chains dangling in races.
• High color contrast.	• Paint facilities all the same color and dress people in clothes that are similar in color to the walls.
• Floor drains or changes in flooring.	• Move the drain from the area where animals walk or make the flooring color surface look the same.

Note: [a]See Chapter 5, this volume, for more information, and also Van Putten and Elshof (1978), Grandin (1982, 1996) and Tanida et al. (1996). Information from many specialists can be found in Grandin (2014).

Welfare Issues with Religious Slaughter (Slaughter without Stunning)

Slaughtering animals without stunning is very controversial from an animal welfare standpoint. It is beyond the scope of this book to carry out a complete review of the literature on religious slaughter. There are two main welfare issues: the throat cut without stunning; and how the animal is restrained during the throat cut. Some of the worst problems with ritual slaughter are the restraint methods. The author has observed horrible plants where long lines of thrashing live cattle were hung upside down by one leg. The vocalization score was nearly 100% of the cattle.

For the best animal welfare, the animal should be held in a comfortable upright position during slaughter without stunning (Fig. 9.7).

Use of an upright restraining box resulted in less vocalization compared to a box that inverted cattle on to their backs (Velarde et al., 2014). The rotating box had less struggling, but the authors did not differentiate between struggling that occurred either before or after loss of sensibility. Struggling after loss of sensibility has no effect on welfare. Struggling that occurs after loss of sensibility would be due to reflexes. A poorly designed rotating box that positioned cattle on their backs was more stressful and had more vocalization and higher cortisol levels compared to upright restraint (Dunn, 1990). An improved-design rotating box with an adjustable side and cutting the animal within a few seconds after inversion resulted in vocalization scores that were similar to an upright box. To assess the quality of handling and restraint during religious slaughter, the plant should be scored for vocalization, falling down, and electric goad use in the same way as conventional slaughter. The percentage of cattle vocalizing should be 5% or less in either an upright or a

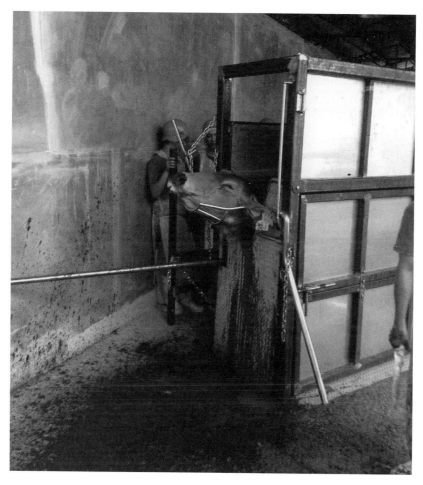

Fig. 9.7. Economical manually operated upright restraint box that can be constructed easily from local materials. The long levers provide a mechanical advantage for manual operation. Designed by Rastro Kolesar.

rotating box, and the standard can be achieved easily (Grandin, 2012; Kristin Pufpaff, 2015, personal communication). People who are implementing animal welfare programs should concentrate on eliminating painful, stressful methods of restraint (see Chapters 5 and 16, this volume). Unfortunately, there has been a recent study carried out in the USA where 47% of the cattle vocalized during handling for kosher slaughter (Hayes *et al.*, 2015). Almost 100% of the animals were moved with an electric prod, and lactate levels were very high (Hayes *et al.*, 2015). This is clearly abusive and not acceptable.

When slaughter without stunning is done correctly, there is little or no reaction from the cattle during the cut (Grandin, 1994). Barnett *et al.* (2007)

reported that only four out of 100 chickens showed a physical response during the cut. When a hand was waved in front of the animal's face, it had a greater behavioral reaction compared to well done Shehita (kosher slaughter) (Grandin, 1994). In most cattle, when the knife first touched the throat, there was a slight twitch. When a hand was waved rapidly within 20 cm of an animal's face, it became agitated because its flight zone had been invaded. The cattle that were observed by the author were extensively raised and they had a large flight zone. Also, the author has observed that when the cut is done poorly, and the wound closes over the knife, the animal may struggle violently. EEG pain measurements indicate that cutting the neck of

Quick Access Information 9.10. Solving problems with stun boxes, head holders, and conveyor restrainers.

Problem	Possible causes	Remedies
• Animal refuses to enter.	• See Quick Access Information 9.9. • "Visual cliff" effect in a conveyor restrainer. Animal can see that the conveyor is above the floor.	• See Quick Access Information 9.9. • Install a false floor to provide the illusion of a floor to walk on (Grandin, 2003b). The false floor should be installed so that it is approximately 15 cm below the feet of the largest animal that is riding on the conveyer.
	• Hold-down rack touches animal's back while entering.	• Raise hold-down rack.
• Animal becomes agitated or vocalizes.	• Slipping on the floor. • Excessive pressure applied by a restraint device.	• Install nonslip flooring. • Reduce pressure. There is an optimal pressure for holding an animal – it must not be too loose or too tight. Install pressure-limiting devices. A head holder will need to have its own separate pressure-limiting device – it requires a much lower pressure than heavy gates.
	• Held too long in the restrainer. • Sudden, jerky motion of the apparatus.	• Stun or ritually slaughter within 10 s. • Install speed-reducing devices so that the device moves with a steady, slow movement. Hydraulic or pneumatic controls should have good throttling control. This enables the operator to move a head holder or other restraint device with a smooth movement, like a car's accelerator pedal (Grandin, 1992).
	• One side of a V-conveyor restrainer runs faster than the other.	• Both sides should run at the same speed. Lengthen the hold down.
	• Hold-down rack does not fully block the animal's vision on a conveyor restrainer.	• Lengthen hold-down rack.
	• Animal sees people and other moving things in front of the stun box.	• Install a solid shield in front of the stun box. Experiment with cardboard in different positions.

109- to 170-kg calves with a sharp 24.5-cm-long knife causes pain (Gibson *et al.*, 2009a,b). Petty *et al.* (1994) reported that catecholamines were higher after kosher slaughter compared to captive bolt stunning. Another welfare concern after slaughter without stunning is aspiration of blood into the lungs. This can vary from 36 to 69% of the cattle (Gregory *et al.*, 2008).

Many Muslim religious authorities will permit stunning prior to the throat cut for halal slaughter (Nakyinsige *et al.*, 2013). In New Zealand, electrical head-only stunning is used for halal slaughter.

Other Muslim religious authorities will permit the use of a nonpenetrating captive bolt because they define death when the heart stops (Farouk, 2013). After both penetrating and nonpenetrating captive bolt, the heart continues to beat for several minutes (Vimini *et al.*, 1983; Jeristrom, 2014). For poultry, electrical water-bath stunning is often permitted. Gas stunning is not permitted. For kosher slaughter, some rabbis will permit captive bolt stunning immediately after the cut. Some people who oppose the use of pre-slaughter stunning will cite research that shows that massive amounts of stress hormones are released after captive bolt or electric stunning (Van der Wal, 1978; Shaw and Tume, 1992). When the stun is applied correctly, it has no effect on welfare because the animal is insensible and unconscious when the hormones are released. Some good sources of additional reading are Daly *et al.* (1988), Grandin (1992, 1994, 2006), Levinger (1995), Rosen (2004), Shimshony and Chaudry (2005), Gregory (2007, 2008) and Johnson *et al.* (2015). See the earlier section on how to determine insensibility for additional information on the time required for loss of sensibility. The OIE, the EU, the USA, and many other countries allow religious slaughter so that people are free to practice their religious beliefs (Rosen, 2004; Shimshony and Chaudry, 2005). Since the first edition of this book, slaughter without stunning is becoming increasingly controversial, and some European countries have banned it. Slaughter without stunning has greater welfare problems with cattle compared to sheep. Sheep lose insensibility more quickly and they are easier to restrain in a low-stress manner. The first sign of loss of consciousness is LOP and the ability to stand (EFSA, 2013). Loss of spontaneous natural blinking is another sign of loss of conscious (EFSA, 2013).

Fig. 9.8. Proper handling of a nonambulatory pig in a plastic sledge. Downed animals should never be dragged.

Handling Nonambulatory Animals and Factors that Make Animals Difficult to Handle

Nonambulatory animals must NEVER be dragged. This is a violation of OIE (2014) codes and EU and USDA regulations for humane slaughter. Pigs, sheep, goats, and other small animals can be moved easily on a sledge or other conveyance. Figure 9.8 shows a simple sledge for moving downed nonambulatory pigs and Fig. 9.9 shows a handcart that can be constructed easily for moving nonambulatory pigs, sheep, or goats. Large animals such as dairy cows that have become nonambulatory are very difficult to be moved in a manner that does not cause suffering. One of the best things the USDA did in the USA was to ban the slaughter of nonambulatory cows in federally inspected plants. Marketing the cows for slaughter when they are in a better condition can prevent most nonambulatory dairy cows. Nonambulatory pigs and sheep can still be slaughtered in an USDA plant. The emphasis has to be on preventing the occurrence of high numbers of nonambulatory animals.

Answers to Frequently Asked Questions

Handling and meat quality problems caused by beta-agonists

Pigs fed high doses of ractopamine (beta-agonist) may be more difficult to move at the plant (Marchant-Forde *et al.*, 2003). James *et al.* (2013) reported that pigs fed ractopamine had more stress-related problems if they were handled roughly. Lairage managers and staff have reported that large market-weight pigs that weigh 125–130 kg (275–285 lb) that have been fed high doses of ractopamine are difficult to move and may become nonambulatory, as they become fatigued and too weak to move. These pigs do not have the symptoms of PSS. See Chapter 5, this volume, and Marchant-Forde *et al.* (2003) for more information. In cattle, a low dose of ractopamine of 200 mg/day for 28 days had only a slight effect on handling (Baszczrak *et al.*, 2006). There were no adverse effects during a short 60-min trip to slaughter.

Fig. 9.9. Handcart for moving a nonambulatory animal. It can be slid under the animal easily. This can be constructed in the plant maintenance shop. Nonambulatory animals must never be dragged. In some countries, animals that arrive downed and nonambulatory must be shot on the truck with a captive bolt.

Cattle fed excessive doses of beta-agonists such as ractopamine or zilpaterol will have tougher meat and less marbling. The author has observed that they may also have symptoms of heat stress and lameness. Longeragan *et al.* (2014) reported that beta-agonists increased death losses during the hot summer months. Their feet will usually look normal but they may be sore footed. Pigs have a higher incidence of hoof lesions when they are fed ractopamine (Poletto *et al.*, 2009). In both cattle and pigs, moderate doses of ractopamine or zilpaterol will increase the toughness of both pork and beef (Carr *et al.*, 2005; Hilton *et al.*, 2009; Gruber *et al.*, 2010). One of the most effective ways to stop the practice of feeding excessive doses of beta-agonists is to charge a fine for moving nonambulatory animals or for tough meat. When a producer sees money deducted from his or her pay check, he or she either will stop using beta-agonists or will use them in a more responsible manner. These substances are banned in many countries.

Tame intensively raised animals versus extensively raised animals

A single-file race or stun box is not required if a few halter-broke animals that have been trained to lead are processed. A tame animal can be led on to the slaughter floor and stunned. Small animals such as sheep or goats can be held by a person and either slaughtered ritually or stunned. Simple systems that may be adequate with tame intensively raised animals that are accustomed to people will be highly detrimental to animal welfare if they are used with extensively raised animals. Systems that may be adequate for tame Middle Eastern sheep are not adequate for imported Australian sheep that are not accustomed to handling. For these sheep, a single-file race is required. For small plants, a simple race made by a local welder could be used. The sheep could be lined up in the race and removed one at a time by a person. For large Middle Eastern plants, a conveyor restrainer would need to be installed at the end of the race.

T. Grandin

Effect of blood on animal behavior

In some systems, especially in developing countries, the animal can see another animal being bled or is required to walk over a floor covered with blood. The author has observed that cattle will enter a kosher restraint box that is covered with blood voluntarily, provided that the previous animals had remained calm. If a previous animal had been jammed in malfunctioning equipment for 10–15 min, the other cattle would refuse to enter. Research shows that a stressor applied for 15 min, such as multiple shocks from a goad, will trigger the secretion of stress pheromones (Vieuille-Thomas and Signoret, 1992; Boissy *et al.*, 1999). If animals remain behaviorally calm, then having blood on the floor may not affect welfare. If an animal becomes agitated for several minutes, the floor may need to be washed before the next animal is dealt with. Observations indicate that stunning an animal in the sight of other animals has almost no effect on behavior, but the sight of a decapitated head may cause the next animal to panic. They do not appear to understand what has happened provided that the fallen animal's body remains intact.

Bruises from restraint equipment or head holder

An important part of a welfare evaluation is inspecting carcasses for bruises or injuries caused by restraint equipment. If a restraining device is causing a bruise or injuries, the problem must be corrected.

Methods for cleaning dirty livestock

The emphasis should be on keeping cattle clean at the farm. Harsh methods such as spraying animals with high-pressure fire hoses should be avoided. Spraying water directly into the face of animals should also be avoided. Some plants have sprayers built into the floor of a holding pen to clean legs and bellies. This is acceptable. Washing should be avoided if the animals are in an environment with below-freezing temperatures. The practice of clipping or shaving live cattle should be stopped because it is stressful to the animal and dangerous for people. Some plants that get very dirty cattle wash the carcasses after stunning and bleeding. From a welfare standpoint, this is much better than removing soil from live animals. Another good alternative is to use sheep shears to clip the areas where the carcass will be opened. This is done after stunning and bleeding.

Stun-to-bleed interval

Regulations in some countries require that animals are bled within 20 or 60 s after stunning. IF an animal shows any sign of return to sensibility, it must be re-stunned BEFORE it is hoisted and hung on the bleed rail. If re-stunning is required, the interval is timed after the second stun. A short stun-to-bleed interval is most important for head-only electrical and nonpenetrating captive bolt.

Procedures during work breaks

DO NOT leave animals in the stun box, restrainer, or restraint box during coffee and lunch breaks. During lunch and other long breaks, the lead-up race should also be empty. In poultry plants, do not leave birds hanging on the line during breaks.

Problems with understaffing of handling and stunning

Understaffing of stunning boxes and handling systems can cause severe welfare problems because workers quickly become fatigued or rushed. The worst problems occur when a particular job can almost be done correctly by one person. At one large beef plant, they had problems with gates being slammed on cattle in the stun box. Adding a second person to operate the gates eliminated this problem.

How stressful is slaughter?

A review of the literature indicates that cortisol levels in cattle during handling on the farm and after slaughter are similar. It varies from very high to very low in both places. Animals that remain calm will have lower cortisol. Many studies are reviewed in Grandin (1997b, 2014).

Principles of Fish Stunning and Assessing Fish Welfare

European fish retailers are demanding higher standards for fish welfare (Lines and Spence, 2014). The OIE (2014) has requirements for effective stunning of fish that produces absence of consciousness. Producers and processors of fish have installed systems for stunning fish and rendering them insensible to pain. There is a discussion on research on pain perception in fish in Chapter 2, this volume. Research indicates that the process of live chilling of fish may be highly stressful (Lines *et al.*, 2003; Roth *et al.*, 2009). A review of the literature indicates that many fish-farm managers have installed stunning systems. An Internet search revealed many recent patents for fish-stunning equipment. Electrical stunning equipment must induce epileptiform activity in the brain to render fish insensible. Information on electrical systems and electrical parameters can be found in Lambooij *et al.* (2007, 2008a,b, 2013) and Branson (2008). Percussion stunning is also used in fish. The technology for stunning fish is evolving rapidly.

Data will have to be collected to determine the numerical scores for an acceptable level of welfare. When auditing of fish processing and farms first begins, the cut-off for an acceptable score should allow the top 25% of the best operations to pass (see Chapter 2, this volume).

Conclusion

The implementation of an auditing program that uses objective numerical scoring will improve animal welfare greatly. Another benefit will be reduced bruises and severe meat quality defects such as PSE and dark cutters. To be effective, the program should audit CCPs such as the percentage of animals stunned correctly on the first attempt, the percentage of animals rendered insensible, the percentage that fall, the percentage moved with no electric prod, and the percentage that vocalize. Numerical scores enable a plant to determine if their practices are improving or deteriorating.

References

Ahsan, M., Hasan, B., Algotsson, M. and Sarenbo, S. (2014) Handling and welfare of bovine livestock at local abattoirs in Bangladesh. *Journal of Applied Animal Welfare Science* 17, 340–353.

Anil, A.M. and McKinstry, J.L. (1992) The effectiveness of high frequency electrical stunning in pigs. *Meat Science* 31, 481–491.

Anil, M.H. and McKinstry, J.L. (1998) Variation in electrical stunning tong placements and relative consequences in slaughter pigs. *Veterinary Journal* 155, 85–90.

Atkinson, S., Velarde, A. and Algers, B. (2013) Assessment of stun quality at commercial slaughter in cattle shot with captive bolt. *Animal Welfare* 22, 473–481.

AVMA (American Veterinary Medical Association) (2013) *Guidelines for Euthanasia of Animals*, 2013 edn. American Veterinary Medical Association, Schaumberg, Illinois.

Baldwin, B.A. and Bell, F.R. (1963a) The anatomy of the cerebral circulation of the sheep and ox. The dynamic distribution of the blood supplied by the carotid and vertebral arteries to cranial regions. *Journal of Anatomy* 97, 203–215.

Baldwin, B.A. and Bell, F.R. (1963b) The effect of temporary reduction in cephalic blood flow on the EEG of sheep and calf. *Electroencephalography and Clinical Neurophysiology* 15, 465–475.

Barnett, J.L., Cronin, G.M. and Scott, P.C. (2007) Behavioral responses of poultry during kosher slaughter and their implications for the bird's welfare. *Veterinary Record* 160, 45–49.

Baszczrak, J.A., Grandin, T., Gruber, S.L., Engle, T.E., Platter, W.J., *et al.* (2006) Effects of ractopamine supplementation on behavior of British, Continental, and Brahman crossbred steers during routine handling. *Journal of Animal Science* 84, 3410–3414.

Becerril-Herrera, M., Alonso-Spilsbury, M., Lemus-Flores, C., Guerrero-Legarreta, I., Hernandez, A., *et al.* (2009) CO_2 stunning may compromise swine welfare compared to electrical stunning. *Meat Science* 81, 233–237.

Bedanova, I., Voslarova, E., Chloupek, P., Pistekova, V., Suchy, P., *et al.* (2007) Stress in broilers resulting from shackling. *Poultry Science* 86, 1065–1069.

Benjamin, M.E., Gonyou, H.W., Ivers, D.L., Richardson, L.F., Jones, D.J., *et al.* (2001) Effect of handling method on the incidence of stress response in market swine in a model system. *Journal of Animal Science* 79 (Supplement 1), 279 (Abstract).

Benson, E.R., Alphin, R.L., Rankin, M.K., Caputo, M.P., Kinney, C.A., *et al.* (2012) Evaluation of EEG based determination of unconsciousness vs. loss of posture in broilers. *Research in Veterinary Science* 93, 960–964.

Berghaus, A. and Troeger, K. (1998) Electrical stunning of pig's minimum current flow time required to induce epilepsy at various frequencies. *International Congress of Meat Science and Technology* 44, 1070–1073.

Blackmore, D.K. (1984) Differences in behaviour between sheep and cattle during slaughter. *Research in Veterinary Science* 37, 223–226.

Blackmore, D.K. and Newhook, J.C. (1981) Insensibility during slaughter of pigs in comparison to other domestic stock. *New Zealand Veterinary Journal* 29, 219–221.

Blackmore, D.K. and Newhook, J.C. (1983) The assessment of insensibility in sheep, calves, and pigs during slaughter. In: Eikelenboom, G. (ed.) *Stunning of Animals for Slaughter*. Martinus Nijhoff, Boston, Masachusetts, pp. 13–25.

Blackmore, D.K., Newhook, J.C. and Grandin, T. (1983) Time of onset of insensibility in four- to six-week-old calves during slaughter. *Meat Science* 9, 145–149.

Boissy, A., Terlouw, C. and Le Neindre, P. (1998) Presence of cues from stressed conspecifics increases reactivity to aversive events in cattle, evidence for the existence of alarm substances in urine. *Physiology and Behavior* 63, 489–495.

Boissy, A., Bouix, J., Orgeur, P., Poindron, P., Bibe, B., *et al.* (2005) Genetic analysis of the emotional reactivity in sheep effects of genotype of lambs and their dams. *Genetic Selection Evolution* 37, 381–401.

Bourguet, C., Deiss, V., Tannugi, E.C. and Terlouw, E.M.C. (2011) Behavioral and physiological reactions of cattle in a commercial abattoir and relationships with organizational aspects of the abattoir and animal characteristics. *Meat Science* 88, 158–168.

Branson, E. (2008) *Fish Welfare*. Blackwell Publishing, Oxford, UK.

Carr, S.N., Ivers, D.J., Anderson, D.B., Jones, D.J., Mowrey, D.H., *et al.* (2005) The effects of ractopamine hydrochloride on lean carcass yields and pork quality characteristics. *Journal of Animal Science* 83, 2886–2893.

Cattaruzzi, B. and Cheek, H. (2012) Method for humanely stunning and slaughtering poultry using controlled low atmospheric pressure. US Patent EP 2055191B1 or CA2701499A1. US Patent Office, Washington, DC.

Coenen, A.M., Lankhaar, J., Lowe, J.C. and McKeegan, D.E. (2009) Remote monitoring of electroencephalogram, electrocardiogram, and behavior during controlled atmosphere stunning in broilers: Implications for welfare. *Poultry Science* 88, 10–19.

Croft, P.S. (1952) Problems with electrical stunning. *Veterinary Record* 64, 255–258.

Daly, C.C., Kallweit, E. and Ellendorf, F. (1988) Cortical function in cattle during slaughter: conventional captive bolt stunning followed by exsanguination compared to sheehita slaughter. *Veterinary Record* 122, 325–329.

Dewell, G.A., Simpson, C.A., Dewell, R.D., Hyatt, D.R., Belk, K.E., *et al.* (2008) Risks associated with transportation and lairage on hide contamination with *Salmonella enterica* in finished beef cattle at slaughter. *Journal of Food Protection* 71, 2228–2232.

Dokmanovic, M., Baltic, Z.M., Markovic, R., Boskovic, M., Loncina, M., *et al.* (2014) Relationships among pre-slaughter stress, rigor mortis, blood lactate, and meat and carcass quality in pigs. *Acta Veterinaria-Beograd* 64, 124–137.

D'Souza, D.N., Dunshea, F.R., Leury, B.J. and Warner, R. (1999) Effect of mixing boars during lairage and pre-slaughter handling on meat quality. *Australian Journal of Agricultural Research* 50, 109–113.

Dunn, C.S. (1990) Stress reactions of cattle undergoing ritual slaughter using two methods of restraint. *Veterinary Record* 126, 522–525.

Dwyer, C.M. (2004) How has the risk of predation shaped the behavioral responses of sheep to fear and distress. *Animal Welfare* 13, 269–281.

Edwards, L.N., Grandin, T., Engle, T.E., Porter, S.P., Ritter, M.J., *et al.* (2010) Using exsanguination blood lactate to assess the quality of pre-slaughter pig handling. *Meat Science* 86, 384–390.

EFSA (2013) Scientific opinion on monitoring procedures at slaughterhouses for bovines. EFSAAHAW Panel on Animal Health and Welfare. *EFSA Journal* 11(12), 3460, 65 pp, doi.10.2903j.efsa.2013.3460.

Farouk, M.M. (2013) Advances in the industrial production of halal and kosher red meat. *Meat Science* 95, 805–820.

Forslid, A. (1987) Transient neocortisol, hippocampal and amygdaloid EEG silence induced by one minute inhalation of high concentration CO_2 in the swine. *Acta Physiologica Scandinavica* 130, 1–10.

Gerritzen, M.A., Reimert, H.G.M., Hindle, V.A., Verhoeven, M.T.W. and Veerkamp, W.B. (2013) Multistage carbon dioxide gas stunning of broilers. *Poultry Science* 92, 41–50.

Gibson, T.J., Johnson, C.B., Murrell, J.C., Hulls, C.M., Mitchinson, S.L., *et al.* (2009a) Electroencephalographic responses of halothane-anaesthetized calves to slaughter by ventral-neck incision without prior stunning. *New Zealand Veterinary Journal* 57, 77–85.

Gibson, T.J., Johnson, C.B., Murrell, J.C., Chambers, J.P., Stafford, K.J., *et al.* (2009b) Components of electroencephalographic responses to slaughter in halothane-anesthetized calves: effects of cutting neck tissues compared to major blood vessels. *New Zealand Veterinary Journal* 57, 84–89.

Grandin, T. (1982) Pig behavior studies applied to slaughter plant design. *Applied Animal Ethology* 9, 141–151.

Grandin, T. (1986) Cardiac arrest stunning of livestock and poultry. In: Fox, M.W. (ed.) *Advances in Animal Welfare Science 1985*. Martinus Nijhoff, Boston, Massachusetts.

Grandin, T. (1988) Possible genetic effect on pig's reaction to CO_2 stunning. *Proceedings of the 34th International Congress of Meat Science and Technology*, Brisbane,

Australia. Commonwealth Scientific and Industrial Research Organization (CSIRO), Brisbane, Australia, pp. 96–97.

Grandin, T. (1992) Observations of cattle restraint devices for stunning and slaughter. *Animal Welfare* 1, 85–91.

Grandin, T. (1994) Euthanasia and slaughter of livestock. *Journal of the American Veterinary Medical Association* 204, 1354–1360.

Grandin, T. (1996) Factors that impede animal movement in slaughter plants. *Journal of American Veterinary Medical Association* 209, 757–759.

Grandin, T. (1997a) *Survey of Stunning and Handling in Federally Inspected Beef, Veal, Pork, and Sheep Slaughter Plants*. United States Department of Agriculture (USDA)/Agricultural Research Service Project 3602-32000-002-08G. USDA, Beltsville, Maryland.

Grandin, T. (1997b) Assessment of stress during handling and transport. *Journal of Animal Science* 75, 249–257.

Grandin, T. (1998a) Objective scoring of animal handling and stunning practices at slaughter plants. *Journal of American Veterinary Medical Association* 212, 36–39.

Grandin, T. (1998b) The feasibility of vocalization scoring as an indicator of poor welfare during slaughter. *Applied Animal Behaviour Science* 56, 121–128.

Grandin, T. (2000) Effect of animal welfare audits of slaughter plants by a major fast food company on cattle handling and stunning practices. *Journal of the American Veterinary Medical Association* 216, 848–851.

Grandin, T. (2001a) Cattle vocalizations are associated with handling and equipment problems in slaughter plants. *Applied Animal Behaviour Science* 71, 191–201.

Grandin, T. (2001b) Solving return to sensibility problems after electrical stunning in commercial pork slaughter plants. *Journal of the American Veterinary Medical Association* 219, 608–611.

Grandin, T. (2002) Return to sensibility problems after penetrating captive bolt stunning of cattle in commercial beef slaughter plants. *Journal of the American Veterinary Medical Association* 221, 1258–1261.

Grandin, T. (2003a) The welfare of pigs during transport and slaughter. *Pig News and Information* 24, 83N–90N.

Grandin, T. (2003b) Transferring results from behavioral research to industry to improve animal welfare on the farm, ranch, and slaughter plant. *Applied Animal Behaviour Science* 81, 215–228.

Grandin, T. (2005) Maintenance of good animal welfare standards in beef slaughter plants by use of auditory programs. *Journal of the American Veterinary Medical Association* 226, 370–373.

Grandin, T. (2006) Improving religious slaughter practices in the US. *Anthropology of Food*, 5 May 2006. Available at: http://aof.revues.org/document93.html (accessed 28 April 2015).

Grandin, T. (2012) Developing measures to audit welfare of cattle and pigs at slaughter. *Animal Welfare* 21, 351–356.

Grandin, T. (2013) *Recommended Animal Handling Guidelines and Audit Guide*, 2007 edn. American Meat Institute Foundation, Washington, DC. Available at: www.animalhandling.org (accessed 9 May 2015).

Grandin, T. (ed.) (2014) *Livestock Handling and Transport*, 4th edn. CAB International, Wallingford, UK.

Grandin, T., Curtis, S.E., Widowski, T.M. and Thurmon, J.C. (1986) Electro-immobilization versus mechanical restraint in an avoid–avoid choice test for ewes. *Journal of Animal Science* 62, 1469–1480.

Gregory, N.G. (2007) *Animal Welfare and Meat Production*, 2nd edn. CAB International, Wallingford, UK.

Gregory, N.G. (2008) Animal welfare at markets and during transport and slaughter. *Meat Science* 80, 2–11.

Gregory, N.G. and Wotton, S.B. (1984a) Sheep slaughtering procedures, I. Survey of abattoir practice. *British Veterinary Journal* 140, 281–286.

Gregory, N.G. and Wotton, S.B. (1984b) Sheep slaughtering procedures, II. Time to loss of brain responsiveness after exsanguination or cardiac arrest. *British Veterinary Journal* 140, 354–360.

Gregory, N.G. and Wotton, S.B. (1984c) Time of loss of brain responsiveness following exsanguination in calves. *Research in Veterinary Science* 37, 141–143.

Gregory, N.G., von Wenzlawowicz, M. and von Holleben, K. (2008) Blood in the respiratory tract during slaughter with and without stunning in cattle. *Meat Science* 82, 13–16.

Gregory, N.G., Spence, J.Y., Mason, C.W., Tinarwo, A. and Heasman, L. (2009) Effectiveness of poll stunning water buffalo with captive bolt guns. *Meat Science* 81, 178–182.

Gregory, N.G., von Wenzlawowicz, M., von Holleben, K., Fielding, H.R., Gibson, T.J., *et al.* (2012) Complications during shehita and halal slaughter without stunning of cattle. *Animal Welfare* 21(Suppl. 2), 81–86.

Grillner, T. (2011) Neuroscience. Human locomotor circuits conform. *Science* 334, 912–913.

Gruber, S.L., Tatum, J.D., Engle, T.E., Chapman, P.L., Belk, K.E., *et al.* (2010) Relationships of behavioral and physiological symptoms of preslaughter stress of beef longissimus muscle tenderness. *Journal of Animal Science* 88, 1148–1159.

Hambrecht, E., Eissen, J.J., Newman, D.J., Smits, C.H.M., den Hartog, L.A., *et al.* (2005a) Negative effects of stress immediately before slaughter on pork quality are aggravated by suboptimal transport and lairage conditions. *Journal of Animal Science* 83, 440–448.

Hambrecht, E., Eissen, J.J., Newman, D.J., Verstegen, M.W. and Hartog, L.A. (2005b) Preslaughter handling affects pork quality and glycocytic potential of two muscles differing in fiber type composition. *Journal of Animal Science* 83, 900–907.

Hansch, F., Nowak, B. and Hartung, J. (2009) Behavioural and clinical response of turkeys stunned in a V-shaped carbon dioxide tunnel. *Animal Welfare* 18, 81–86.

Hartung, J., Nowak, B., Waldmann, K.H. and Ellerbrock, S. (2002) CO_2 stunning of slaughter pigs: effects of EEG, catecholamines and clinical reflexes. *Deutsche Tierarztliche Wochenschrift* 109, 135–139.

Hayes, N.S., Schwartz, C.A., Phelps, K.J., Borowicz, P., Maddock-Carlin, K.R., *et al.* (2015) The relationship between pre-harvest stress and carcass characteristics of beef heifers that qualified for kosher designation. *Meat Science* 100, 134–138.

Hemsworth, P.H., Rice, M., Karlen, M.G., Galleja, L., Barnett, J.L., *et al.* (2011) Human animal interactions at abattoirs. Relationships between handling and animal stress in sheep and cattle. *Applied Animal Behaviour Science* 135, 24–33.

Hilton, G.G., Montgomery, J.L., Krehbiel, C.R., Yates, D.A., Hutcheson, J.P., *et al.* (2009) Effects of feeding zilpaterol hydrochloride with and without monesin and tylosin on carcass cutability and meat palatability of beef steers. *Journal of Animal Science* 87, 1394–1406.

Hoenderken, R. (1978) Electrical stunning of pigs for slaughter. Why? Hearing on preslaughter stunning. Kavlinge, Sweden, 19 May 1978.

Hoenderken, R. (1983) Electrical and carbon dioxide stunning of pigs for slaughter. In: Eikelenboom, G. (ed.) *Stunning of Animals for Slaughter*. Martinus Nijhoff, Boston, Massachusetts, pp. 59–63.

HSA (2000) *Electrical Stunning of Red Meat Animals*. Humane Slaughter Association, Herts, UK. Available at: www.hsa.org.uk (accessed 28 April 2015).

HSA (2001) *Practical Slaughter of Poultry: A Guide for the Small Producer*. Humane Slaughter Association, Herts, UK. Available at: www.hsa.org.uk (accessed 28 April 2015).

HSA (2013) New stunning method shows potential. Humane Slaughter Association, Herts, UK. Available at: http://www.hsa.org.uk/news-events/news/post/8-new-stunning-method-shows-potential (accessed 7 April 2015).

HSA (2014) *Captive Bolt Stunning of Livestock*. Humane Slaughter Association, Herts, UK. Available at: www.hsa.org.uk (accessed 28 April 2015).

James, B.W., Tokach, M.D., Goodband, R.D., Nelssen, J.L., Dritz, S.S., *et al.* (2013) Effects of dietary carnitine and ractopamine HCl on the metabolic response to handling in finishing pigs. *Journal of Animal Science* 91, 4426–4429.

Jeristrom, J. (2014) What and when determines the death of an animal? A study investigating the heart activity during slaughter of farm animals. Thesis, SLU, Skara, Sweden.

Johnson, C.B., Mellor, D.J., Hemsworth, P.H. and Fisher, A.D. (2015) A scientific comment on the welfare of domesticated ruminants slaughtered without stunning. *New Zealand Veterinary Journal* 63, 58–65. doi:10.1080/00480169.2014.964345.

Jongman, E.C., Barnett, J.L. and Hemsworth, P.H. (2000) The aversiveness of carbon dioxide stunning in pigs and a comparison of CO_2 rate vs. the V restrainer. *Applied Animal Behaviour Science* 67, 67–76.

Kannan, G., Heath, J.L., Wabeck, C.J. and Mench, J.A. (1997) Shackling broilers: Effects on stress responses and breast meat quality. *British Poultry Science* 38, 323–332.

Kolesar, R., Villareal, E.J. and Grudnik, T. (2014) Welfare of cattle in pre-slaughter and slaughter. Training DVD of the OIE Improved Welfare Training Program. OIE, World Organization for Animal Health, Paris.

Lambooij, E. (1982) Electrical stunning of sheep. *Meat Science* 6, 123–135.

Lambooij, E. and Spanjaard, W. (1982) Electrical stunning of veal calves. *Meat Science* 6, 15–25.

Lambooij, E. and Van Voorst, N. (1985) Electroanaesthesia of calves and sheep. In: Eikelenboom, G. (ed.) *Stunning Animals for Slaughter*. Martinus Nijhoff, Boston, Massachusetts, pp. 117–122.

Lambooij, B., Merkus, G.S.M., Voorst, N.V. and Pieterse, C. (1997) Effect of low voltage with a high frequency electrical stunning on unconsciousness in slaughter pigs. *Fleischwirtschaft International* 2, 13–14.

Lambooij, E., Pilarczyk, M., Blaiowas, H., van den Boogaart, J.G.M. and Van de Vis, J.W. (2007) Electrical and percussive stunning of the common carp (*Cyprinus carpio* K.): Neurological and behavioral assessment. *Aquacultural Engineering* 37, 117–179.

Lambooij, E., Gerritzen, M.A., Reimert, H., Burggraaf, D. and Van de Vis, J.W. (2008a) A humane protocol for electro-stunning and killing of Nile tilapia in fresh water. *Aquaculture* 275, 88–95.

Lambooij, E., Gerritzen, M.A., Reimert, H.G.M., Burggraaf, D., Andre, G., *et al.* (2008b) Evaluation of electrical stunning of sea bass (*Dicentrarchus labrax*) in seawater and killing by chilling: Welfare aspects, product quality, and possibilities for implementation. *Aquaculture Research* 39, 50–58.

Lambooij, E., Digre, H., Erikson, V., Reimert, H.G.M., Burggraaf, D., *et al.* (2013) Evaluation of electric stunning of Atlantic Cod (*Gagus morhua*) and Turbot (*Psetta maxima*) in seawater. *Journal of Aquatic Food Product Technology* 22, 371–379.

Levinger, I.M. (1995) *Shechita in the Light of the Year 2000*. Maskil L. David, Jerusalem, Israel.

Lines, J.A. and Spence, J. (2014) Humane harvesting and slaughter of farmed fish. *Revue Scientifique et Technique (International Office of Epizootics)* 33, 255–264.

Lines, J.A., Robb, D.H., Kestin, S.C., Crook, S.C. and Benson, T. (2003) Electric stunning: A humane slaughter method for trout. *Aquacultural Engineering* 28, 141–154.

Longeragan, G.H., Thomson, D.U. and Scott, H.M. (2014) Increased mortality in groups of cattle administered to B-Andrenergic agonist ractopamine hydrochloride and Zilpaterol hydrochloride. *PLoS One* 9(3):e91177. doi:10:1371/journalpone.0091177.

McKeegan, D.E.F., Abeyesinghe, S.M., McLemen, M.A., Lowe, J.C., Demmeus, T.G.M., *et al.* (2007a) Controlled atmosphere stunning of broiler chickens, II. Effects of behavior, physiology and meat quality in a commercial processing plant. *British Poultry Science* 48, 430–442.

McKeegan, D.E.F., McIntyre, J.A., Demmers, T.G.M., Lowe, J.C., Wathes, C.M., *et al.* (2007b) Physiological and behavioral responses of broilers to controlled atmosphere stunning: implications for welfare. *Animal Welfare* 16, 409–426.

McKeegan, D.E.F., Sandercock, D.A. and Gerritzen, M.A. (2013) Physiological responses to low atmospheric pressure stunning and implications for welfare. *Poultry Science* 92, 858–868.

Marchant-Forde, J.N., Lay, D.C., Pajor, J.A., Richert, B.T. and Schinkel, A.P. (2003) The effects of ractopamine on the behaviour and physiology of finishing pigs. *Journal of Animal Science* 81, 416–422.

Maria, G.A., Villarroel, M., Chacon, G. and Gebresenbet, G. (2004) Scoring system for evaluating stress of cattle during commercial loading and unloading. *Veterinary Record* 154, 818–821.

Miranda de la Lama, G.C., Leyva, I.G., Barreras-Serrano, A., Perez-Linares, C., Sanchez-Lopez, E., *et al.* (2012) Assessment of cattle welfare at a commercial slaughter plant in northeast Mexico. *Tropical Animal Health and Production* 44, 497–504.

Nakyinsige, K., Che Man, Y.B., Aghwan, Z.A., Zulkifli, I., Goh, Y.M., *et al.* (2013) Stunning and animal welfare from an Islamic and scientific perspective. *Meat Science* 95, 352–361.

Newhook, J.C. and Blackmore, D.K. (1982) Electroencephalographic stunning and slaughter of sheep and calves, Part 1. The onset of permanent insensibility in sheep during slaughter. *Meat Science* 6, 221–233.

OIE (2014) Chapter 7.5. Slaughter of animals. In: *Terrestrial Animal Health Code*. World Organisation for Animal Health, Paris.

Paranhos da Costa, M.J.R., Huertas, S.M., Strappini, A.C. and Gallo, C. (2014) Handling and transport of cattle and pigs in South America. In: Grandin, T. (ed.) *Livestock Handling and Transport*, 4th edn. CAB International, Wallingford, UK, pp. 174–192.

Pascoe, P.J. (1986) Humaneness of electro-immobilization unit for cattle. *American Journal of Veterinary Research* 47, 2252–2256.

Petty, D.B., Hattingh, J., Ganhao, M.F. and Bezuidenhout, L. (1994) Factors which affect blood variables of slaughtered cattle. *Journal of the South African Veterinary Association* 65, 41–45.

Poletto, R., Rostagno, M.H., Richert, B.T. and Marchant-Forde, J.N. (2009) Effects of a 'step up' ractopamine feeding program, sex and social rank on growth performance, hoof lesions, and Enterobacteriaceae shedding in finishing pigs. *Journal of Animal Science* 87, 304–313.

Probst, J.K., Spengler Neff, A., Hillmann, E., Kreuzer, M., Koch-Mathis, M., *et al.* (2014) Relationships between stress-related exsanguination blood variables, vocalization, and stressors imposed on cattle between lairage and stunning box under conventional abattoir conditions. *Livestock Science* 164, 154–158.

Raj, A.B.M. (2006) Recent developments in stunning and slaughter of poultry. *World's Poultry Science Journal* 62, 467–484.

Raj, A.B.M. and Gregory, N.G. (1990) Investigation into the batch stunning/killing of chickens using carbon dioxide or argon induced hypoxia. *Research in Veterinary Science* 49, 364–366.

Raj, A.B.M. and O'Callaghan, M. (2004) Effects of electrical water bath stunning current frequencies on the spontaneous electroencephalogram and somatosensory evoked potentials in hens. *British Poultry Science* 45, 230–236.

Rosen, S.D. (2004) Physiological insights into Shechita. *Veterinary Record* 154, 759–765.

Roth, B., Imsland, A.K. and Foss, A. (2009) Live chilling of turbot and subsequent effect on behavior, muscle stiffness, muscle quality, blood gases, and chilling. *Animal Welfare* 18, 33–42.

Rushen, J. (1986) Aversion of sheep to electro-immobilization and physical restraint. *Applied Animal Behaviour Science* 15, 315–324.

Sandercock, D.A., Auckburally, A., Flaherty, D., Sandilands, V. and McKeegan, D.E.F. (2014) Avian reflex and electro encephalogram responses in different states of consciousness. *Physiology and Behavior* 133, 252–259.

Shaw, F.D. and Tume, R.K. (1992) The assessment of pre-slaughter and slaughter treatments of livestock by measurement of plasma constituents: A review of recent work. *Meat Science* 32, 311–329.

Shimshony, A. and Chaudry, M.N. (2005) Slaughter of animals for human consumption. *Review of Science and Technology* 24, 693–710.

Tanida, H., Miura, A., Tanaka, T. and Yoshimoto, T. (1996) Behavioral responses of piglets to darkness and shadows. *Applied Animal Behaviour Science* 49, 173–183.

Troeger, K. and Wolsterdorf, W. (1991) Gas anesthesia of slaughter pigs. *Fleischwirtschaft International* 4, 43–49.

Van der Wal, P.G. (1978) Chemical and physiological aspects of pig stunning in relation to meat quality, a review. *Meat Science* 2, 19–30.

Van Putten, G. and Elshof, J. (1978) Observations of the effects of transport on the well-being and lean quality of slaughter pigs. *Animal Regulation Studies* 1, 247–271.

Velarde, A., Rodrigues, P., Dalmau, A., Fuentes, C., Llonch, P., *et al.* (2014) Religious slaughter evaluation

T. Grandin

of current practices in selected countries. *Meat Science* 96, 278–287.

Verhoeven, M.T.W., Gerritzen, M.A., Hellebrekers, L.J. and Kemp, B. (2015) Indicators used in livestock to assess unconsciousness after stunning: A review. *Animal* 9, 320–330. doi:10.1017/S1751731114002596.

Vieuille-Thomas, C. and Signoret, J.P. (1992) Pheromonal transmission of aversive substances in domestic pigs. *Journal of Chemical Endocrinology* 18, 1551–1557.

Vimini, R.J., Field, R.A., Riley, M.L. and Varnell, T.R. (1983) Effect of delayed bleeding after captive bolt stunning on heart activity and blood removal in beef cattle. *Journal of Animal Science* 57, 628–631.

Vizzier-Thaxton, Y., Christensen, K.D., Schilling, M.W., Buhr, R.J. and Thaxton, J.P. (2010) A new humane method of stunning broilers using low atmospheric pressure. *The Journal of Applied Poultry Research* 19, 341–348.

Vogel, K.D., Badtram, G., Claus, J.R., Grandin, T., Turpin, S., *et al.* (2010) Head only followed by cardiac arrest electrical stunning is an effective alternative to head only electrical stunning. *Journal of Animal Science* 89, 1412–1418.

Warner, R.D., Ferguson, D.M., Cottrell, J.J. and Knee, B.W. (2007) Acute stress induced by the preslaughter use of electric prodders causes tougher beef meat. *Australian Journal of Experimental Agriculture* 47, 782–788.

Warrington, P.D. (1974) Electrical stunning: A review of literature. *Veterinary Bulletin* 44, 617–633.

Warriss, P.D., Brown, S., Adams, S.J.M. and Corlett, I.K. (1994) Relationship between subjective and objective assessment of stress at slaughter and meat quality in pigs. *Meat Science* 38, 329–340.

Weaver, A.L. and Wotton, S.B. (2008) The Jarvis beef stunner: Effects of a prototype chest electrode. *Meat Science* 81, 51–56.

Webster, A.B. and Fletcher, D.L. (2001) Reactions of laying hens and broilers to different gases used for stunning poultry. *Poultry Science* 80, 1371–1377.

Wenzlawowicz, M.V., Schutte, A., Hollenbon, K.V., Altrock, A.V., Bostelman, N., *et al.* (1999) Field study on the welfare of meat quality aspects of Midas pig stunning device. *Fleischwirtschaft International* 2, 8–13.

White, R.G., DeShazer, J.A., Tressler, C.J., Borcher, G.M., Davey, S., *et al.* (1995) Vocalizations and physiological response of pigs during castration with and without anesthetic. *Journal of Animal Science* 73, 381–386.

Wotton, S.B. and Gregory, N.G. (1986) Pig slaughtering procedures: Time to loss of brain responsiveness after exsanguination or cardiac arrest. *Research in Veterinary Science* 40, 148–151.

Wotton, S.B., Zhang, X., McKinstry, J., Velarde, A. and Knowles, T.G. (2014) The effect of required current frequency combinations (EC1099-2009) on the incidence of cardiac arrest in broilers stunned and slaughtered for the halal market. *PeerJ PrePrints* 2:e255v1.

Zulkifli, I., Goh, Y.M., Norbaiyah, B., Sazili, A.Q., Lotfi, M., *et al.* (2013) Changes in blood parameters and electroencephalogram of cattle as affected by different stunning and slaughter methods in cattle. *Animal Production Science* 54, 187–193.

10 Recommended On-Farm Euthanasia Practices

JENNIFER WOODS[1] AND JAN K. SHEARER[2]

[1]*J. Woods Livestock Services, Blackie, Alberta, Canada;* [2]*Iowa State University, Ames, Iowa, USA*

Summary

Euthanasia means good death in Greek. It is important for people involved with euthanasia to maintain a caring attitude. This chapter covers indications for euthanasia and how to recognize when animals are in pain. It is important to differentiate between a euthanasia method that can be used as a single-step method to one where an additional step, such as exsanguination (bleeding) or pithing is required to ensure death. When firearms or penetrating captive bolts are used, larger caliber bullets for firearms or a special captive bolt with an extended bolt are required for a one-step method. The chapter also covers electrical methods, inhalants, and the use of firefighting foam for poultry. Piglets should be immersed rapidly in high concentrations of CO_2, and poultry should be exposed to lower concentrations that are increased gradually.

Learning Objectives

- Help stockpeople maintain a caring attitude when they have to euthanize animals
- Know when euthanasia of an animal is indicated
- Use of the most effective firearm and captive bolt methods; diagrams show correct shooting positions
- Use of recommended inhalant methods for poultry and piglets
- Confirming onset of death and carcass disposal

Introduction – Attitudes Towards Euthanasia

"Euthanasia is derived from the Greek terms 'eu' meaning good and 'thanatos' meaning death. The term is usually used to describe ending of life of an animal in a way that minimizes or eliminates pain and distress" (AVMA, 2013). The avoidance of pain and distress requires the use of techniques that induce an immediate loss of consciousness, followed by, or in conjunction with, cardiac and respiratory arrest that ultimately results in loss of brain function (AVMA, 2013). For persons performing euthanasia, a certain degree of technical proficiency, knowledge, and appropriate equipment are required.

Livestock owners and others who derive all or a portion of their livelihood from animal agriculture share a moral obligation to ensure the welfare of animals. Part of these obligations is to make certain animals do not experience unnecessary pain and suffering, even in death. Therefore, when disease or injury conditions arise that diminish the quality of life or create pain and suffering that cannot be relieved effectively by veterinary treatment at a reasonable cost, euthanasia is indicated.

Reality is that the situations whereby a person may be required to perform euthanasia are frequently on the farm. In some cases, it is an emergency procedure perhaps associated with some type of traumatic event. In others, it is a decision based on one's assessment of a sick animal's locomotory status, prognosis for recovery, or perceived suffering. Other indications may include loss of productivity, economic restraints of treatment, or a dangerous animal. Regardless of the reason, for the sake of the animal and its caretaker, who may also be the person performing this unenviable task, the ability to induce both a rapid and a humane death is of paramount importance.

This chapter provides guidance on the process of euthanasia, beginning with indications for euthanasia

---

and ending with either disposal of the carcass or processing it for meat. Steps in the process that are covered within this chapter include timeliness of application, animal considerations, human concerns, selection of method, application of technique, methods to ensure a humane death, and assessing your euthanasia system.

A personal experience with euthanasia

As someone (first author) who is not only a livestock owner but also a professional who practises euthanasia through my work in emergency response to livestock truck accidents, training, and research, euthanasia and death is a very large part of my life. It is a job that has to be done and one I have done many times with little hesitation, knowing it is best for the animal. One of the most difficult struggles I have had as a livestock owner was when I had to euthanize my 10-year-old daughter's 4 H (youth project) lamb named Posie. She was different from the other animals on the farm, as not only was she a pet that I was personally attached to but also I was dealing with the emotional distress of my daughter.

As sheep often do, she managed to get her head through two fence panels and consumed a large amount of grain. Though we caught her and another ewe shortly after the incident, the grain overload led to acidosis that ended up going systemic. We treated them around the clock for 3 days and the second ewe pulled out of it, but Posie obviously had consumed the most grain and was slowly dying. Acidosis can take several days to kill an animal, but she was getting progressively weaker and showing no signs of recovery. After the third day, it was obvious she was not going to recover and the winter temperature outside was reaching −25°C.

I struggled with being the one who had to euthanize her, as many other owners do. In my head, I debated having someone else come out and do it, struggled with continuing treatment though I knew she was terminal, and the consideration of taking the easy way and just letting her die a "natural" death. When I returned from the barn after shooting her, my daughter, in tears, told me it was the right thing to do – "it was really hard for us, but the best thing for Posie". The welfare of the animal must always take precedence over our emotions, no matter how hard it may be.

Maintain a caring attitude

People who are really good at caring for animals often have a difficult time making decisions to euthanize them. Blackwell (2004) has found that it was easier for farm stockpeople to euthanize a sick or injured pig if the farm had a written policy that clearly stated the conditions when an animal should be euthanized. Some examples are fractured limbs, emaciated body condition, or a specified number of days with no response to veterinary treatment. The policy helps the stockperson because it makes the difficult decision. It is important to make the criteria for euthanasia strict enough to prevent suffering but not so strict that caring employees are not allowed to nurse sick animals back to health (Grandin and Johnson, 2009). If a caring, good stockperson was told to kill almost every sick pig, they would probably quit. The livestock industry needs caring, good stockpeople because they improve animal welfare and help make animals more productive (see Chapter 7, this volume). Emotional attachment to an animal is just one of the many factors that affect the timeliness, efficiency, and effectiveness of euthanasia. Other influences include sociodemographic considerations, environmental influences, psychological factors, and management practices. When fully understood, each one of these factors can be managed to ensure timely euthanasia and minimize distress.

Cultural factors

Sociodemographic considerations impact the performance of euthanasia to relieve suffering. They include religion, gender, cultural, personality, and age. A person's religion can affect their attitude and beliefs towards euthanasia. There are religions that do not condone the taking of life under any circumstances; others may consider certain species sacred. Men tend to have a more positive attitude towards euthanasia than women, even though both are equally willing to perform the task. Gender also influences the preferred mode of euthanasia (Matthis, 2004). Different cultures have contrasting views on animal welfare, ranging from very high awareness to no awareness at all. Research in the US pig industry has shown that most Americans have a more positive attitude towards euthanasia, and Mexican employees have a more negative attitude towards euthanasia. Different personalities and temperaments have different attitudes and comfort levels pertaining to euthanasia.

Some people have no feelings with regard to the action, whereas others are not comfortable with it under any circumstance, believing there are always alternatives. This survey also found that the age of the stockperson also influenced attitude – the older the stockperson, the more negative the attitude (Matthis, 2004).

Experience with euthanasia

Other factors that affect a person's attitude include previous experience, management, level of exposure, caring, and social stigma. A stockperson's previous experiences will influence their attitude and approach towards euthanasia. People who are raised around animals and in a farm environment are more likely to be comfortable with euthanasia than those who have not been raised around farm animals. Also, people who have had a previous bad experience with poorly carried out euthanasia are more likely not to be comfortable with it.

The frequency of euthanasia and related tasks can affect a stockperson's attitude negatively. Studies both on both pig farms and in animal shelters have shown that the longer a person has been performing euthanasia tasks, the less willing they become to euthanize (*Swine News*, 2000; Matthis, 2004; Reeve *et al.*, 2004). This also applies to the frequency of the performance of euthanasia and related tasks. This phenomenon explains the difference in attitude of swine barn employees – those in farrowing units, where animals are euthanized more frequently, have a more negative attitude towards euthanasia compared to those in a finishing (fattening) barn, where euthanasia is not as frequent (Matthis, 2004).

Caring and killing paradox

The job of a stockperson is to raise and care for animals. When they have to euthanize them, it creates a "caring and killing paradox" (Arluke, 1994; Reeve *et al.*, 2005). Many people struggle with this, especially in cases where they have put extensive effort into saving a sick or injured animal. Euthanizing animals is also often morally stigmatized in some societies. This "tainted" image of euthanizing animals can put added strain on the handlers themselves, as they personalize the stigma.

Many studies have documented that euthanasia can have negative psychological effects on the people who perform it. The majority of studies have focused on animal-shelter workers, with some limited research

in pigs (Arluke, 1994; *Swine News*, 2000; Matthis, 2004; Reeve *et al.*, 2004; AVMA, 2013). Stockpeople who constantly euthanize large numbers of animals may handle livestock roughly and apply euthanasia carelessly. They may also experience job dissatisfaction, increased missed work days, depression, grief, frustration, sleeplessness, nightmares, isolation, high blood pressure, ulcers, and substance abuse.

Psychology of euthanasia

In large slaughter plants, Grandin (1988) reported that the workers who stunned (euthanized) the animals fell into four psychological patterns. They are: (i) caring; (ii) box stapler; (iii) sadists; and (iv) those who carry out religious rituals, such as kosher or halal slaughter. The box stapler does his work efficiently and is emotionally detached. This is the most common psychological pattern. One very skilled stunner operator in a huge plant talked about the weather and golf as he shot 250 cattle/hour. He was efficient and never abused the animals. The sadist enjoys causing suffering and must be removed. Manette (2004) and Wichart von Holten (2003) also discuss ways to help people who have to kill large numbers of animals deal with the psychological trauma. This is especially a problem after mass kills for disease control purposes. Some people become totally desensitized and others have soothed their grief by performing a funeral ritual for the dead animals (Manette, 2004).

Reaction to euthanasia procedures will vary among individuals, with some experiencing no adverse affects, while others are deeply affected. Experience has shown that the more knowledgeable and aware people are made of the correct procedures and reasons for euthanasia, the more comfortable they become with the job. The greater the level of comfort, the less likely they are to experience stress-related symptoms.

On-farm management of euthanasia

Management will often set the tone for the attitude of the stockpeople. Managers must have a proactive approach to all aspects of animal welfare and demand the same attitude from the staff. A stockperson with a good attitude is essential to ensure proper handling and euthanasia procedures on the farm. Research has revealed sequential relationships between the attitudes of the stockperson towards the animals, the behavior of the stockperson with

J. Woods and J.K. Shearer

the animals, the behavioral response of animals to humans (i.e. fear) and the overall performance and welfare of the animal (see Chapter 7, this volume).

When hiring or selecting employees for positions that include euthanasia, it must be ensured that the person is comfortable performing the expected task. It is detrimental to employee morale, safety, and animal welfare to force people to perform euthanasia who are not comfortable or competent with the procedure. People may be comfortable performing euthanasia, but they may not be comfortable with the modes of euthanasia used within the operation (i.e. blunt force trauma). This must be discussed before they are expected to euthanize animals. Company policy must be clearly posted, and there must be an alternative plan for those people uncomfortable with the designated euthanasia process, as euthanasia must be available 24 h a day.

It is also management's responsibility to ensure that stockpeople are trained in euthanasia. Research has shown that the amount and type of euthanasia training a stockperson receives influences their attitude. When people are provided with comprehensive training that covers all aspects of euthanasia, they will often become more comfortable with the procedures and have a better attitude towards euthanasia (Matthis, 2004; Reeve *et al.*, 2004). Training not only gives them the skill to perform the act but also the confidence to make timely decisions on when to euthanize an animal.

Management can also alleviate much of the strain on their stockpeople by rotating the euthanasia task throughout the staff, especially in operations where euthanasia is a regular part of the routine. Managers must keep the lines of communication open with employees and note any changes in behavior, attitude, frequency of sick days, etc. Support should be provided for employees who request it or appear to need it. This support network should include open lines of communication, task rotation when requested, and counselling if required. Management of euthanasia requires effective farm management. Grandin (1988) observed that the farms or slaughter plants that had the best stunning and animal handling practices had a manager who was involved enough to care, but not so involved that he became desensitized.

Indications for Euthanasia

Indications for euthanasia of livestock include poor health, disease, injury, loss of productivity, economics, and safety. There are three possible treatment options for livestock producers when faced with each of these situations: (i) ship the animal for meat processing IF the animal is fit for transport and human consumption; (ii) provide veterinary treatment; or (iii) euthanasia. Just because there is a chance for recovery does not mean treatment is always the best choice for the producer or the animal. When deciding which option is best, there are several questions a person has to ask in order to make a responsible decision.

Whenever the outcome of disease or injury is in doubt, a veterinarian should be consulted for advice. Simply leaving an animal that is suffering to die of natural causes or, in other words, "letting nature take its course" is *unacceptable*. Furthermore, it is *not* acceptable to prolong an animal's misery by delaying euthanasia for reasons of convenience (i.e. waiting for the veterinarian's weekly visit). It is important that when euthanasia is indicated, it be conducted in a timely manner.

Examples of **poor health or injury conditions** that would justify euthanasia include:

- Emaciation and/or debilitation from disease or injury that may result in an animal being too weak to be transported.
- Paralysis from traumatic injuries or disease that result in immobility.
- Cancerous conditions such as lymphoma and squamous cell carcinomas (cancer eye) of cattle. Animals with severe cancer eyes, where the eye has ruptured and the surrounding tissue is invaded, should be euthanized immediately.
- Disease conditions for which the cost of treatment is prohibitive.
- Disease conditions where no effective treatment is known (Johne's disease in ruminants, Porcine Epidemic Diarrhea Virus), prognosis is poor, or time to expected recovery is unusually prolonged.
- Chronic disease where treatment is unlikely to improve outcome (chronic respiratory disease in cattle and sheep).
- Transmittable diseases (zoonotics) where there is significant threat to human health or other animals (rabies, foot and mouth).
- Fractures of the legs, hip, or spine that are not repairable and result in immobility or an inability to stand.
- Emergency medical conditions that result in excruciating pain that cannot be relieved by treatment (trauma associated with road accidents).

- A wound significantly impacting a critical biological function (i.e. major organ, muscle and skeletal systems, brain injury) – one example is severe injuries that expose internal organs.
- Profuse bleeding.
- Large umbilical hernia in a pig that has become injured from scraping the floor.
- Animals that are nonambulatory for more than 24 h. Green *et al.* (2008) found that dairy cows that remained nonambulatory for more than 24 h were less likely to recover.

Livestock owners are responsible for the welfare of the animals in their care. Although profit is the motivation of most livestock enterprises, the welfare of animals within those operations must always take priority. Whenever it is clear that the cost of therapy outweighs financial return, euthanasia should be considered a potential option. The cost of treating a sick animal includes: cost of medication, expense of increased labour during treatment, veterinary care, and input costs to return the animal to production. Other factors that may enter the decision process might include:

1. Will the animal have to endure a painful and lengthy recovery?
2. Will the animal be likely to return to normal function post-recovery (i.e. restoration of quality of life)?
3. Can the required care be provided during the convalescent period?
4. Is the animal likely to suffer chronic pain or immobility following recovery?
5. Will weather extremes create inhumane conditions for this animal during and/or after recovery? (Woods, 2009).

The answer to these questions is not always clear. None the less, they should be part of the decision-making process whenever livestock owners are faced with the choice of treatment, slaughter, or euthanasia.

Time for recovery from illness or injury

One of the biggest challenges is determining: *How long should an animal be given to recover?* Industry literature and guidelines would suggest that sick animals should show evidence of significant improvement within 24–48 h from the onset of treatment. Depending on the specific disorder encountered, most animals will exhibit evidence of a positive response to therapy within 24 h. Rarely do animals recover when they do not show progress within the first 36 h following treatment. Recovery

time for animals that experience injuries may be more prolonged and harder to estimate, so they must be evaluated on an individual basis.

In cases where animals are unfit for transport or not acceptable to be processed for meat, it is the responsibility of the producer to euthanize the animals. Examples of this would be animals that are still within the required drug withdrawal period, animals that would experience further pain and suffering during transport, animals that have high fevers or ones in which illness may lead to the condemning of the carcass (e.g. cancer eye).

Most livestock are raised for production purposes, whether for raising offspring or providing meat, eggs, or dairy products. When the cost of feed, veterinary care, housing, and labour exceeds productivity income of an individual animal, it is the responsibility of the producer to remove the animal from the production cycle. Producers should cull the animal in a timely manner while the animal is still fit for transport and processing. This will allow for the owner to recover some of the cost and maintain good animal welfare. In cases, though, where the animal is at risk and not fit to transport, on-farm slaughter or euthanasia are the only two options.

There are also some indications for euthanasia not related to health, welfare, or production. When an animal has become dangerous to handlers, family members, staff, or other animals, it must be shipped for meat processing or euthanized. No matter how productive the animal is, the owners or management are responsible for the safety of all those on the farm or within the facility. It is irresponsible and the owner may be liable if the animal is sold to another producer knowing the animal can hurt someone (see the section headed 'Bull Behavior and Safety' in Chapter 5, this volume).

Recognition of pain and suffering that may indicate the need for euthanasia

Pain and suffering in livestock are primary indications for euthanasia, but can be easily misinterpreted. Prey animals instinctively avoid expressions of pain in order to evade the notice of predators. Cattle, for example, may simply become less responsive or depressed, and lame animals will adjust their gait or posture to mask evidence of lameness (Chapters 2 and 11, this volume). Predators, on the other hand, freely express signs of pain or discomfort. Consider the reaction of a dog that has inadvertently had its foot stepped on. The response is verbally loud, its flight reaction swift, and on occasion punctuated by aggression.

This is an important distinction, since failure to conduct euthanasia procedures in timely fashion is often related to a misinterpretation of an animal's response to pain and suffering. A survey of respondents on the euthanasia of pigs found that sick pigs that appeared to be in pain were easier to euthanize than pigs that showed no signs of pain (Matthis, 2004). The survey also revealed that an overwhelming percentage of people agreed that it was more humane to euthanize a sick/injured pig that appeared to be in pain than to let it suffer. The experience and education of the stockpeople are important for accurate assessment of animal behavior, particularly when it is necessary to distinguish distress from normal behavior.

There are several behavioral signs that can alert you to the fact that an animal may be in pain. Any of these indications may appear alone or in conjunction with other signs, and may overlap with signs of distress. If you are unsure if the animal may be experiencing pain or why the animal is in pain, consult an experienced stockperson or professional to assist in assessing the situation.

Indications of pain may include:

- inability or unwillingness to rise
- inability or unwillingness to walk
- reluctance to put weight on a limb
- protection of the painful area
- vocalization, especially when the animal moves or the painful area is touched
- open-mouth breathing
- arched or hunched back
- abdomen tucked up
- drooping head and/or ears
- grinding of teeth
- tail uncurled (pigs whose tails have not been docked)
- lack of interest in food or water
- lack of interest in surroundings
- stays away from other animals
- does not respond to other animals attempting to interact with it
- does not respond when touched or prodded
- standing in a rigid position
- shivering
- trembling or profuse sweating
- animal licks or uses a limb to poke the painful area
- scratching or shaking affected area
- kicking or biting at abdomen

- gets up and lies down frequently
- walks in circles
- excessive rolling (horses)
- grunting
- hiding in bedding
- licking
- biting
- change in personality
- change in eye brightness
- glassy eyes, dilated pupils
- reduced suckling
- tail flicking
- unsettled
- inability to get comfortable
- their head is turned back
- lying on their belly when lying down
- increased blood pressure and heart rate
- rough hair or coat that does not lay flat (easier to see in animals with short hair).

This list includes both chronic and acute indicators of pain and distress. See Chapters 2 and 6, this volume, for indicators of acute pain after procedures such as dehorning and castration.

According to Moberg (1985), an animal experiences distress when the biological cost of coping with the stressor(s) diverts resources away from other primary biological functions (growth, reproduction, etc.). Distress behavior includes, but is not limited to, stress vocalization, attempts to flee, aggressive behavior, rapid or irregular breathing, mouth breathing, salivation, struggling, urination, defecation, rapid heartbeat, sweating, dilation of the pupils, tremors/ shaking, freezing/standing completely still, grinding teeth, and reduced or increased suckling. Just like pain, any of these indicators may appear alone or in combination with others.

It is not uncommon to hear stockpeople say that an animal is "just not acting like itself" when initially diagnosing a problem. Knowing your animals and being cognizant of their behavior will assist in early detection of disease or illness, which in turn will allow for an increased chance of recovery or the minimizing of suffering through timely euthanasia. Having employees who can "read" animals is an asset to the operation.

Pre-euthanasia Considerations and Preparations

It is important that stockpeople do all they can to minimize anxiety, fear, pain, and distress for the animal to be euthanized. If the animal to be euthanized is ambulatory and able to be moved without causing distress, discomfort, or pain, it may be moved to an area where the carcass would be reached more easily by removal equipment. The use of aggressive means to force an animal to drag itself or dragging of a nonambulatory animal is not acceptable, and it is a violation of both international and many national laws. In cases where movement may increase distress or animal suffering, the animal should be euthanized first, and moved following confirmation of death.

Restraint methods

If restraint is necessary, it must occur for the shortest time possible and use the least stressful means available. When choosing a method of restraint, it must be safe for handlers and as stress free as possible for animals. If small animals are being restrained by hand for penetrating captive bolt gun (PCBG), extreme caution must be exhibited to avoid shooting the hands or legs of the person performing the restraint. An animal must be restrained so that it cannot strike out or escape if something goes wrong. It must also be possible to remove the carcass from the area following euthanasia.

Proper restraint for pigs depends on the size and condition of the animal. For neonates and early nursery pigs, the animals should be picked up utilizing two points of contact (i.e. leg and flank), either held firmly or placed in an appropriate restraint device (such as a body sling). Under no circumstances should the animals be picked up by one point of contact (i.e. leg) and swung, thrown, etc. (OIE, 2014a). At a minimum, larger animals should be penned. The use of a properly designed restrainer chute minimizes the opportunity for movement and reduces the likelihood of misapplication of the euthanasia method. Snout snares for pigs should be designed for the age and weight of the animal, and be constructed of rope, or smooth, round cable. They must not cut or injure the animal. Snares MUST NOT be used to drag or lead an animal, and euthanasia must be performed immediately after the snare is applied. Restrainer chutes (races and crushes) must be designed for the age and weight of the animal, and the chute must provide adequate access to the animal to allow for the necessary procedure (i.e. euthanasia). If animals are euthanized in a restrainer chute, the design must allow for animal removal once it is confirmed dead. There are many useful pictures in Sheldon *et al.* (2006).

For cattle, squeeze chutes are acceptable, but animals can be very difficult to remove. A headgate (stanchion, head bale) on the end of a single-file race will often work well. Tame ambulatory animals can be led to the place where euthanasia will be performed with a halter (head collar). Horses and sheep can be restrained in restraint chutes or with halters. If you use a halter, make sure it has enough lead to untie following euthanasia so the animal will not be hanging. For elk/farmed deer, a restraint chute (race) designed for their species is recommended. On the farm, poultry can be restrained with the use of cones or by hand.

Euthanasia of fractious animals (such as an aggressive bull or cow) may require capture in a race where the animal can be tranquilized and released to a pen that is accessible by removal equipment. Once the tranquilizer has taken effect and the animal is immobilized, it may be euthanized by an appropriate method that is safe for handlers and assistants.

The safety of other animals in the immediate area must always be considered. If necessary, other animals should be removed from the immediate area to avoid the chance of ricochet, getting hit by a stray bullet, or injury due to startling. Biosecurity is also a safety issue for other animals as the sick animal may release contaminated body fluids or blood following euthanasia.

Determination of the most appropriate euthanasia method

Selection of the best method of euthanasia should include:

- concerns for human safety
- animal welfare
- emotional comfort with the procedure by the person performing the task as well as bystanders
- ability to restrain the animal for proper application of the procedure
- skill of the person performing the procedure
- cost
- rendering and carcass disposal considerations
- potential need for brain tissue for diagnostic purposes.

Some methods of euthanasia are more costly than others. Tools such as captive bolt guns have a high initial cost, but are relatively inexpensive to use. Anesthetic overdose has a relatively high cost since a veterinarian must perform the procedure and carcass disposal is complicated by drug residue.

J. Woods and J.K. Shearer

The number of animals euthanized within the operation is also a consideration. If you only euthanize an occasional animal, cost is not a significant factor compared with a larger operation that euthanizes animals more frequently, or maybe in the event of conducting mass killing of livestock.

Each method of euthanasia requires a certain level of skill and training. This is an extremely important consideration as the skill and efficiency of the person performing the task is vital for the proficiency of the task. Improper use of the tool will not only jeopardize the safety of the handler but also the welfare of the animal. *The vast majority of failed euthanasia is the direct result of human error.*

Individual preference of methods

Individual preference for the method of euthanasia must be considered. People are often more comfortable with one means of euthanasia rather than another. The more comfortable a person is with the chosen acceptable method, the more proficiently they will perform the act. Factors that can determine the level of comfort include religion, background, gender, education/training, previous experience, and aesthetics.

Each method of euthanasia has its own degree of aesthetic displeasure. Some means have a more peaceful appearance (i.e. anesthetic overdose) than others (i.e. blunt force trauma). Aesthetic effectiveness may be affected by blood loss and physical trauma to the animal. Consideration must be given to the comfort level of the person performing the task and that of bystanders (i.e. general public or media). Just because one method is less aesthetically appealing than another does not mean it is less humane. Barbiturate overdose is very effective, peaceful and humane when it is done correctly, but it is not practical on most livestock operations. Animals that are euthanized with anesthetics cannot be used for food for either people or animals. In many countries, rendering plants may not accept animals that are euthanized with barbiturates.

There are also legal restrictions with certain methods of euthanasia. In many countries, lethal injection can only be delivered by or under the supervision of a licensed veterinarian because barbiturates are a controlled substance and have potential for drug abuse. When prescription drugs are used for euthanizing animals, you should find out what the laws are in your country, state, or province. In most countries, the use of firearms or a captive bolt does not require supervision by a veterinarian. However, some countries require the gun to be registered, and the operator must hold a valid firearms licence. In other countries, people are not allowed to own firearms.

Biosecurity considerations include the spread of further disease and clean-up of the euthanasia site. Euthanasia techniques that must be followed by exsanguination lead to the possible contamination of facilities with blood. Contamination of the area where euthanasia is conducted with brain tissue is a possible risk with gunshot, captive bolt gun, and blunt force trauma. Other body fluids may also be released during the euthanasia process, which can contaminate facilities.

When determining which of the various methods of euthanasia should be used, consideration must be given to the type of animal, its size, and age. Manual blunt force trauma is only recommended for poultry, piglets, and infant lambs and kids. Cervical dislocation is only acceptable for small poultry such as chickens. The skulls of boars, sows, bison, and mature bulls are much thicker and more difficult to penetrate with firearms and captive bolt guns, thus limiting the selection of effective tools.

The mode of disposal of the carcass must also be considered in making euthanasia decisions. Such concerns include local and national laws regulating the disposal of dead livestock. If it is likely that a carcass may be consumed by scavengers (buzzards, coyotes, etc.), the animal cannot be euthanized with drugs. Animals that are going to require diagnostic evaluation after death (such as for rabies diagnosis) must be euthanized in a way that will not damage or destroy the brain.

Induction of death

Death is a process and does not occur immediately. Animals are first rendered insensible and then the body begins to die as the brain stops, the heart quits beating, the lungs stop breathing, and the blood quits circulating. There is no such thing as immediate death of the entire body. For example, gunshot or captive bolt will destroy the brain instantly, but the heart may beat for several minutes (Vimini *et al.*, 1982). However, a well-placed gunshot has started the process of death and the animal is instantly rendered insensible to pain and will not recover.

Death may be induced by one or more of the following mechanisms: direct depression of the central nervous system (CNS), hypoxia, and physical disruption of brain activity. *Direct depression of the CNS* is achieved through the injection of an overdose of barbiturates. Though not commonly used in a livestock-production setting, inhalant anesthetics such as ether or halothane will also induce death through depression of the CNS, but they have significant human safety concerns. *Hypoxia*, or lack of oxygen, is achieved by exposing animals to high concentrations of gases such as carbon dioxide (CO_2) or argon, or through rapid blood loss (exsanguination). Gunshot, blunt force trauma, or captive bolt stunning devices induce death by *physical damage to the CNS that results in a disruption of brain activity*. Death occurs as a result of respiratory and cardiac failure (AVMA, 2013).

Lethal Injection

For on-farm euthanasia of livestock, one may choose lethal injection with barbiturates. This requires a veterinarian to be available to perform the intravenous injection, which can make its use restrictive. Other restrictions include the need to restrain the animal, safety of the handler, especially with large animals, disposal of livestock following death, and higher cost and practicality in the euthanasia of several animals. Both the American Veterinary Medical Association (AVMA, 2013) and the British Veterinary Medical Association (BVMA) have specific guidelines. The AVMA guides are available free on the Internet and may be accessed by typing "AVMA Guidelines for the Euthanasia of Animals" into Google. To access the British guidelines will require either a subscription to or membership of the BVMA.

When lethal injection is not available or practical, euthanasia options include gunshot, captive bolt gun, CO_2, electrocution, blunt force trauma, and cervical dislocation. Carbon monoxide and maceration are both considered acceptable means of euthanasia. Due to high human safety and animal welfare issues with carbon monoxide and the restriction of maceration to day-old poultry and embryonated eggs only, they will not be included in this chapter.

Firearms

Gunshot is an acceptable means of euthanasia for all livestock. In the USA, it is the most popular method for on-farm euthanasia (Fulwider *et al.*, 2008). Gunshot euthanizes by mass destruction of the brain. A basic principle is that a larger, more

powerful cartridge and firearm is required when there is no additional step such as bleeding. The most powerful firearms are required for long-distance shots. The degree of brain damage inflicted by the bullet is dependent on the firearm, nature of the bullet (or shotshell), and accuracy of the shot. For euthanasia purposes, handguns are limited to close-range shooting of less than 5–25 cm (2–10 inches). Shotguns are appropriate for a distance of 1–2 m (1–2 yards), and rifles for long-distance shots.

The correct selection of ammunition is vital for successful euthanasia. The energy of a bullet and its destructive capacity are normally described in terms of muzzle energy (i.e. joules or foot pounds of force). It is recommended that a minimum of 407 J (300 ft lb) of muzzle energy be available for firearms used in the euthanasia of animals up to 180 kg (400 lb), and 1356 J (1000 ft lb) of muzzle energy for animals heavier than 180 kg (400 lb) (USDA, 2004). This recommendation is for a single-step method in the field. The Humane Slaughter Association recommends a minimum muzzle energy of 200 joules for large or aged animals (HSA, 2014).

Shotgun 28, 20, 16, 12 or 0.410 gauges

Shotguns are very effective for the euthanasia of livestock. They are best used at a distance of 1–2 m (1–2 yards). The 20, 16, and 12 gauge can be used on all weight and species classes. The smaller 28 and 0.410 gauge shotguns should not be used on larger and mature animals due to skull mass. The first author prefers either the 0.410 or 0.20 gauge loaded with a slug. This works effectively at 6 m (20 ft) or less. Number 4, 5, or 6 birdshot is appropriate ammunition for close range only because, as birdshot leaves the barrel of the gun, it begins to disperse or spread, lessening impact and destruction as the distance from the animal increases. Slugs (special shotgun shells) are considered the best choice for euthanasia as a slug is a solid mass that does not disperse as it leaves the barrel. In a situation where animals are loose or where the handler is unable to get close to the animal, a shotgun with solid slugs should be used.

Rifles

The 0.22 caliber long rifle is one of the most popular firearms, but with an average muzzle energy of only approximately 100 ft lbs, it does not meet the current energy requirements for euthanasia of livestock. Animals up to 180 kg (400 lbs) require a minimum of 300 ft lb, while animals heavier than 180 kg (400 lb) will require a minimum of 1000 ft lb to ensure death (USDA, 2004). The 0.22 must be followed by either pithing or bleeding. A 0.22 should NEVER be used on bison, elk, fully mature bovine bulls, large mature rams, or mature boars because it is not reliable in causing a humane death (AVMA, 2013). If a 0.22 is utilized, a long rifle, round-nose, solid-point lead bullet should always be used. A 0.22 long hollow-point bullet fired from a rifle does not reliably destroy the brainstem (Thomson *et al.*, 2013). Many slaughter plants use hollow points to improve safety because a hollow point is less likely to ricochet and injure people. They also often use cartridges that are smaller than recommended. In a slaughter plant, it is easy to assess the indicators of instantaneous insensibility, and shooting is always followed by bleeding (see Chapter 9, this volume). Out in the field, it is more difficult to assess insensibility, and animals are often shot from a greater distance. In this situation, using a less powerful cartridge is more likely to result in suffering. Hollow points will require the use of larger cartridges compared to solid round points. Rifles are more effective than pistols because the longer barrel increases bullet velocity. In slaughter plants, a single-shot, bolt-action or break-open rifle is recommended to improve safety.

For longer distances of up to 274 m (300 yards), the recommended rifle cartridges would include those shown in Table 10.1. Note the reduction in muzzle energy as it reaches 274 m (300 yards).

Larger caliber rifles such as the 0.308 are not appropriate for close-range shots due to the chance of pass through.

Handguns

Handguns can be used for close-range euthanasia at a distance of 5–25 cm (2–10 inches), with the appropriate round-nose, lead bullet ammunition. Common handgun cartridges that exceed the recommended 300 ft lb of muzzle energy include those shown in Table 10.2. A 9 mm pistol firing a 124 grain cartridge is not effective for cattle (Thomson *et al.*, 2013). The first author prefers the use of a pistol if animals have to be euthanized in a confined space, such as a truck that has tipped over. In other situations a rifle is recommended because it has greater muzzle velocity. Small 0.22, 9 mm pistols are not recommended for large animals due to lower muzzle velocity compared to a rifle of the same caliber.

Application of gunshot for euthanasia

FIREARMS MUST NEVER BE HELD FLUSH TO THE ANIMAL'S HEAD OR BODY. The pressure within the barrel when fired may cause the barrel of the gun to explode. Ideally, the gun should be angled so the bullet follows the angle of the neck or the spine. When angled

Table 10.1. Specifications for rifles. (From USDA, 2004.)

Cartridge	Muzzle energy (J) (ft lb in parentheses)	Muzzle energy at 274 m (300 yards) (ft lb in parentheses)
0.357 Magnum (rifle)	1593 (1175)	457 (337)
0.223 Remington	1757 (1296)	778 (574)[a]
30–30 Winchester	2579 (1902)	883 (651)
0.308	3590 (2648)	1617 (1193)
30–06 Springfield	3852 (2841)	1973 (1455)

Note: [a]The 0.223 Remington is used with a 5.56 mm NATO rifle cartridge.

Table 10.2. Specifications for handguns. (From USDA, 2004.)[a]

Cartridge	Average muzzle energy (J) (ft lb in parentheses)
0.40 Smith and Wesson	553 (408)
0.45 automatic	557 (411)
0.357 Magnum	755 (557)
0.41 Remington Magnum	823 (607)
10 mm automatic	880 (649)
0.44 Remington Magnum	988 (729)

Note: [a]When higher muzzle energy is required, it is recommended you use a rifle.

properly, the bullet will travel through the brain and ideally end up near the top of the spine or brainstem area. The angle of the firearm is important for safety reasons, since it is desired that the bullet remain within the body. Even though the World Organisation for Animal Health (OIE) recognizes a shot to the neck as acceptable for killing, it is only in a situation of mass killing for disease control purposes, not for euthanasia of individual animals. The AVMA (2013), the Humane Slaughter Association (HSA, 2014) and the European Food Safety Authority (EFSA, 2004) does not recommend gunshot to the heart or neck as it does not immediately render the animal unconscious. Firearms used for continuous multiple shots should be rotated to prevent overheating.

Cattle and calves

There are two acceptable points of entry for gunshot in cattle – the front of the head or behind the poll of the animal. If the animal is lying down, has horns, or has a thick skull, behind the poll is the better choice over the front of the head (Gregory et al., 2008).

The frontal target area is high up on the head of the animal, NOT BETWEEN THE EYES. For horned cattle, Fig. 10.1 shows the correct shot location. Draw an imaginary X from the top of the eye to the center of the opposite horn. The shot is placed in the middle of the X. For polled (hornless) cattle, Fig. 10.2 shows the correct shot location. Draw an imaginary horizontal line at the top of the eyes. The correct shot location is midway between the horizontal line and the top of the poll. The firearm should be positioned so that the muzzle is perpendicular to the skull and the bullet will enter the front of the head and travel towards the tail of the animal. There may be some differences in the location of the shot, based on the skull shape and horn mass of an animal.

Pigs

The anatomical site in pigs is slightly more complicated to identify because the pig's brain is smaller relative to the size of its skull. As a pig ages, the skull continues to grow more relative to the brain, which makes the target even more difficult to identify accurately. This is complicated further by the fact that different breeds of pigs can have very different skull shapes, and the shape will change as the pig matures. Mature sows have a large sinus cavity located right in front of the brain, setting the location of the brain even further back into the skull, thus requiring deeper penetration with a bullet or penetrating captive bolt. Similarly, boars form a

Fig. 10.1. Correct location for shooting horned cattle with a captive bolt or firearm. Compared to previous diagrams, the center of the X is slightly raised because the X goes from the top of the eye to the middle of the opposite horn. The ideal shot location is on the intersection of the two lines drawn from the lateral canthus (outer corner of the eye) to the base of the opposite horn. (From Shearer and Nicoletti, 2011; AVMA, 2013.)

Fig. 10.2. Correct location for shooting polled (hornless) cattle with a captive bolt or firearm. An alternative method to identify the ideal shot location is on the midline of the skull midway between two horizontal lines: one drawn from the lateral canthus to lateral canthus and another drawn horizontally across the top of the poll.

ridge down the front of their skull as they mature, which also complicates penetration to the brain. Due to human safety issues with restraint and ricochet, gunshot is not a practical means of euthanasia for piglets under 5 kg (12 lb).

The ideal location for shooting a market-weight pig (100–135 kg) is approximately 2.5 cm (1 inch) above their eyes, in the middle of their forehead (Fig. 10.3). For older boars and sows, the shot should be located 3–4 cm (1.5–2 inches) above eye level and just to the side of the ridge on their skull. Older, mature animals are best euthanized with shotguns using slugs.

Horses

The brain of the horse is located high on the forehead so the shot will need to be slightly above the intersection of an X drawn on the forehead, with each line beginning at the inside corner of the eye and progressing to the opposite ear. The specific anatomical site is approximately 2 cm (1 inch) above the center of the X (Fig. 10.4). It is also important to note that horses may lunge forward or rear up when shot, so the shooter must beware of this when shooting a standing animal and position themselves accordingly.

Sheep and goats

There are three acceptable points of entry on sheep: the front of the head, the top of the head, and the back of the poll. When using the landmarks for a forehead shot, the gun should be positioned so that

Fig. 10.3. Correct position for shooting swine with a captive bolt or a firearm. Many old diagrams show a position that is too low. (Diagram by J.K. Shearer.)

it enters 2 cm above the eyes. Euthanasia of sheep and goats by gunshot can be difficult as different breeds have different skull shapes, which will require some alteration to the target area for gunshot.

The ideal shooting position for sheep is a shot to the top of the head. The shot should be placed in the midline of the skull, pointing straight down at the throat. This shot will take the projectile through the brain. A shotgun is best to use for this shot to avoid the projectile passing through an animal and injuring other people and animals.

In horned sheep, the most effective shot is behind the poll. The projectile will not have to pass through the extra bone mass created by the horns that cover the area in front of the brain. It is best to make this shot with a shotgun, to eliminate the chance of the bullet passing through and into unintended objects or in the direction of other animals or people. The shot should be aimed towards the throat and mouth (Fig. 10.5).

Poultry

Gunshot is an acceptable means of euthanasia for poultry when the shot is directed perpendicular (at a right angle) to the frontal bone of the bird. Though acceptable, this is not a practical method due to the small size of the bird and the high chance of ricochet.

Farmed elk/deer

A deer's brain is situated high in the skull. The shot will need to be 2 cm above the intersection of an imaginary X drawn on the forehead, with each line starting at the inside corner of the eye and the base of the antler on the opposite side (Fig. 10.6).

Farmed bison

Bison MUST be shot with a high-powered rifle or a handgun that is a 0.357 Magnum or larger. Due to the massive skull structure of a bison, the best place to position the firearm is on the side of the head between the base of the horn and the ear. Ideally, the bullet will be delivered on a level line with the base of the horn when the head is carried in a normal walking position. It is best if the shot can be delivered when the bison is standing still, to allow the bullet to travel into the skull cavity.

If the shot must be taken from the front, the point of entry should be right above the intersection of the X made between the eye and the base of the

Fig. 10.4. Correct position for shooting equines. The correct angle is essential. (From Alberta Equine Working Group, Alberta, Canada.)

opposite horns (see Fig. 10.1, same as for cattle). A high-powered rifle is required. It is best to deliver this shot with the bison's head lowered down towards the ground. This can be achieved by offering the bison feed on the ground. Ricochet is a major hazard of shooting a bison in the front of the head.

Penetrating Captive Bolt Gun (PCBG)

Penetrating captive bolt guns (PCBGs) consist of a steel bolt, with a flange and piston at one end, which is housed in a barrel. On firing, the expansion of gases propels the piston forwards and forces the bolt out of the muzzle of the barrel. The bolt is retained within the barrel by a series of cushions that absorb the excess energy of the bolt and keep it within the barrel. The bolt is retracted back into the gun either automatically or manually, depending on the design of the PCBG. To be effective, the PCBG must have sufficient bolt velocity. For cattle, the PCBG will be most effective with volt velocities ranging from 55 to 58 m/s. A slower bolt velocity

J. Woods and J.K. Shearer

of 41–47 m/s is less effective (Daly *et al.*, 1987). Other important research studies on captive bolt devices are Blackmore (1985), Daly and Whittington (1989) and Gregory *et al.* (2007). Captive bolt guns can be designed for either stunning, where an additional step such as bleeding is required, or more powerful extended bolt versions for single-step euthanasia. No additional step such as bleeding is required.

Previous research involving the use of the penetrating captive bolt gun on sheep found there were three levels of brain damage caused by the application of a captive bolt gun (Daly and Whittington, 1989). First, there was tissue damage in the brain from the passage of the bolt through the tissue. Second, the high-speed penetration of the bolt within the cranium creates pressure waves within the fluid medium of the brain, creating additional damage at sites distant from the bolt trajectory. Last, the impact of the bolt on the cranium contributes to the concussion (Daly and Whittington, 1989). PCBGs are powered by gunpowder or compressed air, which must provide sufficient energy to penetrate the skull of the species on which they are being used. Accurate placement, energy of bolt (bolt velocity), and depth of penetration determines effectiveness. Bolt velocity is dependent on maintenance (in particular, cleaning) and storage of the cartridge charges. Many manufacturers of captive bolt guns can also supply a test stand for measuring bolt velocity. Use of this device ensures that the captive bolt is working correctly. Cartridges must never be stored in a damp location as damp cartridges are often less effective. Soft-sounding shots that may be caused by damp cartridges are less effective (Grandin, 2002; Gregory *et al.*, 2007).

Fig. 10.5. Correct location for shooting a lamb or a goat. Some sheep breeds have extremely thick skulls and shooting from the rear position may be more effective. The correct angle is essential to ensure hitting the brain. (From Shearer and Nicoletti, 2011; AVMA, 2013.)

Fig. 10.6. Correct location for shooting deer with a captive bolt or firearm. (From Canadian Agri-Food Research Council.)

PCBGs come in 9 mm, 0.22 caliber and 0.25 caliber. Styles include in-line (cylindrical) and pistol grip (resemble a handgun). Pneumatic PCBGs (air powered) must be supplied with sufficient air pressure and air volume to operate, and for the most part are limited to slaughter plant environments.

Captive bolt technology for single-step euthanasia

In 2007, a captive bolt gun was designed to meet the needs of stockpeople for on-farm euthanasia of livestock. Unlike traditional captive bolt guns that are designed for only stunning animals and requiring a secondary step such as pithing or exsanguinations, the euthanizer unit is more powerful and a secondary step is not required. The gun has a longer bolt length and has sufficient force to kill all species and weight classes of livestock. This device is designed to cause sufficient destruction of the brain (cerebral cortex) to cause instantaneous loss of consciousness and destruction of the brainstem (specifically the medulla oblongata), leading to the loss of life functions with no chance of recovery.

This gun is especially useful for farms that house multiple age/weights of animals (i.e. farrow-to-finish swine farms), farms with mixed species of livestock, or for people that deal with a variety of species and weight classes of animals such as veterinarians, animal control, auction marts, transporters, and livestock rescue crews. A single gun will be effective on both large and small animals. A single shot reliably euthanizes pigs under 200 kg (440 lbs) (Woods, 2012). A study involving 489 sheep found it was effective as a single-step method when the shot was assessed as properly targeted based on the trajectory of the bolt (Gibson *et al.*, 2012). Cattle have thinner skulls than pigs, and a single shot from the euthanizer captive bolt gun will reliably kill them without a secondary step such as exsanguination.

PCBGs should only be applied by trained users. The shooter should wear both protective ear and eye gear. The PCBG must never be cocked until the shooter is ready to fire, and the shooter should make sure that the safety catch or other safety mechanism that will prevent accidental firing is set until it is ready to be discharged. The muzzle end of the captive bolt should always be kept pointed towards the ground. In-line, cylinder-shaped guns are often the cause of injury due to holding the gun with the bolt end up when the handler confuses the discharge end with the top of the gun.

Animals that are upright and mobile may be difficult to euthanize safely with a captive bolt without restraint. The shooter should wait until the animal stops moving its head before firing and should not chase the animal's head. One well-placed shot should readily cause unconsciousness. If there is any doubt, a second shot should be applied. If a second shot is required, it should be administered in a different location from the first shot. If the first shot was placed correctly, the second shot should be applied slightly above and to the side of the first hole. If the first shot was placed incorrectly, the second shot should be placed in the correct position.

It is highly recommended that high-powered captive bolt guns designed for on-farm euthanasia be utilized, as traditional PCBGs designed for stunning may only stun the animal and therefore may require a secondary kill step. Secondary kill steps include exsanguination and pithing. When applying the PCBG, it MUST be held firmly against the animal's head. If it is not held flush to the animal's head, the bolt may not penetrate the skull. It may also cause the bolt to slide up the head of the animal. The animal may be approached from the front. However, it is likely to be easier to approach from the rear and off to one side, thus keeping out of the line of sight of the animal. The handler can then reach over the poll of the animal, placing the PCBG on the front of the head. There are two basic types of trigger mechanisms. One type causes the gun to fire when it is tapped on the head and the other type has to be pulled by the operator to fire the gun. The recoil of the PCBG will vary depending on caliber, buffer configuration, manufacture, and the cartridges used. If necessary, the handler should use two hands when firing the PCBG. The style of gun will also make a difference in recoil and noise associated with firing.

Maintenance of captive bolt guns

The most common cause of misfires and ineffective kills is poor maintenance (Grandin, 1998). PCBGs must be cleaned and maintained in order to operate effectively. PCBGs should be cleaned and inspected at the end of each day of use. PCBGs that are not used regularly still need to be cleaned and oiled. All parts must be put back into the PCBG and lost or damaged parts replaced before use. In the case of a misfire, the stunner breech should remain closed for a minimum of 30 s in case the problem is a "hang fire" due to slow primer ignition.

Refer to the manufacturer's instruction manual for cleaning instructions. Make sure the PCBG is unloaded before cleaning. PCBGs and cartridges must be stored in a dry area. Exposure to humidity will affect both the gun and the effectiveness of the ammunition. Quick Access Information 10.1 has tips for maintaining captive bolt guns.

Location of shot for captive bolt

Figure 10.1 shows the recommended location that is "at the intersection of two lines drawn from the top of the eye to the opposite horn or top of the opposite ear" based on new research focusing specifically on the placement of the captive bolt gun on cattle (Gilliam *et al.*, 2012, 2014). Figure 10.2 shows the correct location for polled cattle. There is also variation in placement, based on the length and shape of the skull. As an example, Holstein dairy cows have much longer skulls than Hereford beef cows. For animals with heavy skulls, such as water buffalo or Brahman cattle, shooting in the hollow behind the poll can be effective (Gregory *et al.*, 2008). The angle must be precise to ensure that the brain is penetrated.

The current recommendations for the point of entry on pigs are the same frontal location as for gunshot (Dr Lyndi Gilliam, Oklahoma State University, 2009, personal communication). Commercial

Quick Access Information 10.1. Tips for maintaining captive bolt guns. (From C. Bildstein, 2009.)

- It must be cleaned every day if it has been fired. Treat it the same way as a firearm.
- Use gun oil and gun cleaner. Do not use machinery grease, white oil, or WD-40.
- Do not get the captive bolt gun wet.
- Use a wire brush to polish the piston end of the bolt every time the stunner is cleaned.
- Replace the bumpers if they are cracked or stiff. This will prevent the bolt from sticking in the head.
- Number the guns and do not mix the parts. Each gun wears differently.
- Have plenty of spare parts for replacement of worn or broken parts.
- Store cartridges in a dry place.
- Captive bolts used in large slaughter plants should be rotated frequently to prevent overheating.

application, though, has shown that the captive bolt gun needs to be placed approximately 2–5 cm higher than the placement for gunshot due to trajectory differences. PCBG placement is difficult with pigs because, in order to achieve the ideal projection path into the brain, the PCBG must be slightly angled so the bolt travels straight ahead – towards the tail of the pig. If the PCBG is angled too much, it may not penetrate the head, but instead ride up the skull.

Also, as a pig ages, the skull continues to grow but the brain does not, making the target even more difficult. Mature sows have a large sinus cavity located right in front of the brain, setting the location of the brain even further back into the skull, requiring deeper penetration. Boars form a ridge down the front of their skull as they mature, making penetration difficult. Even though the point of entry is the same as a gunshot, the captive bolt gun should be placed slightly to one side of the ridge. Initial research has shown the PCBG a reliable method for single-step euthanasia for pigs weighing less than 200 kg. However, for pigs that weigh 200 kg or more, the effectiveness and reliability of this method is not consistent (Woods, 2012). The lead researcher, though, noted that the effectiveness was influenced heavily by placement – correct location combined with the gun being held firmly and flush to the skull. Almost all failed single steps in mature animals observed were noted as not placed correctly or not held flush to the head. For this reason, it is highly recommended that pigs be restrained with snares for application of the PCBG (Woods, 2012).

On the farm, horses must be restrained when applying the captive bolt gun, which makes its use very limited. When shot, horses often make a small forward movement (leap into the air), making it dangerous for the handler to stand in front of the horse, holding the gun on the horse's head (see Fig. 10.4). The captive bolt gun is best used on horses that are down on the ground and immobile, or restrained in a chute as a human safety precaution. The point of entry will be the same frontal location as with gunshot.

For sheep, the point of entry is the highest point/ top of the head, the same as with firearms. Poor aim and shooting sheep in the wrong location is a major cause of failure of captive bolt euthanasia (Gibson *et al.*, 2012). For goats, the captive bolt is best placed behind the poll of the animal. Captive bolt guns are acceptable only on large, mature turkeys, with the shot applied to the head while the bird is restrained by the beak. Appropriate placement for farmed elk and deer is the same frontal location as gunshot.

Captive bolt guns are not recommended for euthanasia of bison.

A small captive bolt gun is available for euthanizing broiler chickens. To be effective, it has to have a 6 mm bolt driven with an air pressure of 827 kPa. The penetration depth has to be 10 mm (Raj and O'Callaghan, 2001). The gun must be placed on top of the animal's head.

Controlled Blunt Force Trauma by Captive Bolt

Controlled blunt force trauma induces euthanasia by physical disruption of the brain. A blow to the head can be an effective stand-alone means of euthanasia to small or young animals with thin craniums. A single, sharp blow must be delivered to the central skull bones with sufficient force to produce immediate depression of the CNS and destruction of brain tissue, without breaking open the skull.

Acceptable tools for controlled blunt force trauma currently include cartridge and pneumatic nonpenetrating CBGs that deliver a blow to the head with either a metal, blunt, flat head or a mushroom-shaped bolt. Like the PCBG, the tool is held flush to the animals head and when fired will deliver a concussive force, resulting in brain hemorrhage, shearing strains and stresses within the brain, and tissue deformation. Recent research supports the use of nonpenetrating CBG as a single-step method of euthanasia. The *Cash Dispatch* with the nonpenetrating head was found extremely dependable with a single shot on piglets up to 10 kg (22 lbs) (Woods, 2012), with preliminary trials showing brain damage to be more extensive on pigs up to 23 kg (50 lbs) than with a standard penetrating captive bolt (Jennifer Woods, 2010, personal communication). Casey-Trott *et al.* (2013) found the *Zephyr E* to be very effective for 3-day-old piglets, while Erasmus *et al.* (2010a,b) determined that the same gun was effective for euthanizing 11- to 13-kg (22–28 lb) turkeys.

Effectiveness is dependent on animal size, skull structure and mass, as well as the skill of the person using the gun. There are several nonpenetrating CBGs currently on the market and each one has been designed for use on specific species, ages, and weight classes. Currently, there are guns for use on piglets, poultry, lambs, neonate calves, and rabbits. **It is critical that the nonpenetrating CBG utilized is a gun that has been designed specifically for the species and weight class/age group being euthanized.** A gun designed only for small animals such as

poultry, rabbits, or neonate piglets would not be effective on a neonate calf or larger weaned piglet. Manufacturers weight, age, and species restrictions must be strictly adhered to, along with the recommend head, cartridge size, and PSI. **Failure to follow the instructions and recommendations will most likely result in failed euthanasia.**

When using a non-PCBG on pigs, poultry, and lambs, the animal must be restrained and the gun applied to the top of the head. Restraint will be dependent on animal type and size (see the section on restraint methods). If there is ever a question of insensibility, a second shot should be applied immediately to the skull of the animal.

Maintenance requirements for nonpenetrating CBGs are the same as for PCBGs. Follow all manufacturers instructions for cleaning and maintenance.

Manual Blunt Force Trauma

Manual blunt force trauma induces death by physical disruption of the brain. A blow to the head can be an effective means of euthanasia for small animals with thin craniums. For livestock, manual blunt force trauma is currently an acceptable means of euthanasia for piglets 5 kg (12 lb) or under, infant lambs under 9 kg (20 lb), infant kids under 7 kg (15 lb), and poultry (chicken and turkeys). A single, sharp blow must be delivered to the central skull bones with sufficient force to produce immediate depression of the CNS and destruction of brain tissue, without breaking open the skull. Even though manual blunt force trauma is an acceptable method, the American Veterinary Medical Association recommends replacing it with other methods (AVMA, 2013). Videos appearing on the Internet of blunt force trauma of small piglets have caused a very negative reaction from the public.

For manual blunt force trauma, the object must be brought to the animal's head, not the animal to the object. Striking the animal to the object significantly decreases the animal welfare standard. If animals are swung during the application of blunt force trauma, they will experience high stress and a much greater chance of injury with dislocated joints, broken legs, etc. Common acceptable tools used for manual blunt force include ball peen hammers, steel rods, wooden clubs, and pipes.

Blunt force trauma to piglets and infant goats must be applied to the top of the animal's head. The top or back of the head is the appropriate application spot for lambs. Poultry will respond violently to blunt force trauma with severe wing flapping and handlers must take precautions to ensure their safety. It is essential that the blow be administered accurately and with absolute determination.

Consistency of delivery is a challenge; therefore, manual blunt force trauma is questionable in terms of reliability and effectiveness. Manual blunt force trauma is *not* an acceptable means of euthanasia for cattle, horses, elk, or bison.

One of the big problems with blunt force trauma is that caring stockpeople who are good at looking after infant animals often do not want to use this method. One way to solve the problem is to have another person come in and do it when the primary stockperson is absent, or instead use gas euthanasia. Practical experience has shown that many people would prefer to put an animal in a "box" (i.e. euthanasia container) instead of hitting it on the head. One person said, "I prefer this, the box does the deed".

Inhalant Methods

Carbon dioxide

The time of exposure and concentration level of CO_2 determines its effectiveness and humaneness. CO_2 is approved in many countries for use on pigs, poultry, neonatal sheep, and goats. On the farm, CO_2 may only be practical for lighter-weight animals because the animal must be placed in an appropriate container. There is currently significant international debate concerning the use of CO_2 for the euthanasia of animals. The use of CO_2 euthanasia for neonate animals (along with domestic waterfowl) requires increased exposure levels and times because they are more resistant to hypoxia.

There are two primary CO_2 techniques utilized in animal production: the high-level precharge system and the gradual-fill system. Each technique must be designed and managed properly to ensure a humane death and minimize stress on the animal. It is recommended that for on-farm euthanasia, the animal be exposed to the target minimum level of $\geq 80\%$ for at least 5 min, and all animals evaluated for indicators of death on removal from the euthanasia container.

Precharged systems – CO_2 systems for piglets

Precharged systems utilize a container in which a high level of CO_2 (>80%, preferably 90%) is injected prior to the animal being placed in the container.

This method is recommended for pigs because it will reduce distress, squealing, and agitation before loss of sensibility when the pig falls over (Sadler *et al.*, 2014b) (see Chapter 9, this volume). The animal should be fully immersed within the gas as quickly as possible, in order to minimize the opportunity for aversive reactions to the gas prior to the animal becoming unconscious. The CO_2 must be maintained at the target level until the animal is confirmed dead. Newborn piglets lose consciousness more quickly in CO_2 than older weaned piglets (Sadler *et al.*, 2014a). A mixture of half argon and half CO_2 provides no advantages (Sadler *et al.*, 2014a).

The euthanasia container should be designed to accommodate the behavioral and physical attributes of the target species, including such factors as door design, nonslip flooring, adequate lighting, etc. It is critical that the container is not overloaded with animals. All animals must be provided with adequate space to stand and lie down. It is not acceptable to pile animals into the container as this will jeopardize their welfare and may cause animals to die of suffocation. The lid or door must also seal and shut securely.

For a gas to fill the chamber, the air in the chamber must be displaced by the gas that is being injected. This is the exchange rate. As CO_2 is heavier than air and will stratify within the euthanasia container, a multiple-port injection system should be utilized to ensure even levels of CO_2 within the container. Gas flow rates must be regulated and monitored to ensure proper flow. Unregulated release of gas from a canister is not recommended due to the potential to chill the animals, freeze the supply lines, or cause lower than desired chamber exchange rate, failed euthanasia, or wasting of CO_2, adding to the cost of the process. Exhaust valves must be designed to avoid pressuring the system during gas injection. Gas flow rates must be slow enough to avoid a high-pitched whistle, which is distressing to livestock. The system should be equipped with a CO_2 monitor or at minimum an alarm that indicates a low CO_2 level within the euthanasia container.

Gradual-fill systems

The specifications for the euthanasia chamber and delivery of the gas has already been covered in the previous section.

Poultry lose consciousness at a much lower level of CO_2 (20–25%); therefore, in addition to the traditional precharged and gradual-fill systems, a low-level gradual-fill system can be utilized. A low-level gradual-fill system utilizes a target level of 40% CO_2 concentration but requires a 20 min exposure period. Gerritzen *et al.* (2004) contains additional information on the use of different gas mixtures for euthanizing chickens.

Due to the aversiveness of CO_2, combinations of CO_2 and various inert gases (argon, nitrogen, etc.) at differing concentrations have also been utilized for euthanasia of livestock and poultry. There is controversy among scientists on the best gas mixtures. Additional information can be found in Raj and Gregory (1995), Raj (1999), Meyer and Morrow (2005), Christensen and Holst (2006) and Hawkins *et al.* (2006). Chapter 9, this volume, contains more scientific studies. On the farm, the most practical way to evaluate gas mixtures is to watch the behavior of the animals. It is absolutely not acceptable to use a gas mixture or methods that cause the animals to attempt to climb out and escape from the container. Gas mixtures and methods should be changed if animals or birds struggle, vocalize, or flap vigorously BEFORE they fall down and lose posture. After loss of posture, when an animal loses sensibility, vigorous bodily reactions will have no effect on welfare.

Electrocution

Electrocution is considered humane when adequate current passes through the brain to induce a grand mal seizure and fibrillation of the heart. Cardiac fibrillation leads to cardiac arrest and ultimately death (see Chapter 9, this volume). The animal is either euthanized by an initial passage of current through the brain, followed by a second application of the current through the heart of the unconscious animal to induce cardiac fibrillation, or through simultaneous induction of unconsciousness and cardiac ventricular fibrillation. Cardiac fibrillation normally leads to cardiac arrest and an interruption in blood flow to the brain and other vital organs.

The amount of current delivered will depend on the voltage and total impedance in the pathway, which is affected by the species, size of the electrodes, applied pressure, location placement, distance between electrodes, and phase of respiration during application of shock. There are a multitude of combinations of voltage, current, frequency, and length of application that have been utilized successfully as a euthanasia technique. However, the selection of these factors should be based on scientific evidence and verified in commercial conditions. The absolute minimum of currents applied through the brain to induce instantaneous insensibility are shown in Table 10.3.

The correct current should be attained within 1 s of initiation and applied for a minimum of 3 s (OIE, 2014a).

Care must be taken that the electrodes are in full contact with the animal before the electrical shock is applied. When the wand is electrified before placement on the animal, it is called "hot wanding", and it will cause pain. Hot wanding of cattle or pigs will cause the animal to vocalize the instant the electrode touches it. Electrocution of cattle should only be used in a slaughter plant because of the specialized equipment and high amperage that are required (see Chapter 9, this volume).

For pigs, electrocution is acceptable on animals over 5 kg (11 lb) when applied properly. The process may be conducted by first applying the electrodes on the opposite sides of the head so that the current travels through the brain. When a commercially available pig stunner is used, the tongs should be placed on the inner base of the ears. After the current has been passed through the head, the tongs are applied a second time across the chest to induce cardiac fibrillation (Vogel *et al.*, 2011). A commercially available pig stunner is recommended. Homemade devices are often dangerous and may not be effective (see Chapter 9, this volume, for further information).

There are no current guidelines or recommendations for on-farm application of electrocution of sheep, horses, elk, or bison. Sheep are problematic, due to the insulating effects of wool. Specialized equipment that is used in a slaughter plant is not practical or safe for on-farm use.

Table 10.3. Minimum currents[a] necessary for killing. (From OIE, 2014a.)

Animal	Current (A)
Cattle	1.5
Calves	1.0
Sheep/goats	1.0
Lambs/kids	0.6
Piglets	0.5
Pigs	1.3

Note: [a]The frequency of current should be no greater than 100 Hz. Standard 50–60 Hz is recommended.

J. Woods and J.K. Shearer

Cervical Dislocation of Poultry

Cervical dislocation induces euthanasia by physical disruption of the brain and spinal cord as the procedure dislocates the vertebral column from the skull, which causes damage to the lower brain region and near-immediate unconsciousness. To ensure death, cervical dislocation should be followed by exsanguination.

Cervical dislocation is only acceptable for small poultry and requires a higher level of skill than other methods. Proper training is vital to ensure the proper application and is best achieved using a stretching method rather than crushing the vertebrate. Small birds can be dislocated by applying rotational movements to the neck. Adult poultry should be held by the shanks (legs) with one hand, and the head grasped immediately behind the skull with the other hand. The neck is then extended and dislocated using a sharp downward and backward thrust. The bird may react violently with wing flapping. It is best to restrain the wings before euthanasia.

Cervical dislocation should be limited to poultry weighing 3 kg (7 lb) or less as it is very difficult on larger, heavily muscled poultry such as broiler breeders and turkeys. Cervical dislocation of 11- to 13-kg (22–28 lb) turkeys failed to abolish the nictitating membrane (eye blink) reflex (Erasmus et al., 2010a,b). For birds that are too large and difficult for cervical dislocations, the use of a nonpenetrating captive bolt euthanasia device is strongly recommended (see the section headed "Controlled Blunt Force Trauma by Captive Bolt"). When cervical dislocation is used on smaller birds, the head should not be pulled off. Decapitation is not an acceptable poultry euthanasia method (AVMA, 2013).

Methods that Should Never be Used

When intravenous methods are used to euthanize animals, the animal must be rendered insensible with barbiturates or other drugs that are true anesthetics. Nonanesthetic agonists must never be used in a sensible animal. If nonanesthetic drugs are used, to ensure death the animal must be rendered insensible with an anesthetic drug before the nonanesthetic agent is administered. Drowning, strangulation, or air embolisms are also methods that should never be used. Both authors of this chapter and the editor of this book have struggled to decide the most practical humane way to euthanize large animals in countries where firearms and cartridge-fired captive bolts cannot be owned by members of the general public. In these countries, the slaughter plants use a pneumatic-powered captive bolt gun that does not use a powder charge. Anesthetic drugs are not a reasonable alternative due to their high cost. The only alternative is to use methods that would not be approved by the professional veterinary associations in many developed countries where there is easy access to captive bolt guns. In this situation, one alternative would be to hit the bovine on the forehead (at X on Fig. 10.1) with a very heavy hammer and then exsanguinate immediately. If exsanguination has to be used as the sole method with no blow to the head, the animal should be cut with a very sharp knife that is twice the width of the neck. The cut should use either the halal or kosher cutting method. The cut should move from the outside of the throat towards the spine and simultaneously cut all four major blood vessels. This method of bleeding should only be used on animals that have been rendered insensible to pain by methods described in this chapter. In mammals, these methods should only be used if a firearm, captive bolt, lethal injection, or other approved method is not available. In poultry, cervical crushing devices or pliers should not be used. Crushing cervical vertebrate is not an acceptable single-step method of poultry euthanasia (AVMA, 2013). Erasmus et al. (2010a,b) reports that cervical dislocation fails to abolish eye reflexes in birds over 11 kg (22 lb).

Secondary Kill Steps

There are several reasons why a secondary kill step may be required. In some cases, the euthanasia tool is only intended for or capable of stunning the animal, not killing it. Many captive bolt guns are designed to stun an animal temporarily. If the first shot with a firearm or captive bolt fails, the animal must be shot again immediately. The secondary kill step should occur within 30 s after the last shot. A secondary kill step can include either exsanguination or pithing.

Exsanguination

Exsanguination is recognized by most welfare guidelines as only a secondary kill step, not a stand-alone kill method. This is especially true if dull knives (that do not have a sharp edge) or stab methods are used. It is inhumane because it causes the animal great pain, suffering, and distress. It has been demonstrated that it may take several MINUTES for the animal to die from exsanguination. During the period when the bleeding animal is still conscious, serious welfare problems are highly likely because the animal can feel anxiety, pain, distress, and other suffering (EFSA, 2004). There have been several research projects documenting the length of time it takes an animal to become unconscious, with one project on calves finding a range of 35–680 s (Bager *et al.*, 1992). Research shows that sheep become insensible more quickly than bovines (see Chapter 9, this volume, for more references).

Exsanguination should be performed using a pointed, very sharp knife with a rigid blade at least 15.2 cm (6 inches) in length. The knife should be fully inserted through the skin just behind the point of the jaw and below the neck bones. From this position, the knife is drawn forward, severing the jugular vein, carotid artery, and windpipe. Properly performed, blood should flow freely with death occurring over a period of several minutes. *Exsanguination is not recognized as a sole method of euthanasia and the animal must be stunned prior to bleeding.* This procedure can be very disturbing to observers due to the large volume of blood loss, which also raises biosecurity concerns.

In cattle, the brachial vasculature may also be severed by lifting a front leg and inserting a knife deeply into the axillary area at the point of the elbow and cutting the skin, blood vessels, and surrounding tissue until the limb can be laid back away from the thorax of the animal.

For pigs, a sharp knife not less than 12 cm (5 inches) should be used. The knife is inserted in the midline of the neck at the depression in front of the breastbone and the skin is raised with the knifepoint using light pressure and a lifting movement. When penetration has been made, the knife handle should be lowered so that the blade is in a near vertical position and pushed upward to sever all major blood vessels that arise from the heart. The brachial vasculature can also be severed in pigs as described above for cattle. Great care must be practiced while performing exsanguination on all animals because, although they are unconscious, they are still capable of making violent, involuntary muscle contractions that cause injury to the handler.

Pithing

Pithing is a technique to bring about death by causing significant destruction of the brain and spinal cord tissue. It is performed by inserting a slender pithing rod or tool through the entry site produced in the skull by the penetrating captive bolt stunner or free bullet. The operator manipulates the pithing tool to destroy both brainstem and spinal cord tissue, which results in death. This procedure is sometimes used in advance of exsanguination procedures to reduce involuntary movement in stunned animals.

A pithing rod can be made from a variety of materials such as a discarded cattle insemination gun, high-tensile wire, a steel rod, or other similar device. Some vendors have disposable pithing rods for purchase. The rod itself must be somewhat rigid, yet flexible. It must be long enough to reach the brain and spinal column through the hole in the skull produced by the projectile from a gun or PCBG. Appelt and Sperry (2007) recommend that pithing should be performed if exanguination is not performed after a captive bolt. It will help prevent return to sensibility.

Confirmation of Death in Euthanized Animals

Understanding the process of death is important. Death is a process and does not occur immediately, even though sensibility and the ability to feel pain are lost instantly after correctly applied gunshot, captive bolt, or electrocution. The animal is still sensible and may be suffering if it is vocalizing (squealing or bellowing), attempts to get up, lifts up its head, or is blinking like a fully alive animal. If any of these signs is present, it must be shot again. Animals are first rendered insensible and then the body begins to die as the brain stops, the heart quits beating, the lungs stop breathing, and the blood quits circulating. There is no such thing as immediate death. This process can take upwards of 10 min to occur when euthanasia is delivered correctly and in the case of barbiturates it may take even longer. Woods (2012) found the mean time to last heartbeat in pigs was 236 s (3:56 m), with a range of 474 secs (7:54 m).

When a person is not knowledgeable about what to expect, he or she may mistakenly interpret any movement as sensibility and a lack of movement as loss of consciousness. Reflex motor activity or muscle spasms may follow the loss of consciousness – this is a normal part of the death process and should not be perceived as the animal being in pain or distress. Certain species and modes of euthanasia will have greater involuntary muscle movement than others (see Chapter 9, this volume).

After gunshot or captive bolt, intense muscle spasms in cattle will last on average only 5–10 s, with neonatal lambs lasting for several minutes. One study recorded an average time to last movement in pigs of 180 s (3 min), with a range of 527 s (8:47 min). The average number of kicks tallied on one hind leg was 116, with a range from 14 to 361 kicks (Woods, 2012). If the animal has an extended period of movement or "flopping", it may only be stunned and should be reshot. All species may have weak, slow kicking movements for several minutes.

It is also important to recognize that these involuntary movements can be distressful for people inexperienced with these procedures. A forewarning to bystanders ahead of time helps reduce the amount of explanation that may be necessary afterwards. Human safety must also be considered when confirming death. The handler must be aware of the involuntary movements that the animal may make in order to avoid being kicked. It is best to approach the animal along the back/spine in order to avoid contact with the legs or head of the animal.

Confirmation of the onset of death (insensible or unconscious) should occur within the first 30 s following completion of the euthanasia procedure; death should be confirmed within 10 min. Confirmation of death can be challenging outside of a veterinarian's office or laboratory. The cessation of a steady, rhythmic heartbeat and absence of respiration are the most reliable modes for determining death, but are extremely challenging to confirm in a field environment, due to background noise. If there is any uncertainty over whether or not an animal is dead, the procedure should be repeated or a second method applied.

Absence of rhythmic breathing is determined by visual observation of movement of the chest or palpation for chest movement. However, animals that are unconscious may have very sluggish and irregular breathing, making it difficult to determine the existence of chest movement. Gasping and random breaths are not considered rhythmic. The lack of chest movement is an indication that the animal is no longer breathing, because their respiratory system has shut down (see Chapter 9, this volume).

Along with the absence of rhythmic breathing, two of the following five other modes of confirmation must be met before an animal is confirmed technically dead: cessation of rhythmic heartbeat; absence of pulse; lack of palpebral reflex; dilation of the pupils; loss of capillary refill. Animals must not be moved until death is confirmed. It is recommended

to wait at least 20 min before moving the animal for disposal.

Cessation of regular heartbeat is evaluated accurately using a stethoscope, electrocardiograph (ECG), or electroencephalogram (EEG), though you may be able to palpate the heartbeat in some circumstances. These measures indicate the animal's heart has stopped pumping, ceasing the flow of blood through the body and to the brain. Death is confirmed when you are unable to detect the beating of the heart for more than 3 min. However, this procedure can be very challenging in a field environment. For detection with a stethoscope, you must have a stethoscope and the heartbeat can be difficult to hear because of background noises. In order to utilize an ECG or EEG, you must have the proper equipment, which is extremely costly and not commonly used outside a veterinary clinic.

Absence of pulse is detected through palpation of an artery and is an indication that the animal's heart is no longer beating. This technique is very challenging in a field environment, requiring operator skill, and is utilized primarily on larger species. A pulse is not always palpable, so it cannot be solely relied on for determining cessation of heartbeat or confirming death.

Palpebral reflex is checked by running a finger along the eyelashes and is to be performed before corneal reflex to avoid pain of touching the eye of a sensible animal. There should be no eye movement or blinking when eyelashes are touched. Lack of palpebral reflex indicates the animal is insensible. This should be done prior to evaluating the corneal reflex by touching the cornea or the surface of the eye. There should be no eye movement or blinking when touched. If the animal is dead, the eye will remain open and the lid does not move.

Visual observation of corneas will determine if there is dilation of the pupils. The pupils will dilate fully when blood is no longer being pumped through the body due to the cessation of heartbeat, stopping the supply of fluid to the eyes. Other tests that can be used are a needle prick on the nose or shining a high intensity flashlight in the eyes. The animal MUST NOT respond to these stimuli.

You can visually observe the mucous membranes in the mouth for loss of color and capillary refill. When animals die, the mucous membranes become pale and mottled. The loss of capillary refill is indicated when there is no return of color or refill of blood after pressure is applied. The membranes will become dry and sticky because blood is no longer being pumped through the body due to the cessation of the heartbeat.

An animal is not dead if it lifts its head off the ground or attempts to right itself, vocalizes after application, has the presence of eye movement or blinking, has a palpable pulse, pupils are constricted, responds to painful stimuli (i.e. pinching the nose), or fails at any of the above tests.

No animal should be moved or disposed of for a minimum of 20 min following confirmation of death (Quick Access Information 10.2). This will ensure that the animal is truly dead and will not regain sensibility. Attempts must be made, though, immediately following euthanasia to confirm the method was effective. The onset of rigor mortis and bloating will vary among the species. In pigs, it will normally set in within 4–8 h, sheep 8–12 h, and cattle 12–24 h. Rigor mortis should not be used as a means to confirm death due to the extreme delay of onset.

Mass Killing for Disease Control

The OIE (2014b) has guidelines for killing large numbers of animals or birds for disease control purposes. The specifications for gunshot, captive bolt, or electrical methods are the same as those specified in Chapters 9, this volume. The following methods should NEVER be used:

- buried alive in a pit
- thrown in on top of each other in a pit, bag, or rubbish bin and allowed to be crushed and suffocated under other animals or birds
- drowning
- strangulation
- poisoning with substances that would cause a painful death.

Quick Access Information 10.2. Confirmation of death – all signs must be absent before carcass disposal.

- All eye reflexes and movements including corneal reflex and palpebral reflexes must be absent.
- Rhythmic breathing and pulse must be absent.
- Vocalization must be absent.
- Righting reflex and the ability to stand must be absent.
- After all of the above are absent, wait 20 min before disposal of the carcass to ensure that the animal will not regain consciousness.

The OIE states:

> When animals are killed for disease control purposes, methods used should result in immediate death or immediate loss of consciousness lasting until death, when loss of consciousness is not immediate, induction of unconsciousness should be non-aversive and should not cause anxiety, pain, distress, or suffering in animals.
>
> (OIE, 2014b)

The destruction of large numbers of poultry in a low-stress humane manner is difficult. Filling an entire house with gas may not kill laying hens housed on upper decks of cages. Killing of poultry for disease control purposes is an area that needs further study. Some of the most promising methods for poultry are putting anesthetics or other agents in food or water and gas stunning in containers (Raj, 2008). Canadian researchers have also used container systems successfully (John Church, 2008, personal communication), and this method complies with OIE 2014 guidelines. US researchers have used firefighting foam to kill broiler chickens (Benson et al., 2007, 2012). Discussion within the poultry industry indicates that the bubble size must be controlled carefully to induce a peaceful induction. For all gas or foam methods, the behavior of the birds prior to loss of posture (falling over) must be evaluated (see Chapter 9, this volume). Gerritzen and Sparrey (2008) reported that firefighting foam generated with bubbles filled with CO_2 might work. The bubbles must be large enough so that the birds are killed by CO_2 inhalation and they are not killed by airway occlusion or drowning. Further research conducted since publication of the first edition of this book indicated that the use of high expansion foam filled with either CO_2 or nitrogen was effective and did not kill the chickens or turkeys by airway obstruction (McKeegan et al., 2013). For ducks, CO_2 should be added to the foam (Benson et al., 2012). Foam containing CO_2 was more consistent than whole-house gassing (Alphin et al., 2010). A major advantage of foam for mass depopulation of chickens or turkeys is that it is much easier to use than whole-house gassing. Raj (2008) have experimented with dry foam filled with nitrogen. As of 2009, foam methods are under study by the OIE.

Carcass Disposal

Euthanasia presents another issue that people frequently fail to consider – disposal of the carcass.

Acceptable means of disposal of dead livestock include scavenging, burial, composting, incineration, rendering, and tissue digestion. Scavenging of dead animals is when animals are left out for other animals to eat and for natural decomposition to occur. Burial is a very common method where animals are buried in the ground following euthanasia. When animals are composted, they are mixed with carbon bulking material such as finely ground straw or sawdust and will break down over time into organic material. Tissue digestors digest carcasses by a process of alkaline hydrolysis. Incineration is when carcasses are burned and rendering is a process whereby the carcasses are cooked to destroy pathogens and yield products such as meat, feather, bone or blood meal. Each one of these methods has its advantages and limitations and will be regulated by local laws. An article by Shearer *et al.* (2008) contains information on composting. Composting does not break down barbituates. Several studies have shown that pentabarbitol still persisted after 180–367 days of composting. An article by Payne *et al.* (2014) has a good review of both composting methods and problems with barbituates.

Eating euthanized animals

In areas of the world where food is scarce, hungry people are going to eat euthanized animals. They may skin and dress the animal on the farm. In places where people are starving, they are not going to throw away food. Common pathogens such as *Salmonella* spp. and *Escherichia coli* and most parasites can be killed by cooking meat until all pink color has disappeared. People must never be allowed to eat meat from animals or poultry with diseases that are transmissible to people, such as tuberculosis, brucellosis, rabies, bovine spongiform encephalopathy (mad cow), or avian influenza (bird flu). In certain areas, there may be additional diseases where the carcasses must be destroyed and not consumed, due to huge risks to public health. Carcasses from animals euthanized with barbiturates or other drugs are not fit for either human or animal consumption.

Personnel Training and Welfare Verification

All farms and ranches are advised to develop personnel training programs for proper instruction of humane euthanasia techniques. The skill and experience of personnel are of paramount importance when conducting euthanasia procedures. Experience has shown that many people (even those experienced in handling livestock) do not understand the physiological parameters required for proper execution of these techniques. Furthermore, persons should be aware that there is significant danger for the operator (or for bystanders with gunshot) whenever these methods of euthanasia are used. Only those who can demonstrate a working knowledge and proficiency with the techniques should be permitted to perform euthanasia procedures. When these methods are not performed properly, animals may become injured, have varying degrees of consciousness, and experience needless pain and distress.

Experienced persons should assist in the training of inexperienced persons and utilize carcasses to demonstrate anatomical landmarks and application of the various techniques. Carcasses or euthanasia dummies (Veterinary Simulator Industries, Alberta, Canada) should be used for practice by trainees until they become competent with the procedures. People must also be aware of how to confirm death. In some cases, this may require specific training with, and observation of, live animals.

Owners and managers must be acutely aware of all the factors outlined in this section, no matter the size of their operation. Whether a farm has a staff of 50 or is owner operated, it is the owner/manager's responsibility to educate themselves on acceptable euthanasia procedures and ensure they assign tasks appropriately to those who are skilled and comfortable with euthanasia procedures.

Training should not only be required for new staff entering the production site, but continual training and assessment of existing staff is essential. The training should be comprehensive and include all aspects of euthanasia, with both literature review and hands-on training. A survey of swine barn workers has shown that handlers prefer to be trained on the farm by managers. Stockpeople should not be allowed to euthanize animals on their own until they have been tested and proved to be proficient on skill, efficiency, and effectiveness.

It is the responsibility of management to verify that all staff are performing euthanasia properly and in a timely manner. Each employee should be verified on a regular basis and barns assessed for active euthanasia practices, utilizing a standard audit program or third-party verification process. An example of a standard animal welfare assessment for on-farm euthanasia (Woods *et al.*, 2008) is shown in Quick Access Information 10.3.

Quick Access Information 10.3. Animal welfare assessment for on-farm euthanasia.

Animal Welfare Assessment for On-farm Euthanasia

Date of assessment:
Location:
Assessed by:
Employee:
Animal type and weight:

Assessment criteria

1. Is the employee trained in the process and technique of euthanasia?
Yes or No

      Name and date of training program: _____

2. Could the employee explain the SOPs for euthanasia correctly?
Yes or No

3. Is the necessary equipment (handling equipment, restraint device, euthanasia tool, etc.) available and in proper working condition?
Yes or No

4. Is the equipment clean and well maintained? Is there a maintenance and cleaning log?
Yes or No

5. Was the appropriate tool chosen for the species and weight class of the animal?
Yes or No

Application of euthanasia

6. Was the euthanasia decision tree followed to ensure euthanasia was conducted in a timely manner?
Yes or No

7. Was the animal properly restrained if required?
Yes or No

8. Was the procedure applied in accordance to the standards outlined in the training manual?
Yes or No

9. Were the indicators of insensibility confirmed within 30 s following application of the euthanasia technique?
Yes or No

If No – were immediate steps to taken to rectify the situation to ensure death?
Yes or No

10. If a secondary kill step (i.e. second shot, exsanguination, pithing, etc.) was required, did it occur in a timely manner necessary to ensure a humane death?
Yes or No or Not required

Post-euthanasia

11. Was death confirmed before the animal was moved?
Yes or No

12. Was the carcass disposed of in accordance to all applicable rules and regulations?
Yes or No

13. Was the location of euthanasia cleaned properly and disinfected?
Yes or No

14. Was the euthanasia equipment clean and restored properly?
Yes or No

If any of these questions were answered NO, it is recommended that the euthanasia training program be reviewed and the euthanasia process re-evaluated at the next three euthanasia opportunities.

Continued

J. Woods and J.K. Shearer

Quick Access Information 10.3. Continued.

Notes or recommendations:

_____
_____
_____

Wilful acts of abuse

Any wilful act of abuse during the euthanasia process is grounds for automatic disciplinary action. Wilful acts of abuse include, but are not limited to: (i) intentionally applying goads to sensitive parts of the animal like the eyes, ears, nose, rectum; (ii) deliberate slamming of gates on livestock; (iii) use of euthanasia tools in a manner that violates the manufacturer's recommendations and international standards for use (examples include using an electrocution instrument as an electric goad, shooting an animal in the leg to render it immobile as to ease proper shot to the brain, etc.); (iv) hitting/beating an animal; (v) dragging of a live animal.

Were any wilful acts of abuse observed?
Yes or No

References

Alphin, R.L., Rankin, M.K., Johnson, K.J. and Benson, E.R. (2010) Comparison of water based foam and inert gas mass depopulation methods. *Avian Diseases* 54(1 suppl), 757–762.

Appelt, M. and Sperry, J. (2007) Stunning and killing cattle humanely and reliably in emergency situations – a comparison between a stunning only and a stunning and pithing protocol. *Canadian Veterinary Journal* 48, 529–534.

Arluke, A. (1994) Managing emotions in an animal shelter. In: Manning, A. and Serpell, J. (eds) *Animals and Human Society*. Routledge, New York, pp. 145–165.

AVMA (American Veterinary Medical Association) (2013) *AMVA Guidelines on Euthanasia*. AMVA, Schaumberg, Illinois.

Bager, F., Braggins, T.J., Devine, C.E., Graafhuis, A.E., Meller, D.J., *et al.* (1992) Onset of insensibility at slaughter in calves: Effects of electroplectic seizure and exsanguination on spontaneous electrocortical activity and indices of cerebral metabolism. *Research in Veterinary Science* 52, 162–173.

Benson, E., Malone, G.W., Alphin, R.L., Dawson, M.D., Pope, C.R., *et al.* (2007) Foam based mass emergency depopulation of floor raised meat type poultry operations. *Poultry Science* 86, 219–224.

Benson, E.R., Alphin, R.L., Rankin, M.K., Caputo, M.P., Hougertogler, D.P., *et al.* (2012) Mass emergency water based foam depopulation of poultry. *Avian Diseases* 56(4 Suppl), 891–896.

Bildstein, C. (2009) Animal Care and Handling Conference, 18–19 March, Kansas City, Missouri.

Blackmore, D.K. (1985) Energy requirements for the penetration of heads of domestic livestock and the development of a multiple projectile. *Veterinary Record* 116, 36–40.

Blackwell, T.E. (2004) Production practices and well-being of swine. In: Benson, G.J. and Rollins, B.E. (eds) *The Well-being of Farm Animals*. Blackwell Publishing, Ames, Iowa, pp. 241–269.

Casey-Trott, T.M., Millman, S.T., Turner, P.V., Nykcamp, S.G. and Widowski, T.M. (2013) Effectiveness of a non-penetrating captive bolt for euthanasia of piglets less than 3 days of age. *Journal of Animal Science* 91, 5477–5484.

Christensen, L. and Holst, S. (2006) *Group-wise Stunning of Pigs in CO_2 Stunning Systems – Technical and Practical Guidelines for Good Animal Welfare: A Danish Perspective*. Danish Meat Research Institute, Roskilde, Denmark.

Daly, C.C. and Whittington, P.E. (1989) Investigation into the principal determinants of effective captive bolt stunning of sheep. *Research in Veterinary Science* 46, 406–408.

Daly, C.C., Gregory, N.G. and Wotton, S.B. (1987) Captive bolt stunning of cattle: Effects on brain function of bolt velocity. *British Veterinary Journal* 143, 574–580.

EFSA (European Food Safety Authority) (2004) *Welfare Aspects of Animal Stunning and Killing Methods*. EFSA AHAW/04-207. EFSA, Parma, Italy.

Erasmus, M.A., Lawlis, P., Duncan, I.J. and Widowski, T.M. (2010a) Using time to insensibility and estimated time of death to evaluate non-penetrating captive bolt, cervical dislocation and blunt trauma for on-farm killing of turkeys. *Poultry Science* 89, 1345–1354.

Erasmus, M.A., Turner, P.V., Nykamp, S.G. and Widowski, T.M. (2010b) Brain and skull lesions resulting from the use of percussive captive bolt, cervical dislocation by stretching, cervical dislocaiton by crushing and blunt trauma in turkeys. *Veterinary Record* 167, 850–858.

Fulwider, W.K., Grandin, T., Rollin, B.E., Engle, T.E., Dalsted, N.L., *et al.* (2008) Survey of management practices on one hundred and thirteen north central and northeastern United States dairies. *Journal of Dairy Science* 91, 1686–1692.

Gerritzen, M.A. and Sparrey, J. (2008) A pilot study to determine whether high expansion CO_2 enriched foam is acceptable for on-farm emergency killing of poultry. *Animal Welfare* 17, 285–288.

Gerritzen, M.A., Lambooij, B., Reimert, M., Stegeman, A. and Sprujt, B. (2004) On-farm euthanasia of broiler chickens: Effects of different gas mixtures on behavior and brain activity. *Poultry Science* 83, 1294–1301.

Gibson, T.G., Ridler, A.L., Lamb, C.R., Williams, A., Giles, S., *et al.* (2012) Preliminary evaluation of the effectiveness of captive bolt guns as a killing method without exsanguination of horned and unhorned sheep. *Animal Welfare* 21(Supl. 2), 35–42.

Gilliam, J.N., Shearer, J.K., Woods, J.K., Hill, J., Reynolds, J., *et al.* (2012) Captive bolt euthanasia of cattle, determination of optimal shot placement and evaluation of Cash Special Euthanizer Kit for euthanasia of cattle. *Animal Welfare* 21(Suppl. 2), 99–102.

Gilliam, J.N., Shearer, J.K., Bahr, R.J., Crochik, S., Woods, J., *et al.* (2014) Evaluation of brain stem disruption following penetrating captive bolt in isolated cattle heads: Comparison of traditional and alternative placement landmarks. International Beef Cattle Welfare Symposium, Iowa State University, Ames, Iowa.

Grandin, T. (1988) Behavior of slaughter plant and auction employees towards animals. *Anthrozoos* 1, 205–213.

Grandin, T. (1998) Objective scoring of animal handling and stunning practices in slaughter plants. *Journal of the American Veterinary Medical Association* 212, 36–39.

Grandin, T. (2002) Return to sensibility problems after penetrating captive bolt stunning in commercial slaughter plants. *Journal of the American Veterinary Medical Association* 221, 1258–1261.

Grandin, T. and Johnson, C. (2009) *Animals Make Us Human*. Houghton Mifflin Harcourt, Boston, Massachusetts.

Green, A.L., Lombard, J.E., Gerber, L.P., Wagner, B.A. and Hill, G.W. (2008) Factors associated with the recovery of nonambulatory cows in the United States. *Journal of Dairy Science* 91, 2275–2283.

Gregory, N.G., Lee, C.J. and Widdicombe, J.P. (2007) Depth of concussion in cattle shot by penetrating captive bolt. *Meat Science* 77, 499–503.

Gregory, N.G., Spence, J.Y., Mason, C.W., Tinarwo, A. and Heasman, L. (2008) Effectiveness of poll stunning in water buffalo with captive bolt guns. *Meat Science* 81, 178–182.

Hawkins, P., Playle, L., Golledge, H., Leach, M., Banzett, R., *et al.* (2006) Newcastle Consensus Meeting on Carbon Dioxide Euthanasia of Laboratory Animals, 27–28 February, University of Newcastle-upon-Tyne, Newcastle-upon-Tyne, UK. Available at: https://www. nc3rs.org.uk/sites/default/files/documents/Events/First%20Newcastle%20consensus%20meeting%20report.pdf (accessed 31 March 2015).

HSA (Humane Slaughter Association) (2014) *Humane Killing of Livestock Using Firearms*, Humane Slaughter Association, Wheathampstead, UK. Available at: http://www.hsa.org.uk/humane-killing-of-livestock-using-firearms-introduction/introduction-2 (accessed 31 March 2015).

McKeegan, D.E., Reimert, H.G., Hindle, V.A., Boulcott, P., Sparrey, J.M., *et al.* (2013) Physiological and behavioral responses of poultry exposed to gas filled high expansion foam. *Poultry Science* 92, 1145–1154.

Manette, C.S. (2004) A reflection on the ways veterinarians cope with death: euthanasia and slaughter of animals. *Journal of the American Veterinary Medical Association* 225, 34–38.

Matthis, J.S. (2004) Selected employee attributes and perceptions regarding methods and animal welfare concerns associated with swine euthanasia. Dissertation, North Carolina State University, Raleigh, North Carolina.

Meyer, R.E. and Morrow, W.E.M. (2005) Carbon dioxide for emergency on-farm euthanasia of swine. *Journal of Swine Health and Production* 13(4), 210–217.

Moberg, G.P. (1985) Biological response to stress: key to assessment of animal well-being? In: Moberg, G.P. (ed.) *Animal Stress*. American Physiological Society, Bethesda, Maryland, pp. 27–49.

OIE (2014a) Chapter 7.5. Slaughter of Animals. In: *Terrestrial Animal Health Code*. World Organisation for Animal Health, Paris.

OIE (2014b) Chapter 7.6 Killing of Animals for Disease Control Purposes. In: *Terrestrial Animal Health Code*. World Organisation for Animal Health, Paris.

Payne, J., Farris, R., Parker, G., Bonhotal, J. and Schwarz, M. (2014) Fate of euthanasia drugs during equine mortality compositing. American Institute for Goat Research, Langston University, Langston, Ohio.

Raj, A.B.M. (1999) Behaviour of pigs exposed to mixtures of gases and the time required to stun and kill them: Welfare implications. *Veterinary Record* 144(7), 165–168.

Raj, A.B.M. and Gregory, N.G. (1995) Welfare implications of the gas stunning of pigs I. Determination of aversion to the initial inhalation of carbon dioxide or argon. *Animal Welfare* 4, 273–280.

Raj, A.B.M. and O'Callaghan, M. (2001) Evaluation of a pneumatically operated captive bolt gun for stunning broiler chickens. *British Poultry Science* 42, 295–299.

Raj, M. (2008) Humane killing of non-human animals for disease control purposes. *Journal of Applied Animal Welfare Science* 11, 112–124.

Reeve, C.L., Spitzmüller, C., Rogelberg, S.G., Walker, A., Schultz, L., *et al.* (2004) Employee reactions and adjustment to euthanasia-related work: identifying turning-point events through retrospective narratives. *Journal of Applied Animal Welfare Science* 7(1), 1–25.

Reeve, C.L., Rogelberg, S.G., Spitzmüller, C. and Digiacomo, N. (2005) The caring killing paradox: Euthanasia-related strain among animal-shelter workers. *Journal of Applied Social Psychology* 35(1), 119–143.

Sadler, L.J., Hagen, C.D., Wang, C., Widowsi, T.M., Johnson, A.K., *et al.* (2014a) Effects of flow rate and gas mixture on the welfare of weaned and neonate pigs during gas euthanasia. *Journal of Animal Science* 92, 793–805.

Sadler, L.J., Karriker, L.A., Johnson, A.K., Schwartz, K.J., Widowski, T.M., *et al.* (2014b) Swine respiratory disease minimally affects responses of nursery pigs to gas euthanasia. *Journal of Swine Health and Production* 22, 125–133.

Shearer, J.K. and Nicoletti, P. (2011) Procedures for humane euthanasia of sick injured and/or debilitated livestock. Iowa State University Extension Publication.

Shearer, J.K., Irsik, M. and Jennings, E. (2008) Methods of Large Animal Carcass Disposal in Florida. University of Florida Extension. Available at: http://edistt.ifas.ufl.edu/pdffiles/VM/VM13300.pdf (accessed 31 March 2015).

Sheldon, C.C., Sonsthagen, T. and Topel, J.A. (2006) *Animal Restraint for Veterinary Professionals*. Mosby/Elsevier, St Louis, Missouri.

Swine News (2000) Euthanasia for hog farms. *Swine News* July.

Thomson, D.U., Wileman, B.W., Rezac, D.J., Miesner, M.D., Johnson-Neilman, J.L., *et al.* (2013) Computed tomographic evaluation to determine efficacy of euthanasia of yearling feedlot cattle by use of various firearm ammunition combinations. *American Journal of Veterinary Research* 74, 1385–1391.

USDA (United States Department of Agriculture) (2004) National Animal Health Emergency Management System Guidelines. USDA, Washington, DC. Available at: www.dem.ri.gov/topics/erp/nahems_euthanasia.pdf (accessed 27 August 2009).

Vimini, R.J., Field, R.A., Riley, M.L. and Varnell, T.R. (1983) Effect of delayed bleeding after captive bolt stunning on heartrate activity and blood removal in beef cattle. *Journal of Animal Science* 57, 628–631.

Vogel, K.D., Badtram, G., Claus, J.R., Grandin, T., Turpin, S., *et al.* (2011) Head only followed by cardiac arrest electrical stunning in pigs is an effective alternative to head only electrical stunning. *Journal of Animal Science* 89, 1412–1418.

Wichart von Holten, S. (2003) Psycho-social stress in humans of mass slaughter of farm animals. *Deutsche Tierärztliche Wochenschrift* 110, 196–199.

Woods, J. (2009) *Swine Euthanasia Resource*. Blackie, Alberta, Canada.

Woods, J. (2012) Analysis of use of CASH Dispatch kit captive bolt gun as a single stage euthanizing process for pigs. Thesis, Iowa State University, Ames, Iowa.

Woods, J.A., Hill, J.D. and Shearer, J.K. (2008) Animal welfare assessment tool for on-farm euthanasia. Presented at the Welfare and Epidemiology Conference: Across Species, Across Disciplines, Across Borders, 14–16 July, Iowa State University, Ames, Iowa.

Further Reading

Articles and papers can be accessed easily by typing the full title into Google or another search engine.

American Association of Bovine Practitioners (2009) Practical Euthanasia of Cattle. (Type full title into Google or other search engine.)

American Association of Swine Veterinarians (2009) On Farm Euthanasia of Swine – Options for the Producer. American Association of Swine Veterinarians, Perry, Iowa. Available at: www.aasv.org/aasv/euthanasia.pdf (accessed 9 May 2015).

Australian Veterinary Association (1987) Guidelines on humane slaughter and euthanasia. *Australian Veterinary Journal* 64, 4–7.

Canadian Food Inspection Agency (2007) *Notifiable Avian Influenza: Hazard Specific Plan*. Canadian Food Inspection Agency, Ottawa, Canada.

Casey-Trott, T.M., Millman, S.T., Turner, P.V., Nykamp, S.G., Lawlis, P.C., *et al.* (2014) Effectiveness of nonpenetrating captive bolt for euthanasia of 3 kg to 9 kg pigs. *Journal of Animal Science* 92, 5166–5174.

Gibson, T.J., Whitehead, C., Taylor, R., Sykes, O., Chancellor, N.M., *et al.* (2015) Pathophysiology of penetrating captive bolt stunning in alpacas (*Vicugna pacos*). *Meat Science* 100, 227–231.

Humane Slaughter Association (2014) *Captive Bolt Stunning of Cattle*. Humane Slaughter Association, The Old School, Brewhouse Hill, Wheathampstead, UK. (Type full title into Google or other search engine.)

Millman, S. (2014) *Caring for the Convalescent Pig and Euthanasia Decisions*. College of Veterinary Medicine, Iowa State University, Ames, Iowa. (Type full title into Google or other search engine.)

Morrow, W.E.M., Meyer, R.E., Roberts, J. and Lascelles, D. (2006) Financial and welfare implications of immediately euthanizing compromised nursery pigs. *Journal of Swine Health and Production* 14, 25–34.

University of California Veterinary Medicine (2009) The Emergency Euthanasia of Horses. University of California Davis, California. (Type full title into Google or other search engine.)

11 Welfare During Transport of Livestock and Poultry

TEMPLE GRANDIN

Department of Animal Science, Colorado State University, Fort Collins, Colorado, USA

Summary

To maintain an acceptable level of animal welfare, animals or poultry must be fit for transport. The OIE has guidelines for transport. Animals that can barely walk or are likely to become nonambulatory should not be transported. A number of different factors can reduce fitness for transport, such as emaciated poor body condition, severe lameness, lack of acclimatization for a different climate, no vaccinations, fed the wrong feed, high doses of beta-agonists, or genetic conditions. Loading ramps and vehicles should have nonslip flooring. Welfare problems during transport of both livestock and poultry can be monitored by measuring bruises, death losses, broken wings, and other injuries. Overloading of transport vehicles may increase death losses or bruising. Careful driving is also essential to enable animals to remain standing. During hot weather, losses from heat stress can be reduced by either keeping vehicles moving or providing fan ventilation.

Learning Objectives

- Determine if an animal is fit for transport
- Preparing animals for transport
- Design of loading ramps and handling facilities
- How to determine if bruises and injuries are occurring on vehicles or in the abattoir
- Research on space requirements and rest stops
- Use of animal-based measurements to assess transport and handling

Introduction

The single, most important factor for the welfare of animals during transport is to put an animal that is fit for transport on the truck (lorry), airplane, or ship. The World Organisation for Animal Health (OIE, 2014a) has specific guidelines on animals that should not be transported (Quick Access Information 11.1).

> **Quick Access Information 11.1. OIE guidelines – animals not fit for travel:**
>
> - those that are sick, injured, weak, disabled, or fatigued
> - those that are unable to stand unaided or bear weight on each leg
> - those that are blind in both eyes
> - those that cannot be moved without causing them additional suffering
> - newborns with an unhealed navel
> - females that have given birth within the previous 48 h traveling without young
> - pregnant animals that would be in the final 10% of their gestation period at the planned time of unloading
> - those whose body condition would result in poor welfare because of expected climatic conditions (land transport only)
> - animals with unhealed wounds from recent surgical procedures such as dehorning.

Risks during transport can be reduced by selecting animals best suited to the conditions of travel and those that are acclimatized to the expected weather conditions.

Animals at particular risk of suffering poor welfare during transport and which require special conditions (such as in the design of facilities and vehicles and the length of the journey) and additional attention during transport may include:

- very large or obese individuals
- very young or old animals
- excitable or aggressive animals
- animals subject to motion sickness
- animals that have had little contact with humans
- females in the last third of pregnancy or in heavy lactation.

Hair or wool length should be considered in relation to the weather conditions expected during transport.

Unfit Cull Animals are a Major Problem Area

Some of the worst welfare problems during transport occur when old cull breeding animals of little economic value are shipped. Cull breeding stock and old animals should be marketed when they are still fit. The OIE (2014a) states that weak, injured, or fatigued animals are unfit for travel. To avoid totally different interpretations by different people, the author recommends using body condition scoring (BCS), lameness scoring (Quick Access Information 11.2), and injury scoring to determine fitness for travel. The advantage of these measures is greater consistency of judgement between the different people

---

Quick Access Information 11.2. Mobility – lameness (difficulty in walking) scoring. Simple 4-point scoring system.

Below is a simple lameness (difficulty in walking) scoring system that can be used by transporters:

1. Normal – walks easily with no apparent lameness or change in gait.
2. Exhibits stiffness, shortness of stride, or slight limp but keeps up with normal animals when the group is walking.
3. Exhibits obvious stiffness, difficulty in taking steps, obvious limp, or obvious discomfort. Score 3 animals lag behind normal animals when the group is walking.
4. Extremely reluctant to move when encouraged by a person. This animal can barely walk and is likely to go down and become nonambulatory.

The author recommends against using a 3-point scoring system because this cannot differentiate degrees of difficulty in walking.

---

who will be making decisions on fitness to travel. BCS charts will need to be developed with photos of the local types of animals in each country (see Chapters 2 and 4, this volume).

Animals that are emaciated or severely lame, with a lameness score of 4, are more likely to fall down and get trampled during transport compared to stronger animals. Audits at slaughter plants that process cull dairy cows indicated that timely marketing was a major problem (Gary Smith, 2008, personal communication). Some dairies waited until their cows had an emaciated body score of 1 before shipping them. When both milk prices and cull cow prices were high, one out of six cull Holstein dairy cows arriving at auctions were emaciated (Kurt Vogel, 2014, personal communication). Similar problems have occurred with sows. Observations by the author in slaughter plants indicate that 90% of the emaciated, weak cull cows and sows originate from about 10% of the worst producers.

Problems in the developing world with old cull animals

In some parts of the developing world, the treatment of old cull animals is atrocious. Cull buffalo in India are beaten and dragged by workers in the markets where they are loaded on to trucks for transport to slaughter (Chandra and Das, 2001). This is in violation of the OIE (2014a) code that states "conscious animals should not be thrown, dragged, or dropped". Forty-three per cent of the bruises were on the hind limbs and 21% were on the abdomen and udder (Chandra and Das, 2001). Bruises in these locations are due to abuse because animals in countries with good practices seldom have bruises in these locations. The incidence of injuries in Nigeria and Columbia was higher during long journeys or on trips with many stops (Minka and Ayo, 2006; Romero et al., 2013).

Effective methods for stopping abuse

One of the most effective ways to get these abusive practices to stop is economic accountability for damage (Strappini et al., 2009; Schwartzkopf-Genswein and Grandin, 2014) (see Chapter 14, this volume, on the effect of economics on animal welfare). It is also obvious from research carried out by Chandra and Das (2001) that loading facilities are not adequate. Two to three people are required to load each animal. A combination of economic accountability and better loading facilities will improve animal welfare in many countries. The author has observed that livestock markets in many countries have more problems with abusive handling compared to farms. This happens because markets are not held accountable for losses.

Proper Preparation of Animals for Transport

Preparing animals for transport will help to prevent sickness or death associated with transport. Below is a list of preparation methods that will help improve welfare and reduce losses caused by lack of animal preparation.

Introduce animals to the ship's feed before loading

This is especially important for sheep traveling long distances on ships. Sheep should be fed the ship's pelleted feed at an assembly yard before transport to get them accustomed to eating it. Failure to train the sheep to eat the new feed may result in high death losses because many animals may refuse to eat on the ship. Sheep need a minimum of 1 week in an assembly yard to get them adjusted behaviorally and physiologically to the ship's pelleted feed (Warner, 1962; Arnold, 1976).

Shear sheep before transport to a hot climate

This is especially important for sheep that will be traveling to a destination with much hotter temperatures. Wool reduces the ability of sheep to dissipate heat from their bodies (Marai et al., 2006).

Ship acclimatized animals

Death losses will be reduced if cold-acclimatized animals are shipped to a cold climate and animals acclimatized to hot weather are shipped to a hot climate. Norriss et al. (2003) reported that death losses in cattle shipped by sea to the hot Middle East were lower when animals from the hot northern part of Australia were shipped. Heat-stress or cold-stress deaths at an animal's destination are more likely to occur in animals that are not acclimatized to the conditions at their destination. The author has observed this problem in many countries where animals were shipped long distances either north or south.

Neonatal calves should walk easily

Young calves will be less likely to get sick and die after transport if they have been fed colostrum within 6 h after birth. Calves should not be shipped until they can walk and stand easily without assistance from a person. The only exception to this rule would be transporting newborn dairy calves to a specialized calf-rearing facility.

Vaccinate and pre-wean weaner calves

Vaccinating and pre-weaning 6-month-old beef calves before they are shipped a long distance to a feedlot reduced sickness after arrival (Swanson and Morrow-Tesch, 2001; Loneragan et al., 2003; Lalman and Mourer, 2005). Unvaccinated calves shipped the same day they were weaned had more death losses and respiratory illness (Fike and Spire, 2006). Calves should be weaned and vaccinated at least 45 days before they leave the ranch (Powell, 2003). Unfortunately, the author has learned that in 2014, 40–50% of US beef calves were weaned the day of shipping. This occurred because there was no economic incentive or accountability for losses caused by this bad practice.

Some animals should be fasted or fed different feed before transport

Cattle from lush green pastures should be brought in and fed dry hay prior to transport. This will prevent runny diarrhoea. A survey of 37 slaughter plants in the EU showed that failure to fast pigs before loading resulted in higher mortalities (Averos et al., 2008). Ideally, no more than 12 h should elapse between feed removal and stunning. For all species, water should never be restricted before or after transport.

Habituate animals to people walking through them

Both pigs and cattle will be easier to drive if they have become habituated to people walking through them (Fig. 11.1). Animals differentiate between a person in their pen and a person in the alley. The person must get in the pen and train the animals to move away in an orderly manner as he or she walks through them. Cattle handled exclusively on horseback may be difficult and dangerous to drive by people on foot. These cattle should get some experience of being moved in and out of pens by people on foot

Fig. 11.1. Pigs housed in this slatted floor pen differentiate between a person in the pen and a person in the aisle. Pigs will be easier to load and unload on to trucks if a person walks through their pens every day during the fattening period. The animals need to learn how to move away from the person in an orderly manner as he or she walks through them. This building in the USA has a well-designed floor where the surface is over 80% solid. It also has lots of natural light. White translucent curtains admit light. Raising pigs in dark or dimly lit buildings may also make them more difficult to handle and load.

before they are transported to a new location (see Chapter 5, this volume, for additional information).

Prevent weakness caused by performance growth promoters

Excessive use of beta-agonist feed additives to increase muscle mass may cause weakness or increased problems with heat stress, unless they are used very carefully. The author has observed healthy pigs and cattle that were lame, stiff, or weak from excessive doses of ractopamine. Heavy 125-kg pigs fed high doses of ractopamine were weak and many animals were not able to walk from the lairage pen to the stunner. Marchant-Forde *et al.* (2003) also reported that pigs on ractopamine were more difficult to handle and more likely to become fatigued. Pigs fed ractopamine were also stressed more easily by rough handling (James *et al.*, 2013). The effects are dose and time dependent. Feeding 200 mg of ractopamine/day for 28 days to beef cattle did not compromise welfare (Baszczack *et al.*, 2006). These cattle were raised in a temperate climate in Colorado, USA, and they were not exposed to extremely hot conditions. They were transported for 60 min to a slaughter

plant. The author has observed that beta-agonists are associated with increased problems with sore feet and stiff walking during hot weather. The detrimental effects often increase as the cattle progress through handling and transport stops to the slaughter plant. It appears to be a fatigue problem (Thomson, Kansas State University, 2014, personal communication). During the summer, beta-agonists increased death losses (Loneragan *et al.*, 2014).

Choose appropriate breeds or cross-breeds for transport to different climates

Ship heat-tolerant breeds or cross-breeds to hot areas and cold-tolerant breeds to cold areas. Heat-tolerant *Bos indicus*-cross cattle had fewer death losses when shipped from Australia to the Middle East (Norris *et al.*, 2003). *B. indicus* cattle have fewer pronounced changes in their physiology when they are exposed to prolonged heat and humidity (Beatty *et al.*, 2006). Highly improved breeds, such as Holsteins, should be crossed with *B. indicus* in hot climates. This is a common practice in tropical parts of Mexico. In extremely cold climates, *B. indicus* breeds will have more difficulty. Brown-Brandl *et al.* (2001) reported that newer lines of lean pigs were more susceptible to heat stress. Their total heat production was up to 20% higher than published standards based on older genetic lines. Black Angus cattle are now very popular for quality beef. Black-hided cattle standing in the sun will get hotter than cattle with light colored hides. Cattle with dark hides may require more access to shade if shipped to a hot climate (Gaughan *et al.*, 2010). A switch to Angus cattle with lighter red hides may help reduce heat stress.

Genetic Problems or Production Practices that Reduce Fitness for Transport

Pigs with the porcine stress syndrome (PSS) stress gene

Pigs that have the PSS stress gene will have much higher transport death losses. Death losses during transport were 9.2% in homozygous positive pigs, 0.27% in carriers, 0.05% in homozygous negative pigs and 0.1% in pigs where the gene was removed (Murray and Johnson, 1998; Holtcamp, 2000). Pigs with the stress gene are often more lean and will often contain Pietrain genetics. When pigs are purchased from large breeding companies, it is recommended to obtain information on whether or not they have the PSS gene.

A US study done by Ritter *et al.* (2009a) of dead-on-arrival and downed, nonambulatory, fatigued pigs showed that over 95% of these compromised pigs were free of the PSS stress gene. A total of 2019 pigs were sampled at a US commercial slaughter plant. The sample of pigs came from over 100,000 animals from over 400 farms. Ritter *et al.* (2009a) concluded that in the USA, the PSS stress gene (HAL-1843 mutation) was not a major cause of losses. Unfortunately, the author observed in 2014 PSS pigs with Pietrain coloration arriving at a European plant. There are still places where these pigs are bred and raised. The author has observed that problems with nonambulatory pigs have worsened since beta-agonists were approved in the USA. The pigs in the Ritter *et al.* (2009a) study were grown to heavy weights and beta-agonists such as ractopamine may have contributed to the pigs becoming nonambulatory.

Weak, fast-growing animals

Some pigs and poultry bred for rapid weight gain are weaker compared to less highly selected animals.

Problems with weaknesses can occur in pigs that are free of the PSS stress gene. Pigs from rapidly growing genetic lines grown on full feed may be weaker and more likely to go down during transport. In one study, heavyweight, 130-kg market pigs had a trend for higher death losses (Rademacher and Davis, 2005). Heavy, fast-growing lines of poultry are also weaker and become fatigued more easily. Genetics and environment interact. Chicken producers have increased the strength and livability of fast-growing poultry genetic lines by changing nutrition so that the younger chicks grow more slowly. In the final days of the growing program, the older birds are put on full feed for maximum growth. When the animals are going to be transported and grown in parts of the world where modern veterinary and nutritional services are not available, the use of hardy local breeds may result in both better animal welfare and fewer death losses.

Animals with poor leg conformation or other defects

Animals with structurally poor legs are more likely to become nonambulatory during transport. Animals that are being selected for breeding should be evaluated for both lameness and leg conformation. Animals with strong legs will have better welfare during transport. Simple leg conformation charts that show examples of good and bad legs can be used for choosing breeding animals (National Hog Farmer, 2008; see Chapter 2, this volume).

Genomic sequencing methods now make it possible to select for desired traits more quickly. Fan *et al.* (2011) have used genomic methods to help select pigs for structural soundness. Breeders using these new tools should appraise these animals carefully to ensure that they have not accidentally selected for problems that may be linked to desirable traits.

Loading Facilities for Cattle, Pigs, and Sheep

Vehicles in Europe and some developed countries have either adjustable decks or loading ramps built into the vehicle. These systems either have a drop-down tailgate, a hydraulic tailgate or decks that move up and down. They work extremely well in technologically advanced countries. In many parts of the world, these systems are not practical because the local people may have difficulty maintaining them. Even a drop-down ramp is problematic because the springs that are used for raising the tailgate are made from special alloy that may not be available locally. During visits to developing parts of the world, the author has observed many rusting pieces of imported equipment that the local people could not maintain.

A lack of good loading ramps is a serious problem in many parts of the world. This results in animals being thrown or forced to jump off a vehicle that is 1.2 m off the ground (Bulitta *et al.*, 2012). Lack of good ramps to get animals on to trucks is also a major problem. In some facilities, ramps have been built but they are either poorly designed or too steep.

There is a great need for the construction of good ramps for loading and unloading vehicles. Stationary ramps can be built easily from locally available materials such as wood, concrete, or steel (Fig. 11.2). Many people in developing countries are very good carpenters, welders, and masons. Loading ramps can be built easily from the same materials that people use to build houses. Concrete blocks, cement, steel reinforcing rods, and wood are available in developing countries. In some situations, a ramp can be made by piling up an earthen berm for the vehicle to back up against. Another alternative is a portable ramp that can be moved from farm to farm. In the developed world, portable ramps are commercially available. In the developing world, the axles and wheels for the portable ramp could be salvaged from an old car. The author has observed that the welders and ironworkers in the developing world are masters at turning old pieces of cars into carts pulled by a motorbike and many other things. A smart welder could make a portable ramp easily if he was given a few pictures printed from the Internet of a commercially made portable livestock ramp.

Recommended design of truck loading ramps

For cattle, a nonslip ramp built on a 20-degree angle or less is recommended (Grandin, 1990; Fig. 11.2). Sheep can move up and down steeper ramps easily. For pigs, a ramp with a 15-degree or less angle is recommended (Berry *et al.*, 2012). The best nonslip footing on concrete ramps is provided by a stair-step design (Fig. 11.3). Steps are more effective than a grooved ramp and they provide nonslip footing even when they are old and worn. Grooves wear down quickly and become slick. Stair steps are a proven design used on many ramps around the world. Figure 11.3 shows the recommended step dimensions for a concrete ramp for cattle, horses, buffaloes, camels, and other large livestock. The recommended dimensions are a maximum 10-cm rise and a 30-cm or longer tread length. Two grooves at least 2.5 cm deep should be made in each step. For pigs, a shorter 8-cm rise is recommended.

Fig. 11.2. This excellent truck loading ramp was built with locally available materials. In South America, most cattle ranches have their own loading ramps. This ramp has a level dock at the top which helps prevent slipping and falling during unloading. Adding completely solid sides may improve cattle movement (see Chapter 5, this volume).

Fig. 11.3. When a loading ramp is constructed from concrete, stair steps are recommended. The steps should have a 10-cm rise and a 30- to 60-cm tread length. Do not exceed a 10-cm rise. The minimum tread length is 30 cm. The steps must be grooved to provide nonslip footing. These dimensions are recommended for cattle, water buffalo, horses, and other large animals. For pigs, a smaller 8-cm rise is recommended. (From Grandin, 2008.)

Fig. 11.4. This is the correct cleat spacing for all livestock. The cleats should be spaced so that the hooves fit easily between them. (From Grandin, 2008.)

Fig. 11.5. Wrong cleat spacing for a loading ramp – the cleats are too close together and the foot may slide and slip over the top of the cleats. (From Grandin, 2008.)

For ramps constructed from either steel or wood, cleats are normally used. Figures 11.4, 11.5, and 11.6 show the correct and the wrong spacing of cleats to provide nonslip traction. Spacing the cleats too far apart is a common mistake and can cause slipping. When cleats are too close together, the hooves slide over the top and may provide poor traction. During unloading, cleats spaced too far apart caused dew-claw injuries in piglets. The feet of the piglets slide and the dewclaws catch on the cleats. For both small and large livestock species, the cleats should be spaced so the hoof of the animal will fit easily between them (Fig. 11.4). For cattle, 20 cm of clear space between the cleats is recommended. Smaller animals require closer spacing. The stair step and cleat spacing recommendations are the same both for animals trained to lead with a halter and for extensively raised animals with a large flight zone.

Cleats can be made from many different materials. On wooden ramps for cattle and other large livestock, 5 cm × 5 cm cleats made from hardwood work well. Use 2.5 cm × 2.5 cm cleats for smaller

Fig. 11.6. Wrong cleat spacing for a loading ramp – the cleats are too far apart and the animal will slip. Animals are more likely to fall down if the cleats are too far apart. (From Grandin, 2008.)

species. Steel concrete reinforcing rods make good cleats on steel ramps. Steel reinforcing rods are readily available in all countries where homes are built from masonry or concrete block. Pipe or square steel tube can also be used. Creative local people may invent other surfaces that are effective and provide good nonslip footing.

T. Grandin

Evaluation of loading and unloading facilities

Ramps and other parts of unloading and loading facilities should be evaluated with numerical scoring. If more than 1% of the animals fall down (body touches the floor) during either loading or unloading, the ramp or floor is either not adequate or the animals are being handled roughly (OIE, 2014a). If more than 1% continues to fall after handling is improved, then the ramp or floor surface will need to be improved to provide better nonslip footing. When facilities are being evaluated, use animal-based numerical scoring instead of input-based engineering standards.

Solving Bruise Problems on Livestock

Animals that are moved through many dealers and markets will have more bruises. Both Hoffman et al. (1998) and Weeks et al. (2002) found that cattle moved through markets had more bruises than cattle sold for slaughter directly from the farm or ranch. Rough handling that causes cattle to run and become excited will almost double the amount of bruising (Grandin, 1981). Handling animals more quietly will reduce bruising. One study showed that 27% of bruises occurred during either loading or unloading (Strappini et al., 2013). Many people do not realize that the hide does not have to be damaged to have a bruise underneath. An animal can have a huge bruise under a hide that has completely normal hair and no sign of injury. Grazing animals have tough hair and hide on the outside but underneath they are as fragile as soft fruit.

Several surveys have shown that horned cattle have twice as many bruises compared to polled or dehorned cattle (Ramsey et al., 1976; Shaw et al., 1976). A few horned cattle can increase bruising greatly. Contrary to popular belief, tipping (cutting off parts of the horn) does not reduce bruising (Ramsey et al., 1976). Tipping horns on heavy-weight steers is painful and reduces live weight gain (Winks et al., 1977). West African cattle, with massive horns, had more bruises and injuries compared to breeds with smaller horns (Minka and Ayo, 2007). Ranchers should either remove horns from young calves or breed polled cattle.

Bruises and injuries are still a big problem in the developing countries (Romero et al., 2013). Data collected in 2005 in the USA indicated that 9.4% of feedlot cattle had multiple bruises (Garcia et al., 2008). The author has observed that slaughter plants that have an effective bruise prevention program have a much lower level of bruises. Scoring of bruises should be kept simple to avoid problems of inter-observer reliability that were found by Strappini et al. (2012). The size of a bruise had the best agreement between different people. Bruises have been reduced from 35% of fed beef carcasses in 2005 to 23% in 2011 (McKeith et al., 2012). In 2014, a study carried out at Kansas State University, USA, showed an increase in back bruises in fed cattle (Dan Thomson, 2014, personal communication). This was due to Holstein steers becoming taller and backs were bruised unless great care was taken during the loading and unloading of US double-deck trailers.

Equipment problems

Edges with a small diameter such as angle irons, channels, truck doors, and the exposed end of a sharp edge of a pipe are most likely to bruise cattle. Bumping into a smooth, wide surface such as a flat wall or a large, round, 15-cm-diameter post is less likely to bruise. Broken boards and protruding gate latches can also cause bruises. Gates should be equipped with tiebacks to prevent them from swinging out into an alley. It is also important to provide nonslip flooring. Slick floors that cause animals to fall cause many bruises and injuries.

Determine origin of bruises

When bruising is observed at a slaughter plant, a determination must be made to find out where the bruises are occurring. If the bruises occur on animals from many different origins, then it is likely that the problem is at the slaughter plant. If bruises occur on animals from one origin, then the problem is likely to be occurring on the truck or on the farm. A sudden occurrence of bruises is usually caused by a change in personnel or broken equipment. If faulty equipment is causing the bruises, there will be sharp edges that have been rubbed shiny from animals hitting them. Determining the precise age of a bruise is difficult, but it is possible to differentiate old from fresh bruises. Fresh bruises are red, and bruises a few days old often have yellowish secretions on them. This secretion will not be present on fresh bruises.

Loin bruises

A major cause of loin bruises is rough handling during truck loading or unloading. This causes cattle that are too excited to wedge themselves in the truck doorway. Loin bruises can be reduced by having full-width doors on trucks. Horns also greatly increase loin bruises. Loin bruises can also be caused by slamming gates on cattle. The loin becomes bruised when the animal becomes wedged between the end of the gate and an alley fence. Protruding gate latches also cause many loin bruises.

Shoulder bruises

Shoulders are often bruised on sharp edges such as protruding gate latches or broken boards. If the bruises are occurring at the slaughter plant, broken parts in the restrainer conveyor entrance may be to blame. Another cause of shoulder bruises is rolling out of the stunning box. Stunned cattle can still be bruised until they are bled (Meischke and Horder, 1976). Other causes are a sliding gate track that is not recessed and broken sideways one-way flapper gates in a single-file chute. Rough handling and horns also cause many shoulder bruises.

Back bruises

Back bruises are most likely to be caused by equipment problems. Being hit with a stick will also bruise backs (Weeks et al., 2002). One common cause is tall cattle hitting the bottom of the top deck when they exit from the bottom compartment of a truck. These bruises can be reduced by unloading cattle at a slow walk. Animals that jump up are more likely to hit the upper deck. Another major cause is careless use of vertical slide gates. The bottom edge of a vertical slide gate should be constructed from 10-cm-diameter round pipe covered with rubber. Another cause of back bruises is improperly adjusted one-way gates. Cattle can get severe back bruises if they back up under a one-way gate that is set too high.

Hind legs, udder, and abdomen bruises

Abusive beating, dragging, or kicking is usually the cause when this occurs on fully ambulatory cattle that have not fallen in a truck.

Bruising over large areas of the body

The major cause of extensive bruising where large areas of the body are bruised is due to the animal being trampled in a truck. If a truck is overloaded, an animal may fall down and it will not be able to get back up. In sheep, extensive bruising is often caused by grabbing sheep by their wool. Handlers should never grab or lift sheep by their wool.

Hide damage and branding

Branding damages the hide. A rib brand put on a small calf will ruin an entire side of leather because the brand grows with the calf. Freeze branding is less painful than hot iron branding (Lay et al., 1992), but unfortunately it still damages the hide. The leather is more likely to crack where the hide has been freeze branded. In countries where animals are extensively raised and have to be branded to prevent theft, both hide damage and welfare can be improved by moving rib brands to the rear of the animals. There has been much concern from animal advocacy groups about branding animals on the face. This stressful practice has been stopped in some countries. The face is a sensitive part of the animal, and moving brands to the less-sensitive hindquarters is strongly recommended. Branding on the most posterior part of the hindquarters will minimize hide damage.

Poking animals with sticks studded with nails will ruin the top grain surface of the hide. Unfortunately, this is still a problem in many countries. When an animal absolutely refuses to move, a brief shock from a battery-operated goad is preferable to a stick with a nail on the end. The author has seen hides where the entire side was ruined by pointed sticks. Other things that are unrelated to transport and handling can damage hides severely. Lice will ruin the top grain of the leather. If a young animal gets lice, the hide may still have a damaged surface.

Transporting Tame Animals

Miriam Parker has done extensive work in the developing world on the handling and transport of cattle, water buffalo, donkeys, and other tame animals that can be led with a head collar (Ewbank and Parker, 2013). If the animal has a nose ring or a lead threaded through its nose (Fig. 11.7), it must NEVER be tied in a vehicle by the nose.

Fig. 11.7. A typical nose lead that is used in Asia and India. This design incorporates the nose lead into a neckband so that the animal can be tied up safely when grazing. It must never be tied up with just the nose lead. However, when the animal is being transported in a vehicle, the use of a halter (head collar) in addition to this rig is strongly recommended to prevent pulling on the sensitive nose during the journey. (Photograph courtesy of Miriam Parker.)

If it falls or becomes frightened, it may tear its nose. If it is tied for transport, a head collar (halter) must be used. Animals of similar size should travel together. If the animals are tied, each animal should have its own lead rope and not be tied to another animal (Fig. 11.8). Figure 11.9 shows tame cattle being transported on the deck of a ship in the Philippines.

Handling and Transport of Poultry

One study showed that machine catching of chickens reduced stress (Duncan *et al.*, 1986). One of the reasons is that in a well-designed, automated catching machine, the birds are not turned upside down. Nijdam *et al.* (2005), at eight commercial farms, found no difference in the percentage of bruises between hand-caught and machine-caught birds. Nijdam *et al.* (2005), Chauvin *et al.* (2011) and unpublished industry data from the USA showed that machine catching had a higher number of deaths. In the USA, chickens are grown where it is very hot, and machine catching increases the number of deaths because it takes longer to catch the birds, unless two machines are used simultaneously. Longer catching times resulted in

the ventilation system being disabled for a longer period of time. In Europe, catching machines are popular due to high labor costs. When hens at the end of lay were transported, factors that were associated with higher death losses were rain, wind, cold weather, and birds with poorer feather cover and lower weight (Chauvin *et al.*, 2011; Weeks *et al.*, 2012).

In areas of the world where labour is inexpensive, the lowest level of broken wings that the author has observed is in Brazil. Each bird was lifted carefully by the body and placed in the coop (crate). Only 0.25% of these carefully hand-caught birds had broken wings. In the USA, the best hand catching of heavy (>3 kg) birds) produced 0.86% of birds with broken wings. Before restaurant companies started auditing their chicken suppliers, broken or dislocated wings were present on 5–6% of the birds. Both hand catching and machine catching can be done with an acceptable level of welfare. To prevent catching from becoming rough and sloppy, the percentage of deaths, bruises, and broken wings should be measured continuously. Putting workers on incentive pay for low levels of broken wings, deaths, and bruises will also help improve handling.

CORRECT   WRONG

Fig. 11.8. Training materials designed for training people who cannot read, showing the right and wrong way to tie up animals. The two illustrations on the left side show the correct way and the two illustrations on the right side show the wrong way. (Reproduced with permission of Miriam Parker.)

Fig. 11.9. Tame cattle that are trained to lead are being transported on the deck of a ship in the Philippines. They remain calm because they have been exposed to many novel new things long before they got on the ship (see Chapter 5, this volume). The animals' welfare is good on this ship. Cattle that are trained to lead but have not been exposed to many new things would probably panic and jump overboard. The animals on this ship have lived most of their lives grazing along a busy road. This is an example of a situation where under one set of conditions the animals' welfare would be bad and under another set of conditions it is good.

Solving Bruise and Injury Problems in Poultry

Bruised legs

The author has observed that rough shackling is a major cause of bruised drumsticks. The people doing the shackling squeeze the legs too hard when they put the birds on the shackles. An understaffed shackle line where people have to hurry is one cause of bruised legs.

Bruised chicken breasts

There are two major causes of bruised chicken breasts. One cause is jamming birds too quickly through the small opening in the top of the coop. The other cause results from machine catching systems when birds are jammed against the coop door by a conveyor that is out of alignment.

Smashed heads

This injury is most likely to occur when drawer systems are used for transporting chickens, where the birds are put in trays that are slid into rack-like dresser drawers (Fig. 11.10). Drawer systems that have been redesigned so that there is a gap between the top of the drawer and the rack frame seldom have this problem. This gap prevents the head of a chicken from being smashed when the drawer full of chickens is slid back into the rack.

Broken wings

The number one cause of this problem is rough handling during catching. Chickens should NEVER be picked up by a single wing. Measuring the percentage of broken wings is a sensitive measure of how people are handling the birds. Broken wings can also occur in systems where birds have to be removed from the coop through a small door. When individual coops are used, the best ones have a small door for loading the birds and the entire top opens to remove the birds for shackling at the plant.

Table 11.1 shows a big difference in the percentage of chickens with broken wings between the best and the worst plants. Some ethicists are hesitant to state an acceptable level of broken wings because that translates into thousands of birds with broken wings. The author has observed that when numerical standards were introduced for measuring broken wings, they were greatly reduced. In the USA before measurements started, 5–6% of the birds had broken wings. With the present data today the maximum acceptable level would be 1%, and plants with 3% are clearly not acceptable.

Broken legs

In spent hens, weak bones are a major contributor to fractures (Webster, 2004). Old hens have high percentages of keel bone fractures due to osteoporosis (Sherwin *et al.*, 2010; Wilkins *et al.*, 2011). In broiler chickens, rough handling is a major cause. There is much controversy among welfare specialists

(a) (b)

Fig. 11.10. Drawer system for transporting chickens in Denmark and Canada. In the left-hand photograph (a), chickens are being conveyed into the drawers from a mechanical catching machine. When the drawer is full, the catcher must slide the drawer into the rack carefully to prevent the heads of the chickens from being caught. A gap between the top edge of the drawer and the rack helps prevent heads or wings from being damaged (b). This is an important design feature that helps prevent injuries.

on the correct way to pick up chickens during catching. Some specialists state that they should never be picked up by a single leg. In some countries, this is the normal catching method. The author has observed one-legged catching where the number of birds injured was very low. The coops were brought into the barn close to the catcher. The catcher never walked more than 3 m to load the chickens into the coops.

Recommendations on poultry catching methods

It is the author's opinion that instead of arguing over whether poultry are hand-caught or machine-caught, caught by one leg or by two legs, or lifted by the entire body, the best approach is to measure injuries and deaths.

Table 11.1. Percentage of chickens with broken or dislocated wings in 22 US and Canadian broiler chicken processing plants.[a]

Birds with broken wings (%)[b]	Number of plants	Plants (%)
≤1	8	36
1.01–2	6	27
2.01–3	6	27
>3	2	9

Notes: [a]Average bird weight = 2.75 kg (6 lb). Birds were handled in the dump module system and all data were collected when the birds still had their feathers on before picking (defeathering equipment). Broken wings were scored after the live birds were hung on the shackles. All birds were stunned electrically. All data were collected in 2008. [b]The average percentage of birds with broken wings in all plants was 1.67%. The best plant had only 0.20% of birds with broken wings and the worst plant had 3.80% of birds with broken wings.

Quick Access Information 11.3. Outcome-based measures for handling and transport of poultry.

- Percentage of birds dead on arrival.
- Percentage with broken or dislocated wings – count with feathers on to avoid counting damage from feather-removal equipment.
- Percentage of broken legs.
- Percentage of bruised breasts.
- Percentage of keel bone fractures (laying hens).

These percentages will measure the outcome of poor handling. Broken wings should be counted when the feathers are on to prevent counting breakage from the feather-removal picking machine. The broken wing score should include both breaks and dislocations.

Space Requirements for Livestock and Poultry

For all modes of transport – land, sea, and air transport – the OIE code states:

> The amount of headroom necessary depends on the species of animal. Each animal should be able to assume its natural standing position for transport (including during loading and unloading) without coming into contact with the roof or upper deck of the vehicle and there should be sufficient headroom to allow adequate airflow over the animals. For poultry, overcrowding in transport crates increased deaths (Chauvin et al., 2011).

(OIE, 2014b)

For sea journeys where animals have to be able to lie down, the OIE (2014b) code states: "When animals lie down, there must be enough space for every animal to adopt normal lying posture". The OIE transport code applies to cattle, horses, sheep, camelids, pigs, and goats. The code does not apply to poultry. Practical experience, industry guidelines, and research show that all species on sea journeys should have sufficient space for all animals to lie down at the same time without lying on top of each other. Poultry are routinely transported in lying positions in containers where they are not able to stand and they must have sufficient space to all lie down at the same time without being on top of each other.

During truck transport, most large animals such as cattle, horses, and water buffalo will remain standing. Unfortunately, there are no space allowance charts in the OIE (2014a) code for animals that remain standing on a truck. This is probably due to the fact that there is controversy among researchers on the correct amount of floor space for each animal. Most researchers will agree that jamming the maximum number of animals that can be forced on to a vehicle is wrong. Death losses, bruises, and injuries will increase in severely overloaded trucks (Eldridge and Winfield, 1988; Tarrant et al., 1988; Valdes, 2002; Schwartzkopf-Genswein et al., 2012). Too many birds loaded into each compartment is a major cause of deaths in chickens (Nijdam et al., 2005). For large animals such as cattle, bruises and injuries

T. Grandin

increase on an overloaded truck because an animal that goes down is not able to get up. This results in it getting trampled. Cattle and other large animals usually remain standing in a truck. Pigs need space to lie down, unless they are hauled a very short distance (Ritter *et al.*, 2007). A study by Ritter *et al.* (2008) under average commercial conditions showed that for large 129-kg pigs transported for under 4 h, there was a critical space requirement to prevent death losses and greater numbers of nonambulatory pigs. Each large pig needed 0.462 m^2 per pig or greater (Ritter *et al.*, 2008). Pigs were transported at six different space allowances ranging from 0.396 to 0.520 m^2. Lambooij (2014) recommends a maximum loading density of 235 kg/m^2, which enables all the pigs to lie down. Quick Access Information 11.4 shows practical market weight pig loading densities for the transport of pigs for short distances of 4 h or less. To comply with European (EU) rules, the 235 kg/m^2 guidelines should be used, and more space will be required. Good information on space requirements for weaner pigs can be found in Sutherland *et al.* (2009). For people who are implementing an animal welfare program, the author recommends that local, national, or industry regulations and guidelines for space allowances on

trucks are used, and outcomes such as bruises, death losses, and nonambulatory animals are measured carefully.

Many countries have either legislation or guidelines for space requirements for truck floor space when animals are being transported. The OIE (2014b) codes are a minimum standard that all countries should follow to provide a basic level of animal welfare. The legislation and standards in many countries will be much stricter. Reviews of scientific studies on transport can be found in *Livestock Handling and Transport* (Grandin, 2014) and in Miranda de la Lama *et al.* (2014). Some of the other major journal article reviews on transportation are: Knowles (1999) (cattle); Fike and Spire (2006) (cattle); Schwartzkopf-Genswein *et al.* (2012), Knowles *et al.* (1998) (sheep); Hall and Bradshaw (1998) (sheep and pigs); Warriss (1998) (pigs); Ritter *et al.* (2009b) (pigs); Weeks (2000) (deer); and Warris *et al.* (1992) (chickens).

Length of Journey

Another contentious issue within both the research community and the animal advocacy non-governmental organizations (NGOs) is the length of time an

---

Quick Access Information 11.4. Recommended space allowances on trucks for pigs. (From Federation of Animal Science Societies, 2010.)

	Weight kg	Lbs	m^2	ft^2	Winter (cold weather)		Summer (hot weather)	
					m^2	ft^2	m^2	ft^2
Small pigs	4.54	10	0.06	0.70				
	9.07	20	0.084	0.9				
	13.60	30	0093	1.00				
	22.70	50	0.139	1.50				
	31.20	70	0.167	1.80				
	36.30	80	0.177	1.90				
	40.80	90	0.195	2.10				
Market pigs and sows	45	100			0.22	2.4	0.30	3.0
	91	200			0.32	3.5	0.37	4.0
	114	250			0.40	4.3	0.46	5.0
	136	300			0.46	5.0	0.55	6.0
	182	400			0.61	6.6	0.65	7.0

---

animal is allowed to travel by road transport before it has to have a rest stop. People who are implementing an animal welfare program should follow the government regulations and industry guidelines in their country. The OIE (2014a) code does not address this issue because the member countries could not agree on standards. For further information on long-distance transport issues, refer to Appleby *et al.* (2008).

In many countries, 48 h is the maximum length of time that animals can remain on a vehicle before they have to be unloaded. The EU (2005) has strict standards. Cattle are allowed to travel for 14 h and if they have a 1 h rest stop, they are allowed to travel for another 14 h. If the second part of such a trip (i.e. that following the rest stop) brings them within 2 h of their destination, the second half of the trip can be extended to 16 h. For longer trips, a 24-h rest stop is required. Pigs can travel for 24 h without stopping if they have continuous access to water. For chickens, 12 h is the maximum time they should be held in the transport containers (Mohan Raj, 2009, personal communication). Further information can be found in Miranda de la Lama *et al.* (2014). There are lots of rules and they are constantly being changed. Transporters who ship livestock or poultry out of their own country must make sure they know the latest regulations. Sometimes, very short trips are more stressful because animals do not have sufficient time to calm down before they are unloaded. The incidence of nonambulatory fatigued pigs was greater after a short journey of less than an hour compared to trips lasting 3 h (Pilcher *et al.*, 2011).

Rest stop research

Rest stops that are too frequent may be "stress" stops instead of rest stops. This is more likely to be a bigger problem when extensively raised animals with large flight zones are being handled. These animals are more likely to become stressed during loading and unloading compared to tame livestock. Sheep need long rest stops to gain a benefit. They often will not drink until they have eaten. They need a longer rest stop compared to cattle in order for them to eat and drink. A short 3-h rest period does not provide enough time for drinking. On a trip that took 24 h, sheep arrived in better condition with no rest stop (Cockram *et al.*, 1997). Australian research showed that sheep could tolerate easily

trips up to 48 h (Ferguson *et al.*, 2008). In cattle, the maximum length of a nonstop road trip should be 24–32 h for 6-month-old to yearling calves (Grandin, 1998; Schwartzkopf-Genswein, *et al.*, 2012). This recommendation is for transportation when the temperature is less than 30°C. Grain-fed cattle should definitely be rested after 48 h.

Methods to Improve Welfare and Prevent Losses during Transport

Measure losses

One of the most effective ways to bring about improvements in transport is to measure losses. Bruises, injuries, death losses, sickness, and other losses should be counted and tabulated. This makes it possible to identify drivers and producers who are having high losses.

Hold sellers accountable for losses

Sellers should receive bonuses or fines based on the condition of the animals and the incidence of sickness a few weeks after arrival at their destination (see Chapter 14, this volume, for further information).

Better management of livestock markets

Lack of refrigeration is one reason why cull animals are transported long distances in the developing world. This may make it impossible to eliminate long-distance transport in some parts of the world. Both improved facilities and management expertise will be required to correct the atrocious conditions reported by Chandra and Das (2001) because sellers have no economic interest in old cull spent animals. The author has observed that markets are the most difficult segment of the industry to improve because middlemen and dealers pass on losses to the next segment in the marketing drain (see Chapter 14, this volume).

Change insurance policies

Insurance policies on livestock should only cover catastrophic losses such as a truck rollover. They should have a high deductible fee to prevent careless or abusive treatment of animals from being rewarded. For example, a driver should not receive an insurance payment for two or three dead pigs.

Careful driving

Smooth starts and careful braking will help prevent animals from falling down during transport. Careful driving is essential. Cockram *et al.* (2004) reported that 80% of the time that an animal lost its balance followed a driving event such as cornering or braking. Unpublished industry data indicated that certain truck drivers had higher percentages of bruised cattle. McGlone (2006) reported on data collected on 38 drivers who transported over a million pigs. The best drivers had 0.3% dead or nonambulatory pigs and the worst drivers had twice as many.

Driver fatigue – a major cause of accidents

A survey conducted by Jennifer Woods in Canada showed that a high percentage of livestock truck rollovers were due to driver fatigue. A total of 415 commercial truck wrecks were tracked in the USA and Canada. Of these, 85% of the wrecks were due to driver error (Woods and Grandin, 2008). Most people assumed that cold, icy weather would be a major cause of truck wrecks. However, the highest percentage of the wrecks occurred in October, when the roads were clear of ice. Another indicator that fatigue was the cause of many of the accidents was that only 20% of the livestock truck wrecks involved another vehicle and 59% of the wrecks occurred during the early morning hours between 12:00 midnight and 9:00 in the morning. In North America, vehicles are driven on the right-hand side of the road and when a driver falls asleep, the vehicle usually drifts to the right. Of the trucks that tipped over, 84% fell on their right side. A Spanish study also showed that driver fatigue was a major cause of accidents (Miranda de la Lama *et al.*, 2014).

Never pay transporters based on livestock weight

This method of payment provides an economic incentive to overload trucks. Transporters should be paid a negotiated rate for each trip. The rate should specify the number of animals of different types to put on the truck.

Never pay livestock or poultry handlers based on speed of handling

Rewarding people for handling more and more animals per hour will motivate abusive handling. Provide incentive pay for reducing bruises and deaths. Unpublished industry data showed that providing financial bonuses greatly reduced the number of broken wings in poultry and deaths in pigs.

Farm of origin effects

The author has observed that about 10% of farms are responsible for about 90% of the worst emaciated cull dairy cows (Grandin, 2001). Both a Spanish and a US study had similar conclusions. Fitzgerald *et al.* (2009) found that out of nine farms, the worst one had 0.93% more losses on each load of market pigs. In Spain, pigs from certain farms had more deaths and carcass problems (Gosalvez *et al.*, 2006). A large Canadian study also showed that farm of origin was a major factor in predicting the percentage of dead and nonambulatory market pigs (Tina Widowski, 2009, personal communication). To reduce problems caused by poor producers, a system of economic penalties and rewards should be implemented. At one large slaughter plant, problems with weak pigs caused by feeding excessive doses of ractopamine was stopped by charging a US$25 handling fee for each nonambulatory pig. A survey of slaughter houses indicated that conditions caused by neglect at the farm of origin created some of the worst welfare problems (Grandin *et al.*, 1999).

Delays at country borders

Some of the most serious delays that are detrimental to animal welfare are delays at the borders between different countries. Government officials, livestock producer organizations, and NGO advocacy groups should work to solve this problem. In some cases, more customs officials may need to be hired. In other cases, paperwork procedures may need to be changed, or a special lane for livestock could be introduced. Forcing a truckload of livestock or poultry to wait hours at a border is a problem that needs to be corrected.

Documentation for Transport

The OIE (2014a) code for land transport lists the documentation that should accompany every consignment of livestock. Each country or region within a country may require additional documents to prove ownership or enable trace-back after a disease outbreak. The OIE code states:

Article 7.3.6.

Documentation

1. Animals should not be loaded until the documentation required to that point is complete.

2. The documentation accompanying the consignment should include:

 a. journey travel plan and an emergency management plan;

 b. date, time and place of loading and unloading;

 c. veterinary certification, when requested;

 d. animal welfare competencies of the driver (under study); [The author recommends using training materials developed in each country.]

 e. animal identification to allow animal traceability to the premises of departure and, where possible, to the premises of origin;

 f. details of any animals considered at particular risk of suffering poor welfare during transport (point 3e of Article 7.3.7);

 g. documentation of the period of rest, and access to feed and water, prior to the journey;

 h. stocking density estimate for each load in the consignment;

 i. the journey log – daily record of inspection and important events, including records of morbidity and mortality and actions taken, climatic conditions, rest stops, travel time and distance, feed and water offered and estimates of consumption, medication provided and mechanical defects.

3. When veterinary certification is required to accompany consignments of animals, it should address:

 a. fitness of animals to travel;

 b. animal identification (description, number, etc.);

 c. health status including any tests, treatments and vaccinations carried out;

 d. when required, details of disinfection carried out.

(OIE, 2014a)

At the time of certification, the veterinarian should notify the animal handler or the driver of any factors affecting the fitness of the animals to travel on a particular journey. The documentation required for sea and air transport is very similar.

Training of Handlers and Drivers

Many livestock associations, federal governments, and state and provincial governments have developed programs for training handlers and drivers. Training programs can reduce bruises and damage to both livestock and poultry (Pilecco *et al.*, 2013; Paranhos da Costa *et al.*, 2014). Some large slaughter plants require all the drivers that enter the premises to be trained. In many countries, training documents for animal welfare are part of industry-sponsored livestock quality assurance programs that also cover disease control and food safety. These types of programs are already in place in Europe, the USA, Australia, Canada, and many other developed countries. Argentina, Chile, Uruguay, and Brazil have also developed programs. A good driver-training program should cover the following areas:

- Behavioral principles for moving animals – such as explanation of the flight zone and point of balance. This can be omitted if all animals are tame and trained to lead (see Chapter 5, this volume).
- A description of the distractions that can frighten animals – drivers need to understand how shadows, reflections, or the sun's position may affect animal movement (see Chapter 5, this volume).
- Biosecurity rules for disease control.
- Health regulations for transport across the borders of countries, regions, state, or provincial lines.
- Methods to prevent bruises and injuries.
- How to determine an animal's fitness for travel.
- Emergency procedures after an accident.
- Proper use of electric goads and other implements for moving animals.
- How to protect animals from heat stress or cold stress – these must be written specifically for the breeds of animals and climatic conditions in each country.

Emergency Plans

Truck drivers should be equipped with an emergency plan so they will know what to do after an accident. Often, police and fire departments do not know how to handle livestock or poultry that get loose on a highway. Sometimes, they respond in an inappropriate manner, which results in terrified cattle or sheep charging down a highway. Chasing scared, frightened animals is the worst thing to do. When police arrive at an accident scene, they should concentrate on directing traffic instead of chasing animals. In some countries, police and fire departments receive specific training on how to respond to livestock truck accidents. The author recommends that programs for training police and fire departments should be implemented. People with livestock expertise should be called to the accident scene to gather up loose animals. The driver should always carry with him the telephone numbers of both police and people who know how to handle animals. Telephone numbers for getting another vehicle should also be included on the list of emergency numbers. The driver needs to have contact information for different people along the route who can assist. The locations of places where animals can be unloaded in an emergency should also be carried with the driver. These locations could be farms, stockyards, or auction markets. The telephone numbers of people who can bring a portable loading ramp and portable fence panels quickly should also be carried by the driver. Emergency response plans work best if plans on how to respond in the event of an accident have been made before an accident happens.

Euthanizing Animals After an Accident

Chapter 10, this volume, covers euthanasia of different types of animals. Animals that are severely injured should be euthanized at the site of the accident to stop continued suffering. A plan for making euthanasia decisions should be made ahead of time. In some cases, there may be legal problems with the owner of an animal when it has survivable injuries and it has been euthanized. Large corporations who transport their own animals should have policies that state that animals that are severely injured in an accident will be euthanized immediately at the accident site. Some examples of injuries where animals should be euthanized at the accident site are a broken leg, if the internal organs are exposed, or the animal is not able to stand.

Weather Stress

Heat stress

Different animals have greater tolerance for heat stress than others. In poultry, researchers in the Czech Republic found that death losses were greater in summer (Vecerek et al., 2006). Heat stress may be responsible for up to 40% of chicken mortalities (Bayliss and Hinton, 1990). When the temperature was over 23°C, the percentage of chicken mortalities was almost seven times higher in modern broilers. Heat stress is a much greater problem for chickens than cold stress, unless the birds are subjected to extreme cold. McGlone and colleagues (McGlone, 2006; McGlone et al., 2014) analysed data from 2 million pigs that arrived at a large slaughter plant in the Midwestern USA. Death losses started to increase

when the temperature was over 23°C (75°F), and at 32°C (90°F) death losses doubled.

Breeds of cattle, sheep, and other animals that come from the hot regions of the world can tolerate more heat compared to breeds developed in more temperate climates. Animals fed beta-agonists may be more prone to heat stress. Beta-agonists fed during the summer increase death losses and heat stress (Macias-Cruz et al., 2010; Loneragan et al., 2014). Heat can build up rapidly in a parked vehicle. This is especially going to be a problem if the vehicle has solid sides and a solid roof. Vehicles and ships where animals are dependent on mechanical ventilation can heat up to lethal levels in less than 1 h if the ventilation system fails. Failure of the ventilation system is a major cause of death losses in animals that are transported below the deck on ships.

When vehicles have to be parked during hot weather, there should be provisions for heat relief. When the temperature is 29°C (85°F), the air inside the vehicle can quickly reach 35°C (95°F) and 95% relative humidity (McGlone, 2006). At poultry plants, trucks loaded with chickens are parked in a shed equipped with a bank of fans. Some pork slaughter plants also have banks of fans for cooling parked trucks. One of the simplest ways to keep a hot truck ventilated is to keep it moving.

Assessing heat stress

Panting in both mammals and birds is a sign that the animals are becoming heat stressed. In cattle, panting scores are a reliable indicator of heat stress (Mader et al., 2005). Mader and colleagues (Mader et al., 2005) developed a simple five-point panting score for cattle:

0. Normal respiration.
1. Elevated respiration.
2. Moderate panting or presence of drool.
3. Heavy open-mouth breathing.
4. Severe open-mouth breathing with protruding tongue. Neck may be extended forward.

The animal's welfare would be severely compromised when the score reached 3 or 4. At score 2, provide heat relief. Under very hot conditions, purebred Brahmans had lower respiration rates than crosses with Herefords (Gaughan et al., 1999). For sheep and cattle, additional information on assessing heat stress can be found in Mader et al. (2005) and Gaughan et al. (2008). For all livestock, one of the easiest to use heat stress charts is shown in Fig. 11.11. A combination of high heat and high humidity can be dangerous to both pigs and cattle. The chart provides guidelines on the combinations of heat and humidity where extra precautions to prevent heat stress must be taken. Pigs are more sensitive than other livestock. For pigs, the emergency zone starts at 32°C (90°F) for all humidity levels. In pigs, respiration is a major indicator of heat stress (Brown-Brandl *et al.*, 2001). An in-depth view of stress physiology during transport can be found in Knowles *et al.* (2013).

Cold Stress

Transporters must be trained to be careful to avoid the effects of wind chill. Freezing rain at 0°C that wets an animal's coat can be deadly because the animal's coat loses the ability to insulate. Driving a truck through freezing rain is often more dangerous for livestock than colder, dry weather where animals do not get wet. In countries with cold weather, guidelines should be developed on when animals in vehicles need to be protected from cold by covering a vehicle with tarps or boards. When the temperature is below –6°C, aluminium trailers transporting pigs should be 96% covered. When the Celsius temperature warms to over 16°C, all coverings should be removed (McGlone *et al.*, 2014). These recommendations are for market-weight pigs. In the northern regions of Europe and North America, great care has to be taken to prevent frostbite or death losses due to cold stress in pigs. For pigs, deep straw bedding will help protect the animals. The author has observed that pigs can get frostbite easily if they lie against a bare metal floor or side of a vehicle, when the temperature is –18°C. Piglets weighing 7 kg should be transported in a warm vehicle, but the maximum temperature should not exceed 30°C (Lewis, 2007). Most transporters know that animals are likely to become dehydrated during hot weather, but Miranda de la Lama *et al.* (2014) report that problems with dehydration during cold weather may be similar to hot weather. Now that animal welfare is a concern all over the world, published standards for acceptable ranges of temperature may be different due to both animal genetics and different conditions in tropical countries.

Vehicle requirements

In the EU, there are no regulatory requirements for truck design for trips under 8 h. For trips over 8 h, fans for ventilation are required. These systems

Fig. 11.11. Heat-stress chart for livestock. A combination of both high heat and high humidity is dangerous for animals. When the temperature and humidity is in the danger and emergency zone, it is recommended to transport the pigs at night or in the very early morning. Temperature conversions: 75°F (24°C); 80°F (27°C); 85°F (29°C); 90°F (32°C); 95°F (35°C); 100°F (38°C). The emergency zone starts at 32°C (90°F) for pigs, at all humidity levels. The alert zone on this chart is correct for pigs.

T. Grandin

Fig. 11.12. Shipping facility for a large feedlot market. Diagonal pens provide easy, one-way flow of cattle because they enter through one end of the pen and leave through the other end. Sharp 90-degree corners are eliminated.

should be designed so that the animals are not totally dependent on the mechanized ventilation system. If the mechanical ventilation fails, the truck or trailer should be designed so that passive systems such as removable panels or curtains can be opened. This would provide ventilation to prevent death losses from heat stress. During cold weather, curtains or panels can be used to close up ventilation openings to prevent frostbite and deaths from wind chill. Many livestock associations have recommendations for the use of panels and curtains at different temperatures. It is important to develop specific guidelines for your local area that are based on collecting data from many truckloads. The vehicle shown in Fig. 11.2 is a typical truck used all over South America and many other countries. In the parts of the world where the temperature is warm year-round, a vehicle like the one shown in Fig. 11.2 provides lots of natural ventilation.

Stockyard Layout

Angled pens in the shipping area at large feedlots, slaughter plants, and markets allow the movement of animals by eliminating sharp 90-degree corners. Figure 11.12 shows an angled pen layout with a curved loading ramp. This design works really well with extensively raised or feedlot animals that are not trained to lead. It is especially suited for large operations where many trucks have to be loaded quickly.

Databases for Welfare Information

Research on animal transport, animal welfare, and animal behavior can be accessed on the Internet at the following databases:

- PubMed (good for veterinary research)
- www.science.direct (good for behavior research)
- Google Scholar
- www.vetmedresource.org (good for author searches)
- CAB Abstracts.

Summaries of research can be acquired for free. The easiest way to find these databases is to type their names into a search engine such as Google. When you get to the correct web page, a new search box will appear on the page. Type keywords into the new search box and you will locate abstracts (summaries) of scientific studies. Use every possible keyword to find information. For example, many

---

Quick Access Information 11.5. Outcome-based measures for transport of livestock.

- Percentage of animals dead on arrival.
- Percentage of sickness (morbidity after arrival).
- Percentage of livestock that arrive nonambulatory and not able to walk.
- Percentage of bruises on each animal. Classify as (1) none, (2) surface, (3) deep, and (4) many bruises over large areas of the body.
- Percentage of livestock with surface hide damage.

---

different words are used for cattle, such as cows, bulls, cattle, steers, bullocks, oxen, and calves. Do a separate transport search with each word on each database. Some full-text papers will be available for free and others will only give you a free abstract (summary). Often, the full papers can be obtained for free if you go to a library at a major university, or you can e-mail the author and request a copy.

Outcome-based Measures for Assessing Transport

People will often do a better job of managing if they measure performance continuously. The outcome-based measurements shown in Quick Access Information 11.5 will make it possible to determine if practices are improving or becoming worse. All measurements are based on the percentage of animals or poultry that have a deficit.

Conclusion

To have an adequate level of animal welfare during transport, fit animals must be loaded in the vehicle. Transporting unfit animals is a cause of major welfare problems during transport. To maintain transport practices at a high standard requires continuous measurement with animal-based outcome measures and accountability for losses.

References

Appleby, M.C., Cussen, V., Garces, L., Lambert, L.A. and Turner, J. (2008) *Long Distance Transport and the Welfare of Farm Animals*. CAB International, Wallingford, UK.

Arnold, G.W. (1976) Some factors influencing feeding behaviour of sheep in pens. In: *Proceedings of the*

T. Grandin

Sheep Assembly and Transport Workshop. Western Australian Department of Agriculture, Perth, Western Australia.

Averos, X., Knowles, T.G., Brown, S.N., Warriss, P.D. and Gosalvez, L.F. (2008) Factors affecting the mortality of pigs transported to slaughter. *Veterinary Record* 163, 386–390.

Baszczack, J.A., Grandin, T., Gruber, S.L., Engle, T.E., Platter, W.J., *et al.* (2006) Effects of ractopamine supplementation on behaviour of British, Continental, and Brahman crossbred steers during routine handling. *Journal of Animal Science* 84, 3410–3414.

Bayliss, P.A. and Hinton, P.A. (1990) Transportation of broilers with specific reference to mortality rates. *Applied Animal Behaviour Science* 28, 93–118.

Beatty, D.T., Barnes, A., Taylor, E., Pethick, D., McCarthy, M., *et al.* (2006) Physiological responses of *Bos taurus* and *Bos indicus* cattle to prolonged, continuous heat and humidity. *Journal of Animal Science* 84, 972–985.

Berry, N.L., Johnson, A.K., Hill, J., Loneragan, S., Karriker, L.A., *et al.* (2012) Loading gantry versus traditional chute for the finisher pig. Effect on welfare at time of loading and performance measures and transport losses at the harvest facility. *Journal of Animal Science* 90, 4028–4036.

Brown-Brandl, T.M., Eigenberg, R.A., Nienaber, J.A. and Kachman, S.D. (2001) The thermoregulatory profile of a newer genetic line of pigs. *Livestock Production Science* 71, 253–260.

Bulitta, F.S., Gebresenbet, G. and Bosona, T. (2012) Animal handling during supply for marketing and operations at an abattoir in developing countries: The case of Gudar Market and Ambo Abattoir Ethiopia. *Journal of Service Science and Management* 5, 59–68.

Chandra, B.S. and Das, N. (2001) The handling and short haul road transportation of spent buffaloes in relation to bruising and animal welfare. *Tropical Animal Health Production* 33, 153–163.

Chauvin, C., Hilton S., Balaine, L., Michel, V., Peraste, J., *et al.* (2011) Factors associated with mortality of broilers during transport to the slaughter house. *Animal* 5, 287–293.

Cockram, M.S., Kent, J.E., Jackson, R.E., Goddard, P.J., Doherty, O.M., *et al.* (1997) Effect of lairage during 24 hour transport on the behavioural and physiological responses of sheep. *Animal Science* 65, 391–402.

Cockram, M.S., Baxter, E.M., Smith, L.A., Bell, S., Howard, C.M., *et al.* (2004) Effect of driver behaviour, driving events and road type on the stability and resting behaviour of sheep in transit. *Animal Science* 79, 165–176.

Duncan, I.J.H., Slee, G., Kettlewell, P.J., Berry, P. and Carlisle, A.J. (1986) A comparison of the effects of harvesting broiler chickens by machine or by hand. *British Poultry Science* 27, 109–114.

Eldridge, G.A. and Winfield, C.G. (1988) The behavior and bruising of cattle during transport at different space allowances. *Australian Journal of Experimental Agriculture* 28, 695–698.

Ewbank, R. and Parker, M. (2013) Handling cattle in close association with people. In: Grandin, T. (ed.) *Livestock Handling and Transport*, 4th edn. CAB International, Wallingford, UK, pp. 76–89.

Fan, B., Onteru, S.K., Du, Z.Q., Garrick, D.J., Stalder, K., *et al.* (2011) Genome-wide association study identifies loci for body composition and structural soundness traits in pigs. *PLoS ONE* 6(2):e14726, doi:10 1371/Journal.pone.0014726.

Federation of Animal Science Societies (2010) *Guide for the Care and Use of Agricultural Animals in Research and Teaching*, 3rd edn. Federation of Animal Science Societies, Champaign, Illinois (www.fass.org).

Ferguson, D.M., Niemeyer, D.P.O., Lee, C. and Fisher, A.D. (2008) Behavioral and physiological responses in sheep to 12, 30, and 48 hours of road transport. In: Boyle, L., O'Connell, N. and Hanlon, A. (eds) *Proceedings of the 42nd Congress of International Society for Animal Ethology (ISAE)*, 5–9 August, Dublin, Ireland, p. 136.

Fike, K. and Spire, M.F. (2006) Transportation of cattle. *Veterinary Clinics of North America, Food Animal Practice* 22, 305–320.

Fitzgerald, R.F., Stalder, K.J., Matthews, J.O., Schultz-Kaster, C.M. and Johnson, A.K. (2009) Factors associated with fatigued, injured, and dead pig frequency during transport and lairage at a commercial abattoir. *Journal of Animal Science* 87, 1156–1166.

Garcia, L.G., Nicholson, K.L., Hoffman, T.W., Lawrence, T.E., Hale, D.S., *et al.* (2008) National Beef Quality Audit – 2005: Survey of targeted cattle and carcass characteristics related to quality, quantity, and value of fed steers and heifers. *Journal of Animal Science* 86, 3533–3543.

Gaughan, J.B., Mader, T.L., Holt, S.M., Josey, M.J. and Rowan, K.J. (1999) Heat tolerance of Boran and Tuli crossbred steers. *Journal of Animal Science* 77, 2398–2405.

Gaughan, J.B., Mader, T.L., Holt, S.M. and Lisle, S.M. (2008) A new heat load index for feedlot cattle. *Journal of Animal Science* 86, 226–234.

Gaughan, J.B., Bonner, S., Loxton, I., Mader, T.L., Lisle, A., *et al.* (2010) Effect of shade on body temperature and performance of beef steers. *Journal of Animal Science* 88, 4056–4067.

Gosalvez, L.F., Averos, X., Valdeivira, J.J. and Herranz, A. (2006) Influence of season, distance and mixed loads on the physical and carcass integrity of pigs transported to slaughter. *Meat Science* 73, 553–558.

Grandin, T. (1981) Bruises on Southwestern feedlot cattle. *Journal of Animal Science* 53 (Supplement 1), 213 (Abstract).

Grandin, T. (1990) Design of loading facilities and holding pens. *Applied Animal Behaviour Science* 28, 187–201.

Grandin, T. (1998) Handling methods and facilities to reduce stress on cattle. *Veterinary Clinics of North America, Food Animal Practice* 14, 325–341.

Grandin, T. (2001) The welfare of cattle during slaughter and the prevention of nonambulatory (downer) cattle. *Journal of the American Veterinary Medical Association* 219, 1377–1382.

Grandin, T. (2008) Engineering and design of holding yards, loading ramps and handling facilities for land and sea transport of livestock. *Veterinaria Italiana* 44, 235–245.

Grandin, T. (2014) *Livestock Handling and Transport*, 4th edn. CAB International, Wallingford, UK.

Grandin, T., McGee, K. and Lanier, J.L. (1999) Prevalence of severe welfare problems in horses that arrived at slaughter plants. *Journal of the American Veterinary Medical Association* 214, 1531–1533.

Hall, S.J.C. and Bradshaw, R.H. (1998) Welfare aspects of the transport by road of sheep and pigs. *Journal of Applied Welfare Science* 1, 235–254.

Hoffman, D.E., Spire, M.F., Schwenke, J.R. and Unruh, J.A. (1998) Effect of source of cattle and distance transported to a commercial slaughter facility on carcass bruises in mature beef cows. *Journal of the American Veterinary Medical Association* 212, 668–672.

Holtcamp, A. (2000) Gut edema: Clinical signs, diagnosis and control. In: *Proceedings of the American Association of Swine Practitioners*, 11–14 March, Indianapolis, Indiana, pp. 337–340.

James, B.W., Tokoch, M.D., Goodband, R.D. Nelssen, J.L., Dritz, S.S., *et al.* (2013) Effects of dietary L-Carnitine and ractopamine HCl on the metabolic response to handling in finishing pigs. *Journal of Animal Science* 91, 4426–4439.

Knowles, T.G. (1999) A review of the road transport of cattle. *Veterinary Record* 144, 197–201.

Knowles, T., Warriss, P. and Vogel, K. (2013) Stress physiology during transport. In: Grandin, T. (ed.) *Livestock Handling and Transport*, 4th edn. CAB International, Wallingford, UK, pp. 312–328.

Knowles, T.G., Warriss, P.D., Brown, S.N. and Edwards, J.E. (1998) The effects of stocking density during the road transport of lambs. *Veterinary Record* 142, 503–509.

Lalman, D. and Mourer, G. (2005) *Effects of Preconditioning on Health, Performance, and Prices of Weaned Calves.* Oklahoma Cooperative Extension Service ANSI-3529, Oklahoma State University, Stillwater, Oklahoma. http://osufacts.okstate.edu (accessed 5 May 2015).

Lambooij, E. (2014) Transport of pigs. In: Grandin, T. (ed.) *Livestock Handling and Transport*, 4th edn. CAB International, Wallingford, UK, pp. 280–297.

Lay, D.C. Jr., Friend, T.H., Randel, R.D., Bowers, C.L., Grissom, K.K., *et al.* (1992) Behavioral and physiological effects of freeze and hot iron branding on crossbred cattle. *Journal of Animal Science* 70, 330–336.

Lewis, N. (2007) Transport of early weaned piglets. *Applied Animal Behaviour Science* 110, 128–135.

Loneragan, G.H., Dargartz, D.A., Morley, P.S. and Smith, M.A. (2003) Trends in mortality ratios among cattle in US feedlots. *Journal of the American Veterinary Medical Association* 219, 1122–1127.

Loneragan, G.H., Thomson, D.U. and Scott, H.M. (2014) Increased mortality in groups of cattle administered the B-adrenergic agonist ractopamine hydrochloride and zilpaterol hydrochloride. *PLoS ONE* 9(3):e91177. doi:10.1371/journal.pone.0091177.

McGlone, J. (2006) Fatigued pigs: The transportation link. *Pork*, 1 February 2006.

McGlone, J.J., Johnson, A.K., Sapkota, A. and Kephart, R.K. (2014) Transport of market pigs improvements in welfare and economics: In: Grandin, T. (ed.) *Livestock Handling and Transport*. CAB International, Wallingford, UK, pp. 298–314.

McKeith, R.O., Gray, G.D., Hales D.S., Kerth, C.R., Griffin, D.B., *et al.* (2012) National Beef Quality Audit 2011: Harvest floor assessments of targeted characteristics that affect quality and values of cattle carcasses and by-products. *Journal of Animal Science* 90, 5135–5142.

Mader, T.L., Davis, M.S. and Brown-Brandl, T. (2005) Environmental factors influencing heat stress in feedlot cattle. *Journal of Animal Science* 84, 712–719.

Marai, I.F.M., El-Darawany, A.A., Fadiel, A. and Abdel-Hafez, M.A.M. (2006) Physiological traits as affected by heat stress in sheep: a review. *Small Ruminant Research* 71, 1–12.

Macias-Cruz, U., Alvarez-Valenzuela, F.D., Torrentera, N., Velazquez-Morales, J.V., Correa-Calderon, A., *et al.* (2010) Effect of zilpaterol hydrochloride on feedlot and carcass characteristics of ewe lambs under heat stress condition. *Animal Production Science* 50, 983–989.

Marchant-Forde, J.N., Lay, D.C. Jr., Pajor, J.A., Richert, B.T. and Schinckel, A.P. (2003) The effects of ractopamine on the behavior and physiology of finishing pigs. *Journal of Animal Science* 81, 416–422.

Meischke, H.R.C. and Horder, J.C. (1976) Knocking box effect on bruising in cattle. *Food Technology, Australia* 28, 369–371.

Minka, N.S. and Ayo, J.O. (2007) Effects of loading behaviour and road transport stress on traumatic injuries in cattle transported by road during the hot dry season. *Livestock Science* 107, 91–95.

Miranda de la Lama, G.C., Villarroel, M. and Maria, G.A. (2014) Livestock transport from the perspective of the pre-slaughter logistic chain: A review. *Meat Science* 98, 9–20.

Murray, A.C. and Johnson, C.P. (1998) Importance of halothane gene on muscle quality and preslaughter death in western Canadian pigs. *Canadian Journal of Animal Science* 78, 543–548.

National Hog Farmer (2008) Poster. Available at: www.NationalHogFarmer.com/posters (accessed 5 May 2015).

T. Grandin

Nijdam, E., Delezie, E., Lambooij, E., Nabuurs, M.J., Decuypere, E. and Stegeman, J.A. (2005) Comparison of bruises and mortality, stress parameters, and meat quality in manually and mechanically caught broilers. *British Poultry Science* 84, 467–474.

Norris, R.T., Richards, R.B., Creeper, J.H., Jubb, T.F., Madin, B., *et al.* (2003) Cattle deaths during sea transport from Australia. *Australian Veterinary Journal* 81, 156–161.

OIE (2014a) *Transport of Animals by Land, Terrestrial Animal Health Code*. World Organisation for Animal Health, Paris.

OIE (2014b) *Guidelines for the Sea Transport of Animals, Terrestrial Animal Health Code*. World Organisation for Animal Health, Paris.

Paranhos da Costa, M.J.R., Huertas, S.M., Strappini, A.C. and Gallo, C. (2014) Handling and transport of cattle and pigs in South America. In: Grandin, T. (ed.) *Livestock Handling and Transport*, 4th edn. CAB International, Wallingford, UK, pp. 174–192.

Pilcher, C.M., Ellis, M., Rojo-Gomez, A., Curtis, S.E., Wolter, B.F., *et al.* (2011) Effects of floor space during transport and journey time on indicators of stress and transport losses in market weight pigs. *Journal of Animal Science* 89, 3809–3818.

Pilecco, M., Almeida Paz, I.C.L., Tabaldi, L.A., Naas, I.A., Garcia, R.G., *et al.* (2013) Training of catching teams and reduction of back scratches in broilers. *Revista Brasileira de Ciencia Avicola* 15(3), http://dx.doi.org/10.1590/S1516-635X2013000300017.

Powell, J. (2003) Preconditioning Programs for Beef Calves. Cooperative Extension Service, University of Arkansas. Available at: www.uaex.edu (accessed 5 May 2015).

Rademacher, C. and Davis, P. (2005) Factors associated with the incidence of mortality during transport of market hogs. In: Allen, D. (ed.) *Proceedings of the Leman Swine Conference*. University of Minnesota, Minneapolis, Minnesota, pp. 186–191.

Ramsey, W.R., Meischke, H.R.C. and Anderson, B. (1976) The effect of tipping horns and interruption of the journey on bruising cattle. *Australian Veterinary Journal* 52, 285–286.

Ritter, M.J., Ellis, M., Bertelson, C.R., Bowman, R., Brinkman, J., *et al.* (2007) Effects of distance moved during loading and floor space on the trailer during transport on losses of market weight pigs on arrival at the packing plant. *Journal of Animal Science* 85, 3454–3461.

Ritter, M.J., Ellis, M., Bowman, R., Brinkman, J., Curtis, S.E., *et al.* (2008) Effect of season and distance moved during loading in transport losses of market weight pigs in two commercially available types of trailer. *Journal of Animal Science* 86, 3137–3145.

Ritter, M.J., Ellis, M., Hollis, G.R., McKeith, F.K., Orellana, D.G., *et al.* (2009a) Frequency of HAL-1843 mutation of the ryanodine receptor gene in dead and non-ambulatory, non-injured pigs on arrival at the packing plant. *Journal of Animal Science* 86, 511–514.

Ritter, M.J., Ellis, M., Berry, N.L., Curtis, S.E., Anil, L., *et al.* (2009b) Review: Transport losses in market weight pigs. A review of definitions, incidence and economic impact. *The Professional Animal Scientist* 25, 404–414.

Romero, M.H., Uribe-Velasquez, L.F., Sanchez, J.A. and Miranda de la Lama, G.C. (2013) Risk factors influencing bruising and high muscle pH in Columbia cattle carcass due to transport and pre-slaughter operations. *Meat Science* 95, 256–263.

Schwartzkopf-Genswein, K.S. and Grandin, T. (2014) Transport of cattle by road. In: Grandin T. (ed.) *Livestock Handling and Transport*, 4th edn. CAB International, Wallingford, UK, pp. 143–173.

Schwartzkopf-Genswein, K.S., Faucitano, L., Dadgar, S., Shand, P., Gonzalez, L.A., *et al.* (2012) Road transport of cattle, swine, and poultry in North America and its impact on animal welfare, carcass, and meat quality: A review. *Meat Science* 92, 227–243.

Shaw, F.D., Baxter, R.I. and Ramsey, W.R. (1976) The contribution of horned cattle to carcass bruising. *Veterinary Record* 98, 255–257.

Sherwin, C.M., Richards, G.J. and Nicol, C.J. (2010) Comparison of welfare of layer hens in four housing systems in the UK. *British Poultry Science* 51, 488–499.

Strappini, A.C., Metz, J.H., Gallo, C. and Kemp, G. (2009) Origin and assessment of bruises in beef cattle at slaughter. *Animal* 3, 728–736.

Strappini, A.C., Frankena, K., Metz, J.H.M. and Kemp, B. (2012) Intra- and interobserver reliability of a protocol for post mortem evaluation of bruises in Chilean beef carcasses. *Livestock Science* 145, 271–274.

Strappini, A.C., Metz, J.H., Gallo, C., Frankena, K., Vargas, R., *et al.* (2013) Bruises on culled cows, when, where, and how are they inflicted. *Animal* 7, 485–491.

Sutherland, M.A., Bryer, P.G., Davis, B.L. and McGlone, J. (2009) Space requirements of weaned pigs during a sixty minute transport in summer. *Journal of Animal Science* 87, 363–370.

Swanson, J.C. and Morrow-Tesch, J. (2001) Cattle transport: Historical, research, and future perspectives. *Journal of Animal Science* 79 (E Supplement), E102–E109.

Tarrant, P.V., Kenny, F.J. and Harrington, D. (1988) The effect of stocking density during 4 h transport to slaughter, on behaviour, blood constituents and carcass bruising in Friesian steers. *Meat Science* 24, 209–222.

Valdes, A. (2002) *Efectos de los Densidades de Carga y dos Tiempo de Transporte Sobre el Peso Vivo, Rendimiento de la Canal y Presencia de Contusiones s en Novillos Destinados al Faenamiento*. Memoria de Titulo para Optar al Titulo de Medico Veterinario Fac. Ciencias Veterinarias, Universidad Austral del Chile Valdivia, Chile.

Vecerek, V., Grbalova, S., Voslarova, E., Janackova, B. and Maiena, M. (2006) Effects of travel distance and the season of the year on death rates of broilers transported to poultry processing plants. *Poultry Science* 85, 1881–1884.

Warner, A.C.I. (1962) Some factors influencing the rumen microbial population. *Journal of General Microbiology* 28, 129–146.

Warriss, P. (1998) Choosing appropriate space allowances for slaughter pigs transported by road: a review. *Veterinary Record* 142, 449–454.

Warriss, P.D., Bevis, E.A., Brown, S.N. and Edwards, J.E. (1992) Longer journeys to processing plants are associated with higher mortality in broiler chickens. *British Poultry Science* 33, 201–206.

Webster, A.B. (2004) Welfare implications of avian osteoporosis. *Poultry Science* 83, 184–192.

Weeks, C.A. (2000) Transport of deer: A review with particular relevance to red deer (*Cervus elaphus*). *Animal Welfare* 9, 63–74.

Weeks, C.A., McNally, P.W. and Warriss, P.D. (2002) Influence of the design of facilities at auction markets and animal handling procedures on bruising in cattle. *Veterinary Record* 150, 743–748.

Weeks, C.A., Brown, S.N., Richards, G.J., Wilkins, L.J. and Knowles, T.G. (2012) Levels of mortality in hens at the end of lay on farm and in transit to slaughter in Great Britain. *Veterinary Record* 170, 647.

Wilkins, L.J., McKinstry, J.L., Avery, N.C., Knowles, T.G., Brown, S.N., *et al.* (2011) Influence of housing system and design on bone strength and keel bone fractures in laying hens. *Veterinary Record* 169, 414.

Winks, L.A., Holmes, A.E. and O'Rourke, P.K. (1977) Effect of dehorning and tipping on live weight gain of mature Brahman crossbred steers. *Australian Journal of Experimental Agriculture and Animal Husbandry* 17, 16–19.

Woods, J. and Grandin, T. (2008) Fatigue is a major cause of commercial livestock truck accidents. *Veterinaria Italiana* 44, 259–262.

T. Grandin

12 Why Are Behavioral Needs Important?

Tina Widowski

University of Guelph, Guelph, Ontario, Canada

Summary

The performance of some aspects of natural behavior is important even for domesticated animals, and understanding their behavioral biology is necessary for ensuring their welfare. Some natural behaviors are more important to the animals than others, and when those "behavioral needs" are not satisfied, animals may suffer negative consequences or develop abnormal behaviors. Scientific methods are used to measure objectively what animals specifically want and the strength of their motivation to obtain resources in order to perform natural behavioral patterns. Preference tests allow an animal to choose between different things such as feeds or flooring in housing. Aversion tests measure an animal's reluctance to re-enter a place where an aversive (unpleasant or painful) event occurred. Strength of motivation tests measure how hard an animal is willing to work to get to (or away) from something. Cognitive bias tests indicate whether an animal is in a negative (pessimistic) or positive (optimistic) emotional state. In this chapter, the student will learn some of the basic principles of animal motivation and the techniques scientists use to identify which aspects of behavior are most important for good animal welfare.

Learning Objectives

- Understand the basics of motivation for resources that support important sequences of behavior
- Discover that how much an animal wants something can be measured objectively
- Understand the different types of tests for measuring how animals feel
- Measure the outcomes of problem behaviors such as aggression or feather pecking

Introduction

For a number of reasons, animal behavior has always been a key component in discussions about animal welfare. One reason is that many people hold the viewpoint that good welfare means that animals should be able to lead relatively natural lives, or at least to behave in ways that are consistent with the nature of their species (Fraser, 2003). Although performing all natural behavior patterns is not necessarily a requirement for good welfare, from a scientific perspective, a species' behavioral biology – its sensory capacities and its general behavioral traits – do determine how the animals perceive and adapt to the ways we house and handle them (Spinka, 2006). Therefore, it is important that the care we provide matches up with their behavioral traits. Further to this is the notion that the performance of some specific behavior patterns, for example nesting by hens or rooting by pigs, may be important to animals and that they may suffer if they do not have the opportunity to perform them (Dawkins, 1990; Duncan, 1998). This view has led to some of the most difficult and contentious issues regarding the welfare of farm animals.

Importance of feelings

Another reason that behavior plays an important role comes from the viewpoint that animal welfare is mainly to do with how an animal feels (Duncan, 1996; Rushen, 1996). According to this viewpoint, animals should be housed and handled in ways that prevent negative feelings such as pain, fear, and frustration, and that may even promote positive feelings such as pleasure or contentment. Chapter 8, this volume, contains information on the biology of emotions. Although animals' feelings cannot be measured directly, a variety of experimental techniques have been developed using behavior so that we now have a number of well-established ways for gauging animals' perceptions and emotions scientifically (Kirkden and Pajor, 2006; Murphy *et al.*, 2014). Some techniques rely on animal preferences and the fact that an animal will work very hard to obtain something that it wants and will avoid or try to escape from something that it finds unpleasant. Other behavioral techniques aimed at assessing welfare quantify postures or vocalizations that animals have evolved to use as signals to communicate emotions such as fear, pain, or distress (Dawkins, 2004). Mellor (2015) states that preference tests and other methods that assess behavioral motivation may help to identify management practices that should be avoided.

Problem behavior patterns

A final reason that behavior plays a key role in animal welfare is that in many commercial settings, animals perform behavior patterns that are aggressive or that cause injury to other animals, which directly reduces the welfare of the victims. Tail biting in pigs and feather pecking in poultry are two examples of behavior problems that can have devastating effects on the welfare of the target animals. In addition, many animals develop behavior patterns that appear to be abnormal, and these are often interpreted as indicators of poor welfare (Mason and Latham, 2004). These include repetitive abnormal behaviors such as sham chewing and bar biting in sows and tongue rolling in cattle. Substantial research has been aimed at identifying the environmental, genetic, and neurophysiological mechanisms underlying these types of behavior problems in order to help us understand why they develop, what they mean for welfare, and how we can prevent them (see Mason and Rushen, 2006).

Use outcome-based measures

In practical settings, brief snapshots of animal behavior can be difficult to interpret, and therefore most behavioral measures used to assess welfare are often better suited to experimental studies.

Animals' natural patterns of behavior, preferences, and aversions are usually determined from controlled experiments, and this information can then be used to make recommendations on input-based indicators of welfare for housing or management practices. Some behavioral indicators of fear, pain, or thermal discomfort such as vocalizations, gait, or postures have been validated by laboratory studies and can be used as outcome-based measures of welfare (see Chapters 1, 2, and 4, this volume). Many of these are recognized readily and used in day-to-day practice by knowledgeable stockpeople, and some can be applied to welfare audits. Behavior that results in either poor condition or injury to either the animal performing them or to groupmates can be indirectly assessed by using outcome-based measures such as wound scores (Turner *et al.*, 2006), feather scores (Bilčík and Keeling, 1999; LayWel, 2006; Welfare Quality, 2009), or body condition scores (see Chapter 4, this volume). However, understanding the underlying causes of those conditions is important for interpreting them in relation to an animal's welfare and for troubleshooting problems.

In this chapter, some of the scientific concepts and approaches used to measure animal behavior in relation to animal welfare will be discussed. This will include how an understanding of behavior and motivation can provide us with an insight as to how animals perceive the environments we provide for them and help us to develop best practices for managing their physical and social environment. It will also include some discussion of indirect measures of behavior and how those can be used to identify and solve behavior problems.

The Importance of Understanding the Behavioral Biology of the Animal

Behavior is a central part of the mechanisms that every animal has for maintaining its health, survival, and reproduction. Throughout its natural history, each species has evolved complex strategies and coordinated sets of responses for obtaining nutrients, protecting itself from predators and other harm, finding mates, and taking care of its young. To a large degree, behavior is under genetic control since genes determine the development of sensory organs, the muscles, and the neural systems that coordinate them. All of our livestock species have been domesticated for thousands of years, and the process of domestication as well as intensive genetic selection

T. Widowski

for production traits has led to some behavioral changes in modern breeds compared to their wild ancestors. However, most of the changes in behavior that have occurred are quantitative rather than qualitative (Price, 2003). What this means is that domesticated breeds retain many (if not most) of the behavioral traits of their wild ancestors, but that those traits vary in the degree that they are expressed.

Comparison of domestic and wild animal behavior

Jensen and colleagues have conducted a comprehensive set of studies comparing the behavior of White Leghorn layers selected for egg production with that of jungle fowl, which is the wild ancestor of domestic fowl (Jensen, 2006; Jensen and Wright, 2014). The birds were incubated, hatched, and reared under identical conditions and then observed in a semi-natural environment (outdoors but with food and some shelter provided) and during a series of behavior tests. The modern egg-laying breed showed all of the same motor patterns and the same social and sexual signals as their wild ancestors, but generally they were less active, less fearful of humans and novel objects, and performed less exploratory and less anti-predator behavior (Shütz et al., 2001). The modern strain of birds fed more intensively – they were less apt to search for food and ate more from a localized food source, and they engaged in more sexual behavior. Although they tended to space themselves more closely together, the modern strain was surprisingly more aggressive than the wild type after new groups were formed. There were no differences between wild and domestic stock in either the nature or the amount of what the authors called "comfort behaviors", which included preening and dust bathing (Shütz and Jensen, 2001). Using modern genomic techniques, Jensen and colleagues (see Jensen, 2006) also investigated the underlying genetic mechanisms for changes in behavior that might be associated with changes in the genome due to selection for production traits. Their results using quantitative trait loci (QTL) analysis suggest that changes in just a few regulatory genes may be responsible for many of the behavioral effects considered to be part of the domestic phenotype (Jensen, 2006; Jensen and Wright, 2014). An interesting study with reindeer shows results similar to poultry. Herds of reindeer that had increased vigilance and flight responses had greater genetic dissimilarity with domestic herds (Reimers et al., 2012).

These results from the studies on laying hens and jungle fowl are in agreement with the idea that domestication and artificial selection for production traits generally tend to increase the frequency of behavior that supports greater feed efficiencies and higher reproductive rates, and makes the domesticated animal easier for humans to handle than wild animals, but that it has not changed many of their basic behavioral predispositions. A number of studies have also been conducted on feral (populations of domesticated animals that have gone back to the wild) and on domesticated chickens (Wood-Gush and Duncan, 1976), pigs (Jensen and Recén, 1989) and cattle (Rushen et al., 2008) held in wild or semi-natural environments in order to determine how animals spend their time, organize their social groups, and what resources or features of the physical environment seem important for supporting their "natural" behavior patterns. These types of studies do not necessarily imply that keeping animals in a natural environment is a requirement for their welfare. "Natural behavior" is actually quite flexible, and variation in natural behavior allows animals to adapt to variations in geography, environmental conditions, and food supplies. In addition, natural conditions (and natural behavior) are not always good for welfare because they can cause animals to suffer from stress or injury (Spinka, 2006). However, some consideration of species typical behavior can provide insight into their physical and social requirements and help us to understand some of the welfare problems that can occur in commercial agricultural settings (Rushen et al., 2008).

What controls behavior?

Motivation is the concept used to describe the internal process or state that determines an animal's behavioral responses to environmental stimuli at different times. In other words, concepts of motivation explain why an animal does what it does when it does it. For example, a sleeping pig gets up, goes to the feed, and begins to eat. What is it about the pig that has caused it to switch from sleeping to eating? Certainly, the mechanism for this difference involves some complex interaction of neuronal, hormonal, and physiologic processes (see Chapters 1 and 8, this volume, for research findings on the core emotional systems that drive behavior). In this situation, we usually say that the animal has simply "become hungry" or that its "feeding motivation" has changed. Hunger, thirst, and libido are the common terms that we use to describe the motivational states for feeding, drinking, and sexual behavior, respectively.

Early theorists on motivation classified different systems along a continuum with behavior motivated primarily by internal physiological factors on the one end and those motivated primarily by external factors or environmental stimuli on the other. Although we know now that most motivational systems depend on much more complex interactions between internal and external factors, it is often helpful to think about the different mechanisms for the control of behavior as lying somewhere along this continuum (Fig. 12.1). For behavioral systems that clearly serve in regulating homeostasis – feeding, drinking, thermoregulatory behavior – internal factors such as blood glucose, osmotic pressure of the blood, or core body temperature, respectively, can predominate. However, there are other nonregulatory behaviors that also appear to depend greatly on physiological factors. Nesting by hens and nest building in domestic sows are examples.

Most complex behaviors also involve a sequence of activities or motor patterns. The early parts of the behavior sequence often involve exploration or goal-seeking behavior; for example, searching for and courting a mate or foraging for food. Ethologists refer to these as appetitive behavior patterns. Once the goal has been reached, the appetitive parts of the sequence feed forward to performance of consummatory aspects; for example, copulation or swallowing of feed (Fig. 12.2). For some types of behavior, the performance of the appetitive aspects of the behavior sequence increase motivation and appear to be rewarding to the animal in and of themselves (Hughes and Duncan, 1988). Performance of the consummatory behavior patterns can also increase motivation for the entire sequence of behavior; for example, when an animal first tastes a highly palatable food. Later, motivation decreases as the animal becomes satiated. Although the sequence of appetitive and consummatory motor patterns may belong to a single motivational system, for example eating, these different parts of the sequence can be controlled by different neural substrates in different parts of the brain (Schneider *et al.*, 2013), and the appetitive motor patterns, for example foraging, may have behavioral and physiological consequences of their own. In many of the feeding and management systems that we use for livestock and poultry today, the appetitive aspects of behavior have been reduced significantly or even eliminated, which can lead to behavior and welfare problems.

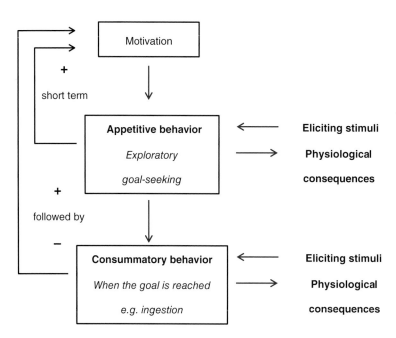

Fig. 12.1. The concept of motivation describes the internal state that determines why an animal does what it does at any given time and generally depends on some combination of internal and external factors. Some behaviors are motivated more by changes in internal stimuli so that the animal wants to perform the behavior regardless of environmental cues.

T. Widowski

Nesting behaviors are highly motivated

Because of their size and general lack of physical development, many newborn or newly hatched animals have very different physical requirements from their older counterparts. In nature, neonates are vulnerable to hypothermia and predation. Therefore, most species of birds and some species of mammals construct a nest for the incubation of eggs and early care of offspring. Even though maternal nests are no longer necessary for the survival of chicks or piglets, hens and sows are still highly motivated to perform nesting behavior before egg laying and farrowing, respectively. Both hens and ducks prefer nest boxes that provide enclosure (Makagon *et al.*, 2011; Ringgenberg *et al.*, 2014) (Fig. 12.3).

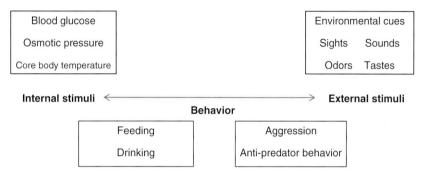

Blood glucose		Environmental cues	
Osmotic pressure		Sights	Sounds
Core body temperature		Odors	Tastes

Internal stimuli ←——————————→ **External stimuli**

Behavior

| Feeding | | Aggression |
| Drinking | | Anti-predator behavior |

Fig. 12.2. Complex behaviors usually involve an integrated system of appetitive and consummatory motor patterns. The performance of the different aspects of the sequence can feed back to either increase or decrease motivation. Each part of the sequence of behaviors can be elicited by different stimuli, may be rewarding to the animal, and have consequences of their own. Thus, it may be important for the animal to perform the entire sequence of behavior. In many modern feeding and management systems, we have reduced or eliminated opportunities for appetitive behavior.

Fig. 12.3. Furnished colony cage that provides a nest box where the hens have a secluded area to lay their eggs behind the plastic curtain. Both hens and ducks prefer an enclosed area for laying their eggs.

Possibly, this is an innate behavior that helps wild nesting birds avoid predation. Hens also preferred a nest with an Astro Turf floor compared to a bare wire floor (Riber and Nielsen, 2013). Both hens and sows perform elements of nest building in the confines of cages and crates, and show behavior changes that have been interpreted as signs of frustration when not given the resources to nest. In addition, many animals engage in bathing or grooming as part of their natural behavioral repertoire. For example, domestic fowl regularly perform dust bathing when provided with a dusty substrate. In wire cages, they show "sham dust bathing", going through all of the motions of dust bathing in the absence of any dusty material. The appearance of these "vacuum activities" has led to the concept of behavioral deprivation and the question of whether animals suffer when they do not have the opportunity to perform some patterns of behavior that may be important to them. Scientists have used studies of motivation to address these questions.

Sow nest building is motivated by internal stimuli

When given the opportunity, sows build nests before giving birth to their piglets. The day before farrowing, the sow seeks out a nest site, excavates a shallow depression, and then gathers various nesting materials, such as branches, leaves, and grasses, which she uses to line the nest. Within a few hours after nest completion, the sow delivers her young in the nest. In intensive swine production systems, many pre-parturient sows kept in farrowing crates still exhibit the motor patterns associated with nest building. During the 16 h or so prior to farrowing, she becomes restless and roots and paws at the floor or bars of the crate. These changes in behavior occur consistently and predictably during a fairly narrow period regardless of the environment that the sow is in (Widowski and Curtis, 1990). Nest building in sows presumably has a strong internal component, because it appears in what seems to be a complete absence of appropriate external stimuli in the farrowing crate – little space and no materials for building. Some of the hormone changes that are also associated with labor and delivery, namely the release of prostaglandins, stimulate the onset of this behavior (Widowski *et al.*, 1990; Gilbert *et al.*, 2002). In a natural environment, the external stimuli – grass, substrate, location of the nest site – then serve to guide the sequence of behavior patterns, which

finally results in nest construction (Jensen, 1993). Considering the function of the maternal nest – a warm, protected place that is necessary for offspring survival – and the critical importance of timing it right, it makes sense that a strong internal control system has evolved. The nest must be completed *before* the young are born, and its being triggered by hormonal changes otherwise associated with reproductive events of parturition ensures that a nest will be completed at the appropriate time.

Behaviors stimulated by external stimuli

In contrast, some behavior patterns are stimulated primarily by external stimuli. Anti-predator behavior – alarm calls, protective postures, avoidance behavior – is the classic example because it generally occurs only in the presence of external stimuli that signal the approach of predators or something perceived as a predator. Animals may perceive rapid movement or a large looming object as threatening (see Chapter 5, this volume). Aggression is also usually due to primarily environmental factors since the presence of a competitor is necessary, although physiological factors such as testosterone in males or hunger may increase the tendency of an animal to engage in fighting. At any time, animals may be stimulated to do different things simultaneously so that different motivational systems are, in essence, competing for control. The behavior that is exhibited is the one that is motivated most strongly at the time or the one that is not inhibited by other factors.

Lack of foraging opportunities in intensive production systems

Within intensive production systems, there are several areas of housing and management that deviate substantially from livestock species-typical behavioral biology. These include the ways we feed them, the ways we manage maternal behavior and mother–young relationships, the sizes and composition of the groups that we keep them in, and the lack of physical resources (e.g. vegetation, soil, or substrate) that support some behavior patterns. With regard to feeding, individual species use a wide variety of strategies for searching, finding, preparing, and finally ingesting their daily nutrient requirements. Examples include grazing and ruminating in cattle and sheep, scratching and pecking in poultry, and rooting and chewing in pigs. In natural feeding systems, foraging and feeding behavior often occupy a large part of an animal's daily activities (Figs 12.4 and 12.5). For example, the observed time spent grazing for pastured dairy cattle ranged from 8.6 to 10 h/day depending on the strain of dairy cow and the level of concentrate feeding (McCarthy *et al.*, 2007), and the time spent rooting and foraging ranged from 12 to 51% of daylight hours in gestating sows managed outdoors, even though the sows were fed a standard daily allotment of feed in the morning (Buckner *et al.*, 1998). This is in contrast to the grazing and foraging opportunities provided for dairy cows in zero-grazing systems and sows in most indoor housing systems. In young mammals, ingestion of milk usually involves contacting the mother in ways that stimulate milk let down, like nuzzling or butting, in addition to sucking behavior. In free-living systems, calves have been observed suckling cows until 7–14 months (see Rushen *et al.*, 2008), and piglets will nurse for anywhere between 10 and 17 weeks (see Widowski *et al.*, 2008). Most dairy calves are fed from buckets from the first day of age, while commercial piglets are commonly weaned from sows to dry diets at around 21–28 days of age,

or even earlier in some systems. These wide departures in the feeding system can lead to behavior problems such as cross-sucking in calves and belly nosing and belly sucking in piglets, and may contribute to the development of oral stereotypies in adult ungulates.

There is some evidence that genetic selection for increased production may lead to more abnormal behaviors related to feed intake. Research indicates that egg-laying hens selected for high egg production have more feather pecking (Muir and Wei Cheng, 2013). Recent studies show that feather pecking may be misdirected foraging behavior instead of aggression (Dixon *et al.*, 2008; de Haas *et al.*, 2010). High-producing livestock and poultry have to eat large amounts of food. The combination of genetic selection for high production and a lack of materials to forage in may contribute to the occurrence of abnormal oral behavior.

Do Animals Have Behavioral "Needs"?

Emotional states are thought to play a role in motivation, because feelings such as fear or frustration or

Fig. 12.4. Hens housed in this cage-free system with litter can perform their natural behaviors of scratching, pecking, and foraging. This system provides opportunities for natural behavior but it has to be managed carefully to avoid air-quality problems.

Fig. 12.5. Furnished colony house enables birds to stand in their full height posture and use perches that help improve bone strength. The one hen is standing on a smooth plastic mat where she can perform natural scratching behavior. This mat will stay cleaner than an Astro Turf or textured mat. Hens cannot perform scratching on a wire floor.

even pleasure make it more likely that the animal does the right thing at the right time (Dawkins, 1990). For well over 40 years, scientists have grappled with the problem of behavioral deprivation and whether animals have behavioral needs (Duncan, 1998). The term "behavioral (or ethological) need" first arose in response to a report by a committee struck by the British government to address public concerns about the welfare of farm animals. The Brambell Committee report proposed that animals have "natural, instinctive urges and behavior patterns" and that animals should not be kept in conditions that suppressed these behavior patterns (Brambell, 1965). From its inception, the term was highly debated and often criticized for its lack of both clear definition and scientific foundation (Dawkins, 1983).

Over time, there was some consensus that the term "behavioral need" should refer to specific behavior patterns that might be important for animals to perform and that, when prevented, would result in frustration or some negative psychological state that would cause suffering and impair welfare (Dawkins, 1983; Hughes and Duncan, 1988; Jensen and Toates, 1993). Behavioral deprivation is generally considered to be more likely to reduce

welfare when the factors motivating the behavior are primarily internal and when the performance of the behavior itself seems to be important to the animal (Duncan, 1998). Dawkins (1990) and Fraser and Duncan (1998) suggested that behavioral "need situations", that is, behavior associated with intense negative emotions, such as fear or frustration, probably evolved for those behaviors where immediate action was necessary to cope with a threat to survival of the individual (e.g. escape from a predator) or survival of offspring (e.g. nesting). They also suggested that other types of behavior that were less immediately critical to survival and that could be performed when opportunity arose (e.g. play, grooming) were more likely to be associated with positive emotional states, such as pleasure and contentment.

Boissy *et al.* (2007) have written a highly cited paper that states, "It is now widely accepted that good welfare is not simply the absence of negative experiences, but rather is primarily the presence of positive experiences such as pleasure", and most current views are in agreement (Mellor, 2015). In order to do this, we must be able to determine objectively what animals want (Dawkins, 2004).

T. Widowski

Scientific Approaches to Measuring Animals' Feelings

Animal welfare scientists have developed a number of different approaches for assessing how animals feel about specific aspects of housing and handling (Kirkden and Pajor, 2006). One way is to offer animals some control over specific resources or specific experiences, either by offering them choices or by providing opportunities to get to or get away from some option, and then observing what decisions the animals make. Basically, these are methods of asking animals what they want and how much they want (or do not want) it. Examples of the sorts of options that can be tested are types of flooring or stall design, different substrates such as straw, wood shavings, or sand, items such as nest boxes, perches, or social companions, and different types of handling and restraint. Three types of standardized tests are used: preference tests, tests for strength of motivation, and aversion tests, which will be explained in more detail. Another approach to measuring an animal's feelings is to keep it in a particular environment (e.g. a wire cage) or expose it to a specific type of experience (e.g. freeze branding) and then observe carefully and measure the animal's responses in order to identify indications of negative feelings such as frustration, fear, or distress. Behavioral responses might include things such as trying to escape (Schwartzkopf-Genswein et al., 1998), behavior that appears out of context to the situation, or stereotypic pacing (Yue and Duncan, 2003). The most recent approach to be developed involves determining whether animals interpret neutral information in a positive or negative way by using cognitive bias tests (Mendl et al., 2009).

Tests for preference, aversion, and strength of motivation

In standardized preference tests, scientists use either a Y-maze or a T-maze, with different options offered at the ends of the maze. Animals are trained to use the maze, they learn what options are available at the ends of the maze, and then their choices of going to the different options are counted over a series of tests. Alternatively, the animal may be given continuous access to different options over a longer period, and the frequency of accessing the options and amount of time spent with the different options is measured by live observation or by video recording. Preference tests have been used to study a whole variety of different design features of housing in all species, including environmental temperature for sows (Phillips et al., 2000; Fraser and Nicol, 2011), different types and intensities of lighting for poultry (Widowski et al., 1992; Davis et al., 1999), different types of bedding and flooring for dairy cows (Tucker et al., 2003), and even different levels of ammonia in the air (Wathes et al., 2002). Preference tests can provide good information about housing alternatives, but the tests have to be designed carefully to ensure that the animals are making choices for the options of interest; for example, not just always choosing the left side. They also have to be interpreted carefully because a number of factors can affect the results, such as the previous experience of the animals (they might be hesitant to enter an area with a novel type of flooring, even though it is more comfortable to lie on), or the motivational state that the animal is in during the test (a pig might choose straw to lie on when it is cold but choose bare concrete when it is hot). Recent research by the author's research group has indicated that there are complex relationships between cage floor area and next area that affect both preferences and agonistic behavior significantly (Hunniford et al., 2014). Another concern with preference tests is that animals do not always make choices that are best for their long-term health and welfare, and this has to be considered when using the results to make housing recommendations. One example may be choosing a floor type that may be comfortable to lie on but may be bad for long-term foot health.

Lighting preferences for poultry

One of the advantages of using preference tests is that they allow us to ask the animals directly about whether they perceive differences in the environment and which ones they find more comfortable rather than relying on human perceptions. In this regard, preference tests have sometimes led to some surprising results. One example comes from a question regarding the types of lighting to use in poultry houses. The visual system of birds is very different from that of humans. Birds have very rich color vision and they can see different wavelengths of light than those seen by humans. Birds also have different abilities for motion detection and are able to perceive flickering light at a much higher frequency than humans – a feature called the critical fusion frequency (CFF). The CFF is the frequency at which motion can no longer be detected or the frequency at which a discontinuous light source (flicker) appears to be continuous. In humans, the CFF is 60 Hz. This means that at frequencies higher than 60 Hz, images blend together and appear to be continuous; this is the phenomenon on which the motion picture depends. Birds generally have much better motion detection than humans. The domestic chickens's CFF has been estimated to be at around 105 Hz (Nuboer *et al.*, 1992).

In the early 1990s, when compact fluorescent bulbs were first developed, there were questions regarding the welfare implications of using them in poultry houses. Common fluorescent light sources powered by magnetic ballasts flicker at twice the frequency of the electricity supply. In North America, the supply frequency is 60 Hz; in Europe, it is only 50 Hz. This means that fluorescent lights flicker at 120 and 100 Hz, respectively, in those places. Most humans cannot detect this flicker because it is well above our CFF. Because birds' CFF is near the flicker frequency of fluorescent light sources, it was speculated that birds might be able to see the flicker of fluorescent lighting and might find it aversive if they had to live in an environment with flickering light (Nuboer *et al.*, 1992). In order to determine whether laying hens found fluorescent lighting aversive, they were allowed to choose between two rooms illuminated respectively by either standard incandescent lighting or a compact fluorescent source during 6 h long preference tests (Widowski *et al.*, 1992). The rooms were identical except for the type of illumination. Light intensities were similar to those used in laying-hen houses.

Contrary to what we expected from what we knew of the bird's CFF, the hens spent around 73% of the time in fluorescent light and only 27% in incandescent light. This meant that either the birds did not perceive the flicker of fluorescent lights (at least in North America) or, if they did, they did not avoid it. In fact, they found some aspect of the fluorescent light more attractive than incandescent

light and spent most of the time in it. Similar results have been obtained from preference tests conducted in the UK on turkeys (Sherwin, 1999). It is not clear why birds prefer fluorescent lighting over incandescent, but it is probably due to differences in the wavelengths of light emitted. The compact fluorescent source used emits more energy in the shorter wavelengths (UV and blue region of the spectrum), and domestic fowl are more sensitive to wavelengths in this region than are humans. The results of these studies on the lighting preferences of birds emphasize how the sensory capacities and perceptions of animals differ from those of humans, and that it can be difficult for people to predict how animals may respond. One of the best ways to obtain answers is to use an experimental technique that allows us to ask the animal directly.

Aversion tests

Aversion tests are based on the idea that unpleasant feelings or negative emotions evolved to help animals learn to avoid situations that might cause them harm. When an animal experiences something that causes fear, pain, or discomfort, it usually behaves in ways that serve to remove the source of those feelings (avoid and escape). If an animal repeatedly experiences something unpleasant, painful, or distressing, it will learn to avoid the place or the situation associated with those unpleasant feelings. A number of studies have shown that cattle will learn to avoid rough handling by either hesitating or refusing to go down a runway where hitting and prodding have been experienced (Pajor *et al.*, 2000), or by choosing the arm of a maze associated with gentle handling rather than rough (Pajor *et al.*, 2003). Similar tests have shown that sheep quickly learned to avoid an arm of a maze in which they were subjected to electro-immobilization as a means of restraint (Grandin *et al.*, 1986). These tests indicate that cattle and sheep do, in fact, distinguish between different types of handling and restraint, and that they find some types more aversive than others.

Although the majority of aversion tests have been used in the context of handling and restraint, the approach has also been used to determine the aversiveness of some gases that are used for stunning at slaughter or euthanasia. Raj and Gregory (1995) trained pigs to put their heads in a chamber in order to eat apples from a box in the chamber. Over the course of a series of test days, the chamber was filled with either air (control), 90% argon gas in air, 30% carbon dioxide (CO_2) in air, or 90% CO_2 in air for a 3 min exposure period. On test days when the chamber was filled with air or argon, all of the pigs remained in the chamber for the majority of the 3 min test period eating the apples, and none of the pigs hesitated to enter the chamber on the subsequent test day. On days when the box was filled with 90% CO_2, on the other hand, nearly all of the pigs withdrew their heads immediately and spent very little time feeding in the box, even on test days when the pigs were fasted for 24 h. The response to 30% CO_2 was intermediate. A number of the pigs hesitated, and one even refused to enter the box on the day following the experience of the 90% CO_2, when the chamber was subsequently filled with air, suggesting that the high concentration of CO_2 was highly aversive (see Chapters 9 and 10, this volume, for further research on gas stunning and euthanasia).

Strength of motivation tests

Although preference tests do provide us with good information about the relative merits of different choices to the animal, they do not give us any indication how important those choices are. When given the option, a hen may prefer to dust bathe in peat moss over wood shavings, so if we were going to offer them a substrate, we know the one that she values more. However, the results of the preference test tell us nothing about *how much she wants* a dust-bathing substrate. This information is particularly important when we consider the more difficult questions regarding whether animals suffer from behavioral deprivation. Dawkins (1983) pioneered the concept of using another type of test, the demand test, to gauge the strength of motivation an animal has for a particular resource. Demand tests are based on techniques used by economists to determine what types of things people view as needs or luxury items based on their purchasing behavior. Demand tests used for animals are designed to ask animals to pay for a resource through their behavior. They may have to incur a cost to obtain a resource, for example, forgoing an opportunity to feed in order to spend time in a larger cage or with a social companion. More often, the animal is asked to perform an operant technique – peck at a key, press a lever with their hoof or snout – in order to obtain the reward. Once the animals learn the task, the price they have to pay for getting to the

item is increased – they have to pay more pecks at the key in order to obtain the reward. Animals' willingness to pay for a resource such as litter or straw is often compared with their willingness to pay for food; food is considered to be the gold standard of what an animal wants. Other types of tasks, referred to as obstruction tests, have also been used to determine how much an animal wants to get a resource – for example, pushing through weighted doors that can be made heavier and more difficult to open (Duncan and Kite, 1987), or squeezing through gaps that are made narrower and more difficult to squeeze through (Cooper and Appleby, 1996). Sometimes, animals are tested before and after being deprived of a resource to determine whether the value of the resource increases after the animal has been without it for some time or, alternatively, whether "out of sight is out of mind" (Duncan and Petherick, 1991).

Using studies of motivation to determine nest-box and dust-bathing needs in laying hens

The conventional cage system for laying hens has been highly criticized, in part because of the lack of opportunities for hens to perform nesting and dust bathing. Literally hundreds of studies have been conducted to help us understand the motivation for these behaviors, and there are several comprehensive reviews of that work (Cooper and Albentosa, 2003; Olsson and Keeling, 2005). Comparisons of what we know about what controls behavior and how hard hens are willing to work to get to a nest box or a dust bath indicate that there are some important differences between these behaviors, with implications for hen welfare.

The pre-laying behavior of hens has been studied extensively for over 50 years, beginning with the studies of Wood-Gush and Gilbert, who showed that nesting was stimulated by the release of hormones during ovulation, resulting in an organized sequence of behavior prior to laying the egg on the following day (see Wood-Gush and Gilbert, 1964). Hens begin to show signs of searching for a nest site several hours prior to egg laying, when they increase locomotion and examine potential nesting sites. The searching phase is followed by a period of sitting at the final site where the egg is laid; there, the hen constructs a nest by rotating her body to form a hollow and arranges nesting materials (Duncan and Kite, 1989).

Most hens prefer to lay their eggs in a discrete enclosed nest box, and the strength of hens' motivation to access a nest box has been demonstrated in a variety of ways. Hens have been shown to be willing to squeeze through narrow gaps (Cooper and Appleby, 1996), push open heavily weighted doors (Follensbee *et al.*, 1992), and pass through cages occupied by unfamiliar or dominant hens in order to gain access to a nest box (Freire *et al.*, 1997). Cooper and Appleby (2003) showed that hens' work rate (by pushing through a locked door) for a small pen furnished with a wooden nest box at 40 min before egg laying was equal to their work rate to return to their home pen after 4 h of confinement without food, and the work rate to access the nest was double that amount at 20 min prior to egg laying. Hens can be highly persistent in their efforts to get to a nest box. When hens were trained to push through a door to get to a nest box and then were prevented from doing so, light hybrid hens averaged 150 attempts to push through the door in an hour (Follensbee, 1992). Figure 12.6 shows a door apparatus that is for measuring the strength of motivation.

When a nest box is not available, hens are more active, engage in walking for a longer duration before laying their eggs, and often perform what has been described as stereotyped pacing; behavioral differences that have been interpreted as signs of frustration (Wood-Gush and Gilbert, 1969; Yue and Duncan, 2003). Yue and Duncan (2003) found that hens in cages without access to a nest box

Fig. 12.6. A hen is trained to push through this door to measure how much she is willing to work to get to a resource, such as a nest box. After the hen has learned to open the door, increasingly heavy weights are attached to the ends of the horizontal crossbars. The amount of weight that the hen is willing to lift to open the door provides a measure of the strength of her motivation.

spent over 20% of the hour before egg laying walking and pacing, compared to around 7% by hens with a nest box. Nesting in a secluded place is a high-priority behavior that can be accommodated by providing the hen with a nest box. Hens will use the nest box in an enriched colony cage the vast majority of the time, with nest use depending on design (Cook *et al.*, 2011; Hunniford *et al.*, 2014).

Measuring an animal's use of different resources

The type of nest box a hen prefers can be determined by counting the number of eggs she lays in different types of boxes. Hens will avoid a box with a wire floor, and they prefer one with Astro Turf (Riber and Nielsen, 2013). Astro Turf (artificial grass) is difficult to keep clean. Wall and Tauson (2013) have found that artificial grass can be replaced with plastic netting that allows manure to fall through. In commercial nest-box systems, hens have definite preferences for different locations (Riber and Nielsen, 2013).

Another useful resource use test is measuring the amount of time an animal housed in a barren environment will spend with a resource after it gains access. To obtain a more complete picture of behavioral needs, both resource use and strength of motivation tests are required, and this may help tease out how animals prioritize resources, as well as appetitive and consummatory aspects of behavior. For example, Elmore *et al.* (2012) report that access to food (consummatory behavior) is the most highly motivated for the hungry sow, but sows will spend more time manipulating straw (the appetitive behavior). There is also considerable variation between individual animals on the use of a resource. A study with dairy cows showed that some cows made greater use of cooling showers than others (Legrand *et al.*, 2011).

How important is dust bathing?

Domestic chicks begin dust bathing during the first weeks after hatching. The behavior involves a sequence of distinct lying and fluffing and rubbing motor patterns that distribute dust under the feathers. Dust bathing is controlled by an interaction of internal and environmental factors, and the motivational system is quite different from that of nesting. Dust bathing occurs on average every 2 days and follows a diurnal rhythm, with most dust bathing

T. Widowski

occurring in the late morning and early afternoon (Lindberg and Nicol, 1997). A number of environmental factors also affect hens' motivation to dust bathe, including the sight of a dusty substrate (Petherick *et al.*, 1995), ambient temperature, light, and a radiant heat source (Duncan *et al.*, 1998). External factors (like sunlight and peat moss) can be very potent stimuli for dust bathing, and the behavior can essentially be "switched on" under the right conditions.

The willingness of hens to work for a dust-bathing substrate after a period of deprivation has been tested using a variety of operant and obstruction tests, but the results have been highly variable (see reviews by Cooper and Albentosa, 2003; Olsson and Keeling, 2005). Widowski and Duncan (2000) found considerable individual variation in hens' willingness to push through a weighted door to access a dust bath in peat moss. Although the majority of hens tested pushed more weight after they had been deprived of litter, some hens pushed more weight after they had just completed a bout of dust bathing in their home pen. Others pushed through the door and then did not perform any dust bathing. The hens' behavior in the test runway was clearly different in this study than in previous studies with nest boxes. We concluded that these results did not support a "needs" model of motivation for dust bathing, but rather that hens dust bathed when the opportunity presented itself. However, the behavior is likely to be rewarding to the hens.

When birds have been kept without a substrate for dust bathing, they begin dust bathing more quickly and they perform longer and more intense dust bathing bouts when they are finally presented with a substrate. This is often referred to as a rebound effect, and is considered to be evidence of a "build up" of dust-bathing motivation after a period of deprivation (Cooper and Albentosa, 2003). However, any behavior thought to be associated with frustration, such as pacing, head flicking, or displacement preening, has rarely been reported in studies of dust-bathing deprivation. Although the appearance of vacuum dust bathing raises concern, hens in enriched cages supplied with a dust bath are often seen to dust bathe on the wire floors. Olsson and Keeling (2005) argue that "need" versus "opportunity" models for dust bathing are more likely to depend on the internal state of the hen. If a hen is presented with a highly attractive substrate, she may perform the behavior even if she has recently dust bathed. However, for a hen deprived of substrate

for some time, the need to dust bathe will supersede any external factors and she will perform sham dust bathing in the absence of substrate.

The various studies on nesting and dust bathing have led a number of scientists to conclude that lack of a suitable secluded nest box is one of the biggest welfare concerns in conventional cage systems, and that hens may be frustrated when they do not have access to a nest (Duncan, 2001; also see Cooper and Albentosa, 2003; Weeks and Nicol, 2006). Nesting in a secluded place is a hard-wired behavior pattern that helped the survival of chicks in the wild. Dust bathing, on the other hand, seems to be less of a priority for laying hens (see Chapters 1 and 8, this volume). Nesting motivation can be satisfied by providing an enclosed nest box lined with substrate or artificial turf (Struelens *et al.*, 2005, 2008). A variety of models of cages furnished with nest boxes, perches, and scratch mats have been developed that provide opportunities for hens to engage in a variety of behavior and at the same time provide the health and hygiene benefits of cages (Tauson, 2005).

Cognitive bias tests

The most recent approach for determining how animals feel is based on the notion that the way that animals (and humans) process and interpret information depends on whether they are in a positive (optimistic) or negative (pessimistic) emotional state (e.g. is the glass half empty or half full?) (Mendl *et al.*, 2009). There is now a growing body of scientific literature using cognitive bias tests in animals (Douglas *et al.*, 2012; Düpjan *et al.*, 2013; Daros *et al.*, 2014). In a cognitive bias test, the subject learns a simple task such as to press a red video screen to get a reward and not to press a white screen, which provides no reward. After the task is learned, the subject is tested with an ambiguous cue – in this case, a pink screen. If the animal is in a more positive affective state, it will expect a positive outcome and press the ambiguous screen to obtain a reward. If the animal is in a more negative state, it will not respond – interpreting the ambiguous cue as negative and not expecting a reward. Calves in pain from horn disbudding were more likely to judge the ambiguous pink screen as negative and not respond (Neave *et al.*, 2013). A similar negative cognitive bias occurred after calves were separated from their mothers (Daros *et al.*, 2014). Positive cognitive bias can be induced by environmental enrichment for pigs (Douglas *et al.*, 2012).

Cognitive bias tests are more useful for determining if a particular environmental enrichment does more than just preventing abnormal behavior. These tests may help to answer a simple question – Does the animal feel good?

Consequences of Behavioral Deprivation

Although studies of motivation can tell us a great deal about what types of behavior are highly motivated and may be important for animals, much less is known about the physiological and health consequences of depriving them. Some of the best examples of the link between behavioral deprivation and any effects on physiology or the development of abnormal behavior have to do with ingestive behavior. Feeding and drinking are "regulatory" behavior patterns in that they directly impact on body function. Performance of the motor patterns associated with feeding and drinking contribute to feelings of satiety and help signal to the brain that the animal has eaten or had a drink well before digestion has occurred or osmotic balance of the blood is re-established. In production settings, there is often a mismatch between how the animals are fed and their natural feeding behavior.

Cross-sucking problems and roughage preferences in calves

Because their survival depends on it, young mammals are highly motivated to perform sucking behavior. Veal calves and dairy replacement heifers are usually weaned from their dams within a day or so after birth and are fed milk (or milk replacer) from buckets. When calves are kept in groups, they often develop cross-sucking, which involves sucking on the ears, muzzle, tail, prepuce, udder, or scrotum of other calves. Cross-sucking can be detrimental to the health of calves. In some cases, cross-sucking can lead to urine drinking and reduced body weight. In the absence of other calves, sucking is often directed at pen fixtures.

Numerous studies have been done to determine the factors that control sucking behavior in calves as well as the physiological consequences (de Passillé, 2001; also see Rushen *et al.*, 2008). In studies investigating the control of sucking in calves, de Passillé and colleagues (2001) found that cross-sucking within groups of calves occurred most often *after*

calves had ingested a meal, and that cross-sucking was reduced when calves were provided with an artificial teat. Their studies also indicated that sucking behavior was motivated by the taste of milk. Even giving a small amount of milk or milk replacer (5 ml) by syringe in the mouth triggers sucking, and the more concentrated the milk replacer, the more sucking is performed. Increasing the volume of milk at a meal does not reduce the motivation for sucking, although the behavior is not completely independent of hunger. Calves that missed a meal or were fed low levels of milk did increase non-nutritive sucking after the subsequent meal. Sucking motivation diminishes spontaneously within 10 min after feeding, whether or not calves have been able to suck. Also, the *performance* of sucking behavior appears to be linked to satiety signals and digestive function of the calves. Providing an artificial teat after feeding milk from a bucket significantly increased plasma cholecystokinin (CCK) and serum insulin concentrations in the hepatic portal vein compared to calves that could not suck. Calves given a dry teat to suck after a meal also had lower heart rates and rested more. Providing either a dry teat or feeding calves from a bottle with a nipple reduces cross-sucking. Many progressive producers have found that feeding calves with a nipple (teat) bottle has many advantages compared to using buckets. The studies on sucking behavior in calves are a good illustration of the close links between appetitive and consummatory behavior patterns and the physiological systems that they support.

Young calves also have a need for access to roughages. Holstein dairy calves that had access to either chopped lucerne (alfalfa) hay or rye grass hay spent less time performing non-nutritive oral behaviors (Castells *et al.*, 2012). Both the preferences and strength of motivation tests can be used to determine the preferred types of roughages. Webb *et al.* (2014) found that calves preferred long hay instead of chopped. Further studies with older heifers have also clearly indicated that more coarsely chopped straw with larger particle size is preferred (Greter *et al.*, 2013). Together, these studies clearly show that ruminant animals need roughages that they can chew.

Weaning problems in piglets

Weaning age and feeding systems in commercial production can also cause problems for young piglets. Early-weaned piglets often have trouble

T. Widowski

getting on to solid feed, spend a great deal of time at the drinker, and develop belly nosing, a rhythmic massage directed at the bellies or navels of other piglets (see Widowski *et al.*, 2008). The behavior appears very similar to massaging the sow's udder, which piglets use to stimulate milk let down. Belly nosing and belly sucking can cause lesions on the recipient pigs (Straw and Bartlett, 2001) (Fig. 12.7). The incidence of belly-nosing behavior is affected significantly by weaning age, with younger-weaned piglets performing significantly more belly nosing. It appears to be a result of confusion among the control systems for sucking, feeding, and drinking. Providing an artificial teat or artificial udder to massage can reduce belly nosing significantly, as well as time spent at the drinker. Drinker design can also affect belly nosing: piglets with push-bowl drinkers perform significantly less belly nosing than piglets with "nipple" drinkers (Torrey and Widowski, 2004). Even though the name "nipple drinker" implies that the drinker allows for sucking, water actually pours into the piglet's mouth and it does not appear to satisfy sucking motivation. Although we know less about the physiological consequences of sucking (or sucking deprivation) in piglets than in calves, we do know that piglets who develop high levels of belly nosing grow more poorly and, in extreme cases, appear to be wasting. In a field study on commercial farms, we found that even short 2 h video samples of weaned piglets are enough to detect when belly nosing is occurring (Widowski *et al.*, 2003). Many US and Canadian producers have stopped extreme early weaning and have switched back to weaning at

21–28 days. Weaning at a later age will help reduce belly nosing and other abnormal behavior in pigs.

Oral stereotypies in adult ungulates

In adult ungulates, restricted feeding and the lack of opportunities to forage appear to contribute to the development of oral stereotypies (Bergeron *et al.*, 2006). The behavior often peaks around feeding time or shortly thereafter, and factors that increase satiety appear to reduce them. Several hypotheses have been developed to explain why oral stereotypies occur. One hypothesis is that the diets provided simply are not satisfying – the diets are deficient in calories or some other nutrient. In restricted-fed sows, increasing feed allotment or fiber and bulkiness of the diet reduces behavior patterns such as sham chewing and bar biting. Another hypothesis for the development of stereotypies is that in horses and ruminants, the behavior fulfils some function on the gut. For example, tongue rolling in cattle and cribbing in horses has been suggested to result in increased saliva production, which buffers the gut, and some studies do support this. Finally, there is the notion that the lack of opportunities to engage in foraging and processing of feeds (reduced rumination time) can contribute to oral stereotypies. While our understanding of the mechanisms of oral stereotypies is still not complete, it does suggest that more attention needs to be given to matching the way livestock are fed with their natural feeding systems. Providing hay straw or other fibrous roughages can help prevent these abnormal oral behaviors in sows, cows, and horses (Castells *et al.*, 2012).

Putting Outcome Measures of Behavior in Context

Solutions to some welfare problems can be accomplished by assessing behavior on farms. Although behavior observations are often too time-consuming for practical welfare assessment, aggression and some other behavior problems that cause injury can be assessed indirectly. Scoring systems for injuries due to fighting in pigs (Séguin *et al.*, 2006; Turner *et al.*, 2006; Baumgartner, 2007) or feather loss due to aggression and feather pecking in hens (Bilcík and Keeling, 1999) have been shown to be

Fig. 12.7. Belly nosing is abnormal behavior that can cause damage to other pigs. Piglets that are weaned too early will often engage in more belly nosing.

correlated to observations of actual fighting and pecking behavior. Attention to the nature of the injury (where on the body it occurs) or the timing of injury in relation to management can help determine what type of behavior and when the behavior is occurring.

In pigs, for example, injuries and scratches to the shoulders and head were correlated with reciprocal fighting and were more prevalent in aggressive animals, while lesions to the flank or back end of pigs were more prevalent in pigs who were observed being attacked by other pigs (Fig. 12.8) (Turner *et al.*, 2006). Baumgartner (2007) also found that lesions on the head and shoulders were more prevalent in mixed pigs (pigs from different pens who had been fighting) than in non-mixed pigs, while lesions on the tails or ears were not. In group-housing systems for sows, aggression can be a problem. Some fighting is almost inevitable when new groups are mixed, because pigs engage in vigorous fighting for several hours in order to establish their dominance hierarchies. Chronic aggression can also occur if sows are competing for food or space.

Fig. 12.8. An aggressive sow is biting the hindquarters of another sow. Some sows are subordinate and they are more likely to be bullied and attacked by other animals. Sows and pigs with scratches on the rear flank are being attacked by other animals.

Scratch scoring determines if fighting continues after initial mixing

Tracking the timing of lesions relative to mixing may help to determine when aggression is occurring and whether chronic aggression over a long period is a problem. In order to determine the welfare implications of converting part of the gestation barn from stalls to group housing, Séguin *et al.* (2006) scored superficial skin scratches on sows each week for 5 weeks after mixing in a group-housing system (Fig. 12.9). Tracking the percentages of sows with severe scratch scores was most useful at identifying lesions due to fighting at mixing. Over 30% of sows had severe scores the day after mixing, but these decreased over time, and by 3 weeks after mixing only around 5% had scratches. Although sows were fed on the floor (Fig. 12.10), body condition scores were good and did not vary over time. Therefore, scratches were due to fighting at the time of mixing sows in the group and there was no evidence of chronic aggression during feeding. When using lesion scores for assessing aggression and welfare in group-housing systems, it is important to consider the scores in

Fig. 12.9. Sow with severe scratches on the shoulder. Shoulder scratches due to fighting can be scored with a simple four-point scoring system: 0 for no scratches; 1 for less than five scratches; 2 for five to ten scratches; and 3 for more than ten scratches. Scratches on the shoulder are correlated with reciprocal fighting. Aggressive animals will usually have more shoulder scratches.

relation to the severity and the time frame from when the sows are mixed. Practical experience in the industry has shown that larger groups of over 60 sows may fight less (Chapter 14, this volume).

Fig. 12.10. Group-housed sows with a feeding system that drops feed on the floor. This type of feeding system may have increased levels of fighting due to competition for feed. The pens have partial partitions and large floor-space allowances, which possibly have a beneficial effect on providing places to escape. This farm had good management and the sows had larger litters in these pens compared to sow gestation stalls. When new facilities are being built, other types of feeding equipment that reduces competition over food is strongly recommended.

Conclusion

The motivation for different types of behavior can be measured in a very scientific and objective manner. Research clearly shows that some natural behaviors are more highly motivated than others and that when animals are unable to perform them, there can be behavioral, physiological, and welfare consequences. Most researchers agree that these types of highly motivated behaviors should be accommodated when livestock or poultry are housed in intensive systems.

References

Baumgartner, J. (2007) How to deal with complex data of skin lesions in weaner pigs. *Animal Welfare* 16, 165–168.

Bergeron, R., Badnell-Waters, A.J., Lambton, S. and Mason, G. (2006) Stereotypic oral behaviour in captive ungulates: Foraging, diet and gastrointestinal function. In: Mason, G. and Rushen, J. (eds) *Stereotypic Animal Behaviour Fundamentals and Application to Welfare*, 2nd edn. CAB International, Wallingford, UK, pp. 19–57.

Bilcík, B. and Keeling, L.J. (1999) Changes in feather condition in relation to feather pecking and aggressive behaviour in laying hens. *British Poultry Science* 40, 444–451.

Boissy, A., Manteuffel, G., Jensen, M.B., Moe, R.O., Spruijt, B., *et al.* (2007) Assessment of positive emotions in animals to improve their welfare. *Physiology & Behavior* 92, 375–397.

Brambell, R.W.R. (chairman) (1965) *Report of the Technical Committee to Enquire into the Welfare of Animals Kept Under Intensive Livestock Husbandry Systems.* Command Paper 2836. Her Majesty's Stationery Office, London.

Buckner, L.J., Edwards, S.A. and Bruce, J.M. (1998) Behaviour and shelter use by outdoor sows. *Applied Animal Behaviour Science* 57, 69–80.

Castells, L., Bach, A., Araujo, G., Montoro, C. and Terré, M. (2012) Effect of different forage sources on performance and feeding behavior of Holstein calves. *Journal of Dairy Science* 95, 286–293.

Cook, N.J., Schaefer, A.L., Korver, D.R., Haley, D.B., Feddes, J.J.R., *et al.* (2011) Minimally invasive assessments of the behavioural and physiological effects of enriched colony cages in laying hens. *Open Agriculture Journal* 5, 10–18.

Cooper, J.J. and Albentosa, M.J. (2003) Behavioural priorities of laying hens. *Avian and Poultry Biology Reviews* 14, 127–149.

Cooper, J.J. and Appleby, M.C. (1996) Demand for nest boxes in laying hens. *Behavioural Processes* 36, 171–182.

Cooper, J.J. and Appleby, M.C. (2003) The value of environmental resources to domestic hens: A comparison of the work-rate for food and for nests as a function of time. *Animal Welfare* 12, 39–52.

Daros, R.R., Costa, J.H.C., von Keyserlingk, M.A.G., Hötzel, M.J. and Weary, D.M. (2014) Separation from the dam causes negative judgment bias in dairy calves. *PLoS One* 9(5):e98429, doi:10.1371/journal.pone.0098429.

Davis, N.J., Prescott, N.B., Savory, C.J. and Wathes, C.M. (1999) Preferences of growing fowls for different light intensities in relation to age, strain and behaviour. *Animal Welfare* 8, 193–204.

Dawkins, M.S. (1983) Battery hens name their price: consumer demand theory and the measurement of ethological 'needs'. *Animal Behaviour* 31, 1195–1205.

Dawkins, M.S. (1990) From an animal's point of view: motivation, fitness and animal welfare. *Behavioral and Brain Sciences* 13, 1–9.

Dawkins, M.S. (2004) Using behaviour to assess animal welfare. *Animal Welfare* 13, 3–7.

de Haas, E.N., Nielsen, B.L., Buitenhuis, A.J. and Rodenburg, T.B. (2010) Selection on feather pecking affects response to novelty and foraging behaviour in laying hens. *Applied Animal Behaviour Science* 124, 90–96.

de Passillé, A.M. (2001) Sucking motivation and related problems in calves. *Applied Animal Behaviour Science* 72, 175–187.

Dixon, L., Duncan, I.J.H. and Mason, G.L. (2008) What's in a peck? Using fixed action pattern morphology to identify the motivational basis of abnormal feather pecking behaviour. *Animal Behaviour* 76, 1035–1042.

Douglas, C., Bateson, M., Walsh, C., Bedue, A. and Edwards, S.A. (2012) Environmental enrichment induces optimistic cognitive bias in pigs. *Applied Animal Behaviour Science* 139, 65–73.

Duncan, I.J.H. (1996) Animal welfare defined in terms of feelings. *Acta Agriculturae Scandinavica (Section A – Animal Science)* 27 (Supplement), 29–35.

Duncan, I.J.H. (1998) Behavior and behavioral needs. *Poultry Science* 77, 1766–1772.

Duncan, I.J.H. (2001) The pros and cons of cages. *World's Poultry Science Journal* 57, 381–390.

Duncan, I.J.H. and Kite, V.G. (1987) Some investigations into motivation in the domestic fowl. *Applied Animal Behaviour Science* 18, 387–388.

Duncan, I.J.H. and Kite, V.G. (1989) Nest site selection and nest-building behaviour in domestic fowl. *Animal Behaviour* 37, 215–231.

Duncan, I.J.H. and Petherick, J.C. (1991) The implications of cognitive processes for animal welfare. *Journal of Animal Science* 69, 5017–5022.

Duncan, I.J.H., Widowski, T.M., Malleau, A.M., Lindberg, A.C. and Petherick, J.C. (1998) External factors and causation of dustbathing in domestic hens. *Behavioural Processes* 43, 219–228.

Düpjan, S., Ramp, C., Kanitz, E., Tuchscherer, A. and Puppe B. (2013) A design for studies on cognitive bias in the domestic pig. *Journal of Veterinary Behavior: Clinical Applications and Research* 8, 485–489.

Elmore, M.R.P., Garner, J.P., Johnson, A.K., Kirkden, R.D., Paterson-Kane, E.G., *et al.* (2012) Differing results for motivation tests and measures of resource use: The value of environmental enrichment to gestating sows housed in stalls. *Applied Animal Behaviour Science* 141, 9–19.

Follensbee, M. (1992) Quantifying the nesting motivation of domestic hens. MSc thesis, University of Guelph, Guelph, Ontario, Canada.

Follensbee, M.E., Duncan, I.J.H. and Widowski, T.M. (1992) Quantifying nesting motivation of domestic hens. *Journal of Animal Science* 70(1), 164.

Fraser, D. (2003) Assessing animal welfare at the farm and group level: The interplay of science and values. *Animal Welfare* 12, 433–443.

Fraser, D. and Duncan, I.J.H. (1998) 'Pleasures', 'pains' and animal welfare: Toward a natural history of affect. *Animal Welfare* 7, 383–396.

Fraser, D. and Nicol, C.J. (2011) Preferences and motivation research. In: Appleby, M.C., Mench, J.A., Olsson, I.A.S. and Hughes, B.O. (eds) *Animal Welfare*. CAB International, Wallingford, UK, pp. 183–199.

Freire, R., Appleby, M.C. and Hughes, B.O. (1997) Assessment of pre-laying motivation in the domestic hen using social interaction. *Animal Behaviour* 54, 313–319.

Gilbert, C.L., Burne, T.H., Goode, J.A., Murfitt, P.J. and Walton, S.L. (2002) Indomethacin blocks pre-partum nest building behaviour in the pig (*Sus scrofa*): Effects on plasma prostaglandin F metabolite, oxytocin, cortisol and progesterone. *Journal of Endocrinology* 172, 507–517.

Grandin, T., Curtis, S.E., Widowski, T.M. and Thurmon, J.C. (1986) Electro-immobilization versus mechanical restraint in an avoid-avoid choice test for ewes. *Journal of Animal Science* 62, 1469–1480.

Greter, A.M., Prinsen, M., Duffield, T.F., McBride, B.W., Widowski, T.M., *et al.* (2013) Growing dairy heifers prefer supplementary long straw when fed a nutrient dense ration in a limited amount. *Journal of Dairy Science* 96, 3950–3958.

Hughes, B.O. and Duncan, I.J.H. (1988) The notion of ethological 'need'. Models of motivation and animal welfare. *Animal Behaviour* 36, 1696–1707.

Hunniford, M.E., Torrey, S., Bédécarrats, G., Duncan, I.J.H. and Widowski, T.M. (2014) Evidence of competition for nest sites by laying hens in large furnished cages. *Applied Animal Behaviour Science* 161, 95–104.

Jensen, P. (1993) Nest building in domestic sows – the role of external stimuli. *Animal Behaviour* 45, 351–358.

Jensen, P. (2006) Domestication – from behaviour to genes and back again. *Applied Animal Behaviour Science* 97, 3–15.

Jensen, P. and Recén, B. (1989) When to wean – observations from free-ranging domestic pigs. *Applied Animal Behaviour Science* 23, 49–60.

Jensen, P. and Toates, F.M. (1993) Who needs 'behavioural needs'? Motivational aspects of the needs of animals. *Applied Animal Behaviour Science* 37, 161–181.

Jensen, P. and Wright, D. (2014) Behavioral genetics and animal domestication. In: Grandin, T. and Deesing, M.J. (eds) *Genetics and the Behavior of Domestic Animals*, 2nd edn. Academic Press (Elsevier), San Diego, California, pp. 41–80.

Kirkden, R.D. and Pajor, E.A. (2006) Using preference, motivation and aversion tests to ask scientific questions about animals' feelings. *Applied Animal Behaviour Science* 100, 29–47.

LayWel (2006) Available at: www.laywel.eu (accessed 9 May 2015).

Legrand, A., Schültz, K.E. and Tucker, C.B. (2011) Using water to cool cattle: Behavioral and physiological changes associated with voluntary use of cow showers. *Journal of Dairy Science* 94, 3376–3386.

Lindberg, A.C. and Nicol, C.J. (1997) Dustbathing in modified battery cages. Is sham dustbathing an adequate substitute? *Applied Animal Behaviour Science* 55, 113–128.

McCarthy, S., Horan, B., Rath, M., Linnane, M., O'Connor, P., *et al.* (2007) The influence of strain of Holstein-Friesian dairy cow and pasture-based feeding system on grazing behaviour, intake and milk production. *Grass and Forage Science* 62, 13–26.

Makagon, M.M., Tucker, C. and Mench, J.A. (2011) Factors affecting nest choice by Pekin ducks. *Applied Animal Behaviour Science* 129, 121–128.

Mason, G.J. and Latham, N.R. (2004) Can't stop, won't stop: Is stereotypy a reliable indicator of welfare? *Animal Welfare* 13, S57–S69.

Mason, G. and Rushen, J. (2006) *Sterotypic Animal Behaviour Fundamentals and Application to Welfare*, 2nd edn. CAB International, Wallingford, UK, 367 pp.

Mellor, D.J. (2015) Positive animal welfare states and reference standards for welfare assessment. *New Zealand Veterinary Journal* 63, 17–23.

Mendl, M., Burman, O.H.P., Parker, R.M.A. and Paul, E.S. (2009) Cognitive bias as an indicator of animal emotion and welfare: Emerging evidence and underlying mechanisms. *Applied Animal Behaviour Science* 118, 161–181.

Muir, W.M. and Wei Cheng, H. (2013) Genetic influences on the behavior of chickens associated with welfare and productivity. In: Grandin, T. and Deesing, M.J. (eds) *Genetics and the Behavior of Domestic Animals*, 2nd edn. Academic Press (Elsevier) San Diego, California, pp. 317–359.

Murphy, E., Nordquist, R.E. and van der Staay, F.J. (2014) A review of behavioral methods to study emotion and mood in pigs (*Sus scrofa*). *Applied Animal Behaviour Science* 159, 9–28.

Neave, H.W., Daros, R.R., Costa, J.H.C., von Keyserlingk, M.A.G. and Weary, D.M. (2013) Pain and pessimism: Dairy calves exhibit negative judgment bias following hot-iron disbudding. *PLoS One*, doi: 10.1371/journal.pone.0080556.

Nuboer, J.F.W., Coemans, M.A.J.M. and Vos, J.J. (1992) Artificial lighting in poultry houses: Do hens perceive the modulation of fluorescent lamps as flicker? *British Poultry Science* 33, 123–133.

Olsson, I.A.S. and Keeling, L.J. (2005) Why in earth? Dustbathing behaviour in jungle and domestic fowl reviewed from a Tinbergian and animal welfare perspective. *Applied Animal Behaviour Science* 93, 259–282.

Pajor, E.A., Rushen, J. and de Passillé, A.M. (2000) Aversion learning techniques to evaluate dairy cattle handling practices. *Applied Animal Behaviour Science* 69, 89–102.

Pajor, E.A., Rushen, J. and de Passillé, A.M. (2003) Dairy cattle's choice of handling treatments in a Y-maze. *Applied Animal Behaviour Science* 80, 93–107.

Petherick, J.C., Seawright, E., Waddington, D., Duncan, I.J.H. and Murphy, L.B. (1995) The role of perception in the causation of dustbathing behaviour in domestic fowl. *Animal Behaviour* 49, 1521–1530.

Phillips, P.A., Fraser, D. and Pawluczuk, B. (2000) Floor temperature preference of sows at farrowing. *Applied Animal Behaviour Science* 67, 59–65.

Price, E.O. (2003) *Animal Domestication and Behavior*. CAB International, Wallingford, UK, 297 pp.

Raj, A.B.M. and Gregory, N.G. (1995) Welfare implication of the gas stunning of pigs 1. Determination of aversion to the initial inhalation of carbon dioxide or argon. *Animal Welfare* 4, 273–280.

Reimers, E., Røed, K.H. and Colman, J.E. (2012) Persistence of vigilance and flight response behaviour in wild reindeer with varying domestic ancestry. *Journal of Evolutionary Biology* 25, 1543–1554.

Riber, A.B. and Nielsen, B.L. (2013) Changes in position and quality of preferred nest box: Effects on nest box use by laying hens. *Applied Animal Behaviour Science* 148, 185–191.

Ringgenberg, N., Fröhlich, E.K.F., Harlander-Matauschek, A., Würbel, H. and Roth, B.A. (2014) Does nest size matter to laying hens? *Applied Animal Behaviour Science* 155, 66–73.

Rushen, J. (1996) Using aversion learning techniques to assess the mental state, suffering, and welfare of farm animals. *Journal of Animal Science* 74, 1990–1995.

Rushen, J., de Passillé, A.M., von Keyserlingk, M.A.G. and Weary, D.M. (2008) *The Welfare of Cattle*. Springer, Dordrecht, the Netherlands, 310 pp.

Schneider, J.E., Wise, J.D., Benton, N.A., Brozek, J.M. and Keen-Rhinehart, E. (2013) When do we eat? Ingestive behavior, survival, and reproductive success. *Hormones and Behavior* 64, 702–728.

Schwartzkopf-Genswein, K.S., Stookey, J.M., Crowe, T.G. and Genswein, B.M. (1998) Comparison of image analysis, exertion force, and behavior measurements for use in the assessment of beef cattle responses to hot-iron and freeze-branding. *Journal of Animal Science* 76, 972–979.

Séguin, M.J., Barney, D. and Widowski, T.M. (2006) Assessment of a group-housing system for gestating sows: Effects of space allowance and pen size on the incidence of superficial skin lesions, changes in body condition and farrowing performance. *Swine Health and Production* 14, 89–96.

Sherwin, C.M. (1999) Domestic turkeys are not averse to compact fluorescent lighting. *Applied Animal Behaviour Science* 64, 47–55.

Shütz, K.E. and Jensen, P. (2001) Effects of resource allocation on behavioural strategies: A comparison of red junglefowl (*Gallus gallus*) and two domesticated breeds of poultry. *Ethology* 107, 753–765.

Shütz, K.E., Forkman, B. and Jensen, P. (2001) Domestication effects on foraging strategy, social behaviour and different fear responses: A comparison between the red jungle fowl (*Gallus gallus*) and a modern layer strain. *Applied Animal Behaviour Science* 74, 1–14.

Spinka, M. (2006) How important is natural behaviour in animal farming systems? *Applied Animal Behaviour Science* 100, 117–128.

Straw, B.E. and Bartlett, P. (2001) Flank or belly nosing in weaned pigs. *Journal of Swine Health and Production* 9, 19–23.

Struelens, E., Tuyttens, F.A.M., Janssen, A., Leroy, T., Audoorn, L., *et al.* (2005) Design of laying nests in furnished cages: Influence of nesting material, nest box position and seclusion. *British Poultry Science* 46, 9–15.

Struelens, E., Van Nuffel, A., Tuyttens, F.A.M., Audoorn, L., Vranken, E., *et al.* (2008) Influence of nest seclusion and nesting material on pre-laying behaviour of laying hens. *Applied Animal Behaviour Science* 112, 106–119.

Tauson, R. (2005) Management and housing systems for layers – effects on welfare and production. *World's Poultry Science Journal* 61, 477–490.

Torrey, S. and Widowski, T.M. (2004) Effect of drinker type and sound stimuli on early-weaned pig performance and behaviour. *Journal of Animal Science* 82, 2105–2114.

Tucker, C.B., Weary, D.M. and Fraser, D. (2003) Effects of three types of free stall surfaces on preferences and stall usage by dairy cows. *Journal of Dairy Science* 86, 521–529.

Turner, S.P., Farnworth, M.J., White, I.M.S., Brotherstone, S., Mendl, M., *et al.* (2006) The accumulation of skin lesions and their use as a predictor of individual aggressiveness in pigs. *Applied Animal Behaviour Science* 96, 245–259.

Wall, H. and Tauson, R. (2013) Nest lining in small group furnished cages for laying hens. *Journal of Applied Poultry Research* 22, 474–484.

Wathes, C.M., Jones, J.B., Kristensen, H.H., Jones, E.K.M. and Webster, A.J.F. (2002) Aversion of pigs and domestic fowl to atmospheric ammonia. *Transactions of the American Society of Agricultural Engineers* 45, 1605–1610.

Webb, L.E., Jensen, M.B., Engel, B., van Reenin, C.G., Gerrits, W.J.J., *et al.* (2014) Chipped or long roughage: What do calves prefer? Using cross point analysis of double demand functions. *PLoS One* 9(2):e88778, doi:10:1371/journal.pone 0088778.

Weeks, C.A. and Nicol, C.J. (2006) Behavioural needs, priorities and preferences of laying hens. *World's Poultry Science Journal* 62, 296–307.

Welfare Quality (2009) www.welfarequalitynetwork.net

Widowski, T.M. and Curtis, S.E. (1990) The influence of straw, cloth tassel, or both on the prepartum behavior of sows. *Applied Animal Behaviour Science* 27, 53–71.

Widowski, T.M. and Duncan, I.J.H. (2000) Working for a dust-bath: Are hens increasing pleasure rather then reducing suffering? *Applied Animal Behaviour Science* 68, 39–53.

Widowski, T.M., Curtis, S.E., Dziuk, P.J., Wagner, W.C. and Sherwood, O.D. (1990) Behavioral and endocrine responses of sows to prostaglandin F2alpha and cloprostenol. *Biology of Reproduction* 43, 290–297.

Widowski, T.M., Keeling, L.J. and Duncan, I.J.H. (1992) The preferences of hens for compact fluorescent over incandescent lighting. *Canadian Veterinary Journal* 72, 203–211.

Widowski, T.M., Cottrell, T., Dewey, C.E. and Friendship, R.M. (2003) Observations of piglet-directed behaviour patterns and skin lesions in eleven commercial swine herds. *Swine Health and Production* 11, 181–185.

Widowski, T.M., Torrey, S., Bench, C.J. and Gonyou, H.W. (2008) Development of ingestive behaviour and the relationship to belly nosing in early-weaned piglets. *Applied Animal Behaviour Science* 110, 109–127.

Wood-Gush, D.G.M. and Duncan, I.J.H. (1976) Some behavioural observations on domestic fowl in the wild. *Applied Animal Ethology* 2, 255–260.

Wood-Gush, D.G.M. and Gilbert, A.B. (1964) The control of the nesting behavior of the domestic hen, II. The role of the ovary. *Animal Behaviour* 12, 451–453.

Wood-Gush, D.G.M. and Gilbert, A.B. (1969) Observations on the laying behaviour of hens in battery cages. *British Poultry Science* 10, 29–36.

Yue, S. and Duncan, I.J.H. (2003) Frustrated nesting behaviour: Relation to extra-cuticular shell calcium and bone strength in White Leghorn hens. *British Poultry Science* 44, 175–181.

13 Animal Well-being on Organic Farms

HUBERT J. KARREMAN[1] AND WENDY FULWIDER[2]

[1]Rodale Institute, Kutztown, Pennsylvania, USA; [2]Global Animal Partnership, Alexandria, Virginia, USA

Summary

Organic livestock rearing uses common sense that is also applicable to conventional farms. By providing the building blocks to health, via clean, dry bedding, housing, genetic selection for disease resistance, choice of whether to be inside or outside, feeds appropriate to species digestive processes, and managed pastureland, both the health and well-being of farm animals can be achieved. In the relatively rare instances of illness, there are many tools from which to choose. This can both stimulate and free a clinician from the false sense that antibiotics must always be used to treat infections. If more clinicians used alternatives to antibiotics, there would be less resistance, which is also good for society. Many principles used in organic agriculture can also help improve biodiversity. A basic principle of organic animal production is to eliminate or greatly reduce dependence on antibiotics and other synthetic inputs. Animals are allowed to perform their natural behaviors and they have direct outdoor contact with the land. Successful organic farming requires managers to look at the big picture of how their farming system interacts with the environment. It is important to use breeds of livestock and poultry that are hardy and can resist disease. Some genetic lines of high-producing animals are not recommended for organic systems with lower inputs. Organic requirements such as grazing on pasture can improve animal welfare by reducing lameness, but excellent management is required to control disease and parasites.

Learning Objectives

- Animal welfare requirements for organic farms
- How to select genetic lines of animals that are suitable for organic farming
- Grazing more than one species on a pasture
- Understand the differences between EU, US, and Canadian organic programs
- The importance of using natural medical treatments and reducing or eliminating antibiotics

Introduction

As an organic veterinary practitioner (HK), it is evident that animal health and welfare are closely linked, if not inseparable. The two issues of health and animal welfare make a difference in the lives of farm animals. In organic agriculture, factors that contribute to health and welfare are often synergistic, and teasing them apart is difficult. An organic farmer looks at "the big picture" when managing his or her farming operation.

An ideal farm with "big picture" management is both profitable and enhances many aspects of life. In this light, the definition of an ideal farm is: (i) that it is biologically alive and self-sufficient; (ii) has biodiversity within the farm ecosystem; (iii) has soils of high mineral content to provide high-quality feeds; (iv) has management that promotes and maintains healthy animals; (v) creates a nutrient-dense, high-quality finished product; and (vi) has positive interaction with the local community. Such a farm will enhance many different aspects of life, both on a macroscopic level (landscape and animals) as well as a microscopic level (soil life and water quality). Many farms can achieve this, whether it is organic, biodynamic, biological, regenerative, sustainable, conventional, or whatever terminology is applicable. Animal health and animal welfare are merged seamlessly to create animal well-being.

Definition of Animal Welfare on an Organic Farm

One definition of animal welfare is that animals are experiencing a good life if they are producing their maximum possible (milk, eggs, rate of gain, etc.). This definition is held mainly by agricultural economists. It focuses solely on output and is putting the proverbial cart before the horse. Instead, it may be more logical to emphasize the input factors that directly affect the lives of the animals, which then go on to create the products. To maintain a high level of both welfare and production will require conscious application of high management skills (Sundrum, 2012). See Chapter 1, this volume, for more discussion on definitions of animal welfare.

A farm animal could be experiencing a good and healthy life when certain basic factors are provided to satisfy its biological needs. The following factors are necessary:

1. Ample clean and dry bedding since dampness is an initiating factor in many diseases. To avoid skin sores, the bedding should provide a soft cushion for the animals to lie on. Bedding should not be allowed to become damp or soiled.
2. Provide appropriate housing and proper ventilation for periods of inclement weather while avoiding drafts at ground level. Ventilation is essential to prevent high ammonia levels.
3. Animals should *always* have the choice of whether or not to be outdoors to enjoy fresh air and direct sunlight.
4. Land should be managed in a rotational, strip grazing method in order to provide scheduled fresh foraging while simultaneously resting previously grazed land.
5. Provide a variety of feedstuffs proper for the species' digestive processes for biologically robust digestion, a critical component to long-term health. For ruminants this would be a forage-based diet of grasses and legumes (fresh, dry or ensiled) with relatively low grain input (if any). For monogastric animals such as pigs and poultry, this would be a variety of feeds that are based more on grain/starches/seeds, with a choice to forage freely.

Caring for farm animals in a manner that enhances their well-being by providing the building blocks for health is not rocket science; it is actually back to basics. The preceding five factors that satisfy biological needs apply to any type of farm, whether managed organically or conventionally. These factors appear to have a true impact on animal well-being, as shown by a dramatic reduction in calls for veterinary services when farms become organic (private clinic business records of HK). Moreover, organic regulations in various countries specifically *require* certain minimum dietary, housing, and pasture requirements appropriate to the various farm animal species (USDA NOP a,b,c,d; EU Council Regulation, 2012; CAN/CGSB) (Quick Access Information 13.1). In conventional farming, housing is constructed for maximal efficiency of workers, and feed is formulated and fed for maximum production. Certain levels of various maladies are acceptable in conventional settings because they can be overcome easily by various "production tools", oftentimes antibiotics, hormones, and parasiticides, i.e. to grow as many young stock as quickly as possible. Use of antibiotics, beta-agonists, and hormones is forbidden by US organic regulations. In Europe, hormones are banned but limited use of antibiotics is allowed (Quick Access Information 13.1). Therefore, the conscientious prevention of problems is actively employed on well-managed organic farms. This can usually be achieved by having a lower density of animals on the farm and by providing the five basic building blocks to health and well-being already outlined. A goal of most organic farms is to produce food predominantly from within the farm's soil, pasture, cropland, and water resource base rather than "buying" production by purchasing feed, fertilizer, and other inputs from the open market.

Natural Behavior is Important

It is easier to provide and maintain robust animal well-being if management methods approximate the natural biological cycles of farm animals' cousins that live in the wild. For example, allowing dairy calves to suck directly from their dams allows calves to satisfy their innate urge to drink milk frequently, as deer and elk (and beef calves) do. Frequent small sucking bouts provide a steady influx of nutrients while simultaneously allowing calves to learn by watching their mothers, i.e. which plants to eat and which plants to avoid (Provenza and Balph, 1987). This information becomes imprinted on their memories via the taste/olfactory system for use later in life (Galef *et al.*, 1994). Learning to graze very early in life is the natural behavior of wild ruminants. This may help develop a body conformation that is more suitable for adult grazing than calves reared indoors. Calves on pasture with their mothers will

Quick Access Information 13.1. Organic standards for medications, hormones, and feed additives.

	USA	Canada	European Union
Antibiotics	None Animals given antibiotics are removed from organic production	Limited use, 30-day minimum withdrawal time or twice the label withholding if longer than 30 days (refer to standard) Not for growth promotion	Allowed with twice the withdrawal time Not for growth promotion
Chemical (synthetic) Allopathic drugs	See list of allowed synthetic drugs See 7CFR205.603	Allowed under veterinary supervision Restrictions on slaughter animals	Allowed under veterinary supervision
Hormones	Not for growth promotion Oxytocin allowed postpartum emergencies and epinephrine	Not for growth promotion Allowed for therapeutic reasons under veterinary supervision	Not for growth promotion Allowed for therapeutic treatment Not for reproductive induction
Beta-agonists	Banned	Banned	Banned
Feed requirement	Total ration from organic sources certified by NOP See 205.236 (a)(2)(i) for allowed synthetic substances	Feedstuffs produced in accordance with organic standards	Portion of the ration from nonorganic production allowed (see standard)
Vaccinations	Allowed No GMOs	Allowed No GMOs	Allowed No GMOs
Pain relief sedatives	Allowed Xylazine, Butorphanol Flunixin Lidocaine Procaine Aspirin	Allowed under veterinary supervision for physical alterations	Allowed under veterinary supervision
Parasiticides	Prohibited in slaughter stock Allowed in breeding stock with conditions 90-day withdrawal for dairy products (see standards)	Allowed twice the withdrawal times See standard for other restrictions	Allowed See standards for restrictions

chew their cud very early in life, often within 2 weeks of being born. This is biologically relevant because developing the rumen to process fibrous feeds is the goal of sheep, goat, or cattle farmers. Chewing the cud is proof positive of a functioning rumen. In conventional systems, calves are usually separated from their mothers immediately after birth, fed milk replacer and grain twice daily, and offered hay only *after* they are weaned at 6–8 weeks of age. In feeding systems that deny young ruminant animals forage, it is usual to see calves nibbling at their bedding in a desperate attempt to satisfy their instinct to eat fibrous material. These animals have an increased likelihood of ingesting pathogens such as bacteria and parasites. The use of milk replacer and feeds fortified with antibiotics and coccidiostats reduces illness in the conventional farming sector. Raising dairy calves on their dams is increasing in organic dairy farming, but most organic dairy farms still separate calves from cows. See Chapters 8 and 12, this volume, for further information on behavioral needs. However, organic dairy farms usually feed hay and whole milk, thus decreasing the pathogen load in animals that consume fibrous materials by nibbling at bedding. This still approximates Mother Nature more closely than the conventional method of medicated milk replacer and medicated grain feeding to calves.

CowSignals® and Assessing Animal Welfare

A group of Dutch veterinarians have successfully advanced a production-based animal well-being program called CowSignals (Hulsen, 2005; www.cowsignals.com). The CowSignals® method of assessing cattle well-being considers six factors that enhance health or, if lacking, are obstacles to health. In CowSignals®, the concept is based on the six freedoms provided by pasture. They are: (i) feed; (ii) water; (iii) light; (iv) fresh air; (v) adequate space; and (vi) comfortable rest. The overall paradigm is to consider what cows will experience on pasture and provide very similar conditions while they are in a barn. By doing this, dairy cows will lead very comfortable longer lives and thus produce effortlessly. The program focuses on direct observations and comparing them to the pasture ideal, becoming aware of the reasons for what is observed and then taking action to remove anything that is an obstacle to health, and therefore production. Cows can have a great quality of life on any kind of farm (organic or conventional), and this can be assessed easily by the CowSignals® method. Rather than penalizing farmers for problems, the program allows farmers to understand the problems in terms of affecting milk production. This approach makes the program very palatable to farmers. CowSignals unquestionably enhances animal welfare while presented in a farmer-friendly way. It is a clever way to make farmers aware of needed changes for better animal welfare. Significantly, most changes are simple and inexpensive. Animal welfare should also be evaluated using assessment tools such as scoring lameness, body condition, sickness, and other measures described in Chapters 1, 2, and 4, this volume.

Indeed, observation and adapting to what the animals are telling us is already accepted as standard operating procedure within organic livestock production. For instance, adjusting feed by visually noting the amount of pasture eaten and comparing to how much the cows are milking while observing rumen fill (Burfeind et al., 2010) and manure composition. This is in contrast to total control when animals live in confinement, geared to ensure continual maximum output alone with little regard for the animal, i.e. gestation stalls for swine, battery cage systems for poultry, or dairy cows living continuously on concrete, fed homogeneous mixed rations at the feed bunk, and only walking as far as the milking parlor.

Animals Must Have Direct Contact with the Land

Organic regulations differ with respect to geography, animal species, and type of land (pasture, savannah, ley, woodland). All organic livestock in US, EU, and Canadian programs are *guaranteed* direct contact with land (USDA, NOP 7CFR205.230(1)) (Quick Access Information 13.2). This is in stark contrast with modern conventional agriculture, which separates animals from the land by keeping them inside. They can no longer live the way in which they have developed over the eons. Instead of allowing the animals to self-harvest at least part of their feed and randomly distribute their own manure, conventional agriculture will bring all the feed to the animals and then haul away all of the manure in very prescribed allotments. Numerous studies show that cattle that have access to pasture have the benefits of less lameness, death losses, pneumonia, and digital dermatitis (Burow et al., 2011; Haufe et al., 2012; Richert et al., 2012; Fabian et al., 2014).

Regardless of species or breed, there have been many improvements in animal comfort in confinement systems. But any species of animals kept confined indoors is denied the ability to truly express deeply imbedded, "hard-wired", biological activities. The natural eating behavior of cattle and sheep is grazing, goats graze and browse, pigs root and wallow, and poultry scratch and peck. Most of these activities require direct contact between animals and land. By denying direct contact with the land, indoor confinement systems perpetually hinder true animal welfare.

Biodiversity and sustainability land management

There is evidence that well-managed organic farming can increase biodiversity (Maeder et al., 2002; Bengtsson et al., 2005; Tuck et al., 2014). Grazing can improve pasture when it is managed correctly and is bad for the environment if poorly managed (Franzluebbers et al., 2012). Soil organic carbon (SOC) is a key indicator of good grazing practices. Grazing management can be complex, and a practice that works well in one location may work poorly in another (Teague *et al.*, 2013). Many people do not know that animal learning can have a large effect on the plants ruminants will choose to graze (Provenza and Balph, 1987). Continuous light stocking of a pasture is bad because it trains generations of animals to "eat the best and leave the rest" (Provenza *et al.*, 2003). Higher stocking rates and frequent pasture rotation encourage animals to eat a wider variety of plants. When this method is used, pastures must be rested for a sufficient period to allow full recovery.

A common question that is often asked is how can organic agriculture be sustainable since animal productivity is reduced? Output is reduced, but there is also a huge reduction in inputs. Dairy cattle on conventional farms are fed twice as much grain (Stiglbauer *et al.*, 2013). Organic crops have 20% lower production, but fertilizer and energy are reduced 34% and 53%, respectively, and pesticides are reduced 97% (Maeder *et al.*, 2002), but the amount of land required would be increased.

Genetic Selection of Appropriate Animals for Organic Agriculture

Keeping animals in conventional farming is very uniform and homogeneous from farm to farm – a "one size fits all" type of approach. Farms look essentially the same, cattle at a feed bunk under a roof, a set of white hutches where baby calves are raised, and open cropland surrounding the main barn area. Livestock on conventional farms tend to be from a limited set of breeds (i.e. black and white

Holstein dairy cows, hybrids of white Cornish broiler poultry, and pink Yorkshire swine). But animals genetically bred for high production also need intensive management and extra inputs in order to perform well. Some animals, such as the hybrid Cornish crosses, have been bred to gain weight quickly at the expense of using their limbs to move about. Such birds are not suited to being outside on pasture. Organic farmers need to choose strong animals with good legs for walking as a main trait over high production alone. Some farmers will cross-breed experimentally to see how the offspring perform. While lower-producing animals do not produce as much per unit of animal, traditional/heritage breeds and cross-breeds usually need less intense management. This is probably due to better innate disease resistance and genetics more suitable for pasture-based production. Some dairy breeds such as Milking Shorthorn, Lineback, and Normandie (to name a few) are better at maintaining body condition with less grain. Holsteins crossed with these breeds are fairly common on organic dairy farms in the USA. Purebred Heritage breeds (those breeds that are endangered) tend to be well suited to traditional farming systems that include pasture. It is not uncommon to see a farm that has become organic to change slowly to more hardy genetics, and increase use of pasture. When animals are out on pasture, it is likely that a farm is organic.

Research on genetics, disease, and health

Selection of animals for disease resistance and providing proper conditions to prevent disease is important

---

for all systems of animal keeping. Jones and Berk (2012) and Sutherland *et al.* (2013) maintain that hardy breeds should be used for organic production. This is especially important for broiler chickens. Some genetic lines of rapidly growing animals are not recommended for organic programs. Both Sorensen (2012) and Sophia (2014) recommend the use of slower-growing broiler chickens that gain 40–50 g/day. Ricke *et al.* (2012) contains more information on suitable genetics for meat chicken production.

Slower-growing birds require less nutritional inputs in order to thrive. Strains of birds that grow rapidly have higher methionine requirements, and it is difficult to supply their needs with all natural diets (Jacob, 2013). Birds deficient in methionine are more likely to have poor feather growth and more feather pecking. Synthetic methionine is allowed in US organic programs but the European Union (EU) forbids it (Hungerford, 2005). The use of synthetic methionine is controversial, and its utilization facilitates the use of faster-growing broilers. Jankowski *et al.* (2014) review methionine requirements.

In the UK, a survey of 206 dairy farms indicated that the Holstein Friesian breed had more lameness than all other dairy breeds observed (Barker *et al.*, 2010). Stear *et al.* (2012) contains an extensive review of breeding for disease resistance. Breeding for parasite resistance is also important. Several researchers have found that parasites are more likely to be a problem on organic farms unless pastures are rotated carefully (Hovi *et al.*, 2003; Lindgren *et al.*, 2014). Parasites are a bigger problem on outdoor pig and poultry operations (Sutherland *et al.*, 2013). Any animal should do well with ample dry bedding, fresh air, well-managed pastures, high-forage diets (ruminants) or a variety of dietary inputs, direct access to sunshine, and appropriate housing for inclement weather. Animals given the choice to be inside or outside will experience the least health problems since they can get to fresh air and/or come in from bad weather. This is contrasted by animals bred specifically for high production or rates of gain, which tend to fare better in confinement, where many more inputs in their lives can be controlled.

Integrated Animal and Grazing Systems

Diversified animal species such as cows, pigs, and chickens are more likely on an organic farm. The longer farmers are organic, the more likely it is to see different groups of animal species integrated into the grazing operation. For instance, cattle may go through pasture eating swaths of fresh grass, legumes, and weeds. Placing horses or mules into the paddock after cows will level any uneaten vegetation. The horses should be in the paddock no longer than 12 h to avoid overgrazing the vegetation, which horses are known to do. Interestingly, horses will graze fairly close to freshly deposited cow manure. Poultry following cattle or horses in a pasture will scratch and refresh the soil surface by fluffing up areas that may have become compacted. Pigs following the chickens will then act as biological plows by rooting with their very strong and muscular snouts, thus aerating the soil at lower depths. Followed by proper rest time, vibrant pasture regrows and will be ready in a few weeks for the cattle to begin the whole grazing cycle again.

Having both mammals and poultry on pasture has health benefits. Poultry that follow cows on pasture can help reduce parasite build-up because chickens will naturally peck and scratch apart manure paddies in search of insects. This mechanical action to find insects also helps break down the manure paddy faster by exposing it to the sun and drying it out more quickly. Alternatively, pigs can follow cows as they are naturally inclined to root through cow manure and eat it in order to consume any partially undigested grains. Chickens can then follow after the pigs in order to reduce parasites by pecking through manure paddies to get insects, while also exposing the inside of the manure to sun and wind, thereby desiccating parasite eggs (internal eggs deposited or fly eggs laid). Mammals followed by poultry is a symbiotic relationship, similar to wild birds pecking at ticks and flies among large game animals in the wild. While parasites may still be present, this natural cycle will help keep parasite pressures low. This illustrates how mimicking natural biology tends to keep animals in a healthy state, even though there may be some disease pressure or challenge in their environment.

Grazing different species

Integrated pasture-based livestock production allows livestock and poultry the ability to express their natural behaviors to the fullest extent while enhancing soil health. An extension of sequential grazing is co-grazing, where different species graze at the same time in the same paddock. For instance,

H.J. Karreman and W. Fulwider

grazing cows and sheep is possible and complementary since cows will wrap their tongues around a bunch of herbage while sheep will nibble closer to the ground. What the cows do not eat or leave behind, the sheep may still graze. The same holds for cows and horses, since horses have both upper and lower front teeth and will nibble to the ground like sheep tend to. By co-grazing, more use will be made from pasture at that moment, and potentially longer rest periods are achievable for pasture regrowth. Taking into account appropriate fencing needs for the various species is obligatory.

Biosecurity to Protect Organic Animals Against Disease

To maintain both health and animal welfare, managers must prevent diseases from being introduced into the farm by new breeding stock. New breeding cattle, sheep, swine, and other livestock should be quarantined on a remote part of the farm away from the other animals for 30 days. Double fencing that prevents nose-to-nose contact will block the spread of some, but not all, diseases. Visitors who have been on other farms should be provided with boots. Since some organic farms mix different species, care must be taken to avoid spreading diseases between species.

One of the worst diseases from a welfare standpoint is Aujeszky's disease, also called pseudorabies or mad itch. Pseudorabies is not rabies. This disease has been eradicated in the US commercial swine herd but it still exists in feral (wild) swine (Cramer *et al.*, 2011). It is endemic in many other countries. Infected swine can pass Aujeszky's disease to cattle (Thawley *et al.*, 1980). Cattle get a horrible form of the disease where they scratch and mutilate their heads due to intense itching. Further information can be obtained (California Department of Agriculture, 2014, and Iowa State University, 2014). Consult veterinary specialists in your country for recommendations and vaccination protocols for all diseases in your area.

Physical Alterations of Animals

Organic regulations worldwide *require* that management provides animals with resources that fit their natural biology. The regulations prohibit certain physical alterations that would prevent animals from expressing their natural behaviors.

Physical alterations that are banned are nose rings in pigs, docking tails of cows and pigs, mulesing of sheep, and beak trimming of poultry. Some physical alterations such as dehorning, disbudding, and castration require proper anesthesia to be used (USDA NOP 7CFR238(a)(5)). See Chapter 6, this volume, on methods of pain relief after disbudding and castration. Before using medications, refer to the legislation and regulations in your country. However, if an animal injures its tail and it needs amputation, it is allowed (with proper anesthesia). Organic regulations in all countries require animals have access to the outdoors and minimum space requirements, as well as providing enough stalls for every animal. The preceding discussion of animal management on organic farms describes the general daily life of animals on these farms. Regardless of farm management systems, animals should have their biological and behavioral needs met every day.

Antibiotics and Organics

Why are people willing to pay a premium for organic products? The main reason is consumer perception of a wholesome product. This is based on certain rules regarding organic farming, such as any field spray must be environmentally friendly (USC6518, Section 2119(m)). Consumers also know that the biology of the land is respected. The animals are required to be on the land and able to express natural behaviors. Additionally, organic consumers know that the animals have not received growth hormones or antibiotics in their feed. In the USA, consumers are very touchy about the issue of antibiotic or hormone use. Consumer perceptions and expectations are a major factor of the high demand for organic products. US consumers expressed their opinions about organic farming via the public comment process to the National Organic Standard Board. These consumer comments helped determine which materials could and could not be used in US organic farming. In the USA, consumers want antibiotics removed. This is a controversial issue as products move across the globe.

As a US veterinary practitioner (HK) working in the certified organic livestock sector since 1995, the ability to use antibiotics to treat illness in organic livestock has never been easy. The use of antibiotics in a USDA certified organic herd, per the National Organic Program (NOP) Rule, is prohibited and it requires permanent removal of the animal from organic production (USDA, NOP 7USFR205.238(c)(7)).

After the legal withdrawal time, the animal can be sold to a conventional farm to live out its life or be sent to slaughter. The ban on antibiotics has led to the development of natural methods of treating common farm animal diseases in the NOP certified organic sector. Let us explore the reason for the complete prohibition of antibiotics in US organics, and the unintended positive consequences of this prohibition.

Reasons for US antibiotic prohibition

There are many reasons for the complete antibiotic prohibition in the USA. One major reason is the perception (correct or not) by consumers that antibiotics have been overused or abused in the conventional sector. The impression at the time of the debate (mid-1990s) about US organic standards was that prohibiting antibiotics for organic livestock would create a stronger demarcation between conventional and organic. And indeed it has, just as the pasture requirement has. During the national discussion, the Food and Drug Administration (FDA) forced the National Organic Standards Board (NOSB) to make a decision as to whether antibiotics would be allowed with a simple requirement to double the withholding time or to ban them completely. These were the two choices the federal government was proposing to the young organic community. There is reason to believe that the FDA was hoping the NOSB would take the "no antibiotics at all" stance as the FDA thought it would simply be a matter of time until the organic livestock sector would collapse. The NOSB eventually did elect for a complete ban on antibiotics in organic livestock. This was due mainly to the comments submitted from both the public and the organic farmers, who wanted complete prohibition.

Europe and Canada allow antibiotics

The EU, on the other hand, has always allowed the use of antibiotics, with a penalty time of twice the withholding time and a maximum of 2–3 antibiotic treatments per lactation (dairy cows). In animals living less than a year, two or three treatments are allowed.

Canada has chosen a middle path between the NOP and EU standards in regards to antibiotic use. While the majority of Canadian regulations quote directly from both the NOP and the EU, when it comes to antibiotics they have taken a novel approach: twice the withholding time or 30 days withholding time, whichever is greater. Since most antibiotics, especially for lactating dairy cattle, have withholding times of only 36–72 h for conventional use, the 30-day withholding time would make a dairy farmer in Canada think long and hard before using an antibiotic on a lactating cow. The Canadian standards would make it possible to use antibiotics to save a sick baby calf, and the grown animal could still be used for Canadian organic production. Baby dairy calves are a major welfare problem area on poorly managed organic dairies. At one US organic dairy with sloppy practices, 30% of newborn Holstein dairy calves died (T. Grandin, 2010, Colorado State University, personal communication).

Differences between US and European organic systems

In Europe, there is greater emphasis on natural living, and in the USA, there is more emphasis on no antibiotics and avoiding contaminated food. People in Europe are also more concerned about genetically modified organisms (GMOs), but within the past few years, American consumers have become more aware of this issue. Quick Access Information 13.1 outlines the differences between US, Canadian, and European organic standards. The US standard is much stricter, requiring all pasture, hay, and feed to be grown on organically certified land without artificial pesticides or fertilizers. Since this is very expensive, many "natural" programs have been developed in the USA and other countries. They allow conventional feeds to be used, but the use of antibiotics, growth-promoting hormones, and beta-agonists is forbidden. For further information, refer to legislation in your country and the standards that are required by the market to which the animals are being sold.

Alternative Medications for Organic Animals

Strict penalties for using antibiotics and other prohibited materials give farmers incentives to try natural alternatives to antibiotics. Strict rules stimulate caregivers and veterinarians to think "outside the box". It is unfortunate but true that veterinarians are reluctant to steer away from antibiotic and hormone therapies learned in veterinary school. Veterinarians already carry many supplies that are permitted in organic animals to relieve acute pain and suffering immediately. They are dextrose, calcium, hypertonic saline, lactated ringers solution, sodium iodide, injectable vitamins and minerals,

H.J. Karreman and W. Fulwider

flunixin, xylazine, butorphanol, and lidocaine. Naturally occurring biologics such as specific antibodies and vaccines (if not genetically engineered) are allowed. Parasiticides are allowed in all organic programs for breeding animals.

There are many good textbooks on natural treatments that are allowed in US organic and other programs that forbid antibiotics. Before using any treatments, consult with your local veterinarian, the legislation in your country, and the standards of the marketing program your herd or flock is participating in. The following books can be obtained from major academic publishers – Karreman (2004), Vaarst *et al.* (2006), Blair (2007), Wynn and Fougere (2007), and Siragusa and Ricke (2012). One of the authors of this chapter has written an easy-to-use barn guide to natural treatments for dairy cows (Karreman, 2011).

An Organic Program Should Never Compromise Welfare

Withholding an antibiotic from an individual animal when it is truly needed would impact animal welfare negatively. But antibiotics – in and of themselves – do not ensure proper animal welfare.

Antibiotics save lives. Using them in a timely fashion prior to any permanent damage is a critical component in the responsible use of these lifesaving medicines. When an organic farmer is having trouble deciding whether or not to use an antibiotic when it is truly needed, a veterinarian may need to remind him or her that a few animals will leave the herd every year and that the animal truly needing an antibiotic will be one of them. The organic consumer is paying a premium for organic products, and the farmer should do the right thing and not let any animal suffer. Perhaps the farmer should be asked gently what the organic consumer would want done for the animal in the situation if he or she were standing there and had been given all the information and treatment options. A final nudge to use an antibiotic when it is truly necessary would be to state that it is better to have a live conventional animal than a dead organic animal. One of the authors of this chapter (HK) has used each of these rationales separately and in tandem on a few occasions when a farmer simply could not make up his or her mind about using an antibiotic. The farmer was having a difficult time with the reality of having to remove the animal permanently from US organic production. In a certain sense, when needing to reach for an antibiotic, animal welfare on the farm has already been compromised in some manner and most organic farmers feel slightly defeated when an animal has to be given an antibiotic.

Is the neglect/suffering of one organic animal not given antibiotics in time worse than the potential disease transmission of using reproductive hormone injections in synchronization protocols on conventional farms without using new needles on different cows? Does the potential suffering of one animal not given an antibiotic in time offset the continual day in and day out feeding of high grain rations for high production or rate of gain, which creates subacute rumen acidosis? And does the suffering of one animal not given antibiotics in time overshadow cows that spend their entire lives on concrete and never experience walking on actual earth and grazing pasture? Is it always the case that cows that can be given antibiotics on conventional farms are always treated promptly with antibiotics when they really need them? Additionally, livestock do die on farms – just because antibiotics can be used on conventional farms in the USA does not guarantee that animals will never die.

Small Versus Large Organic Farms and Antibiotics

Within the organic sphere, there are generally two ideologies regarding the size of a farm and its place in organics. It is generally assumed that small organic farms (less than 100 dairy cows, less than 5 sows, etc.) allow for developing close relationships and feelings for each individual animal. Unfortunately, the sheer workload may not always translate into best practices being done every day. This is contrasted by large organic farms where there is less "personal touch" due to large numbers of animals. But there are standard operating procedures carried out reliably in a standardized way in well-managed operations.

In regards to animal care, sometimes a farmer operating a small organic farm will want to know everything possible he or she can do for an animal (short of giving an antibiotic). All natural treatments, regardless of scientific merit, will be tried to save "Bessie", but then the farmer realizes only too late that an antibiotic should have been used. This is probably unique to small organic farms, since a large organic farm may give the antibiotic quicker because there is a ready replacement animal for the one to be removed. The animal that must leave a large organic farm does not create as much loss as an animal removed from a small organic farm. Large

US dairies that operate both conventional and organic operations may switch 10% of the baby calves and 2% of the cows from organic to conventional after the animals have been treated with antibiotics.

A conventional farm can always utilize an antibiotic and anti-inflammatory without thinking about other possible options – just give a shot and get on with other work. Whenever antibiotics are easily allowed, people will always reflexively reach for them. There is no real incentive to research functional alternatives to antibiotics unless there are strict prohibitions in place, as with the US National Organic Program.

Different Approaches to Antibiotic Use

There are three situations where antibiotics should be used without hesitation: (i) generalized peritonitis; (ii) bone infections and gangrene; and (iii) when two or more organic systems are involved and the animal is depressed.

By combining allowed synthetic materials from the listing found at 7CFR205.603 along with biological antibodies, botanical antibacterials, and mineral-based antiseptics, relief of pain and suffering can be accomplished. Thus, relief of pain and suffering is not a function of antibiotics alone. It is conceivable, due to the prohibition of antibiotics in US organic livestock, that the USA is the global leader of antibiotic-free treatments against infections in livestock. Traditional thinking is that Europe leads the way in naturopathic medicine – this is quite possibly the case in human practice, but highly unlikely in farm animal practice, due to the very minimal twice the withholding time for antibiotic use (which may sometimes be as little as 4 days withholding). As long as antibiotics can be used without much penalty, they will be reached for quickly, especially when regulations state that they can be used when alternatives will not be effective. It is human nature to take the easy route. Perhaps a good middle ground is to look to the Canadian system of a 30-day withholding time as a world standard – strict enough to be an incentive to try out natural alternatives to antibiotics for common diseases but still allowing antibiotics for truly life-threatening situations.

References

Barker, Z.E., Leach, K.A., Whay, H.R., Bell, N.G. and Main, D.C. (2010) Assessment of lameness prevalence and associated risk factors in dairy herds in England and Wales. *Journal of Dairy Science* 93, 932–941.

Bengtsson, J., Ahnström, J. and Weibull, A.C. (2005) The effects of organic agriculture on biodiversity and abundance: A meta-analysis. *Journal of Applied Ecology* 42, 261–269.

Blair, R. (2007) *Nutrition and Feeding of Organic Pigs.* CAB International, Wallingford, UK.

Burfeind, O., Sepulveda, P., von Keyserlingk, M.A.G., Weary, D.M., Veira, D.M., *et al.* (2010) Evaluation of a scoring system for rumen fill in dairy cows. *Journal of Dairy Science* 93, 3635–3640.

Burow, E., Thomsen, P.T., Sorensen, J.T. and Rousing, T. (2011) The effect of grazing on cow mortality in Danish dairy herds. *Preventative Veterinary Medicine* 100, 237–241.

California Department of Agriculture (2014) Available at: www.cdfa.ca.gov/ahfss/animal_health/pdfs/Pseudorabies.pdf (accessed 20 October 2014).

CAN/CGSB-32.310-2006 6.46.7+6.8.

COMMISSION REGULATION (EC) No 889/2008 of 5 September 2008 Article 24.4

Cramer, S.D., Campbell, G.A., Njaa, B.L., Morgan, S.E., Smith, S.K. II, *et al.* (2011) Pseudorabies virus infection of Oklahoma hunting dogs. *Journal of Veterinary Diagnostic Investigation* 23, 915–923.

EU Council Regulation (2012) EU Council Regulation (EC) No. 834/2007 of 28 March 2012 Article 14(e)(i). Available at: www.ifoam-eu.org/en/organic-regulations/list-eu-organic-regulations (accessed 14 November 2014).

Fabian, J., Laven, R.A. and Whay, H.R. (2014) The prevalence of lameness on New Zealand dairy farms: A comparison of farmer estimate and locomotion scoring. *Veterinary Journal* 201, 31–38.

Franzluebbers, A.J., Paine, L.K., Winston, J.R., Krome, M., Sanderson, M.A., *et al.* (2012) Well managed grazing systems: A forgotten hero of conservation. *Journal of Soil and Water Conservation* 67, 199.

Galef, B.G., Mainardi, M. and Valsecchi, P. (1994) *Behavioral Aspects of Feeding.* Harwood Academic, Chur, Switzerland.

Haufe, H.C., Gygax, L., Wechsler, B., Stauffacher, M. and Friedli, K. (2012) Influence of floor surface and access to pasture on claw health in dairy cows kept in cubicle housing systems. *Preventive Veterinary Medicine* 105, 85–92.

Hovi, M., Sundrum, A. and Thamsborg, S.M. (2003) Animal health and welfare in organic livestock production in Europe: Current state and future challenges. *Livestock Production Science* 80, 41–53.

Hulsen, J. (2005) *Cow Signals: A Practical Guide for Dairy Farm Management.* Roodbont, Zutphen, the Netherlands.

Hungerford, C. (2005) *There is a Synthetic in My Organic Chicken.* Rodale Institute, Kutztown, Pennsylvania.

Iowa State University (2014) *Pseudorabies – PRV.* Veterinary Diagnostic and Production Animal Medical, Iowa State University, Ames, Iowa.

Jacob, J. (2013) Synthetic methionine and organic poultry diets. Available at: www.extension.org (accessed 27 May 2014).

Jankowski, J., Kubinska, M. and Zdunczyk, Z. (2014) Nutritional and immunomodulatory function of methionine in poultry diets – A review. *Annals of Animal Science* 14, 17–32.

Jones, T.A. and Berk, J. (2012) Alternative systems for meat chickens and turkey production health and welfare. In:Sandilands, V. and Hocking, V. (eds) *Alternative Systems for Poultry, Health, Welfare and Productivity*. CAB International Wallingford, UK, pp. 250–296.

Karreman, H.K. (2004) *Treating Dairy Cows Naturally – Thoughts and Strategies*. Acres USA Publishing, Austin, Texas.

Karreman, H.K. (2011) *The Barn Guide to Treating Dairy Cows Naturally. Practical Organic Cow Care for Farmers*. Acres USA Publishing, Austin, Texas.

Lindgren, K., Bochicchio, D., Hegelund, L., Leeb, C., Mejer, H., *et al.* (2014) Animal health and welfare in production systems for organic fattening pigs. *Organic Agriculture* 4, 135–147.

Maeder, P., Fliessbach, A., Dubois, D., Gunst, L., Fried, P., *et al.* (2002) Soil fertility and biodiversity in organic farming. *Science* 296, 1694–1697.

Provenza, F.D. and Balph, D.F. (1987) Diet learning by domestic ruminants: Theory evidence and practical implications. *Applied Animal Behaviour Science* 18, 211–232.

Provenza, F.D., Villalba, J.J., Dziba, L.E., Atwood, S.B. and Banner, R.E. (2003) Linking herbivore experiences, varied diets, and plant biochemical diversity. *Small Ruminant Research* 49, 257–274.

Richert, R.M., Cicconi, K.M., Gamroth, J.M., Schukken, Y.H., Stiglbauer, K.E., *et al.* (2012) Risk factors in clinical mastitis, ketosis, and pneumonia in dairy cattle on organic and small conventional farms in the United States. *Journal of Dairy Science* 96, 4269–4285.

Ricke, S.C., Van Loo, E.J., Johnson, M.G. and O'Bryan, C.A. (2012) Genetics of poultry meat production in organic systems. In: Ricke, S.C., Van Loo, E.J., Johnson, M.G. and O'Bryan, C.A. (eds) *Organic Meat Production and Processing*. John Wiley and Sons, Hoboken, New Jersey, pp. 137–145.

Siragusa, G.R. and Ricke, S.C. (2012) Probiotics as pathogen control agents for organic meat production. In: Ricke, S.C., Van Loo, E.J., Johnson, M.G. and O'Bryan, C.A. (eds) *Organic Meat Production and Processing*. John Wiley and Sons, Hoboken, New Jersey, pp. 329–349.

Sophia, W. (2014) How strain and production system affects chicken welfare and quality of meat. Thesis, SLU, Agricultural Science, Uppsala, Sweden.

Sorenson, P. (2012) Genetics of poultry meat production in organic systems. In: Ricke, S.C., Van Loo, E.J., Johnson, M.C. and O'Bryan, C.A. (eds). John Wiley and Sons, Ames, Iowa, pp. 137–146.

Stear, M.J., Nikbakht, G., Matthews, L. and Jonsson, N. (2012) Breeding for disease resistance in livestock and fish. *CAB Reviews: Perspectives in Agriculture, Veterinary Science, Nutrition and Natural Resources* 7, 1–10.

Stiglbauer, K.E., Cicconi-Hogan, K.M., Richert, R., Schukken, Y.H., Ruegg, P.L., *et al.* (2013) Assessment of herd management on organic and conventional dairy farms in the United States. *Journal of Dairy Science* 96, 1290–1300.

Sundrum, A. (2012) Health and welfare of organic livestock and its challenges. In: Ricke, S.C., Van Loo, E.J., Johnson, M.G. and O'Bryan, C.A. (eds) *Organic Meat Production and Processing*. John Wiley and Sons, Ames, Iowa, pp. 87–112.

Sutherland, M.A., Webster, J. and Sutherland, I. (2013) Animal health and welfare issues facing organic production systems. *Animals* 3, 1021–1035.

Teague, R., Provenza, F., Kreuter, U., Steffens, T. and Barnes, M. (2013) Multi paddock grazing on rangelands. Why the perceptual dichotomy between research results and rancher experience. *Journal of Environmental Management* 128, 699–717.

Thawley, D.G., Wright, J.C. and Solorzano, R.F. (1980) Epidemologic monitoring following an episode of pseudorabies involving swine, sheep, and cattle. *Journal of the American Veterinary Medical Association* 176, 1001–1003.

Tuck, S.L., Winquist, C., Mota, F., Ahnstrom, J., Turnbull, L.A., *et al.* (2014) Land use intensity and the effects of organic farming on biodiversity: A hierarchial meta-analysis. *Journal of Applied Ecology* 51, 746–755.

USDA NOP (National Organic Program) 7CFR205.237(c) All standards available at www.ams.Usda.gov/AMSV1.0/nop (accessed 15 November 2014).

USDA NOP 7CFR205.230(1)

USDA NOP 7CFR205.238(a)(a), (a)(2)

USDA NOP 7CFR205.238(a)(6)

USDA NOP 7CFR205.239(a)

USDA NOP 7CFR205.240

USDA NOP 7CFR205.603

USDA NOP USC6518 section 2119(m)

USDA NOP 7USFR205.238 (c)(7)

USDA NOP 7YSC6509

Vaarst, M., Roderick, S., Byarugaba, D.K., Kobayashi, S., Rubaire-Akiiki, C., *et al.* (2006) Sustainable veterinary medical practices in organic farming a global perspective. In: Halberg, N., Alroe, H., Knudsen, M. and Kristensen, E. (eds) *Global Development of Organic Agriculture: Challenges and Prospects*. CAB International Wallingford, UK.

Wynn, S.G. and Fougere, B. (2007) *Veterinary Herbal Medicine*. Mosby, St Louis, Missouri.

14 The Effect of Economic Factors on the Welfare of Livestock and Poultry

TEMPLE GRANDIN

Department of Animal Science, Colorado State University, Fort Collins, Colorado, USA

Summary

Economic factors can serve as either incentives to improve animal welfare or disincentives that may worsen animal welfare. High standards set by large meat buyers can be powerful motivators to improve animal treatment. An objective scoring system that was developed for slaughter plants has been used successfully by major retail buyers to enforce better standards. The financial penalty was removal from the approved supplier list. Holding transporters and producers financially accountable for bruises, death losses, and dark cutters will reduce them greatly. Animal welfare will suffer in segmented markets where producers are not held financially accountable for bruises or sickness. The loss gets passed on to the next segment in the marketing chain. Selling animals through many dealers is usually detrimental to welfare. People handling animals should receive extra payments for low bruising and death losses. Economic factors may decrease welfare by encouraging producers to overcrowd animals or convert pasture-land to crops.

Learning Objectives

- Learn how standards implemented by major meat buyers can be used to improve welfare
- Understand how different marketing systems can either improve welfare or make it worse
- How to use incentive payments to improve welfare
- Biological overload of an animal is detrimental to welfare
- Understand how short-term economic gains obtained by converting pastureland to crops may be detrimental to both welfare and sustainability

Introduction

During a long career spanning over 35 years, the author has learned to understand more and more how economic forces can be used to improve animal welfare. Economic incentives to treat animals better can be very effective. One huge positive force for improving animal welfare is that consumers are demanding that animals be treated better. Corporations, both large and small, can be motivated to improve practices when consumers demand it. Easy-to-use video cameras in everybody's mobile telephone are also great drivers of change. People are shocked when they see videos of animal abuse posted on the Internet. Undercover videos can go viral and be seen by thousands of people.

All of the things the author recommends are based on either scientific papers, first-hand experience of implementing welfare auditing programs, observations during extensive travel to many countries, or research and interviews with other individuals who have implemented effective programs. In the first section of this chapter, the effects of economic factors on improving animal welfare will be covered. The second half of the chapter will cover economic factors that are detrimental to good animal welfare.

Economic Incentives Provided by Buyers Can Improve Animal Welfare Greatly

Alliances between meat companies and producers

In these systems, ranchers and farmers produce animals that must meet specific criteria for animal

---

welfare, food safety, or other requirements. Alliance programs have standards that exceed regular commercial practices. Some examples of alliance systems are branded programs for organic, natural, or high-welfare production. Producers are often eager to join these programs to get higher prices. Participation in an alliance is voluntary.

Standards set by importing countries

The demand for meat, milk, and eggs is rising rapidly in the developed world. The European Union (EU) has high welfare standards. Since the first edition of this book, South American meat companies have built many new slaughter plants and have initiated programs to market pasture-raised cattle to Europe. This has resulted in improvements in handling, transport, and slaughter.

Welfare and food safety supply chain auditing by major meat buyers

The programs that have been implemented by supermarkets and restaurants to inspect farms and slaughter facilities have resulted in great improvements in how animals are treated (Grandin, 2005, 2007). These audits have resulted in tremendous improvements in plant facilities. The most noticeable changes are much better repair and maintenance of equipment such as stunners, races, and pens. Out of 75 beef and pork plants in the USA that were on the McDonald's approved supplier list, only three had to build totally new systems. Most plants made simple economical improvements, which are discussed in Chapters 5 and 9, this volume. At three other plants, no improvements occurred until a new plant manager was hired. This shows the importance of management attitude. In the USA, most plants already had at least adequate facilities. In South America and other parts of the world, many new lairages, races, and stun boxes have been built to replace poor facilities. Tesco and other supermarket buyers from Europe and McDonald's units within each South American country demanded better animal treatment. Socially responsible buying programs from big corporations have the power to stop many bad practices.

Executives Motivated to Improve Operations After Seeing Bad Practices

The author had the opportunity to take upper management people from many different companies on their first trips to farms and slaughter plants. When things were going well, they were happy and when they saw abuses, they became highly motivated to make improvements happen. The executives had to see bad practices with their own eyes to motivate them to make improvements. One executive became really motivated to improve conditions after he saw an emaciated, sick, old dairy cow going into his hamburger. Animal welfare became real and was no longer an abstract concept that was delegated to the public relations or legal department. It is essential to get high-level executives out of the office so they can see bad practices to motivate them to change. The reactions were similar to the reactions of CEOs who went undercover as regular employees in their own corporations. In the USA, there is a television show called *Undercover Boss* and it shows the shocked reactions when the boss sees mistreatment of employees. The author witnessed the same shocked reactions.

Make Producers and Transporters Financially Accountable for Bruises, Poor Meat Quality, Nonambulatory Animals, and Death Losses

Bruises on slaughter cattle were reduced greatly when producers or transporters had to pay for them. In the USA, payment programs where bruises and other losses were deducted from producer payments greatly improved cattle treatment during transport to the plant. Grandin (1981) found that when producers had to pay for bruises, bruising was reduced by half. Paranhos da Costa *et al.* (2012, 2014) in Brazil reported that when supermarkets audited bruises and made deductions from transporters' pay, bruising was reduced from 20% to 1% of the cattle. Bruises cause severe economic losses and large portions of the meat have to be removed from severely bruised carcasses (Fig. 14.1). Carmen Gallo in Chile also reported that bruises were reduced when transporters were fined for damage to the animals (Grandin and Gallo, 2007). In another case, problems with weak pigs that were too fatigued to walk off the truck or move to the stunner were reduced greatly when producers were fined US$20 for each fatigued pig. Producers reduced nonambulatory and weak pigs by decreasing greatly the dose of the beta-agonist, ractopamine (Paylean®). This feed additive makes pigs big and lean, and too high a dose may increase the percentage of nonambulatory pigs.

Fig. 14.1. Rough handling and overloaded trucks can cause extensive bruising and damage to the meat. Bruised meat has been trimmed from this carcass. Careful handling will prevent bruises.

Use objective methods for assessing handling and transport losses

Vague guidelines that use terms such as "adequate space" or "proper handling" are impossible to implement because one person's interpretation of proper handling will be different from someone else's. Loading and unloading of trucks and moving animals through vaccination races should be measured with numerical scoring of variables, such as the percentage of animals that fall, the percentage of animals that are prodded with an electric goad, and the percentage that move faster than a trot. Moving at a walk is preferable. For more information, see Grandin (1998a, 2007, 2013), Maria *et al.* (2004), and Chapters 2 and 4, this volume. Alvaro Barros-Restano (2006, personal communication) reported that in auction markets in Uruguay, continuous monitoring has improved handling greatly. Handling practices need to be measured continuously to prevent them from gradually becoming increasingly rough. Measures of death losses, nonambulatory animals, bruises, injuries, pale soft meat in pigs, and dark cutting dark firm and dry (DFD) beef should be used to provide either bonuses or deductions from transporter or producer pay. In many countries, there is a large payment deduction for meat-quality defects such as DFD in cattle.

Improving meat quality is an incentive for improving animal treatment

When the author first started working with pork slaughter plants in the early 1980s, the handling of pigs was terrible and every pig was shocked multiple times with an electric goad. In the late 1980s, the USA started exporting pork to Japan and the Japanese grader who worked in the plant rejected pale, soft exudative (PSE) meat. The author visited many different plants and stopped the excessive use of electric goads in the race that led up to the stunner. The next day, 10% more pork was suitable for export to Japan because it was not pale, soft and watery due to rough handling. This provided a motivation to handle pigs more carefully. Today, several research studies clearly show that how the pigs are handled in the last few minutes in the stunning chute has a big effect on pork quality. Excitement and electric goad use in the stunning race increases both PSE and blood lactate (Grandin, 1985; Hambrecht *et al.*, 2005; Edwards *et al.*, 2010). One study in feedlot cattle showed that cattle that have a large flight zone and become agitated when people approach have tougher meat (Gruber *et al.*, 2006). An earlier study done by Voisinet *et al.* (1997) also found that excitable cattle had tougher meat and more dark cutters.

Economic Advantages of Improving Handler Safety, Efficiency, and Value of Old Cull Animals

The author has convinced many slaughter plants, feedlots, and farms to improve cattle handling to reduce accidents and injuries to people. Marcos Zapiola in South America has used the same approach to promote quiet handling methods.

Understanding cattle behavior may also help people to avoid injuries. In dairies, handlers were more likely to get injured when moving cows into a hoof-trimming facility compared to moving them for milking (Lindahl, 2014). Hoof trimming was aversive to cows and they did not want to move into the facility. This is more likely to result in a cow kicking or knocking a person over. There is a need to use comfortable restraint devices and to take steps to make hoof trimming less aversive. Training workers in good stockmanship can increase milk production (Sorge *et al.*, 2014). Unfortunately, 50% of dairy managers were not interested in more stockmanship training (Sorge *et al.*, 2014). It is difficult for some people to understand fully that animal behavior is important. Stockmanship is also an excellent way to reduce accidents.

The author has also used the safety of people as a major selling tool to eliminate live shackling and hoisting of animals by one back leg in US slaughter plants. Douphrate *et al.* (2009) studied 10 years of worker injury data in the USA. Accidents while handling cattle were a major cause of severe debilitating injuries that required costly medical care. In Australia, the most common of injuries to veterinarians was being kicked or struck by cattle (Lucas *et al.*, 2013). In the USA, shackling and hoisting is still legal due to religious exemptions. When a veal calf slaughter plant replaced shackling and hoisting with an upright restraint for kosher slaughter, there was a huge reduction in accidents. For an 18-month period, before installation of the upright restrainer, there were 126 working days of lost time due to accidents involving employees being kicked or trolleys falling on them. Three workers were absent for more than 3 weeks. After the shackle hoist was replaced with the restrainer, during the next 18-month period one employee was absent for 2 days due to a bruised hand (Grandin, 1988).

Reduce labor requirements with animal-welfare-friendly equipment

The author designed and installed many innovative cattle-handling systems that were sold to plant and feedlot management as a way to reduce labor costs. Half the cattle in the USA and Canada are handled in a restrainer system designed by the author (Grandin, 2003). When the author presented her proposal for a totally new handling system at a slaughter plant, many managers bought the system because one or two full-time employees could be removed. The cost savings of reducing labour requirements, improving meat quality, reducing accidents, and reducing bruises were all put on the proposal to emphasize how much money they could save. In Uruguay, cattle are restrained for religious slaughter by three or four employees, who hold them down. Replacing the highly stressful method of restraint with better equipment would improve both animal welfare and employee safety. It would also reduce labor requirements.

Increase the economic value of cull animals

Some of the worst abuses the author has observed in transported animals were animals that were not fit for transport. They were treated badly because they were worth very little money. Emaciated, weak, old cows, sow or ewes should be euthanized on the farm and not loaded on to a vehicle. Published materials for assessing body condition, lameness, and injuries should be used (see Chapters 2 and 4, this volume). The use of pictures and videos to make assessment of the animal's condition more objective is strongly recommended. Livestock quality assurance schemes in many countries have excellent materials for assessing fitness for travel. When programs are implemented and producers receive more money for cows that are in better condition, they will be motivated to sell their cull animals before they become emaciated (Roeber *et al.*, 2001). In the USA, there are several successful programs for improving the value of cull cows. They are fed in a feedlot for 60–90 days to improve meat quality, to make the meat more valuable.

Promote the use of livestock identification and trace-back

In much of the developed world, animals are required to be identified with either an individual identification number or the identity of their farm of origin. Animal identification makes it possible to trace animals back to the farm of origin, which enables customers to determine where their meat comes from. Trace-back also makes it easier to hold producers and transporters accountable for losses.

Economical Strategies for Improving Welfare

In many countries, there is a need for high-quality, small, local slaughter plants in areas where animals are raised. More plants will help reduce the need for long-distance transport. Efforts by the government to build local slaughter plants have failed in some countries because the government did not provide funds to hire experienced managers to operate them. There are many rusting hulks throughout the developing world where this mistake has been made. Some of these plants contained equipment that was too expensive and difficult for local people to maintain. Efforts in building producer-owned cooperative slaughter plants have had mixed results. The successful cooperatives have clauses in their legal documents to prevent a few producers from buying too big a portion and then selling out. This leaves the other coop members at the mercy of a new plant owner who is no longer legally bound by the original cooperative agreements. The author has witnessed this sad fate for three large cooperatives in the USA. Successful cooperative plants must have a strong, highly experienced leader and legal documents that will prevent one or two members from taking over and doing things that are detrimental to other coop members. Fighting between producer board members has destroyed some cooperatives.

Hire and train experienced people who know how to implement practical solutions and pay them well

There is a tremendous need for more educated, experienced people to work on hands-on, practical things. They need to have both scientific knowledge and hands-on experience to bridge the gap between making policy and applying it successfully. In the USA, there are not enough students who want to become large animal veterinarians. A big part of the problem is that a veterinary education is very expensive, and it is difficult for the student to pay back their student loans. There are similar concerns in Europe. Programs need to be developed to expose young students to farm animals, and economic incentives should be used to encourage students to go into careers that will help both animal welfare and sustainable agriculture. In 2014, the author visited both Uruguay and Brazil and there were many eager young students going into beef production. The government has made training programs affordable. Policies and legislation are useless unless there are practical people on the ground who can implement them. Input from practical field researchers will help create policy and legislation that will work. In many fields, there is a shortage of people to work in the "trenches" and make real constructive change. This has happened in many fields, ranging from medicine to agriculture. The author urges governments, non-governmental organization (NGO) animal activist groups, and livestock companies to support and educate skilled fieldworkers and researchers. These people are essential to make real change and improvements take place.

In some developed countries, there is a shortage of qualified truck drivers. Hands-on jobs such as truck driver and animal handler need to receive more recognition and pay. The author has observed effective programs in which handling improved when employees received training, better supervision, higher pay, and recognition with a special animal welfare emblem for their hats. For these programs to work, they must be backed by a firm commitment from upper managers.

Implement simple, practical, and economical ways to improve handling transport in developing countries

Fancy equipment such as hydraulic tailgate lifts or aluminium trailers are often not appropriate in developing countries. The people do not have the equipment or the money to maintain these items. While traveling to many countries, the author observed that simple improvements made a big difference. Nonslip flooring is essential in vehicles used to haul livestock, and on scales, unloading areas, and stun boxes. Floors can be made nonslip by welding readily available steel bars on the floor in a grid pattern. A nonslip floor will prevent many serious animal injuries. There is also a need to build ramps for loading and unloading. Many animals in developing countries are injured when they are forced to jump off a vehicle. See Chapter 11, this volume, on design of loading ramps that can be built easily in a developing country. Training people in animal behavior and low-stress handling methods is also essential. Further information can be found in Chapter 5, this volume, and Grandin (1987, 1998b, 2014), Smith (1998) and Ewbank and Parker (2014). Many people made the mistake of thinking that fancy equipment would solve all the problems. Over the years, the author has learned that good equipment makes good treatment easier, but it is useless without good management.

Corporate NGO and foundation sources of funding for animal welfare projects and student scholarships

In both Europe and the USA, government sources of research funding are drying up. In the future, funding will need to come from private sources. Programs in sustainable agriculture and local production of food are activities that many private sources of funding like to support. Local initiatives will become increasingly important. A good example of an effective project is the mobile slaughter unit that is funded by the Lopez Community Land Trust, a nonprofit group that supports sustainable agriculture (Etter, 2008). This system enables small, local sheep and cattle producers to have their animals slaughtered in a United States Department of Agriculture (USDA)-inspected plant and sell their meat without any restrictions. The mobile unit improves animal welfare because the slaughterhouse comes to the farm. To make this type of program successful, funds must also be provided for hiring competent people to run and operate the system. The project is likely to fail if funds for operating expenses are omitted. In some areas, mobile units that visit "docking stations" with water, sewage, and lairage facilities close to local farms may be more economical.

Private sources of money can also be very effective for supporting veterinary and graduate students who go into the animal welfare field. Students will enter the animal welfare field quickly if money is available for projects and scholarships.

Practical Methods to Use Economic Factors to Improve Welfare

Reward animal handlers with extra pay for low levels of bruises, injuries, and deaths. In the US and British poultry industries, broken wings were reduced from 5% to 1% by paying a bonus to the chicken loaders when there were 1% or less birds with broken wings. The same system has also worked well for people handling pigs and cattle. The worst way to pay animal handlers is based on how many they can handle per hour. This will result in rough treatment of the animals. The author has observed terrible handling of pigs, cattle, and poultry when workers were financially rewarded for high-speed handling. Workers should be rewarded for high-quality handling. It is also essential not to overwork the animal handlers or put them in a situation that is understaffed. Tired people will abuse animals. Internal unpublished data from large pig and poultry companies showed that death and injuries doubled after the truck loading crews had worked more than 6 h.

Combining animal welfare audits with food safety and quality facilitates implementation

When welfare programs are first getting started in a country or region, they are easier to implement if they are combined with food safety programs and health and vaccination programs. In the USA, the same people who do food safety audits do all of the welfare audits in the slaughter plants. This made the programs easy to implement because no additional people had to be hired, as the food safety auditors were already inspecting the plants. In South America, animal welfare is being combined with programs to prevent injection site damage and programs to obey medication withdrawal times. The prevention of bruises is a major part of South American programs. In European countries, where animal welfare is an established concept, there are many people for whom welfare inspection is their only job.

Design insurance programs to reward good practices

Mistreatment of animals and high levels of bruising and death losses will occur if a truck driver or transport company is reimbursed for every dead, injured, or bruised animal. The most effective insurance programs pay for catastrophic losses such as rolling a truck over, but they will not pay for the first five dead pigs. There is a deductable amount of losses that are not covered. This motivates drivers and transporters to handle animals well so they will not have to pay for losses caused by rough handling or carelessness.

Major Problem Areas Where Economic Factors Work Against Animal Welfare

There has also been controversy about beef cattle shipped from Australia to Indonesia. Cattle are fattened in Indonesian feedlots and slaughtered under substandard conditions. Shipping chilled beef will not work because Indonesians cannot pay a sufficient price to pay for shipping empty refrigerated containers back to Australia. Half the land in Australia does not have enough water to grow crops. The use of grazing animals is the only way that food can be raised on this land. Grazing well done is a highly sustainable type of agriculture, which can improve pasture (Franzluebbers et al., 2012). The most practical way to improve animal welfare is to build better slaughter plants and staff them with experienced Australian managers. The Australians need markets outside Australia for their beef and sheep meat that is grown sustainably on the vast outback.

Standards become less strict when demand for the product exceeds the supply

The author has worked with many of the grocery stores that sell either organic or high-welfare standard meat, milk, or eggs. In the beginning, they start out with strict standards. When the product becomes popular, demand exceeds supply. The grocery store cannot get enough products so it is tempted either to lower its standard or to start using less reliable suppliers. The author observed a natural beef company that became lax in enforcing its own standards and the practices became so bad in its overworked little slaughter plant that it got shut down several times by the USDA for humane slaughter violations. Standards can also slip when the original founder of the company leaves and the company is sold to a large corporation. When the founder leaves, the company no longer has the founder's vision.

Old cull livestock and poultry of little economic value

Some of the worst abuse during handling, transport, and slaughter occurs with old cull breeding stock. In the USA, cull dairy cows and old breeding sows often travel greater distances than young animals that have been fattened on either grain or grass. A survey of auctions that sell old dairy cows in Wisconsin, USA, revealed that up to 20% of cows were in an emaciated poor body condition (Kurt Vogel, 2014, personal communication). There is less economic incentive to treat these animals well because they are less valuable than young fed animals. An effective way to reduce abuses is to increase the value of old breeding stock. This provides an economic incentive to treat them better. Producers need to be educated that if they sell animals before they become skinny and emaciated, they will receive more money for them. In the USA and other parts of the developed world, programs have been implemented in some areas to fatten old breeding stock so that they will become more valuable for meat.

Highly segmented marketing chains with many dealers, agents, and middlemen are bad for animal welfare

In the developed world, such as Europe and North America, most high-quality young animals that are fattened for slaughter go directly from the feedlot or farm to a slaughter plant. This makes it much easier to make people accountable for losses. Old breeding stock often passes through a series of auctions or dealers, and the origin of the animals may not be able to be traced. In the developing world, all classes of livestock are often sold through middlemen and dealers. In all countries, the sectors of the livestock market where the animals go through a series of auctions, dealers, or middlemen will be the most difficult to improve. In a highly segmented market system, there is often no accountability for losses. Middlemen and dealers who do not own the animals have little economic incentive to reduce bruises, injuries, and sickness because they are not held financially accountable for losses.

Sacrificing the well-being of the individual animal for greater profits from the entire flock

When pigs or chickens are jammed too tightly into a house, the welfare and productivity of each individual animal usually declines. Unfortunately, there is an economic incentive to do this because the overall output of eggs or meat per house may be greater. The bad economic incentive to overstock a building is most likely to occur in places where both land and buildings are expensive.

Overworked employees and overloaded equipment are detrimental to animal welfare

Many people who are interested in animal welfare assume that slaughter plants with a high line speed are bad. Data collected by Grandin (2005) indicated that when equipment was designed and staffed with enough people, line speed had no effect on the percentage of cattle stunned with one shot from a captive bolt (Table 14.1). The author has observed that the worst problems occur when either equipment or people are overloaded. This occurs when the plant's meat sales exceed the capacity of the plant's equipment and staffing. One plant that slaughtered 26 cattle/h worked really well, but when its line speed was increased to 35 cattle/h with no increase in employees and no equipment modifications, employees repeatedly slammed the stun box door on cattle. The plant ended up having its USDA inspection suspended for violation of the Humane Slaughter Act.

Table 14.1. Effect of line speed on stunning in beef slaughter plants. (From Grandin, 2005.)

Line speed (cattle/h)	Cattle rendered insensible with a single shot from a captive bolt stunner (%)[a]
<50 (16 plants)	96.2
51–100 (13 plants)	98.9
101–200 (10 plants)	97.4
>200 (27 plants)	96.7

Note: [a]Percentages shown are average values for the number of plants.

No financial incentive to vaccinate animals and prevent sickness

In Australia, vaccinating calves and training them to eat from feed troughs reduced sickness (Walker *et al.*, 2006). In the USA, bovine respiratory disease (BRD) is a major problem. Most cases of BRD could be prevented if ranchers pre-weaned and vaccinated their calves before they left the ranch. Half of the ranchers fail to do this because they receive no financial incentives for vaccinating and pre-weaning their calves (Suther, 2006). When calves get sick at the feedlot, they are someone else's problem. When beef prices are high and there is a shortage of calves to fill up the feedlots, producers can receive high prices for calves that have never been vaccinated. In 2014, 42% of US ranchers planned to skip pre-weaning and vaccinations (Rutherford, 2014). The financial incentive to reduce losses is lost when cattle prices are high.

One way to get a financial benefit from preconditioning is for either a producer or a group of producers to develop a reputation for producing quality calves and selling them in special sales (Thrift and Thrift, 2011). To prevent sickness, beef calves should be preconditioned 45 or more days before they are shipped to a feedlot. Preconditioning consists of vaccinations, weaning, and teaching the calves to eat from a feed bunk and drink from a water trough. Preconditioning will reduce sickness and low animal performance significantly (National Cattlemen's Beef Association, 1994; Arthington *et al.*, 2008; Thrift and Thrift, 2011). The best way to motivate ranchers to implement these practices is to pay them a premium for preconditioned calves. Cattle that get sick have lower-quality grades due to less marbling and are worth less money (Texas A&M University, 1998; Waggoner *et al.*, 2006).

Marketing systems that can be developed to replace segmented marketing chains and improve welfare

Marketing systems that can be developed to replace segmented marketing chains and improve welfare include:

- **Special sales where only certified preconditioned calves are sold.** State or provincial cattle associations work with local producers to develop these programs. Buyers will pay more for the cattle (Troxel *et al.*, 2006; Thift and Thrift, 2011). Verification by an independent third party that preconditioning procedures were done increased premiums buyers paid for calves (Bulot and Lawrence, 2006).
- **Producers are enrolled in a contractual agreement** with a supermarket, restaurant, meat company, or other buyers to produce the animals to certain specifications such as cage-free, preconditioned calves, etc. Producers are paid a premium to be in these programs.
- **Producers sell locally grown animals through a cooperative** that has strict standards for welfare. For all types of programs, auditing is required to ensure that people follow the specifications.

Customers are very poor and feeding a hungry family is their top priority over animal welfare

People in this situation will buy the cheapest meat they can get. This is especially a problem in poor, developing countries. Feeding the family is their first priority.

Overloading the Animal's Biology Causes Suffering

Broiler poultry, laying hens, dairy cows, and pigs that are raised in intensive systems have been genetically selected over the years to provide more and more meat and milk. Some of the welfare problems caused by single-minded selection for production traits have resulted in increased lameness and leg problems in dairy cows and chickens (Knowles *et al.*, 2008; Grandin, 2014). These problems have gradually become worse over the years, and some new people entering the industry do not know that high percentages of lame animals are totally abnormal. There are three basic ways that animal biology can be overloaded to the point where welfare will be poor: genetic selection solely for rapid weight gain; large muscle growth; or ever-increasing milk production. Spent laying hens have severe osteoporosis and high percentages of broken keel bones (Sherwin *et al.*, 2010; Wilkins *et al.*, 2011). Changing housing has reduced fractures, but the percentage of broken bones is still way too high. Hen welfare is simply going from atrocious to poor.

Single-minded overselection for production traits has many bad side effects, such as reduced disease or parasite resistance (Meeker *et al.*, 1987; Johnson *et al.*, 2005; Jiang *et al.*, 2013). A study in Belgium showed that the slower-growing genetic types of broiler chickens on organic farms had less fluctuating asymmetry compared to conventional fast-growing genetics (Tuyttens *et al.*, 2007). Fluctuating asymmetry occurs when one side of an animal may be either bigger or smaller than the other side. Animals that are not symmetrical often have genetic defects. A high-producing American Holstein gives twice as much milk, but she may burn out after only two lactations. There are now concerns about loss of functionality (Rodenbury and Turner, 2012). A New Zealand grass-fed Holstein lasts twice as long. Over the years, as milk production rose, the ability to reproduce and produce a calf declined (Spencer, 2013). The producer wants to get the short-term economic gains, but in the long run, economics may be worse due to a lack of disease resistance, poor reproduction, high cost of replacement heifers, and higher feed costs. Since the first edition of this book, some progressive dairy farmers have switched to slightly less productive genetics and the cow now will last for three or more lactations.

Quality and quantity of meat are two opposing goals. Beef cattle and pigs fed too many beta-agonists had greater muscle mass and tougher meat (Xiong *et al.*, 2006; Fernandez-Duenas *et al.*, 2008; Arp *et al.*, 2013). Meat companies who want high-quality meat have banned or greatly restricted these products in their programs. Unfortunately, there are some meat companies who pay a premium for animals with a high percentage of lean meat. This has provided an economic incentive to overuse beta-agonists, which has resulted in many lame cattle. Lameness and muscle stiffness increases during hot weather and progressively worsens as cattle move from the feedlot to the abattoir. They act as if their muscles become easily fatigued. Pigs fed ractopamine are more susceptible to stress during handling (James *et al.*, 2013), and are difficult to handle (Marchant-Forde *et al.*, 2003). During the

hot summer weather, cattle fed beta-agonists had higher death losses (Longeragan *et al.*, 2014). Zilpaterol was removed from the US market in 2014 due to welfare concerns (Lyles and Calvo-Lorenzo, 2014).

The growth hormone, rBST, causes problems for dairy cows unless it is used in a very carefully managed program. It can cause excessive loss of body condition and increased mastitis (Willeberg, 1993; Kronfield, 1994; Collier *et al.*, 2001). People who are assessing welfare in feedlots and farms that use these products should evaluate animal-based outcomes, such as body condition and lameness, and host stress symptoms, such as panting, carefully.

Auditors Must Avoid Conflict of Interest

It is important that the person who is evaluating a farm or slaughter plant does not have a financial conflict of interest. Audits for animal welfare and food safety have the greatest credibility when they are conducted by either a third party, an independent auditor, a representative from the company that is buying the meat, a government employee, or an employee of the meat company that has a contract with the producer to grow livestock or poultry. A farm's regular veterinarian has a conflict of interest. If he or she is too strict, the farm may fire him or her. This situation would be like a traffic policeman giving his boss a fine for speeding. To prevent bribery, auditors must be paid enough so they will be less likely to accept bribes. The farm veterinarian plays a very important role in helping his or her clients to obey the standards. The farm veterinarians should conduct internal audits on a regular basis so that his or her clients will be ready for the external audit. It is also a conflict of interest for an auditor to be profiting from selling equipment, drugs, feed, or any other service to the people who are being audited.

Effect of the Price of Grain and Biofuel Crops on Welfare and the Destruction of Pasture

In the USA, the beef industry started putting cattle in feedlots during the 1960s because grain was cheap. Cheap grain provided an incentive to take cattle off pasture and raise them in feedlots. The pork and poultry industries also expanded greatly due to the availability of cheap grain. Since the early 2000s, the use of grain to make ethanol has increased around the world. This has resulted in high grain prices. The author visited Brazil and Uruguay in 2007, 2008, and 2014, and has learned that the high prices and demands from export markets is an economic incentive for turning pastureland into crops. Up to 30% of the prime pastureland in Uruguay has already been converted to growing soybeans. Growing eucalyptus trees for biofuel may convert more pastureland. Cattle are being moved to areas with poorer soil (La Manna, 2014). New varieties of high-producing crops such as palm oil and eucalyptus may provide economic incentives to cultivate them in new areas (Carrasco *et al.*, 2014). In Brazil, more and more cattle are being moved off pasture into feedlots. A similar situation is also occurring in Argentina and Uruguay. Much of Argentina's grain crop goes to Europe, and the Argentinean treasury receives huge amounts of income in grain taxes. It has been estimated that taxes on exported grain may produce 80% of the country's tax revenue (Nation, 2008). The main factor that prevents pasture conversion to crops in Uruguay are government tax incentives. Ranchers in the USA have also converted more of their land to grain. In the USA, county extension agents in Illinois report that hilly pastureland that is not really suited for crops is being planted with crops. In 2014, the trend slowed due to a drop in maize prices. Unfortunately, large amounts of prime native prairies had been plowed up and converted. The USA still has vast amounts of western ranges where grazing is the only way food can be produced on this land. In New Zealand, dairy cattle pasture has taken over the flatland and the sheep and beef cattle have moved to the hills (Morris and Kenyon, 2014).

To have good animal welfare in a feedlot requires dry ground. The author has observed from traveling all over the USA, Canada, Australia, New Zealand, and Central and South America that, in areas where the rainfall is greater than 50 cm (20 inches) of rainfall/year, it is very difficult to keep the surface of the feedlot pens dry. This is the reason why so many US feedlots are in the high plains area, where rainfall is less. In the 1970s, ranchers tried building dirt feedlots in the rainy south-eastern USA, but they gave it up due to the mud. Some people are moving cattle back to the rainy Midwestern grain-growing areas to take advantage of feeding ethanol by-products, which

are wet and more expensive to ship than dry grain. This has resulted in some cattle being put in indoor feedlots. Feeding cattle high grain rations for long periods when they are housed on concrete has resulted in feet and leg problems (see Chapter 16, this volume).

Welfare Problems Caused by War and Corruption

Unstable or corrupt conditions in a developing country may create economic incentives that motivate people to abuse animals severely. An extension animal scientist, who cannot be named due to concerns for his safety, has worked in a country where corruption and bad decisions by the government have provided an economic incentive to the local people to use atrocious management practices. Water buffalo and local cattle are given so much rBST (growth hormone) and oxytocin that they are milked until they have become emaciated skeletons. The dairies are located on the perimeter of a large city. Due to war and other conditions, there is no economic incentive to rebreed the animals for another cycle of lactation. The animals are just "used up" and then dealers bring in more animals that they purchase from destitute farmers who live in the surrounding countryside. The drugs are cheap copies that are obtained from firms in Asia.

Well-intentioned Legislation with Bad Consequences

Over the years, the author has observed many situations where well-intentioned legislation or activist activities may be detrimental to improving animal welfare. Prime examples of these are the laws in the USA banning horse slaughter for human consumption. The closure of two out of the three US horse slaughter plants has already resulted in unwanted horses being transported even further distances to either Canada or Mexico for slaughter. Some have been transported all the way to Mexico City. Live horses are also being shipped to Japan. When the Humane Society of the USA lobbied the government to pass this law, nobody thought about worse fates that some unwanted horses could suffer. The fates that are worse than slaughter in Texas or Illinois are: (i) longer transport times; (ii) transport under substandard conditions in Mexico; (iii) being neglected and left to starve in the desert (high hay and grain prices have made this problem worse); and (iv) being ridden and worked in Mexico until they become totally debilitated. The author has seen these worse fates and they are awful. Horse slaughter became such an emotional issue that animal advocates chose to ignore the observations of people in the field that indicated that there were worse fates than slaughter in a US plant.

Animal welfare can also become worse when legislation makes standards so strict that an animal industry is shut down in one country and production is transferred to another country that has atrocious standards. Eggs are now being imported from Eastern Europe into Western Europe. The standards for animal welfare in Eastern Europe are poor. Exporters of eggs, milk, and meat should be required to adhere to the standards of the importing country.

Conclusion

An understanding of how economic factors affect the way in which farm animals are treated will help policy makers to improve welfare (Quick Access Information 14.1 and 14.2). Holding people financially accountable for losses or providing incentives for low losses will improve animal treatment greatly. The wise use of the tremendous purchasing power of large meat buyers has already brought about some dramatic improvements. Unfortunately, there are also many situations where economics is detrimental to animal welfare. One of the worst problems is pushing animals past their biological limits by either single-minded selection for production or the overuse of performance-enhancing substances.

Quick Access Information 14.1. Economic factors that improve animal welfare:

- Meat buyers such as importing country or retail buyers enforce welfare standards.
- Paying premiums for high welfare to producers.
- Financial accountability for bruises and death losses during transport.
- Use of livestock identification and trace-back to farm of origin. Enables accountability for losses.

Quick Access Information 14.2. Economic factors that make animal welfare worse:

- Old cull breeding animals have little economic value.
- Segmented marketing chains with many dealers where losses can be passed on to the next segment. No financial accountability for bruises, death losses, or sickness.
- Overloading the animal's biology to increase production.
- Paying handlers based on speed of handling and loading animals or paying transporters based on the maximum weight a truck can carry. This encourages rough handling and overloading of vehicles.
- Well-intentioned legislation with bad consequences.
- Pastureland being converted to crops and biofuels.

References

Arp, T.S., Howard, S.T., Woerner, D.R., Scanga, J.A., McKenna, D.R., *et al.* (2013) Effects of ractopamine hydrochloride and Zilpaterol hydrochloride supplementation on longissimus muscle shear force and sensory attributes of beef steers. *Journal of Animal Science* 91, 5989–5997.

Arthington, J.D., Qiu, X., Cooke, R.F., Vendramini, J.M.B., Araujo, D.B., *et al.* (2008) Effects of preshipping management on measures of stress and performances of beef steers during feedlot receiving. *Journal of Animal Science* 86, 2016–2023.

Bulot, H. and Lawrence, J.D. (2006) The value of third party certification of preconditioning claims in Iowa. Department of Economics Working Papers, Iowa State University, Ames, Iowa.

Carrasco, L.R., Larrosa, C., Milner-Gulland, E.J. and Edwards, D.P. (2014) Conservation. A double edged sword for tropical forests. *Science* 346, 38–40.

Collier, R.J., Byatt, J.C., Denham, S.C., Eppard, P.J., Fabellar, A.C., *et al.* (2001) Effects of sustained release bovine somatotropin (sometribove) on animal health in commercial dairy herds. *Journal of Dairy Science* 84, 1098–1108.

Douphrate, D.I., Rosecrance, J.C., Stallones, L., Reynolds, S.J. and Gilkey, D.P. (2009) Livestock handling injuries in agriculture: An analysis of Colorado workers' compensation data. *American Journal of Industrial Medicine* 52, 391–407.

Edwards, L.N., Grandin, T., Engle, T.E., Porter, S.P., Ritter, M.J., *et al.* (2010) Use of exsanguination blood lactate to assess the quality of preslaughter handling. *Meat Science* 86, 384–390.

T. Grandin

Etter, L. (2008) Have knife will travel: A slaughterhouse on wheels. *Wall Street Journal*, 5 September, p. 1.

Ewbank, R. and Parker, M. (2014) Handling cattle raised in close association with people. In: Grandin, T. (ed.) *Livestock Handling and Transport*, 4th edn. CAB International, Wallingford, UK, pp. 77–93.

Fernandez-Duenas, D.M., Myers, A.J., Scramlin, S.M., Parks, C.W., Carr, S.N., *et al.* (2008) Carcass meat quality and sensory characteristics of heavy weight pigs fed ractopamine hydrochloride (Paylean). *Journal of Animal Science* 86, 3544–3550.

Franzluebbers, A.J., Paine, L.K., Winsten, J.R., Krome, M., Sanderson, M.A., *et al.* (2012) Well-managed grazing systems: A forgotten hero of conservation. *Journal of Soil and Water Conservation* 67, 100A–104A, doi:10.2489/jswc.674.100A.

Grandin, T. (1981) Bruises on Southwestern feedlot cattle. *Journal of Animal Science* 53 (Supplement 1), 213 (Abstract).

Grandin, T. (1985) Improving pork quality through handling systems. In: *Animal Health and Nutrition*. Watt Publishing, Mount Morris, Illinois, pp. 14–26.

Grandin, T. (1987) Animal handling. *Veterinary Clinics of North America, Food Animal Practice* 3, 323–338.

Grandin, T. (1988) Double rail restrainer conveyor for livestock handling. *Journal of Agricultural Engineering Research* 41, 327–338.

Grandin, T. (1998a) Objective scoring of animal handling and stunning practices in slaughter plants. *Journal of the American Veterinary Medical Association* 212, 36–39.

Grandin, T. (1998b) Handling methods and facilities to reduce stress on cattle. *Veterinary Clinics of North America, Food Animal Practice* 14, 325–341.

Grandin, T. (2003) Transferring results from behavioral research to industry to improve animal welfare on the farm, ranch, and slaughter plant. *Applied Animal Behaviour Science* 81, 215–228.

Grandin, T. (2005) Maintenance of good animal welfare standards in beef slaughter plants by use of auditing programs. *Journal of the American Veterinary Medical Association* 226, 370–373.

Grandin, T. (2007) Introduction: Effect of customer requirements, international standards and marketing structure on the handling and transport of livestock and poultry. In: Grandin, T. (ed.) *Livestock Handling and Transport*. CAB International, Wallingford, UK, pp. 1–18.

Grandin, T. (2013) *Recommended Animal Handling Guidelines and Audit Guide*, 2013 edn. American Meat Institute Foundation, Washington, DC. Available at: www.animalhandling.org (accessed 9 May 2015).

Grandin, T. (2014) Animal welfare and society's concerns, finding the missing link. *Meat Science* 98, 461–469.

Grandin, T. and Gallo, C. (2007) Cattle transport. In: Grandin, T. (ed.) *Livestock Handling and Transport*. CAB International, Wallingford, UK, pp. 134–154.

Gruber, S.L., Tatum, J.D., Grandin, T., Scanga, J.A., Belk, K.E., *et al.* (2006) Is the difference in tenderness commonly observed between heifers and steers attributable to differences in temperament and reaction to pre-harvest stress? Final Report, submitted to the National Cattlemen's Beef Association, Department of Animal Sciences, Colorado State University, Fort Collins, Colorado, pp. 1–38.

Hambrecht, E., Eissen, J.J., Newman, D.J., Smits, C.H.M., Verstegen, M.W. and den Hartog, L.A. (2005) Preslaughter handling affects pork quality and glycolytic potential in two muscles differing in fiber type organization. *Journal of Animal Science* 83, 900–907.

James, B.W., Tokach, M.D., Goodband, R.D., Nelssen, J.L., Dritz, S.S., *et al.* (2013) Effects of dietary L-carnitine and ractopamine HCl on the metabolic response to handling in finishing pigs. *Journal of Animal Science* 91, 4426–4439.

Jiang, C., Xing, F., Xing, J., Jiang, Y. and Zhou, E. (2013) Different expressive patterns of PRRV mediator genes in lung tissue of PRRVS resistant and susceptible pigs. *Developmental and Comparative Immunology* 39, 127–131.

Johnson, R., Petry, D. and Lurney, J. (2005) Genetic resistance to PRRS studied. Swine research review. *National Hog Farmer*, 15 December, pp. 19–20.

Knowles, T.G., Kestin, S.C., Haslam, S.M., Brown, S.N., Green, L.E., *et al.* (2008) Leg disorders in broiler chickens: Prevalance, risk factors and prevention. *PLoS ONE* 3(2), doi:10.1371/journal.pone.0001545.

Kronfield, D.S. (1994) Health management of dairy herds treated with bovine somatotropin. *Journal of the American Veterinary Medical Association* 204, 116–130.

La Manna, A. (2014) Intensification of beef production in Uruguay. Contributions of sustainable products and processes. 60th International Congress of Meat Science and Technology, 17–22 August 2014, Punta del Este, Uruguay.

Lindahl, C. (2014) Risk factors for occupational injuries during cattle handling on Swedish dairy farms. Acta Universitatis Agriculture Sueciae (1652-6880), Dissertation, Swedish University of Agricultural Sciences.

Loneragan, G.H., Thomson, D.U. and Scott, H.M. (2014) Increased mortality in groups of cattle administered the B-adrenergic agonists ractopamine hydrochloride and zilpaterol hydrochloride. *PLoS One* 9(3), doi:10.1371/journal.pone.0091177.

Lucas, M., Day, L. and Fritschi, L. (2013) Serious injuries to Australian veterinarians working with cattle. *Australian Veterinary Journal* 91, 57–60.

Lyles, J.L. and Calvo-Lorenzo, M.S. (2014) Practical developments for managing animal welfare in beef cattle: What does the future hold? *Journal of Animal Science* 92, 5334–5344.

Marchant-Forde, J.N., Lay, D.C. Jr, Pajor, E.A., Richert, B.T. and Schinckel, A.P. (2003) The effects of ractopamine on the behavior and physiology of finishing pigs. *Journal of Animal Science* 81, 416–422.

Maria, G.A., Villarroel, M., Chacon, C. and Gebresenbet, C. (2004) Scoring system for evaluating stress to cattle during commercial loading and unloading. *Veterinary Record* 154, 818–821.

Meeker, D.L., Rothschild, M.F., Christian, L.L., Warner, C.M. and Hill, H.T. (1987) Genetic control of immune response to pseudorabies and atrophic rhinitis vaccines. *Journal of Animal Science* 64, 407–413.

Morris, S.T. and Kenyon, P.R. (2014) Intensive sheep and beef production from pasture – A New Zealand perspective of concerns, opportunities, and challenges. Meat Science 98, 330-335.

Nation, A. (2008) High grain prices are creating more confinement grain feeding in Argentina. *The Stockman Grass Farmer*, January, 1–4.

National Cattlemen's Beef Association (1994) *Strategic Alliances Field Study*. Coordination with Colorado State University, Texas A&M University and the National Cattlemen's Beef Association, Englewood, Colorado.

Paranhos da Costa, M.J.R., Huertas, S.M., Gallo, C. and Dalla Costa, O.A. (2012) Strategies to promote farm animal welfare in Latin America and their effects on carcass and meat quality traits. *Meat Science* 92, 221–226.

Paranhos da Costa, M.J.R., Huertas, S.M., Gallo, C. and Strappini, A.C. (2014) Handling and transport of cattle and pigs in South America. In: Grandin, T. (ed.) *Livestock Handling and Transport*. CAB International, Wallingford, UK, pp. 174–192.

Rodenburg, T.B. and Turner, S.P. (2012) The role of breeding and genetics in the welfare of farm animals. *Animal Frontiers* 2, 16–21.

Roeber, D.L., Mies, P.D., Smith, C.D., Belk, K.E., Field, T.G., Tatum, J.D., *et al.* (2001) National market cow and bull beef quality audit, 1999: A survey of producer-related defects in market cows and bulls. *Journal of Animal Science* 79, 658–665.

Rutherford, B. (2014) Keeping 'girls' home. *Beef Magazine*, September 2014, 51, 21–26.

Sherwin, C.M., Richards, G.J.B. and Nicol, C. (2010) Comparison of the welfare of laying hens in four housing systems in the UK. *British Poultry Science* 51, 488–499.

Smith, G. (1998) *Moving 'Em, A Guide to Low Stress Animal Handling*. Graziers, Hui, Kamuela, Hawaii.

Sorge, U.S., Cherry, C. and Bender, J.B. (2014) Perception of the importance of human–animal interactions on cattle flow and worker safety on Minnesota dairy farms. *Journal of Dairy Science* 97, 4632–4638.

Spencer, T.E. (2013) Early pregnancy: Concepts, challenges and potential solution. *Animal Frontiers* 3, 48–55.

Suther, S. (2006) CAB (Certified Angus) drovers survey: Benchmarks for producer practices and opinions. *Drover's Journal*, Vance Publishing, Lenexa, Kansas, pp. 18–19.

Texas A&M University (1998) *Ranch-to-Rail Statistics 1992–1998*. Department of Animal Science, Texas A&M University, College Station, Texas.

Thrift, F.A. and Thrift, T.A. (2011) Review: Update on preconditioning beef calves prior to sale by cow-calf producers. *The Professional Animal Scientist* 27, 73–82.

Troxel, T.R., Barham, B.L., Cline, S., Foley, D., Hardgrave, R., *et al.* (2006) Management factors affecting selling prices of Arkansas beef calves. *Journal of Animal Science* 84 (Supplement 1), 12 (Abstract).

Tuyttens, F., Heyndrickx, M., DeBoeck, M., Moreels, A., Nuffel, A.V., *et al.* (2008) Broiler chicken health, welfare and fluctuating asymmetry in organic versus conventional production systems. *Livestock Science* 113, 123–132.

Voisinet, B.D., Grandin, T., O'Connor, S.F., Tatum, J.D. and Deesing, M.J. (1997) *Bos indicus* cross feedlot cattle with excitable temperaments have tougher meat and a higher incidence of borderline dark cutters. *Meat Science* 46, 367–377.

Waggoner, J.W., Mathis, C.P., Loest, C.A., Sawyer, J.E. and McCollum, F.T. (2006) Impact of feedlot morbidity on performance, carcass characteristics and profitability of New Mexico ranch-to-rail steers. 2006 *Cattle Growers' Short Course Proceedings & Livestock Research Briefs*. New Mexico State Univ. p. 72.

Walker, K.H., Fell, L.R., Reddacliff, L.A., Kilgour, R.J., House, J.R., *et al.* (2006) Effects of yard weaning and behavioural adaptation of cattle to a feedlot. *Livestock Science* 106, 210–217.

Wilkins, L.J., McKinstry, J.L., Avery, N.C., Knowles, T.G., Brown, S.N., *et al.* (2011) Influence of housing system and design on bone strength and keel bone fractures in laying hens. *Veterinary Record* 169, 414.

Willeberg, P. (1993) Bovine somatotropin and clinical mastitis: Epidemiological assessment of the welfare risk. *Livestock Production Science* 36, 55–66.

Xiong, Y.L., Gower, M.J., Li, C., Elmore, C.A., Cromwell, G.L. and Lindemann, M.D. (2006) Effect of dietary ractopamine on tenderness and postmortem protein degradation of pork muscle. *Meat Science* 73, 600–604.

T. Grandin

15 Improving Animal Care and Welfare: Practical Approaches for Achieving Change

HELEN R. WHAY AND DAVID C.J. MAIN

University of Bristol, Langford, Bristol, UK

Summary

There are vast amounts of knowledge that can help farmers improve the care of their animals. Sometimes, it is difficult to encourage farmers to use them. Measuring problems such as the percentage of lame (difficulty in walking) dairy cows is effective for motivating the best producers to improve. Unfortunately, it has less effect on the poorer producers. Various educational approaches can help farmers recognize problems and motivate them to improve facilities. Some of these methods are visits from veterinarians and agricultural facilitators (educators), and charts to track improvements. A social marketing approach can also be used to help farmers take ownership of the problems on their farm.

Another educational approach is farmers visiting other farms, and participation in discussion groups. Change can also be induced with the enforcement of standards by either retailers or legislation. Enforcements are most effective for maintaining minimum standards. One approach is to start with enforcements and, after producers improve, provide incentive payments to reward the best farmers. Benchmarking is often an effective motivator because it enables each farmer to see how he or she compares to other producers. It is recommended for measures that can be quantified easily, such as the percentage of animals that are lame, have poor body condition, are dirty, or have swollen hocks.

Learning Objectives

- Learn how outcome measures can be used to motivate farmers to reduce lameness

- How to use social marketing methods to promote improvements on the farm
- How to use participatory tools such as a seasonal lameness calendar, priority matrix, and lameness transect walk
- Organization of farmer groups where different farms are visited to share ideas
- Effective use of enforcement approaches by either retailers or legislation

Introduction

This chapter addresses the implementation of knowledge to improve animal care and welfare. There is a vast wealth of information available about what management practices, activities, and processes lead to improvements in animal welfare. Such information comes from the work of animal welfare scientists and agricultural scientists, or is captured from the experiences of farmers, livestock hauliers, auctioneers, and abattoir staff, and is published extensively elsewhere. However, this chapter will not cover the technical knowledge of how to meet animals' requirements. This chapter is about the process of taking this vast wealth of knowledge and translating it into actions by those who have responsibility for animals. In short, this chapter is about changing the behavior of humans.

Human behavior change is a science in itself, commonly associated with the term "social medicine". For years, workers in human medicine have been trying to encourage groups within the population to quit smoking, lose weight, drink less alcohol, practise safe sex, or eat more fruit and vegetables. This need to change human behavior has also extended

---

beyond health to encompass activities such as recycling refuse, saving water, reducing speed on roads, wearing seat belts while driving, and so on. The relevance of this for animal welfare is that we have to recognize that the lives of modern domesticated animals, including production animals, are under the total and absolute control of humans. Humans decide what food they have access to, how much, and when. Humans decide what environment they should live in, how much bedding should be available, how clean and dry it will be, and how much light they will have to see it by. Humans control animal health; they choose management strategies that might promote or prevent the transmission of infectious disease, they decide on vaccination programs, and decide when and if treatment is offered. Humans even control how and when an animal will die. Once we accept that animals have virtually no genuine control over their own lives, then it becomes obvious that in order to achieve welfare improvement, we need to work with the humans responsible for their care. This may come as something of a surprise to those of us who entered agriculture or veterinary science because we thought we wanted to work with animals.

Targets for Behavior Change: Who Do We Want to Influence?

If we accept that humans are the mediators of animal welfare, it is worth looking at the various players who have a stake in implementing change. Animal carers, in this case, farmers, transporters, and abattoir staff, are the most obvious targets for animal welfare behavior change interventions. The knowledge generated by science needs to feed through to and be implemented on the ground by those people who have direct control over animal lives. It is important to recognize that within this group there are people who behave in a competitive, entrepreneurial way and who automatically and regularly implement changes to the management of their livestock. There are also those who want to do the best for the livestock under their care and who do spontaneously make some management changes but find keeping up with and filtering current knowledge a challenge, perhaps finding that translating information so that it is relevant to their own situation is difficult, and who often struggle to divert from existing routines in order to try out new ideas. Finally, there is the group who, again, would like to manage

their stock in the best way possible but always seem to be paddling hard just to stay still and find virtually no opportunities to keep up with the latest information or to implement management changes; these are the "crisis management" folks. Beyond the producers there are, of course, many other groups with an interest in implementing behavior change: farm advisors and sales representatives, farm assurers and standard-setting bodies, legislators, animal welfare charities and campaigners, veterinary surgeons and animal health technicians, animal welfare scientists, retailers, and in some cases the purchasers, of the final product. These different interest groups often come with different motivations for wanting to see changes implemented by livestock carers. Most, although possibly to varying degrees, would like to see change implemented to make the lives of animals better. A farm advisor is also looking for the credibility and enhancement of reputation that comes with their advice actually being implemented on the farm. Legislators are interested in seeing legislation properly implemented, not only because laws are introduced for a purpose but also because failure to implement legislation undermines its role within society. Animal welfare scientists hope to see their work taken up and used; but further to that, the process of implementation may allow them to see how well their findings actually work at the farm level. Then, as a final example, there are the all-important, but often overlooked, customers. Customers of animal products have tremendous power to influence animal welfare; they just do not know they have it. Unfortunately only a relatively small proportion of food customers purchase primarily on the basis of the animal welfare provenance of their food. However, interest in purchasing is growing, and a survey carried out by a UK supermarket (Talking Retail, 2008) found that their customers rated animal welfare as a greater purchasing motivation than environmental issues. Many of the interest groups described above can be seen as having a stake in seeing animal carers implement changes on their farms, but it is also worth remembering that these interest groups can also be targets for behavior change interventions themselves.

So far, this chapter has highlighted that changing human behavior is the route to improving animal welfare, and that many groups have an interest in seeing changes implemented. However, if it were as simple as "wish and it be done", then it probably

H.R. Whay and D.C.J. Main

would not merit a chapter in this book. Unfortunately, there is considerable evidence that humans often find changing their behavior extremely difficult; one only has to look at the number of people still smoking cigarettes or trying to lose weight to appreciate the magnitude of the problem. This challenge is then magnified by the fact that implementing animal welfare improvements is about asking people to make changes on behalf of a third party: the animal. If people find it difficult to implement changes such as taking daily exercise or reducing alcohol consumption that protect their own health, it is clearly a challenge to ask people to make changes to improve the lives of animals.

Dairy cow lameness resistant to improvement

An example from the agricultural sector illustrating how knowledge and change have failed to be implemented successfully is dairy cattle lameness in the UK. Since the late 1980s, information about the levels of lameness (difficulty in walking) among the UK herd has been collected regularly (Clarkson *et al.*, 1996; Whay *et al.*, 2003). These data tell us that on any one day, between 21% and 39% of the national dairy herd are likely to be lame. Unfortunately, lameness has worsened since the first edition of this book, despite the huge wealth of available information about methods for reducing lameness. When asked why new lameness management strategies are not being implemented, farmers most commonly report lack of time, skilled labor, or the unpopularity of the task (Leach *et al.*, 2013). Throughout this chapter, examples relating to dairy cattle lameness will feature quite prominently; this is because it is an area where there has been significant investment in behavior change research and an area in which the authors have first-hand experience.

Influencing the best and poorest producers to improve

In order to bring about a change in behavior to address an animal welfare problem, there must be genuine awareness that a problem exists, coupled with knowledge about possible solutions. Unfortunately, awareness of a problem and knowledge of a solution do not necessarily lead to changes on the farm. It is the job of those who are facilitating change to create situations and present options for change that will help people to take the step that leads to actual changes in animal care practices.

Not all people are susceptible to the same types of intervention to bring about improvement. Producers can be divided into three categories according to which type of support they are most likely to respond to. (i) There are producers whose animals have the poorest welfare and who are unlikely to access new knowledge and change poor practices voluntarily. This group is only susceptible to enforcement as a means of promoting change. This might be enforcement of legislation and codes of practice, and using tools such as farm assurance schemes to set minimum standards and ensure compliance before products can reach the marketplace. (ii) In the middle are the majority of producers, who are likely to make changes with a combination of encouragement and enforcement. They are unlikely to actuate change for themselves so require external contact to initiate and, to some extent, sustain the process. (iii) At the other end of the spectrum are the best producers, who are self-motivated and self-actuating to a high degree; they tend to be competitive, prepared to take risks, and are regularly looking for new market opportunities. This group does not have to be forced to change but benefits from receiving access to new knowledge and scientific findings. These top producers are rewarded by being able to produce premium-priced products, cornering market share, or accessing niche marketing opportunities. This group has a tendency not to feel the need to either share or receive information from their peers.

The following sections in this chapter will review some of the currently available approaches for both encouraging and enforcing change.

Use Measurement to Raise Awareness of Lameness

Dairy farmers are often not fully aware that they have a serious problem. When asked to estimate the percentage of lame cows, the percentage is greatly underestimated (Fabian *et al.*, 2014). Estimates may be less than half. Recent research has shown that actually measuring the percentage of lame cows can motivate farmers to reduce lameness. Over a 3-year period, having a trained evaluator score lameness on a four-point scale as cows exited the milking parlor had significant beneficial effects. Lameness dropped from 35% to 22% in a sample of 128 dairy farms (Main *et al.*, 2012).

Measurement motivates the better farmers to improve

The better dairy farmers who took pride in having a healthy herd had a lower percentage of lame cows at the beginning of a 3-year study. When they received feedback from lameness scoring, they were motivated to reduce lameness further (Leach *et al.*, 2013). The poorer dairies that had higher percentages of lame cows also had a higher tolerance for lame cows. They stated higher percentages when asked "what proportion of cows lame at one time do you consider a problem in your herd?" (Leach *et al.*, 2010). They were less likely to make improvements after being provided with lameness measurements.

Effects of additional interventions to motivate improvements

Farmers at 128 dairies were provided with either lameness measuring only or additional supports such as visits from a veterinarian and meetings with other farmers. These additional supports motivated some of the dairies to make improvements to reduce lameness, such as modifying flooring surfaces and better hoof trimming. This chapter will discuss how additional methods can be used to motivate positive improvements.

A recent survey of 108 dairies in Minnesota, USA, showed that stockmanship training was associated with higher milk production (Sorge *et al.*, 2014). Unfortunately, 50% of the dairy managers were not interested in additional stockmanship training (Sorge *et al.*, 2014). There is definitely a group of producers who are eager to improve and another group who are resistant to change.

Encouragement as a Means of Implementing Change

The key principle of encouraging sustainable behavior change is to transfer ownership of both the problem and the solution to the person, in this case the producer or farmer, responsible for implementing change. The golden rule to remember is that telling people what to do, no matter how encouragingly, simply does not work – no matter how wonderful,

strikingly obvious, or just plain sensible the information is, no matter how beautifully and enthusiastically presented the solutions may be, and no matter how simple and straightforward the change you are hoping to see implemented. This point is well illustrated in the excellent book, *Fostering Sustainable Behavior* (McKenzie-Mohr and Smith, 1999), where the authors Doug McKenzie-Mohr and William Smith give the example of some work done by Scott Geller (Geller, 1981). Geller and colleagues ran a series of workshops for householders providing information about the importance of home energy conservation and providing advice on simple measures that could be implemented by each householder to conserve energy. The workshops were designed to give the householders the message that it was entirely possible for them to make these changes in their own homes. The attitude of the participating householders was measured before and after the workshops and it was found that following the workshop, attendees reported a much greater awareness of energy conservation issues and an intention to implement energy-saving measures within their own homes. However, this change in attitude was not translated into a change in behavior. During follow-up visits to 40 of the participating households, it transpired that only one person had followed through on the recommendation to lower the thermostat on the hot-water boiler. Two participants had placed blankets around their boilers as advised, but it transpired that they had done this before

---

Quick Access Information 15.1. Key aims of an encouragement approach to improve conditions on a farm.

- Pass ownership of the problem and finding solutions to the farmer.
- Give farmers the opportunity to rehearse mentally any changes they might make, and even encourage them to try out changes before implementing them fully.
- Encourage farmers to discuss their problems with colleagues on other farms – it is very valuable for them to know that making changes is a normal behavior.
- To increase awareness, measure the percentage of animals that have a problem such as lameness, poor body condition, injuries, or are dirty.

H.R. Whay and D.C.J. Main

attending the workshop. Indeed, the only significant change to be implemented was that eight of the workshop participants had installed low-flow showerheads, though this was not precipitated by the advice alone, as each participant who attended the workshop was provided with a free low-flow showerhead.

How to get farmers to take ownership of problems and be partners in generating solutions

The concept of ownership of change is much more subtle than simply saying to a producer "it is you who has to implement these changes so therefore you are the owner of them". It is also not quite as cynical as trying to make someone think that the changes were their idea in the first place. Giving someone ownership of change is about creating opportunities for them to explore and realize their problem and allowing them to be a partner in generating ideas for possible solutions. Pointing out to a farmer that their animals are suffering from a welfare problem and even going as far as suggesting they may be breaching laws or codes of practice is a challenge to their dignity and professionalism and can easily be seen as a hostile act. To return to the previously used example of dairy cattle lameness: imagine a scenario where a farmer is aware that he or she has lame cows in the herd but has not yet considered the implications of this lameness. By asking a series of questions about what effect it has on the cows, how the cows feel when they are lame, what the time costs are for the farm workers, and in what ways lame cows cost money, the farmer is able to build up a picture of all the potential consequences of having lame cows on the farm and to rationalize how important these pitfalls are. So now the problem belongs to the producer and is personal to their farm. Others may not accept that lameness is a problem until they hear about and discuss the problem with other farmers in their area. Likewise, ownership of solutions is important because there are rarely one-size-fits-all solutions to problems on farms, as no two farms operate in exactly the same way. While considering and discussing possible solutions, farmers need to run through implementing the change on their own farm in their mind; this process of rehearsing changes is an important step in achieving implementation.

In fact, rehearsing change may even extend as far as carrying out a series of tests of the desired change before committing to it fully. Again, it is also often important to hear from others that they too are trying out changes, and often it is more reassuring to hear from a fellow farmer that a particular action worked than to hear it from a farm advisor, veterinary surgeon, or welfare scientist.

In fact, encouraging change seems to be much more about creating the right environment for farmers to realize and address problems for themselves than about being too forceful and hands on in addressing a problem directly. However, it is worth remembering a few cautionary points. It will never be possible to achieve 100% compliance with any intervention; people are individuals and may well have other priorities in their lives that overshadow the purpose of your proposed area of intervention. People are often unpredictable and do not make decisions that we, from our own standpoint, consider logical (bearing in mind that to others we appear equally illogical). In a guide to human behavior change, Kerr *et al.* (2005) wrote: "interventions often make the mistake of assuming human behaviors are straight forwardly controlled by reason, attitude and will", which is a good reminder that people do not behave in a particular way just because we want them to. There also seems to be some basic law of intervention returns that dictates that you only achieve success commensurate with the effort you put in, and this appears to be true both for the facilitator and the implementer, so do not expect changing behavior to be easy and be prepared to commit sufficient resources to get the job done. Finally, the types of intervention described in this chapter are, of course, a form of manipulation, albeit well intentioned and on behalf of animals, which are unable to represent themselves. However, those who develop skills in this type of manipulation can give in to the temptation to manipulate others for their own Machiavellian reasons; this would be unethical and diminish all those who try to promote behavior change for the benefit of other humans and animals.

The following section describes three broad approaches (social marketing, participatory methods, and farmer groups) for encouraging behavior change that are currently being implemented and tested in the agricultural sector. Each will be considered in terms of its potential strengths and weaknesses.

A social marketing (type) approach

Social marketing is widely used to promote change among groups of people. It is an extension of the principles used in marketing and advertising. We are all familiar with the way advertisers try to persuade us to change our habits and purchase products by implying that owning a particular item will make us glamorous, popular, save us time, and so on. Advertisers also attempt to reassure us by letting us know that many others are making a similar purchase and offering us special incentives to encourage us to spend our money. Social marketing is also about trying to persuade people to do something different, and uses a range of tools to achieve this, but the two key differences from commercial marketing are that social marketing is: (i) directed towards encouraging change that affords a benefit to society (or in our case, to animals and their owners); and (ii) works very specifically to identify and overcome barriers to change.

In an animal welfare and agricultural context, we have found that we have been able to apply many of the tools of social marketing but have had to make some modifications to take into account the isolated nature of farmers. Farmers often work alone on their farms, they have very limited contact with others and their days involve completing a lot of repetitive, routine tasks. So, social marketing for farmers needs to include more contact with individuals than would normally be expected. To understand how social marketing works, the following example will illustrate the types of tools available while describing a project currently running in the UK to encourage dairy farmers to take action to reduce lameness in their dairy cattle.

Benefits and barriers

For every desired change in behavior, there will be both perceived benefits and perceived barriers. These may be internal, external, or both:

- **Internal benefits** may include a change that aligns with a person's ethical position (i.e. believing that they should do everything possible for the welfare of their animals), offering a feeling of pride or fellowship with others who are working on a similar problem.
- **External benefits** may include believing that the change will save time, offer economic benefit, or perhaps contribute to making other tasks on the farm easier as well. For example, keeping the feet of cows clean in order to reduce infectious lameness may also result in cleaner udders and faster milking times.
- **Internal barriers** to change may include a fear that the change will be inconvenient, time-consuming, and difficult to implement in existing routines. A further internal barrier might simply be fear of change itself.
- **External barriers** might include a lack of appropriate equipment; for example, the yard scraper may be inefficient and need repair, modification, or replacement. Other external barriers might be losing time on other important jobs, having a farm layout that makes the change difficult, or hearing from others that making changes is a waste of time and will not work.

It is important for anyone promoting behavior change to recognize that these perceptions of benefits and barriers exist. Also, that other potential routes of behavior change may exist that may be similar but will not be of equal animal welfare benefit. For example, a farmer may perceive that it is indeed important to keep the yards cleaner than they currently are but be reluctant to increase the frequency of yard scraping each day, and instead he or she proposes to flood wash the yards. This will keep the yards very clean and exploit the currently unused wastewater from the milking parlor. However, the reason for cleaning the yards was to help reduce digital dermatitis by creating a cleaner, dryer environment for the cows, and the flood washing will regularly soak the cows' feet, increasing their susceptibility to infectious lesions. Ensuring a proper understanding of benefits and barriers from the farmers' point of view is a cornerstone of social marketing, and running focus groups is a good way of doing this. It also allows learning of local colloquial terms that are used and finding phrases and quotes that will make sense when speaking to farmers about making changes.

Facilitation of positive change with farm visits

Each farmer receives a visit from a member of the behavior change project who has a good knowledge of cattle lameness and the types of changes needed to resolve the problem. The visit is not about giving advice but about helping the farmer to generate solutions that are appropriate to his or her own farm. The facilitator will walk around the farm

H.R. Whay and D.C.J. Main

with the farmer and ask questions about particular aspects of the farm that are likely to be risk factors for lameness. The facilitator will address barriers to change presented by the farmer by encouraging him or her to weigh them against potential benefits. The facilitator will also share the experiences of other farmers by describing the actions they have taken, and offer the contact details of other farmers (with their permission) who have found ways of tackling a similar problem. At the end of the facilitated visit, before leaving the farm, the facilitator will compile an action list of the changes the farmer has identified, including notes on who will be responsible for implementing each change (the farm manager, herdsman, tractor driver, etc.) and when the change is going to be implemented, along with a space to tick when the change has been introduced. This list is then left with the farmer for the coming year.

Social "norms": people do things that other people are doing

Norms describe the process of reassuring farmers that others are also making changes (i.e. that it is normal behavior to make changes to reduce lameness). Despite the fact that we mostly perceive ourselves as being individuals who plow our own furrow, social research shows that we are actually more comfortable and reassured when we know that others are doing the same thing. A good example that illustrated this point was an American advertisement run in the 1970s that showed a Native American shedding a tear over the widespread corruption of the environment. This was a powerful and effective campaign. Later, to capitalize on this success, a second advertisement was designed that showed people leaving their rubbish and waste at a bus stop while the tear-shedding Native American looked on. Much to the consternation of the advertisers, this poster precipitated an increase in littering at bus stops. What happened was that this second poster showed people that littering at bus stops was normal and something that everyone else does; in effect, it gave them permission to drop litter at the bus stop.

The way norms were used in the cattle lameness project was first to create an image and name for the project (illustrated in Fig. 15.1) so that all the participants were aware they belonged to a larger project in which others were involved and that they had a group identity they could be proud of.

Healthy Feet Project
Working together to reduce cattle lameness

Fig. 15.1. The lameness project logo, used as a "norm" to create a group identity and to let farmers know they are working together with other farmers to reduce lameness.

As previously described, norms are also created through describing what changes other farmers have made on their farms. This helps to address perceived barriers, but also acts to reassure each farmer that others are also making changes and overcoming problems. The activities of other farmers are relayed using verbal descriptions, photographs of what they have changed (with their permission), and using quotes (e.g. "You can fix things around the farm but you can't fix cows' feet if you damage them"). The project also produces a regular newsletter that features case examples of farms where changes have been implemented.

Commitment to positive change

Commitment on the part of project members is critical for sustaining behavior change. Although farmers who join a project such as the lameness project being described here are showing some degree of commitment simply by joining up, this is often more in the form of a spectator than a fully signed-up member. There are various techniques to encourage more positive commitment. People tend to like to think of themselves as behaving consistently, so by displaying signs of commitment to a project they are more likely to be loyal to the project and continue with any changes they have implemented. Within the lameness project, all participating farmers are given a jacket lapel badge and a car sticker of the project logo (shown in Fig. 15.1), and are encouraged to display them. Although this is a relatively small act, by showing others that they are part of the project, they are more likely to go on to take on the more challenging behavior change. To use an example from a different context (Goldstein *et al.*, 2007), it was found that a group of people who were asked to wear a lapel pin for a charitable cancer society were subsequently nearly twice as likely to donate money to the charity as people who were not asked to wear a lapel pin.

In the lameness project, pictures of changes that have been made are shown to other farmers also considering similar changes. This is not only useful in creating norms and overcoming barriers among the farmers considering change, but the farmer who has already implemented the change is very clearly told that others will be shown the photograph of the change he or she has made. This serves multiple purposes: by agreeing to allow the photograph to be shown to others, the farmer is agreeing to display a further outward sign of commitment to the project; knowing that others are seeing the photograph also encourages the farmer to sustain the change that has been made. This is reinforced further by asking permission to give their contact details to other farmers who might wish to come and view the changes.

A further area where commitment is promoted is through asking farmers to put their signature on the action plan that is drawn up during the facilitation visit. Permission is also sought to compile a list of farmers who have signed up to their action plan so that it can be posted on the project website. Again, this is to encourage as much outward display of commitment to the project as possible.

Prompts with posters and educational materials

Prompts are aides-memoire to help remind people of their intended changes in behavior. Although people's intentions to change a particular practice or habit are generally good, new activities can be forgotten easily or slip the mind, especially when they involve making changes to existing routines or when people find themselves under time pressure. Prompts need to be targeted as closely as possible to the desired behavior change; very generic prompts tend to be less effective. In the lameness project, extremely targeted prompts have so far proved difficult to create because there is such a huge diversity of different behavior changes being initiated. To date, the prompts that have been used (besides the logo and strap line shown in Fig. 15.1) are cartoons targeting specific activities (an example is shown in Fig. 15.2).

Fig. 15.2. A cartoon prompt to remind farmers to work on early detection and treatment of lameness. (Cartoon by Steve Long.)

A catalog of suppliers of equipment, services, and materials that are commonly needed when making lameness-reducing changes is presented to the farmer at the time when the facilitated action list is generated. The catalog is intended to prompt picking up the telephone and placing an order or booking a service, etc., as a common stalling point for action is farmers saying they do not know where to buy a material (e.g. wood shavings to spread on cows' beds to increase their lying comfort); the catalog overcomes this. Farmers whose action list involves introducing or changing foot bathing routines are given a small laminated card that gives the appropriate dilutions for different footbath chemicals and explains in very straightforward measures how to achieve the dilution for a range of footbath sizes. Finally, a poster has been made that highlights the financial losses associated with lameness (Fig. 15.3). This is only used once farmers have begun to initiate behavior changes. Contrary to popular opinion, especially in the agricultural sector, promoting the financial benefit of behavior change rarely precipitates action. However, once behaviors are changed and people start to notice the financial benefit for themselves, it is a useful tool in sustaining behavior change.

Incentives Motivate Improvements

Incentives can be very powerful tools for behavior change; they may take the form of financial rewards or penalties, offers of approval or acceptance, or small "drug-company type" gifts. It is remarkable how motivated people can become to earn relatively small "gifts", which in reality they could purchase for themselves for much less effort. This phenomenon is regularly observed in the relationship between pharmaceutical companies and veterinary surgeons who will work hard to promote a particular drug

COSTS OF LAMENESS

Healthy Feet Project

Healthy Feet Project

Cows produce less milk when they are lame

Milk loss facts for lame cows

Loss per 305 day lactation associated with:

- Medium lameness score = −442.8 kg
- High lameness score = −745.6 kg
- Claw lesions = −360 kg

Milk loss greatest in early lactation
Milk losses for 2 months before a sole ulcer
Milk losses for 5 months after sole ulcer and
white line disease

Lame cows are less likely to get in calf

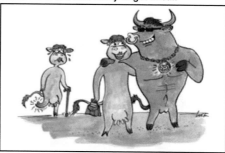

Fertility facts for lame cows

Increased intervals for lame cows

- Calving to conception = 14–50 days longer
- Calving to first service = 4 days longer
- First service to conception = 8 days longer

10% fewer pregnancies/first service
0.42 more services per conception
1.16 times more likely to be treated for anoestrus

Lame cows are more likely to be culled early

Culling facts for lame cows

8.4 times more likely to be culled early

Many cows that become lame have
higher yields before they become lame than
cows that never become lame **SO** lame
cows are culled later in lactation than cows
culled for mastitis or fertility.

Cows that become lame make less money for the farmer

Overall costs of a case of lameness

Reduced fertility = £46.14

Reduced milk yield = £55.05

Culling/replacements = £53.72

Treatment costs = £23.32 (1.4 treatments/cow)

TOTAL COST = £178.23 per lame cow

The poster shows the results of numerous studies that have measured the effects of lameness. The overall costs of lameness are the average cost of lameness taken from the DAISY Research Report No. 5 in 2002. These figures are intended as a guide but will vary from farm to farm.

Fig. 15.3. Poster highlighting the financial losses associated with lameness. (Compiled by Dr Zoe Barker, cartoons by Steve Long.)

to their clients in order to receive a mug or padded waistcoat. To date, the lameness project has not fully exploited the "drug-company type" gift, although occasionally small logoed gifts such as a woolly hat or an insulated mug are handed out to farmers who have implemented a change, to reinforce the approval for the change being voiced by the facilitator. As part of the lameness project, there is a positive control group that is able to access modest but not insignificant financial rewards for having implemented changes outlined in its lameness action plan. Although the final outcome of this incentive scheme has not yet been evaluated fully, the highest level of behavior change appears to be in this group.

Financial incentives from buyers of farm products

Other models of financial incentive already exist within the agricultural sector. One example that is widely used around the world is the payment scheme for bulk somatic cell counts in milk. The milk payment that farmers receive is adjusted according to their levels of bulk milk somatic cell counts. Dekkers and colleagues (1996) reported that dairy farmers in Ontario, Canada, could achieve substantial additional milk revenues through this scheme. An incentive-based intervention approach to improve the welfare of commercially produced broiler chickens in Sweden has been described by Algers and Berg (2001). An animal welfare program for broilers set up by the Swedish Poultry Meat Association monitored the standards of the buildings and equipment used in the rearing of the birds and a score was formulated on the basis of the assessment. This score then determined the future maximum stocking density at which a producer was allowed to keep broilers. The stocking densities at which the birds could be kept ranged from 20 kg/m² to 36 kg/m²; this had obvious economic implications for the producers, and as such was promoted as a reward system for those producers who chose to improve their broiler rearing facilities. Contained within the animal welfare program was the foot health program that monitored and scored the prevalence of footpad dermatitis in the birds at the time of slaughter. As well as being part of the stocking density reward program, the foot health monitoring program was linked to an advisory service that was available to the farmers. Within the first 2 years of running the foot health program, the prevalence of severe footpad lesions had decreased from 11% to 6%.

As demonstrated by the examples above, incentives have a potentially powerful role in encouraging behavior change. The limitation of some of the described approaches is that they can require considerable external investment through government, retailers, and food processors or charities looking for routes to promote animal welfare improvement. Incentives also have a role within enforcement models of behavior change, such as restricting access to markets unless specified animal welfare criteria are met. This aspect of incentive use will be discussed in more detail in the enforcement section of this chapter.

Participatory Approaches to Improve Animal Care

The participatory approaches described in this section were created for use in community development, primarily in developing countries. The approaches described here are powerful and based around empowering the most marginalized and vulnerable people to make changes to improve their own lives. Previously, we have used these approaches in Asia to help the owners of equids used for traction work to make improvements to the welfare of their animals without risking their income. More recently, we have brought some of these approaches on to farms in the UK in order to help farmers become aware of, plan, and own behavior changes to improve the lives of their animals. The approach described here is commonly referred to as either PRA (Participatory Rural Appraisal) or PLA (Participatory Learning and Action), although many variations on these names exist. In this section, we will discuss two main areas of participatory approaches: (i) the principles that guides their use; and (ii) the types of tools available. These principles can be used in both developed and developing parts of the world.

Principles of participatory approaches

The key to the participatory approach is to develop interactive tools or exercises that allow a single farmer or a group of farmers to share, present, analyse, and enhance their knowledge of a problem they may be facing; in this case, our interest is animal welfare. The first principle to be aware of is that of participation. Put most purely, this means that "all people have a right to play an active part in shaping the decisions which affect their lives" (International HIV/AIDS Alliance, 2006). So, for us, this is about

H.R. Whay and D.C.J. Main

getting farmers involved in and wanting to make changes to the lives of their animals because these changes will affect their lives as well. It is also important to recognize that on a farm there may be many voices that need to be included in discussions about change, not just the farm owner or manager. Some of these voices are rarely heard, and often not respected; for example, the relief milker, the farmer's wife, or the son who feeds the calves when he gets home from school. So, our goal is to include as many participants in the process as possible; the more people who have a stake in and support a change, the more likely it is to happen. The second principle is that this approach respects the fact that local people (in this case, farmers) hold much more knowledge about their own situation than we as external parties could ever hope to do. It would be arrogant and presumptuous of us simply to give advice or instruction without recognizing that our understanding of how a particular farm works and why it has developed the routines and practices that it currently uses is very limited. We generally rush to bring new knowledge to farmers without stopping to consider and use the knowledge they already hold. So a participatory approach aims to bring out this knowledge in a way that is visual, engaging, and organized for the farm team to consider and learn from. The third, and often overlooked, principle of using a participatory approach is about the need for proper reflection and analysis of information. There is a common temptation to use a tool and complete an exercise as rapidly and efficiently as possible without pausing to review, think about, and discuss what can be learned from having completed the exercise. In fact, there would be no point conducting the exercise at all if no subsequent reflection or review takes place.

As may be becoming apparent, this process relies on someone bringing these principles and tools to a farm or group of farmers and guiding them through the process. This person is usually the external agent who is interested in helping a farm to implement behavior change and is usually called a facilitator. The success of the process is heavily dependent on the competence of the facilitator. The term "facilitator" translates roughly into "making it easy", and can be interpreted as someone whose role is to find ways to help others through the principles outlined above without bringing their own values or prejudices into the discussion. Good facilitators encourage others to speak, usually by asking questions without offering opinions, and spend the majority of their time listening rather than talking.

Participatory tools

As already mentioned, these tools came about for use among marginalized communities in developing countries. They are designed to be as accessible as possible, so exploit local materials such as dried beans, sticks, stones, etc., and are usually carried out on the ground. Written words are avoided so as not to exclude the illiterate, and pens and paper are not often used, as many people are uncomfortable with this medium. In Europe, we tend to work sitting around a table, use pens and large sheets of paper, and brightly colored, sticky Post-it® notes. We do still try to use pictures and not rely solely on written words. The reason for this is that according to official figures, one in six adults in the UK is functionally illiterate, to the extent that they cannot use a telephone directory to find a plumber. So, drawings avoid exclusion of the less literate and act as prompts to help remind people what a particular question or category referred to. This is also helpful in situations where we would like to include migrant farm laborers from Eastern Europe, but we do not share a common language with them.

The following three tools or exercises are illustrated pictorially and their purpose explained. When facilitating exercises such as these, it is important to break them down into steps and work on completing one step at a time.

Example tool 1: seasonal lameness calendar

The seasonal lameness calendar illustrated in Fig. 15.4 is compiled by the facilitator or agricultural consultant from the farm records. Ideally, this would be done in collaboration with members of the farm team, but we often find that presenting the final product for analysis and discussion is still a useful exercise when the farm personnel have only limited time. This chart allows farmers to consider management changes in relation to peaks of lameness, it helps them to identify potential risks that are specific to a particular type of lameness, and to consider how particular management activities could be timed to pre-empt a rise in lameness levels.

Example tool 2: matrix for prioritizing lameness management activities

The matrix illustrated in Fig. 15.5 is built up in stages. Initially, the participant(s) are asked to

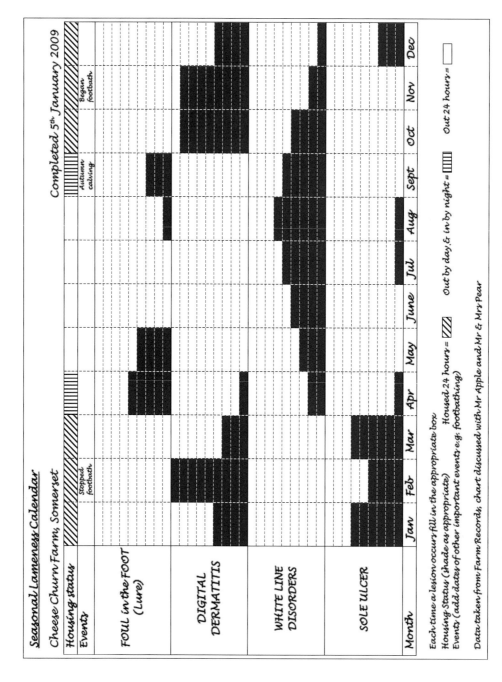

Fig. 15.4. Seasonal lameness calendar compiled from the farm records. This chart illustrates visually how different types of lameness occur at different times, and can be used to promote discussion about potential risk factors for different types of lameness and the timing of lameness management activities. (Compiled with the assistance of Dr Zoe Barker.)

H.R. Whay and D.C.J. Main

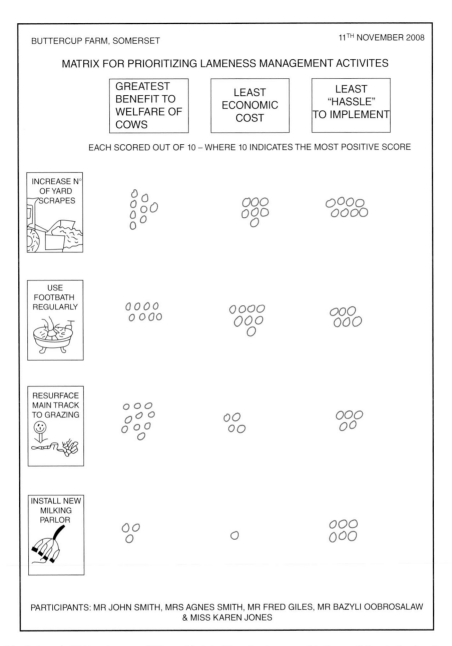

Fig. 15.5. Matrix for prioritizing changes. This matrix is built up in stages and helps participants to view the potential positive and negatives aspects of a change in manageable chunks.

make a list of potential changes they are considering that might contribute to reducing lameness levels among their dairy cattle; these are illustrated on separate cards and laid out in a column. They are then asked to consider each of these possible changes with reference to how much welfare benefit each one would confer to their cattle. Each one is scored out of ten, with ten indicating the greatest benefit. Next, they are asked to score how much each of the changes will be likely to cost. The question is phrased so that again a score of ten represents the most positive outcome (i.e. the least cost). When considering cost, the participant(s) can consider the potential cost of implementing a permanent change to a daily routine versus a one-off capital investment. Finally, the participant(s) are asked to score the "hassle" (extra daily work) factor of implementing the change; once again, a score of ten represents the least hassle. Once the matrix is completed, the starting point for the review is to look where the greatest scores appear. This exercise is useful for prioritizing actions and breaking down considerations of the implications of implementing a change into bite-sized chunks.

Example tool 3: lameness transect walk

The lameness transect walk (Fig. 15.6) is useful for giving structure to a meeting of a group of farmers. At the beginning of the meeting, the group is asked to select the route "following in the footsteps of the cows" for a walk around the host's farm. Each participant is given a farm map on a clipboard and asked to note down any good ideas they see, any potential risks for lameness, and any opportunities for change. After the walk, each member of the group is invited to transcribe their observations on to brightly colored Post-it® notes and stick them on to a giant pre-prepared chart. A facilitator then promotes a discussion around areas that have been highlighted on the chart, often by a high density of stickers. The facilitator's role is to keep the discussion moving along and to bring out as many potential opportunities for change as possible. Although the discussion is based around the farm being visited, all participants benefit from hearing the ideas, suggestions, and experiences of others, and often bring into the discussion the particular problems they are facing.

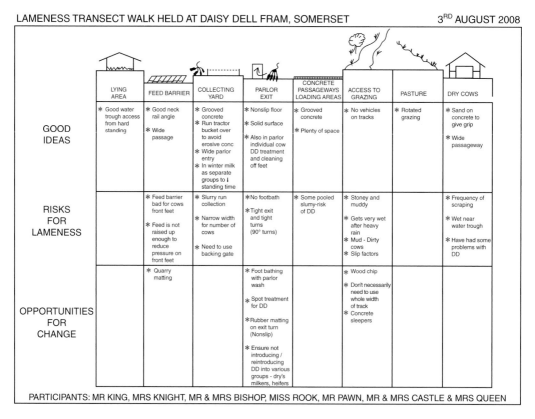

Fig. 15.6. This chart produced from a farm walk is transcribed from a very large chart built up by a group of farmers who wrote their comments on to brightly colored Post-it® notes. The giant chart formed the basis for a facilitated discussion following the walk. A copy of the transcribed chart was then sent to all the farmers who came to the meeting. (Transcribed with the assistance of Miss Clare Maggs.)

H.R. Whay and D.C.J. Main

Farmer Self-help and Discussion Groups

The concept of farmers forming self-help groups or, more commonly, discussion groups is not new, and these groups are often useful for breaking down the isolation associated with farming and for sharing new ideas. This section looks at two approaches to how farmer groups can be managed to achieve optimum likelihood of changing behavior and addressing areas of animal welfare concern. The two models of farmer group structures to look at in this section are the "stable school" and the "monitor farm". The actual formation of groups is often not easy, but is a discussion for another time. For the purposes of this section, we will assume that we have a group of farmers looking for a structured group format.

Stable schools – all farms in the group are open to all farmers

The stable school system is adapted from a system known as Farmer Field Schools running in Uganda and has been used as the Stable School approach in Denmark by Mette Vaarst and colleagues (Vaarst et al., 2007) and in Germany by March et al. (2014). In Denmark, six organic farmers formed a group with the stated goal of using health promotion to reduce and, where possible, phase out antibiotic usage in their cows. To participate in the stable school, each farmer agreed to host a visit from the group and to be open to receiving management recommendations from the group. The visit consisted of a farm walk where the host was asked to present an aspect of the farm he or she was proud of, as well as two problem areas. Once the walk was completed, the group returned to the farmhouse or office and examined the farm's accounts. Following this, a facilitated discussion took place in which the group worked to develop recommendations on changes that could be made to improve the host farm's management and help reduce the need for antibiotics. The goal was to prevent problems arising requiring antibiotic treatment, not to withhold treatment from animals in need. The facilitator's role was to encourage participation by all group members and to act as the recorder and reporter of the outcome of the discussions. After all farms had been visited, each then received a follow-up visit, so that in total each farm was visited twice. This approach produced some notable successes with reference to the common

goal of reducing antibiotic usage in the Danish study and improving cow cleanliness in the German study. However, as well as this common goal, each farmer also formed his or her own "local" goals, which in some cases were animal health oriented and in others were based on improving family or social environments. These were important markers of the success of the project, and many participants felt they benefited very much from joining a group.

Designated monitor farms are opened up to other farmers

The monitor farm approach was developed originally in New Zealand and has been imported to Scotland. The purpose of the farmer groups currently engaged in the monitor farm approach is to improve farm performance and profitability, and it is likely that animal welfare may be implicit within these goals. In the context of this chapter, we will have to extrapolate this approach to consider how it might work when focused explicitly on animal welfare improvement.

The concept of the monitor farm is that one farmer agrees to open up his farm business to a group of farmers, and possibly members of allied industries such as veterinary surgeons and feed specialists. The monitor farm hosts regular visits from the group, during which aspects of the farm's management and potential changes to be made are discussed. The regular meetings then allow the group to observe the progress of the farm business and to see the effect of implementing suggested changes (or not). The monitor farm also hosts an open day once a year for visitors outside the group to view the farm and learn about the process. Once again, this process is supported by a facilitator, who helps with setting the objectives of the meetings, managing the logistics, and where needed, analysis of data.

Clearly, the monitor farm itself is likely to be implementing changes, although it is not obliged to follow the recommendations of the group. The group members themselves are also likely to form the habit of trying to implement changes on their own farms, benefiting from hearing others' experiences and learning from the implementation of changes on the monitor farm, for which they already feel some level of ownership. In some cases, members of the group have been reported also to develop cooperative activity. It is also suggested that the "trickle through" effect can lead to changes being implemented in the wider community. Trickle through is a commonly promoted concept within development, but can be very difficult to demonstrate.

Enforcement as a Means of Implementing Change – Legislation and Government Enforcement

Insisting that farmers take particular actions to improve the welfare of animals under their care is a well-established route to compel behavior change. Traditionally, this approach has centered on setting legal requirements that a farmer must comply with. Failure to meet these legal requirements could result in prosecutions and/or penalties, depending on the local regulatory requirements.

Legislation is clearly a well-recognized mechanism for governments to bring about changes in behavior. However, legislation is just one of a set of tools also known as "policy instruments" used by government to implement policy. The Farm Animal Welfare Council (FAWC, 2008) has reviewed the policy instruments that the UK government could use to improve animal welfare. One example of a policy instrument has been the government's role in producing guidance in the form of species-specific animal welfare codes. The welfare codes promote good animal husbandry and animal care, as well as presenting relevant pieces of animal welfare legislation in a clear and easily accessible format. The animal welfare codes have been influential in guiding both farmers and farm inspectors and advisors. While legislation may remain the most obvious tool available to governments, the FAWC suggested that: "To achieve the levels of animal welfare that people want requires a coordinated approach to the use of policy instruments to achieve desired behavioural change".

Requirement set by retailers

However, more recently, it is the marketplace that is insisting producers maintain minimum animal welfare standards. Failure to meet the criteria of the marketplace could result in a farmer finding himself or herself unable to sell livestock products; this is clearly a very powerful motivator for change. In many countries, these marketplace requirements are achieved by insisting that farmers become members of farm assurance schemes. The farm assurance schemes aim to provide the purchaser of a product an assurance that animal welfare (and food safety) standards are being met. Often, these schemes are driven by retailers who are keen to ensure minimum standards for their products. The retailers' primary motivation here is to avoid negative media publicity associated with problems on the farms that supply them. The minimum these schemes normally aim to achieve is to verify compliance with national legislative requirements. However, some retailers and schemes may formulate additional requirements, either as part of the company's corporate social responsibility (CSR) policy or to obtain a marketing differential from other products, and animal welfare may be included in their remit. The examples describing the influence that the fast-food chain, McDonald's, has had on animal welfare through the requirements they set for slaughterhouses, presented in Chapters 2 and 9, this volume, are based on this principle.

Effectiveness of the enforcement approach

Although the requirement for farmers to comply with specific standards may originate from different sources, for example politicians representing citizen interests or the marketplace representing consumer interests, the net result is the same: a powerful statement to producers that they have no choice other than to comply. Both "enforcement" approaches (legislation and marketplace requirements), therefore, have the same inherent strengths and weaknesses.

The strengths of the enforcement approach are:

1. It can be effective in ensuring the worst extremes of animal welfare problems are avoided.
2. A clear set of requirements can be transparent and communicated to all farmers.
3. Provided a careful development process is followed that includes the incorporation of relevant practical experience and scientific knowledge, then the standards can be fair to producers.
4. Considering the penalties involved (prosecution or inability to access the marketplace), farmers will be highly motivated to comply.

H.R. Whay and D.C.J. Main

5. Once an inspection and certification system has been established, the minimum standards can be supplemented to include additional requirements that can be used in marketing initiatives, and thereby lead to a price premium (i.e. a positive financial incentive to be members of such schemes).

The weaknesses of the enforcement approach are:

1. The inherent "negativeness" of this approach can generate bad feeling with farmers as it conveys the message: "We do not trust you to care for your animals".
2. Farmers may be encouraged to comply up to a minimum standard, but it does not recognize excellence, continuous improvement, or innovation beyond the minimum required.
3. Written requirements are difficult to construct, so are often poorly defined and, therefore, ambiguous. Standards are also often developed in a haphazard manner with insufficient consultation (see Chapter 4, this volume, on writing clear standards).
4. If no marketing claims arise from the marketplace requirements, then farmers required to comply with a standard are at a commercial disadvantage to producers (perhaps from a developing country) who do not have such an approach.

Enforcement approach is weak for chronic long-term problems such as lameness

It seems reasonable that farmers are required to achieve a minimum standard. Indeed, it is in all farmers' interests that unacceptable welfare conditions are regulated, as the actions of a few can have a very significant adverse effect in the media and on public opinion. So the enforcement approach is certainly useful for regulating the extremes.

The experience of Temple Grandin's approach based on setting requirements to achieve a set of welfare outcomes within the slaughterhouse environment is very effective (see Chapters 4 and 9, this volume). The key here seems to be focusing on a few parameters and demonstrating to slaughterhouse managers that achieving change is possible. The challenge will be to replicate this success on the farm, where the welfare problems are often more chronic in nature and directly related to an inherent component of the production system; for example,

resolving the problem of tail biting in pigs, which seems to be intrinsic to some types of production system. In a slaughterhouse environment, welfare improvements due to modifications in handling systems can be demonstrated in the next batch of animals, whereas on the farm it can be a considerable time before the outcome and positive effect of changes can be seen. Once a dairy cow becomes lame with a sole ulcer, her foot is damaged for life; it will not be until the next generation of heifers joins the herd that the outcome of implementing new preventive measures will be seen.

The Use of Standards and Animal-based Outcome Measures in an Enforcement Approach

It is beyond the scope of this chapter to review all the possible legislation and standard options that might influence animal welfare. The OIE (World Animal Health Organisation) and most countries will have legislation that aims to avoid cruelty or "unnecessary suffering". This is a useful absolute baseline standard for animal welfare. However, beyond this baseline, the legal requirements will, of course, vary between countries, although a common minimum standard for farm animal welfare is applied within several countries of the European Community (Directive 98/58/EC). Similarly, penalties for failure to comply with standards will also vary between countries.

In order to understand the potential power of using standards to improve animal welfare, it is useful to examine what types of standards are available. The examples given below cover standards for health planning, treatment of both individuals and groups of animals with welfare problems, and the provision of resources. Quick Access Information 15.2 illustrates how these standards can be applied to the problem of cattle lameness. The potential impact of these standards will, of course, vary. For example, it is likely that standards that require specific actions directly associated with lameness will be more beneficial than actions associated with records and protocols, although, of course, these can have a role. As can be seen from the variety of approaches shown in this chapter, it can be very valuable to link the standards with animal-based outcomes on the farms. Animal-based outcomes are measures of welfare made by looking directly at the animal itself; for example, how many cows are limping, what body condition score the animals are in, or how animals respond behaviorally to an approaching human. Animal-based outcome measures are distinct from resource-based measures that define the environment in which the animal should live but do not describe what the welfare of the animal should be. Animal-based outcome measures may take the form of farmers self-assessing their own animals or involve external observers, inspectors, or assessors. The findings from an animal outcome assessment may lead to a requirement for a farmer to take specific actions once an inspector observing animals with a particular condition such as lameness finds that the prevalence of the problem has gone above a predefined threshold.

The following four examples illustrate how standards may be used to enforce changes intended to improve animal welfare (these have been developed in association with Dr Siobhan Mullan).

Standards associated with health planning activities

Health planning is an important tool for farmers to achieve early recognition of welfare problems including disease, to identify potential risks for their introduction and spread, and to implement management of identified risks (Defra, 2004). These standards can ensure that farmers undertake preventative or corrective health planning activities relevant to a specific parameter. This may involve changes to housing, facilities, routines, or planned use of medication. These standards can be required of the producer at all times. Examples of standards that should be implemented at all times are as follows:

Quick Access Information 15.2. **An illustration of how enforcement standards may be applied to the problem of lameness in cattle.**

Examples of UK standards relevant for dairy cattle lameness

The following are examples from either UK legislation or from the UK welfare codes that are relevant for dairy cattle lameness. Although UK welfare codes are recommendations, farm assurance schemes often require adherence to them as well as UK legislation.

Requirements associated with health planning activities

- The stock-keeper should draw up a written health and welfare plan with the herd's veterinary surgeon and, where necessary, other technical advisors, which should be reviewed and updated each year.[a]
- The written health and welfare plan should … look at … lameness monitoring and foot care.[a]
- A record must be maintained of … any medicinal treatment given to animals.[b]
- … you may find it useful, as part of the health and welfare plan, to note specific cases of … lameness … and where appropriate the relevant treatment given.[a]

Requirements associated with individual affected animals

- Any animals that appear to be ill or injured must be cared for appropriately and without delay. Where they do not respond to such care, veterinary advice must be obtained as soon as possible.[b]

- If a lame animal does not respond to the veterinary surgeon's treatment, you should have it culled rather than leave it to suffer.[a]
- … you must not transport any cattle off-farm that cannot stand up unaided or cannot bear their weight on all four legs when standing or walking.[a]

Requirements associated with herd-/group-level outcomes

- … all keepers of animals must … ensure that the needs of an animal for which he is responsible are met to the extent required by good practice and those needs shall be taken to include … its need to be protected from pain, suffering, injury and disease.[c]

While not currently explicit, standard guidance notes for inspectors could state that the prevalence of lameness above a certain percentage could be used as evidence of failure to provide this need.

Requirements associated with provision of specific resources

- The floor should not slope too steeply … as steeper slopes can cause leg problems, slipping and falling.[a]
- [For slatted accommodation] the gaps between the slats should not be wide enough to cause foot injuries.[a]

Continued

H.R. Whay and D.C.J. Main

- [For cubicles] you need to have enough bedding to … prevent them from getting contact or pressure sores.[a]
- Farmers (and/or their veterinary surgeon) must inspect animals for presence of "a condition" on a regular (e.g. weekly) basis.
- Those inspecting animals for the presence of "this condition" must record such information in a timely manner.
- The records associated with "this condition" must be reviewed by the producer and their veterinary surgeon on a regular basis (e.g. every 6 months).

- A written action plan must be developed that includes preventative husbandry and medicine-related actions that minimize the prevalence of "this condition".
- The written action plan should also include pre-defined farm-specific intervention levels that determine when investigation activities need to be conducted.

Notes: [a]Defra (2003) Welfare Code of Recommendation: Cattle. PB7949; [b]Welfare of Farmed Animals (England) Regulation (2007) SI 2078; [c]Animal Welfare Act (2006), Chapter 45.

Requirements associated with individual affected animals

Standards can be directed at ensuring that a farmer suitably manages the most severely affected individuals. This is applicable for all individual animals with a welfare problem above a predefined severity threshold; this is relevant for more severely affected animals, for example severely lame animals. Requirements could include one or more of the following components:

- All animals affected with "this condition" must receive prompt treatment and nursing care.
- Veterinary advice must be obtained for affected animals that fail to respond to treatment within 3 days.
- Animals affected with "this condition" must not be transported for slaughter.
- Animals that are severely affected by "this condition" must be euthanized/culled humanely.

Requirements associated with herd-/group-level outcomes

Standards can be directed towards ensuring that the farmer appropriately manages welfare problems that occur at the herd or group level. This would be applicable where the prevalence of animals moderately affected by a particular problem exceeds a predefined intervention guideline; for example, the combined prevalence of moderate and severely lame cows exceeds 10% on the day of assessment. Requirements could include one or more of the following components:

- Producers must manage "the condition" so that the overall prevalence does not exceed the pre-defined intervention guideline.
- Producers must develop a farm-specific written action plan in collaboration with their veterinary surgeon when the prevalence of "the condition" exceeds the intervention guideline.
- Producers must demonstrate that appropriate preventative actions have been taken when the herd level exceeds the intervention guideline.
- Additional verification visits will be required if the prevalence of "the condition" exceeds the intervention guideline. Such visits will require an additional fee.

Requirements associated with provision of specific resources

Farm assurance and marketplace requirements include many resource-based standards: animals must be provided with defined facilities, space, bedding, etc. These are sometimes referred to as engineering standards. Animal outcomes can also be used to assist in the interpretation of such requirements. If a standard requires an "appropriate" or "sufficient" resource, for example bedding substrate, then animal outcome measures of cleanliness and body lesions and swellings can help define what is "sufficient" in terms of delivering adequate animal welfare. For example:

- Animals must be provided with sufficient resource (e.g. bedding) to ensure that no animals develop the relevant outcome (e.g. swollen hocks).

Outline of Methods for Use of Animal-based Outcome Measures in Enforcement Approaches

It is clear that animal-based outcome measures are not only a useful tool for farmers but also that they should form part of welfare enforcement mechanisms. In order to implement standards that incorporate animal-based outcome assessments, the following conditions will also need to be considered:

1. **Competency** of inspectors is obviously an important criterion. For most of these standards, it would be important for the inspector to observe the condition among the animals. The inspector, therefore, needs the relevant knowledge, skills, and attitude to conduct these tasks. The standard may also require observation of written records or protocols, and verification that appropriate actions have been taken on the farm. To assess these standards, the inspector would need to have sufficient knowledge of the system to make reasonable judgements.

2. **Guidance notes** are crucial for all standards as there are inevitably opportunities for variable interpretation.

3. **Training** inspectors, producers, or veterinary surgeons in how to carry out the assessment of relevant animal-related outcomes is important, and ensuring consistency of assessment is a key challenge.

4. **Benchmarking** information is essential for setting intervention guidelines and advising farmers. The farmer would need to know how their assessment results compared to other units of a similar type. Also, the product retailer would be keen to have this information in order to describe to interested consumers the welfare status of goods they are selling.

5. **Sampling protocols** are also needed to define how many animals need to be observed. This will depend on the variability and prevalence of parameters in normal units. For example, on pig units, if a parameter were normally relatively consistent in pens or groups, then fewer pens would need to be assessed to give reasonable certainty that the prevalence had been properly counted. If the parameter is generally known to have a relatively low prevalence, then more animals in more pens will need

to be assessed to produce a definitive assertion of the on-farm prevalence. In addition, if the purpose of the assessment is to produce an overall farm average, then sampling from a representative proportion of pens will be satisfactory. However, if a farmer needs to know from where the problem originated, then more pens will need to be sampled.

6. **Advisory information** that allows farmers to manage these animal outcomes needs to be available. Although farm assurance schemes are usually not permitted to provide advice alongside an impartial assessment, there does need to be some mechanism for getting advice to farmers. Often, the veterinary surgeon should be best placed to do this.

The Link Between Enforcement and Incentives

Most farm assurance schemes and marketplace standards are usually perceived by farmers to be a negative enforcement approach because failure to comply can result in inability to access the market. However, some purchasers of livestock products may be prepared to pay an additional premium in recognition that farmers are complying with additional standards. This may, therefore, be perceived as a type of incentive. So, for example, the Royal Society for the Prevention of Cruelty to Animals (RSPCA) Freedom Food scheme aims to deliver a high level of assurance with respect to livestock welfare, and this may be valued by consumers. Any premium generated by these claims would be an incentive for farmers to participate in such schemes. Additional potential incentives for participation in such schemes might be the creation of a niche market for a specialist produced product, receiving contracts with a particular retailer (especially if they have a good track record for dealing fairly with their suppliers), or even ensuring continuation of a strong generic product market.

Another example was the Sustainable Dairy Project, which provided a premium for producers supplying Tesco, a major UK retailer, in return for following some additional welfare requirements, including the requirement for self-evaluation of lameness. Tesco received a Good Business Award for Innovation from the RSPCA in 2007 for recognition of its efforts.

In the same way that incentives may be perceived positively, some financial instruments are quite clearly disincentives. Two useful examples exist here, again for lameness in dairy cattle. First, there is variation in Europe over the interpretation of hygiene regulation for dairy production. Some authorities have interpreted that lame cows have been defined as "unhealthy", and as such, milk from these animals, even if free from relevant medication, cannot enter the food chain. This means that farmers are directly affected by a loss in milk sales for every lame cow; it provides a very strong disincentive for having lame cows on the dairy farm.

Second, in order to gain subsidies under the Single Farm Payment scheme, European farmers are required to demonstrate cross-compliance (i.e. adherence to legal standards for aspects of their farm management such as environmental protection and animal welfare). These requirements are verified by government agencies and failure to comply can result in withdrawal or reduction in scheme payments. Even though the welfare standards are nonspecific, the guidance provided to veterinary inspectors may be much more specific. For example, the European Community Directive 98/58/EC requires that: "Any animals which appear to be ill or injured (a) shall be cared for appropriately without delay; and (b) where they do not respond to such care, veterinary advice shall be obtained as soon as possible". The guidance provided in the UK to the veterinary inspector refers explicitly to lame animals, as it states that:

> If a significant number (over 5% is the current suggested intervention point) of dairy cows or dairy goats are lame at the time of inspection, the Inspector should make enquiries to establish if the cause of the lameness has been recognized and whether appropriate action has been taken and veterinary advice or FHWP [Farm Health and Welfare Plan] protocol has been followed.
>
> (See, for example, http://assurance.redtractor.org.uk/ rtassurance/schemes/resources.eb.)

Conclusion

This chapter has highlighted several different approaches to promoting behavior change among farmers that should lead to animal welfare improvements. Several practical tools for achieving this have been proposed. It is likely that more than one method applied in a coordinated manner will be most effective as different farmers respond to different approaches. It is also important to highlight that behavior change, whether through the route of encouragement, enforcement, or both, rests on the farmers

having a good knowledge of how to farm in a caring and humane manner, i.e. "professional knowledge". Without this knowledge, farmers will be unaware that problems exist and ill equipped to find solutions. In this chapter, we have argued that the majority of farmers do hold a great wealth of professional knowledge, both general and specific to their own situations. In fact, there currently seems to be a danger that farmers often receive such a barrage of information that they find it difficult to sort and filter. It is also important to reiterate that holding this knowledge (although vitally important) is not sufficient on its own to precipitate behavior change. All of the tools described here offer mechanisms to help farmers deal with and act upon the knowledge and skills they already have.

As a final note, we want to point out how crucial it is that any intervention is monitored carefully so that success or failure can be identified and appropriate modifications made to the intervention program. Responsive monitoring not only protects the animals from unintentional harms that may arise from introducing changes but also allows us all to learn more rapidly and efficiently from the process of making change.

References

Algers, B. and Berg, C. (2001) Monitoring animal welfare on commercial broiler farms in Sweden. *Acta Agriculturæ Scandinavica Section A, Animal Science* Supplement 30, 88–92.

Animal Health and Welfare (Scotland) Act (2006) HMSO, London.

Animal Welfare Act (England and Wales) (2006) HMSO, London.

Clarkson, M.J., Downham, D.Y., Faull, W.B., Hughes, J.W., Manson, F.J., Merritt, J.B., *et al*. (1996) Incidence and prevalence of lameness in dairy cattle. *Veterinary Record* 138, 563–567.

Defra (Department for Environment, Food and Rural Affairs) (2003) *Welfare Code of Recommendation: Cattle*. PB7949. Defra, London.

Defra (2004) *Animal Health and Welfare Strategy*. Defra, London.

Dekkers, J.C.M., Van Erp, T. and Schukken, Y.H. (1996) Economic benefits of reducing somatic cell count under the milk quality program of Ontario. *Journal of Dairy Science* 79, 396–401.

European Community (Directive 98/58/EC) Concerning the protection of animals kept for farming purposes. *CELEX-EUR Official Journal* L 221, 8 August 1998, pp. 23–27.

Fabian, J., Laven, R.A. and Whay, H.R. (2014) The prevalence of lameness in New Zealand dairy farms: A comparison of farmer estimates and locomotion scoring. *Veterinary Journal* 201, 31–38.

Farm Animal Welfare Council (FAWC) (2008) *Opinion on Policy Instruments for Protecting and Improving Farm Animal Welfare*. FAWC, London.

Geller, E.S. (1981) Evaluating energy conservation programs: is verbal reporting enough? *Journal of Consumer Research* 8, 331–335.

Goldstein, N.J., Martin, S.J. and Cialdini, R.B. (2007) *Yes! 50 Secrets From the Science of Persuasion*. Profile Books, London, pp. 18–21.

International HIV/AIDS Alliance (2006) *Tools Together Now! 100 Participatory Tools to Mobilise Communities for HIV/AIDS*. International HIV/AIDS Alliance, Hove, UK. Available at: http://www.aidsalliance.org/assets/000/000/369/228-All-together-now_original.pdf?1405520021 (accessed 1 April 2015).

Kerr, J., Weitkunat, R. and Moretti, M. (2004) *ABC of Behaviour Change. A Guide to Successful Disease Prevention and Health Promotion*. Elsevier Churchill Livingstone, Amsterdam.

Leach, K.A., Whay, H.R., Maggs, C.M., Barker, Z.E., Paul, E.S., *et al.* (2010) Working towards a reduction in cattle lameness: 1. Understanding barriers to lameness control on dairy farms. *Research in Veterinary Science* 89, 311–317.

Leach, K.A., Paul, E.S., Whay, H.R., Barker, Z.E., Maggs, C.M., *et al.* (2013) Reducing lameness in dairy herds – overcoming some barriers. *Research in Veterinary Science* 94, 820–825.

McKenzie-Mohr, D. and Smith, W. (1999) *Fostering Sustainable Behavior. An Introduction to Community-based Social Marketing*. New Society Publishers, Gabriola Island, British Columbia, Canada.

Main, D.C.J., Leach, K.A., Barker, Z.E., Sedgwick, A.K., Maggs, C.M., *et al.* (2012) Evaluating an intervention to reduce lameness in dairy cattle. *Journal of Dairy Science* 95, 2946–2954.

March, S., Brinkman, J. and Winckler, C. (2014) Improvement in animal health in organic dairy farms through stable schools: Selected results from a pilot study in Germany. *Organic Agriculture* 4, 319–323.

Sorge, U.S., Cherry, C. and Bender, J.B. (2014) Perception of the importance of human animal interactions on cattle flow and worker safety on Minnesota dairy farms. *Journal of Dairy Science* 97, 4632–4638.

Talking Retail (2008) Co-op Shoppers Put Animal Welfare Above Climate Change. Available at: www.talking retail.com (accessed 1 December 2008).

Vaarst, M., Nissen, T.B., Østergaard, S., Klaas, I.C., Bennedsgaard, T.W., *et al.* (2007) Danish stable schools for experimental common learning in groups of organic dairy farmers. *Journal of Dairy Science* 90, 2543–2554.

Welfare of Farmed Animals (England) Regulations (2007) HMSO, London.

Whay, H.R., Main, D.C.J., Green, L.E. and Webster, A.J.F. (2003) Assessment of the welfare of dairy cattle using animal-based measurements: direct observations and investigation of farm records. *Veterinary Record* 153, 197–202.

Further Reading

ADAS (2008) An Investigation into the Role and Effectiveness of Scottish Monitor Farms. A report prepared for the Scottish Government by ADAS (Agricultural Development and Advisory Service) UK Ltd. Scottish Government Publications, Wolverhampton, UK. Available at: www.scotland.gov.uk/publications/2008/10/29093936/0 (accessed 7 July 2009).

Braun, A. and Doveskog, D. (2008) The farmer field school approach – history, global assessment and success stories. Background Paper for IFAD Rural Poverty Report, 2011. Available at: http://www.ifad.org/rpr2011/background/1.pdf (accessed 1 April 2015).

Hastings, G. (2007) *Social Marketing: Why Should the Devil Have All the Best Tunes?* Elsevier, Oxford, UK.

Kumar, S. (2002) *Methods for Community Participation: A Complete Guide for Practitioners*. ITDG Publishing, London.

McKenzie-Mohr, D. (2013) *Fostering Sustainable Behavior. An Introduction to Community-based Social Marketing*. New Society Publishers, Gabriola Island, British Columbia, Canada.

Main, D. and Webster, J. (2011) Assessment, implementation, and promotion of farm animal welfare. In: Webster, J. (ed.) *Management of Welfare of Farm Animals*. UFAW Farm Handbook. Universities Federation of Animal Welfare, Potters Bar, Herts, UK, pp. 564–584.

Vaarst, M. (2007) Participatory common learning in groups of dairy farmers in Uganda (FFS approach) and Danish stable schools. *DJF Animal Sciences* 78 (special edition). Available at: http://orgprints.org/13731/ (accessed 1 April 2015).

Waddington, H., Snilstveit, B., Hombrados, J.G., Vojtkova, M., Anderson, J., *et al.* (2012) Protocol Farmer Field Schools for improving farming practices and farmer outcomes in low and middle income countries: A Systematic Review. The Campbell Collaboration, The Norwegian Knowledge Center (www.campbellcollaboration.org), Oslo, Norway.

Webster, A.J.F. and Main, D.C.J. (2003) Proceedings of the 2nd International Workshop on the Assessment of Animal Welfare at Farm and Group Level. *Animal Welfare* 12(4), 429–713.

16 Successful Technology Transfer of Behavioral and Animal Welfare Research to the Farm and Slaughter Plant

TEMPLE GRANDIN

Department of Animal Science, Colorado State University, Fort Collins, Colorado, USA

Summary

In any industry, there are good old technologies that get rediscovered and bad old technologies that get reused. The bad old technologies get reused because the younger generation thinks they are new. There is a great need for the results from behavioral and welfare research undertaken in a research setting to be transferred successfully into a commercial setting in the livestock and poultry industry. This chapter describes the successful transfer of the center-track restrainer system, which is used in many large beef plants, from a research project to widespread commercial use. It also provides tips on the successful conversion of individual gestation stalls to group sow housing. Some of the principles of successful technology transfer are choosing enthusiastic early adopters and making sure they do not fail. This may require more time than the original development of the new methods and equipment. Do not allow your technology to get tied up in patent disputes.

Learning Objectives

- How to transfer research results from the laboratory to the field
- Successful transfer of a new technology may require more work and time than the original research
- The importance of behavioral principles in design
- Finding the root cause of a problem

Introduction

In technology, sometimes a good old technology becomes lost and then rediscovered because the younger generation has no knowledge of the past. In other cases, a technology that has failed in the past has been resurrected and reused. In this chapter, the author will discuss ways in which technology is adapted by farmers and the meat industry.

Principles of Successful Technology Transfer From Development to Commercial Use

Ethologists, veterinarians, and animal scientists need to spend more time transferring the results of their research to industry. The author will now outline four steps that are required for the successful transfer of research results from the laboratory to producers. These recommendations are based on years of work on improving livestock handling, inventing new equipment, and implementing animal welfare auditing programs.

It is important to publish your research in peer-reviewed scientific journals so that knowledge does not get lost. Publishing in refereed journals also helps to ensure that the research results are verified by the use of good scientific methods. Researchers also need to publicize their work by giving talks and lectures, writing articles for industry magazines, and creating and maintaining websites. One of the reasons the author was able to transfer cattle-handling designs to the industry is that she wrote over 300 articles on her work for the livestock industry press. Every research project had several articles published about it. She also gave talks at producer meetings, and she posted her designs on her website, where anyone could download them for free (www.grandin.com). People are often too reluctant to give information away. The author discovered that when she gave away a lot of information, she got more consulting jobs than she could handle. She gave the standard designs away for free and made a living by charging for custom designs and consulting to solve problems on farms and in slaughter plants.

Make sure your early adopters are successful

For successful technology transfer, the first people who adopt it have to succeed. Researchers and developers need to choose companies with management that believes in the new technology. Early adopters must be supervised during every step of installation and start-up to make sure that the new method works correctly. Choosing the right farm or plant to test new technology is critical. The place should have a manager who loves the project and is dedicated to making it work. A poor manager with a bad attitude can make a good system fail. The author had a promising technology fail because the plant manager was not interested in the project. The people at the corporate office wanted the project but the local plant manager did not like having the hassle of working on a new project.

Supervise other early adopters to prevent mistakes that could cause the method or technology to fail

Transferring the author's center-track technology out of the first plant that used it successfully was very time-consuming (Fig. 16.1). A lot of time was spent on site at the next nine veal and beef plants to make sure the equipment had been installed correctly. The steel-welding companies often made many poorly designed modifications. They thought they were improving the system, but their modifications made it work poorly. In half the plants the author visited, she found either installation mistakes or modifications that would have caused the system to fail. Transferring the technology from the first plant to the next nine plants required more time and travel than the installation in the first plant.

Do not allow your method or technology to get tied up in patent disputes

There have been cases of companies buying patent rights to good technologies to prevent a new design from being adopted. That happened in the pig industry in the 1970s, when a designer in Ireland developed a humane, low-cost electric stunner for small pork plants. His design was very clever. He made it out of bicycle parts, which were cheap, and he designed it so it ran automatically and you did not have to pay an employee to operate it. That made it even cheaper to run, so small companies could afford it. But they never got a chance to buy

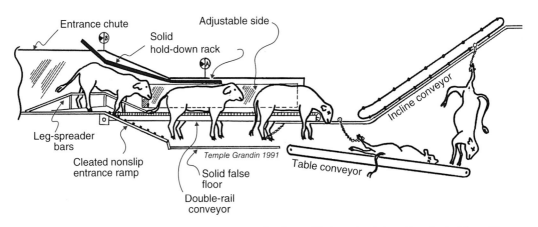

Fig. 16.1. Side view of the centre-track conveyor system that is used in large beef slaughter plants. The cattle straddle a moving conveyor built from metal slats that are 26 cm wide. The outside width of the moving conveyor and the frame must not exceed 30 cm.

T. Grandin

it because one of the big equipment companies that manufactured and sold an expensive stunner for large plants bought the patent rights. They put the new invention on the shelf and removed it from the market (Grandin and Johnson, 2009).

That was a terrible waste because the new design was more humane than the equipment the small plants were using, and it was too small to work in the larger plants. The company that bought the rights was not going to lose any money if the cheaper design for small plants went on the market. They just wanted to get rid of anything that had any possibility of competing with their system (Grandin and Johnson, 2009).

Case Study of Successful Transfer of Technology From the Research Laboratory to the Industry

The original idea for the center-track restrainer came from work done at the University of Connecticut (USA) in the 1970s (Westervelt *et al.*, 1976; Giger *et al.*, 1977). In the early 1970s, the Council for Livestock Protection, a consortium of US animal welfare groups, granted US$60,000 to researchers at the university to develop an alternative method to the cruel shackling and hoisting of conscious calves and sheep by one rear leg for slaughter in kosher plants. The Humane Society of the USA and other major animal non-governmental organization (NGO) groups contributed to this project. The researchers developed and built a plywood prototype, but many more components needed to be invented to make a commercially usable system (Fig. 16.2). With the prototype, the researchers determined that straddling a conveyor was a low stress, comfortable system for restraining an animal (Westervelt *et al.*, 1976). Before the invention of the straddle (center-track) design, all the plants were using a V-restrainer that had two angled conveyor belts on either side of the animal, squeezing the animal as it carried it along. The V-restrainer was invented for pigs and it worked well for fat, plump pigs. It was uncomfortable for a more angular, lean animal like a calf or cow.

In 1985, the Council for Livestock Protection awarded another US$100,000 grant to build an operational system inside a commercial slaughter plant that handled 220 kg veal calves. A major factor that made the transfer of the research prototype to a commercial plant successful was that the same funding agency funded both the scientific research and the development of the first commercial system.

Fig. 16.2. Prototype calf restrainer developed at the University of Connecticut that was used to determine that a straddle-type restrainer was a low-stress method for holding calves and sheep. A space between the bars enables the animal to rest on its armpits and prevents excessive pressure on the sternum (brisket).

The author was hired to design the commercial system and supervise its installation. The council also helped find a plant to build it in – a plant run by an innovator named Frank Broccoli. Frank was very interested in helping to develop a new system. His support and enthusiasm helped make the project successful. The council patented the original straddle idea in the 1970s. They did this so nobody else could patent it, and the design was put in the public domain for everyone to use.

The author had to invent an entrance design so that the cattle or calves could walk on to the conveyor belt and straddle it with their legs in the correct position. The original plywood prototype had no usable entrance design and no way to adjust it for different-sized animals. The author also had to design adjustable sides, to allow for animals of different sizes (Fig. 16.3). To visualize this, you can think of the adjustable sides in a printer feed tray, which you can narrow for envelopes and expand for typing paper. Without those two new parts, the system would not work (Grandin, 1988, 1991) (Fig. 16.3). Figure 16.4 shows a head-holder for the center-track restrainer.

Another hassle is teaching each new equipment company how to build the system correctly. One company built the adjustable sides backwards and another company used lighter gauge steel and many parts broke. Over 20 systems were already being used, but new shops still had to be trained for construction in other countries. The author spent

Fig. 16.3. A large steer is held in the center-track restrainer system. He remains calm because his body is supported fully by the conveyor. Adjustable sides can be adjusted easily for different-sized animals. A gap has been left between the adjustable side and the framework to prevent fingers from being pinched.

Fig. 16.4. Head holder mounted at the end of the center-track restrainer for kosher or halal slaughter. The round 7-cm diameter pipe fits behind the animal's poll. See Grandin (2014) and www.grandin.com for further information.

T. Grandin

hours on the telephone going over pictures and drawings. It was usually possible to fix the problems without traveling to the plant. Today, live video feeds over the Internet would be very helpful in solving problems with equipment.

People need to see it work

Even though the calf system had been operating for 4 years, and had been publicized widely, none of the beef plant managers was able to visualize how this system could be scaled up for cattle twice the size. No one was willing to invest in a large cattle system until they had seen one in operation. Additional grant money was obtained to build a large cattle system. After people saw it work, the technology spread rapidly through the industry. For a 2-year period, the author went to the equipment start-ups at the next seven plants to correct and prevent installation mistakes. Both the calf plant and the beef plant had to sign contracts that enabled viewing access to people who were interested in buying the equipment.

Avoid Premature Transfer of Technology to Industry

A common mistake is to attempt to transfer a new invention or method to the entire industry before it is fully perfected. Before any attempt was made to install the center-track technology in a second plant, the author made sure that EVERYTHING was working perfectly on the first system. Below is an example of a failure of a new technology due to launching it before it was completely perfected.

The USA had a really bad problem in the 1980s with the premature sale of electronic sow feeders for group housing of sows. In these systems, each sow has an electronic key that allows her to enter a feeding stall and get her precise portion of feed. In the early systems, a sow had to back out of the feeding stall and the next sow who was waiting often bit her. Many farmers installed the systems and then ripped them out because they worked poorly. The marketing department at the equipment company was selling systems before the research and development team had finished inventing them. In the 1990s, many of the problems with electronic sow feeders were solved. The innovation that really improved electronic sow feeders was a gate that allowed a sow to exit through the FRONT of the feeding stall. Today,

they work well. When a technology fails, it sometimes takes 10–20 years for people to be willing to try it again.

The disastrous launch of electronic sow feeders for group housing was probably a major contributor to the spread of individual gestation stalls throughout the USA. When the US pork industry underwent a huge expansion in the 1990s, sow stalls were chosen over group systems. Previous bad experiences with electronic sow feeders probably played a role in the decision of the three largest pork companies to build stalls (Grandin and Johnson, 2009). Another factor was that building contractors probably encouraged the construction of individual stalls because they made more money on welded steel compared to group pens.

Successfully Switching to Group Sow Housing

Many farm managers will be supervising the changeover to group housing systems for sows. There are many different systems that are available. A simple very effective system is to lock up the sows in individual stalls for feeding. This system is low tech and can be built and maintained easily in any country. No electricity is required and old sow gestation stalls can be converted easily to feeding stalls. Electronic feeders work really well in developed countries, and the computerized records enable a farmer to determine how much a sow is eating. Electronic feeders will require people with computer skills.

Genetics and sow aggression

Some genetic lines of lean, rapidly growing pigs are very aggressive. Selection for leanness was associated with aggression (Rydmer and Canario, 2014). The author observed that the incidence of tail biting more than doubled when a popular lean genetic line was introduced. Many producers who switched from gestation stalls to group housing found that they needed to change their pig genetics to lines with less aggressive animals. Sows that bite vulvas and inflict serious injuries to other sows should be removed from group housing immediately. One must remember that for over 30 years, some sow herds have had no culling for temperament. Because the sows lived in stalls, producers stopped culling aggressive animals. Aggression does not occur when animals are housed individually.

Group size in sow housing

Some of the most successful systems have sows housed in large groups of over 60 sows/pen. Large groups fight less. When pigs are mixed, small groups of five or six in a small pen will often get in severe fights. One of the reasons this happens is that there is no way for an animal that is being attacked to get away. Sows can be housed in either small, "static" groups of five or six, where they are never mixed again, or large, "dynamic" groups, where animals are constantly being added and removed. Dynamic groups MUST be LARGE groups of 60 or more sows per pen. Quick Access Information 16.1 provides suggestions to help with a successful transition to group housing.

Strategies to reduce fighting

Innovative design features can reduce fighting. When electronic sow feeders are used, an additional electronic lock on the entrance gate of the feeders reduces fighting. It prevents a sow who has finished eating from immediately re-entering the feeder. This stops fights caused by an aggressive sow, who immediately returns to bully another sow out of the feeder.

Another option that may help reduce fighting is providing "hides". McGlone and Curtis (1985) built small half stalls along one wall of a pen. When a pig entered the "hide", its head and shoulders were protected from another pig by a solid partition. Providing "hides" reduced injuries from fighting. When pigs fight, they attack the shoulder using an instinctual, hard-wired behavior pattern. The little half stall covers up the area where pigs instinctually attack. Injuries were reduced in pens equipped with "hides". Innovative managers and scientists need to explore the option of providing protected space. The paper can be downloaded for free from the Internet and it has clear photographs and diagrams that easily could be enlarged.

Welfare Problems Caused by Building Contractors

For years, the author has worked with many building contractors. Building contractors like to construct buildings that are labor-efficient for them to erect. Contractors can estimate precisely the costs of steel fabrication they do in their shop because they know their exact labor costs. When a contractor has to build something complicated on the

---

Quick Access Information 16.1. Tips for successful conversion from sow gestation stalls to group housing.

- Pig genetics may have to be changed. Some lines of lean pigs are very aggressive. Pigs can be genetically selected to be less aggressive (Rydmer and Canario, 2014). Never mix gentle genetic lines of pigs with aggressive genetic lines. The pigs from the gentle genetic line may be seriously injured by the pigs from the more aggressive genetic line.
- IMMEDIATELY remove sows that severely injure other sows from group housing. Quick removal is essential to prevent other sows from learning her aggressive behavior. The aggressive sow and her offspring must be removed from the future gene pool. Cull the aggressive sow after she has her piglets and feed out all her pigs as market pigs. A few gestation stalls may need to be kept for holding the really aggressive sows until they farrow.
- When sows are mixed, put them in a new pen to prevent the problem of newcomers invading another sow's territory.

- Providing sows with straw or other fibrous plant material to manipulate will help reduce fighting during mixing. In addition, try adding more bulk and fiber to sow diets. Adding fiber such as soybean hulls, wheat bran, or sugarbeet pulp to sow diets resulted in faster-growing piglets (Goihl, 2009).
- Do not constantly mix sows housed in small groups. Groups of five or six sows should be kept together as a "static" group. Groups of over 60 sows can be "dynamic", where sows are constantly added or subtracted. Big groups fight less because an animal that is attacked can escape.
- The farm manager must be enthusiastic about changing the system. A manager who is against the change may cause it to fail.
- If possible, use new gilts (young females) instead of sows that have lived their entire life in stalls.
- Provide straw or enrichment devices that provide chewing activities to reduce fighting, belly nosing, or tail biting (Studnitz et al., 2007; D'Eath et al., 2014).

T. Grandin

farm, his costs may vary because bad weather can delay construction.

The weather problem provides an economic motivation to design buildings that can be erected quickly on the farm, which minimizes costly delays caused by rain and storms. These designs are often not best for the animals. A well-designed, naturally ventilated dairy cow barn or covered lairage should have a high-pitched roof and a large ridge vent. Contractors may prefer to build a flatter roof with a smaller ridge vent because less labor on the farm is required to build it. A flatter roof with a small 30 cm ridge vent will result in a hotter barn in the summer. During cold weather, in a naturally ventilated building that lacks a large ridge vent, there can be greater condensation and water dripping from the inside of the roof. The author has visited lairages, stockyards, and dairies in many parts of the world. When a large space has to be covered with a roof and ventilated naturally, the best buildings have a high-pitched roof and a large ridge vent that is 2 m or more wide (Fig. 16.5). This creates a chimney effect that pulls heat and moisture out of the building.

Contractors may also try to persuade farmers to move away from naturally ventilated buildings to more intensively managed, fan-ventilated buildings. The fan-ventilated buildings are faster and easier for a contractor to build because the high-pitched roof and large ridge vent are not required. They also have an economic incentive to make money selling fans and other expensive ventilation components.

In the developing countries in warm regions, complicated mechanized ventilation systems should be avoided. They have high energy costs and are more difficult to maintain. During travels to Brazil, Chile, the Philippines, Mexico, and China, the author observed locally designed, naturally ventilated buildings that worked extremely well. In China, a naturally ventilated broiler chicken house was kept cool by evaporation by sprinkling water on the roof. The building has no mechanical components.

A Bad Old Methodology Returns from the Past

Since the publication of the first edition of this book, there has been increasing reinvention of a technology that failed in the early 1970s. In the USA, there was a boom in the construction of indoor feedlots where cattle were housed on concrete flats. By the late 1970s, many of the companies that built these concrete slatted floor feedlots failed financially because cattle became lame with swollen knees and were unable to reach market weight. After about 180 kg (400 lbs) weight gain, the cattle on concrete slats became lame. These

Fig. 16.5. A roof with a high pitch and a large ridge vent will keep lairages, dairies, and other covered livestock facilities cooler in the summer and well ventilated. The use of this design will reduce or eliminate the need for mechanical ventilation.

animals were fed a high-concentrate diet of 80–90% grain. The industry learned that cattle on a high-grain diet grew from a 204 kg (450 lb) calf to a 545 kg (1200 lb) steer on concrete slats. Today, people are building these same systems. If they add a rubber surface on the slats, they can reach 225 kg (500 lbs) of weight gain on a high-grain ration. If they attempt to put on too much weight gain, they have the same problem that the author observed in the 1970s. To fix the problem will require either a lot of soft bedding or greatly reducing the grain in the diet. Another alternative is to pasture cattle until they reach a heavier weight, before putting them on concrete slats.

Good Old Ideas are Reinvented

Danish sow farms are adopting a farrowing stall that can be opened up when the sow is in the nest-building phase. At this stage, the sow's need for freedom of movement is greatest (Wischner et al., 2009). Then the sow can be closed in when the newborns are small and fragile. Piglet mortality was low when the sow's movements were restricted for the first 4 days. After she has farrowed (Moustsen et al., 2013), then the stall can be reopened to allow her to turn around.

In 2012, the Danish Meat Research Institute described their work on these combi pens (The Pig Site, 2012; Jacobson, 2013). This is a good idea that has been around for a long time. When the author searched patents, a design by Newman (1956) popped up (Fig. 16.6a,b). The design is also similar to an old FAO (Fig. 16.7) drawing and a stall the author saw in Canada over 20 years ago. In technology, it sometimes takes time to reinvent good old ideas. The idea of a stall that could be opened up to allow a sow more freedom of movement has been around for a long time, but the Danish farmers discovered a better way to use it. They opened up the pen to allow freedom of movement during the nest-building and farrowing stage, and then closed it for 4 days after farrowing to protect the tiny newborns.

A Look at Future Welfare Technology

When you watch a dairy cow using the motorized cow brush, most people will say that she is enjoying it (Fig. 16.8). Numerous videos on the Internet show cows grooming all parts of their bodies with this brush. The first reaction of a scientist might be

"It's not scientific to say she likes or enjoys it". Now, imagine you have just downloaded a new app for your mobile telephone that is scientifically designed to determine if the dairy cow is really having a positive experience. After shooting 30 s of video, the app says, "yes".

The relatively new sciences of Qualitative Behavioral Analysis (QBA) shows that words people use to describe animal behavior are correlated with physiological measures. QBA research may some day be used as the basis for my proposed app. When numerous observers watch videos of cattle or sheep riding on trucks or during handling, they assign words to describe behavior such as calm, relaxed, frightened, or anxious. When the evaluations of 40–60 observers are pooled, there is high agreement (Stockman et al., 2011; Wickham et al., 2012). Sheep and cattle that were habituated to transport were more likely to be judged as calm and relaxed, and they had lower physiological indicators of stress (Stockman et al., 2011). At the slaughter plant, animals that were described as nervous or anxious had higher lactate levels (Stockman et al., 2012).

It appears that QBA works well for making a judgement about welfare for relatively short-term handling or transport events. In this situation, the animals are reacting to a specific event. Andreasen et al. (2013) found that having observers rate dairy cows in a barn as relaxed, anxious, or content had little correlation with a Welfare Quality audit that had many animal-based measures. Overall, the welfare of cows resting in their home barn could not be assessed with QBA.

At its present stage of development, QBA is not ready for commercial use. Today, 40–60 people are required to watch the videos and then their responses are pooled and analyzed by sophisticated statistics. In the future, it may be possible to replace the 40–60 observers with an artificial intelligence computer program that has been trained by hundreds of observers. These programs are already commercially available (Yuen et al., 2011). To make this technology work, it MUST be fully tested BEFORE it is released to commercial farms. Otherwise, it could fail like the early sow feeders and years would be required to convince people to try it again.

The Importance of Behavior in Design

One of the most difficult concepts to transfer to the livestock and poultry industry is the importance of design features on equipment that influence the

Sept. 25, 1956   H. C. NEWMAN   2,764,127

FARROWING STALL

Filed Jan. 14, 1953

(a)

(b)

INVENTOR.

Hillis C. Newman

BY

Fig. 16.6. An old US patent for a farrowing pen where the sow can be locked in (a) or allowed to turn around (b). Good, old ideas often get reinvented.

animal's behavior. On both dipping vats and the center-track restrainer system, the author designed an entrance ramp that provided nonslip footing (Grandin, 1980, 1988, 2014). The importance of providing nonslip footing is hard for many people to understand. The author has had to return to five or six beef plants and several dip vats to put the nonslip ramps back in after a user replaced it with a slick, slippery ramp. If an animal refused to enter the system, the user just assumed that he or she needed to change the ramp instead of finding the distractions that have been described in Grandin (1996) and Chapter 5, this volume.

The mistake that people make is that they often attempt to use force instead of using behavioral principles to induce an animal to move through a facility. The first step in fixing an existing facility is to remove distractions. People should use behavioral principles first before resorting to force.

Fig. 16.7. A diagram from the Food and Agriculture Organization for a turnaround farrowing stall that is very similar to a modern combi stall.

It is hard for some people to understand the importance of a nonslip floor in a veterinary race or a stun box. Animals will remain calmer if they have a good footing. Almost every time the author has visited a new beef slaughter plant, problems were observed with a slippery floor in the stun box. In 2014, the author visited three beef plants for the first time. Two out of the three plants had cattle slipping on the stun box floor. The managers are often surprised that cattle will stand still on a nonslip floor. Another technology that has been difficult to transfer is the use of solid shields to control what animals see during handling. At two out of the three beef plants, incoming cattle balked and refused to enter the stun box because they saw people through the front of the box. Installing a shield to block the view of visible people will improve cattle movement (Grandin, 2014). On the center-track restrainer systems, there are two large metal panels that are visual shields to control animal behavior (Grandin, 2003).

The first solid panel is located over the entrance and the first part of the conveyor (Fig. 16.9). The purpose of this panel or hold-down is to block the animal's vision until it is completely in the conveyor and restrainer (Grandin, 2003). If this panel is too short, the animals often become highly agitated. A 0.5 m difference in the length of this panel has a huge effect on cattle behavior. Well-intentioned welding shops shortened this panel on restrainers in two of the beef plants. This resulted in the cattle with large flight zones going crazy. To prove that this piece of metal was required, the author laid a piece of cardboard across the entrance to lengthen the

Fig. 16.8. It is likely cows have positive emotions when they use this motorized brush. Videos on the Internet make it obvious that dairy cows will stretch themselves into many positions to groom many parts of their bodies. (Photograph courtesy of DeLavel.)

Fig. 16.9. A curtain prevents the next animal in line from seeing the movement of the large pneumatic stunner. This curtain should be replaced if it becomes ripped or torn. It is important for managers to pay attention to these design details.

T. Grandin

hold-down. The plant was able to run for the rest of the day with cardboard holding in 560 kg wild cattle. That is the power of behavior. At the end of the shift, the cardboard was replaced with metal.

The second component that some people have removed is a false floor, which prevents the animals from seeing the "visual cliff" effect. The restrainer conveyor is 2.2 m off the floor and cattle refuse to enter if they see the 2.2 m drop-off. A false floor has to be installed to provide the optical illusion that there is a floor to walk on. The false floor is positioned so that it is 15 cm below the hooves of the largest animal. Unfortunately, people often removed the false floor because they did not see the purpose of it. They thought it was extra metal that they did not want to clean and maintain. On many occasions, the author has had to return to a plant to solve problems with animals refusing to enter the restrainer. The problem was fixed by replacing the false floor that had been removed.

Why is it Hard to Understand Behavioral Features of Design?

A person would never remove a mechanical component such as a drive unit on a piece of equipment because they know that a conveyor has to have a drive unit in order for it to move. The problem with parts that are there for behavioral reasons is that some people do not see the purpose of them, so they remove them. This is most likely to occur when the people the author trained have left and new people have replaced them. Animals notice small sensory details that people often fail to notice. Since animals do not have verbal language, they are much more aware of tiny visual details in their environment, which people often fail to notice (Grandin and Johnson, 2005). Figure 16.9 shows a white curtain made from lightweight conveyor belting that prevents the next animal in line from seeing movement of the stunner. Animals will often stop and refuse to move when they see rapid movement or high contrasts of light and dark (see Chapter 5, this volume).

Technology is easier to transfer than management know-how and good stockmanship

During a long career of working with many clients, the author has observed that people are more willing to adopt a new technology to solve a problem than to improve their management methods. For example, they are more willing to purchase new equipment compared to spending the time to train and supervise the employees who will be handling animals. The author's consulting business sells twice as many books on the design of livestock handling facilities compared to materials on the behavioral principles of handling cattle. Good equipment is essential, but it is useless unless good management is combined with it. People often want a quick technology fix instead of concentrating on training employees to take the time to become better managers.

The research presented in Chapter 7, this volume, shows very clearly that good management and good stockmanship will produce healthier animals that have better welfare and production. Learning good stockmanship takes time, dedication, and hard work. A common mistake is to assume that all the problems on a livestock or poultry farm can be fixed by buying new equipment. Good equipment can fix only half of the problems. The other half of the problems have to be fixed by good management practices. Well-designed equipment provides the tools that make good treatment of animals easier, but it is useless if management fails to train and supervise employees.

Problems with getting people to build things correctly

During a long career, the author has designed many curved handling systems for ranches (Fig. 16.10), farms and slaughter plants. The behavioral principles of these systems are explained in Chapter 5, this volume. A curved race works well because it takes advantage of the animal's natural tendency to circle back in the same direction it came from. Another reason why curved races are efficient is that animals entering the race cannot see the people at the other end when they first enter. The systems must be laid out correctly. The most common mistake is bending the race too sharply at the junction between the single-file race and the crowd pen. An animal entering the race must be able to see two to three body lengths up the race before it curves. Even though the principles of layout are clearly stated on the website (www.grandin.com), some people still lay the systems out incorrectly. Layout mistakes can wreck the system. The junction between the crowd pen and the single-file race is very critical. Some people may have a difficult time seeing the importance of the behavioral principles of layout.

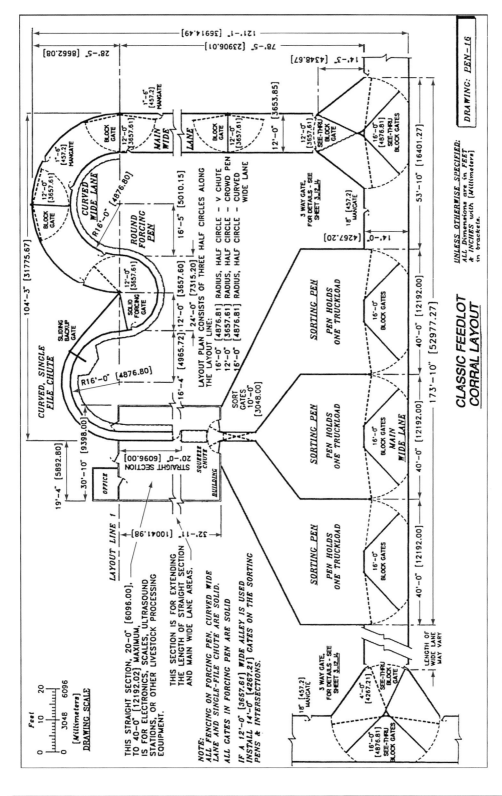

Fig. 16.10. Layout of a curved system for handling cattle in large feedlots. After the squeeze chute, the cattle can be sorted three ways. Correct layout is essential. The junction between the single-file chute (race) and the crowd pen must be laid out correctly. An animal standing at the entrance of the single file must be able to see two body lengths of the race. It must not be bent too sharply.

T. Grandin

Determine the Real Causes of Problems

When a problem occurs, it is important to determine the real cause of the problem. Over the years, poultry and pigs have become larger, more fragile, and difficult to handle and transport. Instead of building more and more technological fixes, maybe a better solution to weak animals would be to breed a hardier animal. A slight decrease in productivity would be more than paid back by improved welfare, less sickness, and fewer losses to death.

The author learned this lesson about the importance of determining the true cause of problems back in 1980. She was asked by a large pork slaughter plant to design a conveyor system to transport pigs that were too weak to walk up a big, long ramp. This total engineering approach to solving this problem was all wrong. A conveyor was installed in the bottom of the single-file pig race and it caused the pigs to jam up and flip over backwards. The entire expensive mess had to be torn out.

When the author started tracking the weak pigs, it was discovered that they all came from a single large farm. Instead of changing all the equipment at the plant, changes should have been made on this one farm. The weak pig problem could have been corrected easily by changing boars and breeding out a genetic defect called "spraddle leg". The farm also had smooth metal flooring that caused the pigs to grow excessively long hooves. Correcting the problems at this one farm would have been much easier and cheaper than rebuilding plant equipment to accommodate animals with genetic defects and excessive hoof growth. The lesson to be learned from this case is that managers, veterinarians, and engineers must find out what the true cause of a problem is before they initiate expensive methods to correct it.

A lot of little changes equal big improvements in animal handling

Many older facilities can often be improved greatly with many small improvements. You would never want to copy the bad old designs for a new facility, but a high standard of animal welfare can usually be attained in many older facilities. Below is a case history from one meat plant.

Step 1. Train the employees in basic behavioral principles (see Chapter 5, this volume). Employees were trained to move smaller bunches of cattle. Moving small groups will require more walking.

Step 2. Repair broken gates and other equipment that directly contacted the animals. Small repairs such as fixing broken or dragging gates was much more important than fixing a leaking lairage roof.

Step 3. Identify and correct all distractions that caused animals to back up or balk. Below is a list of distractions that had to be eliminated in this particular facility:

- Install curtains to block sunbeams shining in the main alley.
- Install a shield to prevent approaching cattle from seeing the person who kept track of cattle ID tags.
- Add a lamp to illuminate a dark race entrance.
- Install a curtain to prevent cattle from seeing people walking by.
- Close in the side of the single-file race to block vision.

All of these distractions had to be located and fixed before the cattle would move easily.

A Summary of Principles for Successful Technology Transfer

1. Knowledge of BOTH the practical and the scientific aspects of your area of speciality in animal welfare is essential. One must keep one's knowledge up-to-date by continually visiting farms and plants. Continuous reading of the scientific literature is also required.
2. Continually writing and communicating ideas and methods – similar lectures and articles must be presented many times to many different audiences. Build a website full of practical information with free access. Document successes in the scientific literature that will be archived permanently in scientific databases.
3. Steady, sustained effort over many years will bring about more lasting improvements than short, intense bursts of activity.
4. The methods or technologies you implement must be practical, and they must work on the farm or in the plant. You need to get out into the farms and plants. Talk to the people who will be using the equipment. These people are often really good at making incremental improvements. However, they often do not have the vision for trying something totally new. They need to see really good videos to be convinced that a new technology will work.

5. Be willing to be criticized, especially when doing new, pioneering things. Accept constructive criticism and be willing to modify what you have been doing.

6. Have the wisdom to work on changing things that can be changed and do not bash your head against the things that cannot be changed.

7. Be positive in your communications and always keep your word. Present the SAME message to the livestock industry, the public, and animal advocacy groups. Never be nasty, mean, or rude.

8. Find innovative farmers and plant managers who want to make your methods work to be the early adopters. To be successful, the early adopters must believe in your methods and research.

9. The time required to transfer your research results successfully to the livestock producers may require more work and time than doing the scientific research.

10. Maintain confidentiality so you can continue with access to farms and slaughter plants. When bad practices were observed, the author would write and talk about them without revealing the location. If the plant or farm were identified, it would have caused the industry to shut its doors on the author's efforts to reform it.

11. Never sign a commercial noncompete clause with a client or technical confidentiality agreement that would interfere with your ability to communicate your ideas and methods.

References

Andreasen, S.N., Wernelsfelder, F., Sandoe, P. and Forkman, B. (2013) The correlation of qualitative behaviour assessments with welfare quality protocol outcomes in on-farm welfare assessments of dairy cattle. *Applied Animal Behaviour Science* 143, 9–17.

D'Eath, R.B., Arnott, G., Turner, S.P., Jensen, T., Lahrmann, H.P., *et al.* (2014) Injurious tail biting in pigs: How can it be controlled in existing systems without tail docking? *Animal* 8, 1479–1497.

Giger, W., Prince, R.P., Westervelt, R. and Kinsman, D.M. (1977) Equipment for low stress animal slaughter. *Transactions of the American Society of Agricultural Engineering* 20, 571–578.

Goihl, J. (2009) Bulky gestation diets help piglets grow faster. *Feedstuffs* 23 March, pp. 12–13.

Grandin, T. (1980) Observations of cattle behavior applied to the design of cattle handling facilities. *Applied Animal Ethology* 6, 19–31.

Grandin, T. (1988) Double rail restrainer for livestock handling. *Journal of Agricultural Engineering Research* 41, 327–338.

Grandin, T. (1991) *Double Rail Restrainer for Handling Beef Cattle*. Paper No. 91–5004. American Society of Agricultural Engineers, St Joseph, Michigan.

Grandin, T. (1996) Factors that impede animal movement at slaughter plants. *Journal of the American Veterinary Medical Association* 209, 757–759.

Grandin, T. (2003) Transferring results of behavioral research to industry to improve animal welfare on the farm, ranch and the slaughter plants. *Applied Animal Behaviour Science* 81, 215–228.

Grandin, T. (2014) Handling and welfare in slaughter plants. In: Grandin, T. (ed.) *Livestock Handling and Transport*. CAB International, Wallingford, UK, pp. 329–353.

Grandin, T. and Johnson, C. (2005) *Animals in Translation*. Scribner (Simon and Schuster), New York.

Grandin, T. and Johnson, C. (2009) *Animals Make Us Human*. Houghton Mifflin Harcourt, Boston, Massachusetts. Sections of this chapter originally appeared in this book.

Jacobson, A.P. (2013) Combi-pens reduce pre-weaning mortality to 9%. Pig Progress. Available at: http://www.pigprogress.net/Breeding/Welfare-Environment/2013/2/Combi-pens-reduce-pre-weaning-mortality-to-9-1167885W/ (accessed 1 April 2015).

McGlone, J.J. and Curtis, S.E. (1985) Behavior and performance of weanling pigs in pens equipped with hide areas. *Journal of Animal Science* 60, 20–24.

Moustsen, V.A., Hales, J., Lahrmann, H.P., Weber, P.M. and Hansen, C.F. (2013) Confinement of lactating sows in crates for 4 days after farrowing reduces piglet mortality. *Animal* 7, 648–654.

Newman, H.C. (1956) Farrowing stall. US Patent 2764127A.

Pig Site, The (2012) Danish Pig Research Centre Annual Report 2012, Farrowing Pens. Available at: www.thepigsite.com/articles/4417/danish-pig-research-centre (accessed 12 October 2014).

Rydmer, L. and Canario, L. (2014) Behavioral genetics in relation to welfare. In: Grandin, T. and Deesing, M. (eds) *Genetics and the Behavior of Domestic Animals*, 2nd edn. Elsevier, San Diego, pp. 397–434.

Stockman, C.A., Collins, T., Barnes, A.L., Miller, D., Wickham, S.L., *et al.* (2011) Qualitative behavioral assessment and quantitative physiological measurement of cattle naïve and habituated to transport. *Animal Production Science* 51, 240–249.

Stockman, C.A., McGilchrist, P., Collins, T., Barnes, A., Miller, A., *et al.* (2012) Qualitative behavioral assessment of Angus steers during pre-slaughter handling and relationship to temperament and physiological response. *Applied Animal Behaviour Science* 142, 125–133.

Studnitz, M., Jensen, M.K. and Pedersen, L.J. (2007) Why do pigs root and in what will they root? *Applied Animal Behaviour Science* 107, 183–197.

Westervelt, R.G., Kinsman, D., Prince, R.P. and Giger, W. (1976) Physiological stress measurement during slaughter of calves and lambs. *Journal of Animal Science* 42, 831–834.

Wickham, S.L., Collins, T., Barnes, A.L., Miller, D.W., Beatty, D.T., *et al*. (2012) Qualitative behavioral assessment of transport naïve and transport habituated sheep. *Journal of Animal Science* 90, 4523–4535.

Wischner, D., Kemper, N. and Krieter, J. (2009) Nest building behavior in sows and consequences for pig husbandry. *Livestock Science* 124, 1–8.

Yuen, M.C., King, I. and Leung, K.S. (2011) A survey of crowdsourcing systems. IEEE Third International Conference on Social Computing. Available at: http://ieeexplore.ieee.org/xpls/abs_all.jsp?arnumber=6113213 (accessed 1 April 2015).

17 Practical Methods for Improving the Welfare of Horses, Donkeys, Mules, and Other Working Draft Animals in Developing Areas

CAMIE R. HELESKI[1], AMY K. MCLEAN[2] AND JANICE C. SWANSON[1]

[1]Michigan State University, East Lansing, Michigan, USA; [2]American Quarter Horse Association, Amarillo, Texas, USA

Summary

Both the welfare and working ability of horses and other draft animals can be improved greatly with simple interventions. In hot climates, horses can become dehydrated easily. Studies showed that 41–50% of working horses were dehydrated. During hot conditions, a horse should be offered 60 l/day (two or three large buckets) and donkeys should receive 20 l (one large bucket). Common mistakes that contribute to sores and injuries are placing the cart shafts too high and incorrect cart balance. The animal should be brushed or washed daily to help prevent harness sores. Many people take good care of the horse's feet, but are not aware that other simple interventions, such as replacing harness padding, will improve both the performance and the welfare of their horse or donkey. The use of correct padding is essential. Never use rough material such as burlap or woven plastic bags.

Learning Objectives

- Understand the importance of working equids from a global perspective
- Learn about common welfare problems faced by these animals
- Recognize realistic solutions to many of these welfare challenges
- Recognize this is an underserved livestock population, in need of more resources, more research, and more outreach

Introduction

One of this chapter's authors (Camie Heleski) has spent the majority of her academic career as Coordinator of the MSU Horse Management Program. One of her primary goals has been to teach students strategies to manage horses optimally to maximize competitive performance, enhance diet formulation, optimize profitability, enhance breeding programs, and select horses based largely on aesthetic parameters.

What a rude awakening for her to become passionately involved with the welfare of working equids, beginning in 2000 with her first of seven international trips focusing on working equids. Since the first edition of this book, she has done extensive work in remote Mexican villages. She now think in terms of what are the 5–15 essential things these owners could do to enhance the welfare of their animals. Realizing that in enhancing the welfare of their carthorse or donkey, they, in turn, enhance the well-being of their family (Starkey, 1997; Kitalyi et al., 2005; Upjohn et al., 2014).

Most of the owners/handlers of these horses, donkeys, and mules are not intentionally cruel; rather, they live in suboptimal environments, and equine husbandry information and outreach is a rare commodity. In some cases, it is the limitation of their financial resources that prevents them from doing more for their animals, but in many cases, it is simply a lack of knowledge on their part.

If you are a professional/paraprofessional working in a developing part of the world, your work is to be commended. You are serving a very important purpose, and one that will improve greatly the welfare of many animals. The suggestions that follow are designed to involve minimal financial inputs. We hope to transition the husbandry of working equids from reactive health care (treating wounds or rehabilitating animals too ill or undernourished to continue working) to proactive husbandry, so that animals can work more productively and have enhanced longevity and well-being. This, in turn, contributes to the well-being of the families who depend on these animals and, as we have seen firsthand, contributes to the pride they have in their working animals.

The Importance of Working Draft Animals in the Developing World

The importance of working draft animals in developing areas of the world is often overlooked (e.g. Pearson *et al.*, 1999; Popescu and Diugan, 2013). Research on these animals in refereed publications is comparatively scarce (though in preparation to write this chapter, it was impressive how much information on draft animals is available if one knows where to look). Based on world equid population statistics (e.g. Pollock, 2009), of the approximately 55 million horses, nearly 84% are used for work in developing countries; of the 41 million donkeys, 98% are used for this purpose; and of the 13 million mules and hinnies, 96% are used for performing work in developing parts of the world. According to some estimates,

98% of the world's equine veterinarians work on a mere 10% of the world's equine populations (Pollock, 2009). It is understandable that most equine veterinarians would work in developed areas of the world; however, there is tremendous working equid suffering taking place that merits our attention (Burn *et al.*, 2010; Riex *et al.*, 2014; Upjohn *et al.*, 2014). For people who have not yet worked with livestock in developing parts of the world, the inevitable question comes … why help the animals when the people are often suffering? It has been our experience that when working animal welfare is enhanced, it subsequently enhances the well-being of their respective families. Furthermore, there are environmentally friendly aspects to supporting the sustenance, possibly even the growth, of draft animal usage in developing countries where efficiency may be best captured through the use of lower rather than higher forms of technology (FAO, 2002; Arriaga-Jordán *et al.*, 2006).

With this in mind, this chapter is written to provide equine professionals/paraprofessionals working in developing regions of the world with fundamental knowledge important to enhancing the husbandry and subsequent welfare of working animals, with a particular emphasis on working equids. Direct experience working with carthorses in southern Brazil, draft donkeys in Mali, West Africa, and working equids in remote villages in Mexico is combined with a review of the literature to form the basis of the insight and

---

Quick Access Information 17.1. Tips to improve both the welfare and the performance of working equids.

1. Check harness fit and modify when necessary and do not use harness materials that are likely to cause sores (Quick Access Information 17.2).
2. Check shaft height and cart balance.
3. Clean the horse or donkey in between work sessions (washing/brushing).
4. Provide clean harness padding where needed.
5. Clean and treat lesions (and other wounds) with antibiotic salve as needed.
6. Provide forage as frequently as possible.
7. Add high quality forage/concentrate when animal is working especially hard or starting to get thin.
8. Try to get the animal into shade during work breaks.

9. Provide salt, preferably trace mineral salt.
10. Provide fresh, clean water multiple times per day. Horses require significantly more than donkeys.
11. Deworm the animal for parasites that are found in your region.
12. If at all possible, vaccinate for critical diseases, based on veterinarian advice.
13. Keep the hooves maintained adequately.
14. If at all possible, do not work lame animals or those with serious injuries/sickness.
15. Treat the animal fairly – no beating! Very few equids need beating to go forward unless they are completely exhausted (in which case, they should be rested) or are completely overburdened.

---

recommendations presented in this chapter. We have tried to make this chapter as realistic and applicable as possible. We recognize that most of the owners/users of working draft animals have very few financial resources to enhance the well-being of the animals in their care. Our suggestions are based on those enhancements that can provide maximal benefit to the animal's comfort and longevity, with minimum inputs, other than time.

The remainder of the chapter discusses 15 recommendations that can enhance greatly the welfare and longevity of working equids and can be implemented with very little financial cost.

Harness Fit, Harnessing, Loading Practices

One of the least expensive ways to enhance working equid welfare is by improving the harness fit, to add padding where needed, to keep the animal clean where the harness is likely to rub, to keep the harness itself clean, and to be certain the cart's load is as balanced as possible. All of these factors will help to minimize the harness lesions that tend to be prevalent in working equids (e.g. Diarra *et al.*, 2007; Sevilla and León, 2007; McLean *et al.*, 2012). Unfortunately, poor condition of the harness is still a major problem (Upjohn *et al.*, 2012). Pressure on the back and withers is affected by harness type (McLean *et al.*, 2012). In addition, animals that are of moderate body condition score (i.e. not thin)[1] will have less friction between body contact points and the harness, generally leading to fewer lesions. If the animal has sweated during the work day, which is more likely in horses than donkeys (Bullard *et al.*, 1970), with mules typically being intermediate, the animal should be rinsed off or sponged off at the end of the work day. If water is a rare commodity, the animal can be brushed off the following day, prior to having its harness put on. In this way, the sweat does not build up under the harnessing area and cause additional friction on subsequent workdays. When dirt or sweat accumulates, it increases the friction of the harness rubbing on the animal. When the animals are thin, there is less natural padding on the body and lesions are even more likely to occur. See Fig. 17.19 for an example of lesions on the withers of a horse.

In one study on Brazilian carthorses (Zanella *et al.*, 2003), we initially found 96% of the horses had harness lesions, most frequently on their withers

or behind their elbows. After an educational workshop for the horse owners that emphasized the previously discussed recommendations, the percentage of lesions dropped to 62% when rechecked 4 months later. A 34% reduction in harness lesions of working horses, as shown in this study, is a significant welfare improvement.

A survey in Mali by Diarra *et al.* (2007) found that >60% of the donkeys had significant harness lesions. It should be noted that with both the horses and the donkeys, many of the lesions were still open sores. Working a full day with a harness rubbing against an open sore is very painful. Improvements in harness fit and harness padding will enhance the welfare of working equids greatly (e.g. Ramswamy, 1998).

Figures 17.1–17.5 show horses and donkeys with generally well-placed harness adjustments. Each country, and sometimes each region, has slightly different modifications to standard harness, saddle, and bridle designs and commonly used carts (see Figs 17.6 and 17.7). A carefully designed harness and cart for all owners of working equids is currently financially unfeasible. In the meantime, persons working with these populations should develop efficient, reasonably comfortable harnesses for the animals they are engaged in helping. Readers are encouraged to visit the *Draught Animal News* (DAN) website (http://www.vet.ed.ac.uk/ctvm/Research/DAPR/draught%20animal%20news/danindex.htm) for downloadable documents containing numerous suggestions, pictures, and drawings for enhancing draft animal harnesses and implements in developing countries.

Our observations of horses in Brazil show many of the horses have the shafts placed too high (Fig. 17.8), which places considerable pressure on to the withers – sometimes referred to as vertical weight (Jones, 2008a). Many of the horses are observed to have lesions in the wither area (see Fig. 17.9) that may cause considerable discomfort. Our experience with Malian donkeys has shown that many of them also have wither lesions and/or fistulous withers (Doumbia, 2008; McLean *et al.*, 2012). It appears this can be due to shaft placement being too high, and/or the harness padding material is too abrasive and contributes to these lesions. In West Africa, SPANA (Society for the Protection of Animals Abroad) has been working hard to distribute appropriate harness backband padding to donkey owners (see Figs 17.10 and 17.11).

C.R. Heleski, A.K. McLean and J.C. Swanson

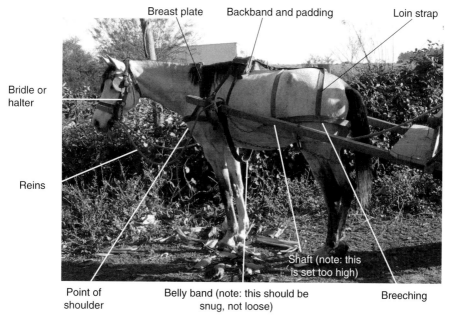

Breast plate   Backband and padding   Loin strap

Bridle or
halter

Reins

Point of
shoulder

Belly band (note: this should be
snug, not loose)

Shaft (note: this
is set too high)

Breeching

Fig. 17.1. Diagram showing well-positioned harness, shafts, and all appropriate parts of a harness to help distribute loading.

A recent study by Burn and others (2008) found that many of the ridden pack donkeys in Jordan had under-tail lesions caused by poorly made/poorly cleaned cruppers under the tails.

Determining Load Capacity of Working Equids

As a general rule of thumb, working equids should not be asked to pull more than their own body weight for any considerable length of time (Jones, 2008a). Setting a specific recommendation for length of time is difficult because it also depends on whether the terrain is flat or hilly; whether the road is hard packed, or sandy, or rocky; whether the wheels have good bearing and turn smoothly or are, themselves a large contributor to friction; what speed the donkey or horse is being asked to travel; how many hours will the draft animal need to pull the load; will there be breaks provided? What fitness/condition is the equid in? Pound for pound, donkeys and mules are generally considered to be stronger than the typical light horses found in most developing parts of the world.

Regarding carried loads, the numbers can be even more difficult to calculate. In developed parts

of the world, 20–25% of the horse's weight has become the recommendation for how much a horse can carry, assuming average condition of the horse, average terrain to be covered, etc. (Wickler *et al.*, 2001). Many working donkeys are observed to carry likely nearly 50% of their own body weight over considerable distances, and often over rough terrain. This does not mean it represents an ideal situation and may be partially why many of these animals have significant lesions on their body.

Assessing Body Condition of Working Equids

Most of the working equids we have observed are thinner than would be desirable (see Figs 17.12 and 17.13). Many of them are dangerously thin, and their body condition is likely to reduce the length of their working life and contribute to significantly reduced welfare. Thin animals often develop more harness lesions and may have subsequently reduced immune systems that make them more likely to become ill. A few different body condition scoring systems have been implemented by professionals/paraprofessionals working in the various countries (e.g. Henneke *et al.*, 1983;

Fig. 17.2. Typical harness set-up for working carthorse in Brazil. The shafts are slightly high and ideally should be longer. The shafts should be level with the point of shoulder in most cases. Notice the harness in the photograph. This is a good example of a harness for working horses. (See Fig. 17.1 if unfamiliar with harness part names.) The breast collar or strap placed over the shoulders of the horse and around the chest is adjusted properly and made of material that is sturdy but will not rub the horse excessively. Also, the horse is equipped with a bellyband and backband that has adequate padding placed above the shoulders and withers. Both of these straps, the backband and the bellyband, help stabilize the load and help keep the shafts in place. This horse is also equipped with breeching, which is needed to help stop the cart and load. Often, primitive harnesses will not include breeching (see Fig. 17.4), but this is needed when animals are pulling heavy loads, for stopping, and especially when working in hilly or mountainous terrain. Traces or trace chains, the metal chains attached to the harness at the breast collar and then to the cart, are also needed, so the animal's power can reach the cart better and be implemented. Again, primitive or crude harnesses often lack traces, and these should be added to assist the animal.

Svendsen, 1997); however, the bottom line is usually something like: overweight (almost never observed in working equids), moderate (see Figs 17.14 and 17.15), thin, and various levels of extremely thin (see Figs 17.12 and 17.13) (Pritchard *et al.*, 2005).

Provide Plenty of Water to Prevent Dehydration

The most important nutrient to the well-being of working equids is water. Many owners are unaware how much water is required for their animal, especially if it is a horse. Horses may sweat up to 15 l/h of trotting work, especially when working in hot, humid climates (Clayton, 1991). It could be expected that a working horse in a warm climate might need 40–60 l/day (10–15 gallons/day) of clean, fresh water (Fig. 17.16). However, a donkey may need only 20 l/day (5 gallons/day) (Fig. 17.17). Donkeys are often the best choice for really hot climates because they sweat less than horses. Water would ideally be offered intermittently throughout the day. In southern Brazil (Zanella *et al.*, 2003), most working horses were only offered water at night, and it was often just one bucket made available. When skin hydration was checked in nearly 5000 working equids (Pritchard *et al.*, 2007), 50% of horses and 37% of donkeys showed significant signs of dehydration. Similar problems with dehydration were also found in Nicaragua, where 41% of working horses were dehydrated (Wilgert, 2010). Recent studies in Africa and Botswana showed that donkeys in poor body condition were more likely to be dehydrated (McLean *et al.*, 2012) and that 66% had poor body condition (Geiger and Hovorka, 2015). Another problem, at least in southern Brazil, is the presence of leptospirosis in still (stagnant) water (R. Zanella, personal communication). Often, horses were turned loose in small paddocks containing small, stagnant ponds as the only water source. Sadly, the seroprevalence of leptospirosis was quite high. In addition to the potential damage this can cause to horses (e.g. moon blindness/uveitis, abortions, and depression), it is also a zoonotic disease, which creates a further concern to the humans living nearby.

Fortunately, donkeys are far more drought-resistant than horses (NRC, 2007). It is not uncommon for wild donkeys to go to water only once every 2–3 days. However, when they do gain access to water, they can drink a large amount in a very short amount of time (Bullard *et al.*, 1970). Working donkeys should be provided water daily. In hot weather, they should be offered water several times per day. As previously discussed, the quality and health status of the water is important.

A major problem with insufficient hydration in working equids is that food intake becomes depressed (NRC, 2007) and leads to an apathetic animal incapable of performing its usual work duties at full capacity. Decreased work performance could lead to owner frustration and increases the likelihood of whipping or beating of the animal.

Despite traditional beliefs to the contrary, repeated studies with heavily worked horses show no ill effects in allowing the horses, mules,

C.R. Heleski, A.K. McLean and J.C. Swanson

Fig. 17.3. Another typical harness set-up for a working horse in Brazil. Again, shafts are somewhat too high and should be longer, extending out to the point of shoulder. Notice there is no breeching on this horse. Breeching is needed when pulling heavy loads, and it helps the horse stop the load, especially when working in hilly or mountainous terrain. Notice, though, that this horse has a leather strap around his tail called a crupper. This, too, can help hold a load off of the horse, or at least distribute the load more evenly.

or donkeys to drink their fill, whenever water is made available, and returning to work (Jones, 2004; Pearson, 2005).

Provide Lots of Opportunities to Forage

Few working equids are able to consume enough calories to keep up with their workload; thus, many are observed as being thin (e.g. Pearson, 2005; Pritchard *et al.*, 2005; Burn *et al.*, 2010). Owners should allow the animal to forage whenever possible. Also, work breaks should be provisioned to allow the animals to eat. In southern Brazil, grass quality is generally good; however, the working horse owners often live in small villages on the edge of the city (peri-urban), where dozens of horses have been staked out and have overgrazed the grass to the ground. The horse may spend the evening grazing, but under these conditions his calorie intake will be minimal. The more informed horse owners observed in this study sent their children to cut bags of grass from neighboring ditch banks. Horses receiving the cut grass were in

better condition than the average horses observed (Figs 17.18 and 17.19). (Note: cut grass should never be stored in bags for any length of time before being given to a horse; it will mold and lead to a sick horse).

In addition to the observations noted above, draft horses identified to be in good condition were supplemented with leftover melons from local fruit stands, leftover bread from local bakeries, and wheat bran and oatmeal obtained from the local market. We discovered that extra calories from even the most unorthodox sources were utilized by the horses. Obviously, the potential for toxic plants should be considered. Most horses, even very thin ones, are good at avoiding toxic plants. This is not an endorsement of unorthodox feed resources, but there are more working equids suffering from lack of food than the rare one that might be lost to toxicity. In these developing parts of the world, everything becomes a risk–benefit assessment. The utilization of locally available food resources, even those that deviate from what is considered "normal" feed, should be considered.

Fig. 17.4. Another typical cart donkey set-up observed in western Africa. Note the appropriate placement of shafts as compared to the point of the shoulder. Also note that, in this case, the donkey is guided primarily by its halter and reins. We observed that when the back padding was made from cotton-type materials, it served its purpose considerably better than when abrasive plastic burlap padding was used. The lack of breeching is a concern with this style of harness.

Working donkeys require fewer daily calories than working horses (NRC, 2007). The donkey is truly amazing with respect to what minimal resources are required for survival. Many of the donkeys we have observed subsisted on seasonal opportunity forage. For example, when cowpea bean hay was occasionally available during the growing season, this was a higher-quality food source for the donkeys that helped them regain body condition lost during the drought season. At other times of the year, donkeys had to be content with only straw or maize stover, both of which are very low in nutritional content. Somehow, the donkey is able to utilize effectively every possible nutrient from feedstuffs (e.g. straw) that would be useless to many animals (Pearson, 2005; NRC, 2007). On rare occasions, donkeys were offered supplemental grain, but this practice was unusual. Donkeys, and other equids, do need some of their nutrient needs met with fresh, green forage in order to accommodate their vitamin requirements.

Co-author McLean, has considerable experience working with mules. In general, the nutrition needs of the mule (or hinny) are intermediate between horses and donkeys.

Salt Requirements

The need for salt is a well-established nutritional requirement for horses (NRC, 2007). Horses should have access to salt, preferably trace mineralized (TM) salt, developed for the region of the world that the animal is living in. For example, there are many parts of the world that are selenium deficient. In those areas, TM salt will often have the addition of selenium. Horses who are not receiving salt are likely to become dehydrated and depressed more quickly than horses with access to salt. Their water intake, in some cases, will become depressed. Salt is a cheap nutrient to supply in most parts of the world. We would like to see all mobile veterinary clinics servicing working equids provide TM salt to their client animals.

Less is known about the salt requirements of the donkey (Pearson, 2005; NRC, 2007). Donkey husbandry guides typically recommend supplying TM

C.R. Heleski, A.K. McLean and J.C. Swanson

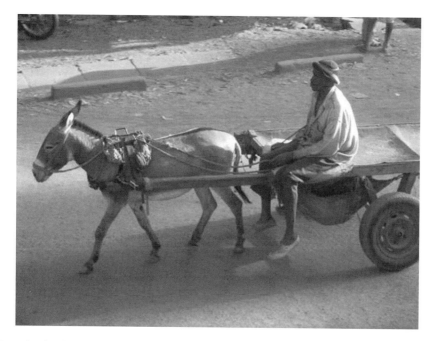

Fig. 17.5. Example of a donkey being driven with a halter and reins. Notice the length of shafts, which is good, and they are at an appropriate height, even with the point of the shoulder. Also, the cart has higher wheels and a lower axle than some shown in earlier pictures. This makes it easier for the donkey to pull the cart forward because the axle and wheels are balanced properly. Also, a synthetic material, probably foam, has been used for backband padding for the donkey. This material is soft, not abrasive, easy to clean, and probably recycled. This donkey is also in good physical condition. Donkeys and horses that receive proper nutrition can be more productive when working. (From Ellis *et al.*, 1980.)

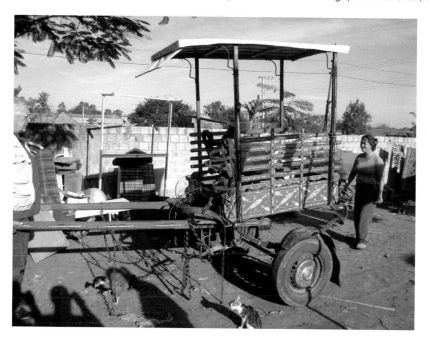

Fig. 17.6 A typical cart used for transporting people and picking up small loads of goods or recyclables (southern Brazil).

Fig. 17.7. A typical working cart donkey harness set-up in western Africa. Note back padding to help minimize harness lesions. A donkey such as this would be guided primarily with a stick. Also, notice that the harness is very crude. There is no halter or bridle for guiding and communicating with the donkey; the bellyband is made of a thin rope, which could lead to lesions. It should instead be made of wider and thicker material. Also, this harness has no traces. The traces could help transfer the power of the donkey pulling to moving the cart. Also, if the wheels were higher on the cart, then the load would roll easier.

Quick Access Information 17.2. Harness padding and repair materials to help prevent sores and lesions.

Good materials	Bad materials
Cotton canvas	Woven plastic bags – the WORST material
Soft leather	Burlap
Sheepskins	Lightweight cloth that wrinkles and forms ridges that rub
Woven wool blanket	Polyethylene sheets – these are too hot
Foam padding	Innertube rubber – this is too hot
Twine or string for quick repairs	Wire for quick repairs – wire should never come in contact with the animal

Cleaning tips for harnesses

Scrape off crusted dirt and sweat
Pick debris out of woven wool
Replace foam monthly, if possible
Use soap and water to clean but ensure animal contact surfaces are allowed to dry completely before using
If saddle soap or oil for leather is available, it should be used to keep leather soft

---

salt to donkeys. If TM salt is not available, even a handful of regular salt or access to rock salt will benefit the donkey in replacing the major electrolytes such as sodium and chloride.

Mules tend to sweat an intermediate amount between horses and donkeys, but little documentation exists that verifies this anecdotal observation. It is assumed their salt requirements

C.R. Heleski, A.K. McLean and J.C. Swanson

Fig. 17.8. These shafts are set significantly too high. You can see the horse trying to brace the load by putting its front legs well under its body. The distribution of weight on the cart could also contribute to the shafts being placed so high. One positive aspect about this photograph is the use of breeching and trace chains. However, if the horse is hitched improperly, then the breeching and traces cannot be used to their fullest to alleviate extra weight being placed on the forehand of the horse. Unevenly distributed loads can lead to lameness and create lesions from rubbing the horse's body.

are likely to be intermediate between that of horses and donkeys.

Parasite Control

In developed countries, parasite control programs have undergone a significant overhaul in the past decade (see the AAEP guidelines at http://www.aaep.org/custdocs/ParasiteControlGuidelinesFinal.pdf.)

In developing parts of the world, any enhancement in parasite control would be considered an improvement. During our owner interviews in southern Brazil (Zanella *et al.*, 2003), we found that less than 5% of horse owners were deworming – despite the animals being staked out in close proximity and eating in areas of high concentration of manure. Twice per year deworming with a low-cost ivermectin product could reduce parasite loads dramatically. An advantage of ivermectin-based products that is sometimes forgotten by people in developed countries is that ivermectin helps get rid of external parasites that may plague these working equids. Many skin conditions have been observed in both working horses and donkeys. Some of these conditions are exacerbated, if not caused, by external parasites.

Fig. 17.9. Wither lesions are very common in working equids worldwide, often because the harness or saddle does not fit well, because the padding is insufficient, or because the horse is quite thin. Additionally, traditional beliefs often mean the equids do not get bathed, so dirty harnesses are often put back on sweat-caked animals, adding to the friction problem.

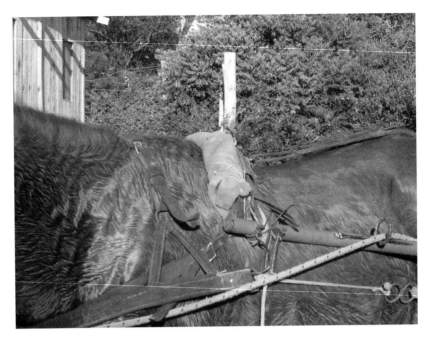

Fig. 17.10. One type of padding (foam) sometimes used to help protect the horse's withers from harness lesions. If the padding is kept clean, it can reduce greatly the number of lesions that a horse might get.

Fig. 17.11. The backband padding in this photograph is made of a cotton canvas and will help alleviate pressure on the donkey's withers and back. Frequently, the material used for padding is abrasive, so try to use cotton or animal skins, ideally a material that will not rub the animal when friction occurs. The padding should also be cleaned and the donkey should be groomed in between work sessions, also to reduce friction.

C.R. Heleski, A.K. McLean and J.C. Swanson

Health Management (Wound Care)

Perhaps nowhere is owner knowledge deficiency more obvious than in wound care and administration of vaccines. Horses are often staked out amid junk and debris, much of which has sharp edges.

Fig. 17.12. This horse is too thin. Note that her ribs and her spine and her hip bones are all quite noticeable; also, her neck is very thin. She would be approximately a "2" on the 1–9 Henneke (Henneke *et al.*, 1983) scoring system. An animal this thin will have trouble working as hard as her owner might want her to. She may also have more trouble resisting disease.

Not surprisingly, this can lead to significant injuries (see Fig. 17.20). Unfortunately, there has been little communication providing information on first aid for working equids (Pearson and Krecek, 2006). Owners have frequently been observed putting mud, manure, and diesel fuel, among other traditional remedies, on to open wounds. While there would be a chance that mud might provide some benefit, e.g. by keeping flies out of the wound, many of the substances will simply prolong healing and add insult to the injury. Furthermore, contracting tetanus is a real concern in developing areas of the world. Tetanus vaccinations would be a relatively inexpensive way to address this problem.

Basic wound cleaning, ideally with soap and clean water, is a significant first step toward enhancing wound care. This, followed by the administration of antibiotic wound ointments, could reduce the number of badly infected wounds dramatically. Bandaging the wounds should be suggested with careful forethought. Inappropriate bandage care will likely create more problems than it solves. It is important to remember that in the developing world, you are basically working in a "triage" scenario. It is rare that we think about optimal care; rather, we need to think of ways to minimize animal suffering and enhance the chances of animal longevity.

Fig. 17.13. This donkey is too thin; note the ribs showing, the hip bones showing, and the thinness of the neck. Also note the various lesions on the donkey; some from harness wear, some from abusive treatment.

Fig. 17.14. This horse is in moderately good shape. He would be a body condition score of "5" on the Henneke (Henneke *et al.*, 1983) system. Note the smooth covering of flesh over the ribs, hip bones, and spine. Also notice that the coat is shiny from receiving better nutrition and grooming.

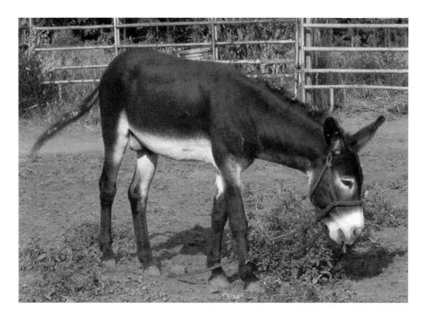

Fig. 17.15. This donkey is in moderately good body condition; note the smoothness of flesh covering the ribs and the smoothness of the neck to wither junction. Also note the shine of the hair coat due to receiving better nutrition and better grooming.

C.R. Heleski, A.K. McLean and J.C. Swanson

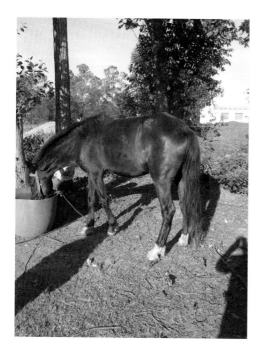

Fig. 17.16. Horses should have frequent access to fresh, clean water. This is especially important for working horses and those who live in hot, humid environments.

Lesions from poorly fitting tack (harnesses, saddles, bridles) are another common source of injuries (see Fig. 17.21). Ways to minimize these injuries are mentioned in the section on harnessing and Quick Access Information 17.2. With respect to the wound, it should be cleaned and dressed. Under the best conditions, the animal should have time off until the wound has healed. This is often not possible; at minimum, sufficient padding needs to be applied to the harness where the trouble spot has been identified. Measurement of leg lesions using a simple checklist reduced the frequency of lesions (Leeb *et al.*, 2013).

Another commonly observed injury in working equids is the rope burn (see Fig. 17.22). In horses, this is typically caused by staking the horse out without allowing it sufficient time to understand the concept of being tied. The horse then gets the rope wrapped around its ankle, panics, and fights until there is a deep, burn-type injury around the pastern area. In developed countries the practice of staking out a horse is discouraged because of the high risk of rope injuries. However, our experience with working equids indicates that almost the only chance the animals have to forage adequately is if they are staked out, and preferably

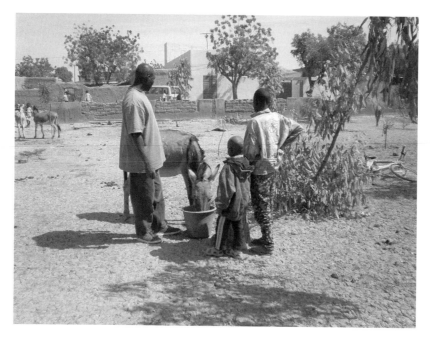

Fig. 17.17. Donkeys are more drought-resistant than horses, but they still require frequent access to clean, fresh water. When donkeys do not receive enough water, their feed intake goes down, which in turn reduces their work output.

often. Surprisingly, we observed many horses with no evidence of rope burns. It is possible that horses under these work conditions are so fatigued at the end of their workday, they do not have the strength to panic and fight the rope. If the opportunity is available, having the rope or chain wrapped within a length of old watering hose is one way to make it safer. Also, if the horse has been trained slowly to accept the concept of giving to rope pressure, this will also assist his learning curve. In donkeys, our experience suggests that

the problem is more often due to the use of abrasive materials, or tethering ropes that are left on for too long.

One other unfortunate cause of injuries for working equids is leaving on certain items of tack/equipment. For example, we have observed donkeys whose halters had been left on so long that the skin was attempting to grow around the rope. We have seen horses wearing crudely made permanent splint boots (pieces of rubber tied on to the legs with binder twine) in which the twine has begun cutting into the leg. Others (Jones, 2008b) have observed harnesses and bridles being fixed with wire, which then often cuts into the animal, especially when left on for too long.

Vaccination

In developed parts of the world, a great deal of money is spent on vaccinating equids for nearly every disease they might encounter. For the most part, this is a good practice. When it comes to the working equid in the developing world, a risk–benefit assessment needs to be considered, and take into account the unlikelihood that the resource-poor owner can afford vaccines.

Fig. 17.18. This horse is eating fresh forage. In this case, the grass was cut from a nearby ditch bank.

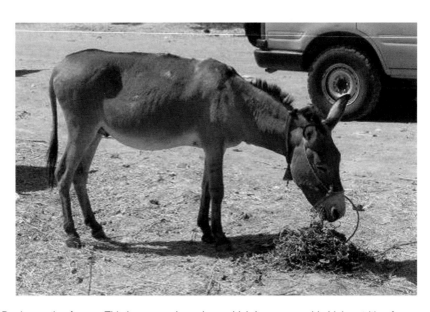

Fig. 17.19. Donkey eating forage. This is cowpea bean hay, which is a reasonably high-nutrition forage available to donkeys in western Africa. Donkey forage is often purchased at the market, especially during the dry season. Also, notice that the donkey's owner has chosen to unharness his donkey and let him eat while he is not being worked. When donkeys and horses are not working, their harness should be removed and they should be allowed to graze or offered food, and especially water. If possible, they should be moved into the shade during breaks.

C.R. Heleski, A.K. McLean and J.C. Swanson

Fig. 17.20. Working equids are frequently kept in areas that have not been checked for junk and debris lying around. Because of this, various wounds are common, such as this wire cut. All owners should have antibiotic salve on hand to help treat wounds after cleaning with soapy water. Ideally, working equids should have their tetanus immunizations up to date.

When minimal cost or donated vaccines can be obtained, it is important to recommend those vaccines that have a high efficacy rate for preventing zoonotic diseases. For example, in southern Brazil, we were able to procure vaccines for many horses to be vaccinated and boostered against tetanus, EEE (Eastern Equine Encephalomyelitis, or sleeping sickness) and WEE (Western Equine Encephalomyelitis, also called sleeping sickness). Workers operating in a geographic area need to consider the endemic disease risks for the region. For example, some regions may need to consider rabies as a potentially important vaccine. Other regions may need to consider WNV (West Nile Virus) or VEE (Venezuelan Equine Encephalomyelitis) as potentially important vaccines. Typically, cheap vaccines, with high efficacy rates, that prevent diseases with significant human risk are the only ones that can reasonably be considered. It is likely that Coggins testing for equine infectious anemia (EIA) should be implemented, but unless replacement equids can be provided for those families who own a positive animal, this process will make no sense to the draft animal owner.

Hoof Care

Surprisingly, during our research work in southern Brazil, we saw few hoof-care problems. The owners were spending a reasonably high percentage of their limited income to have their horses trimmed and reshod every 6–10 weeks. The workmanship of the farriery was quite good. As a consequence, we rarely saw lame horses being worked. At least in southern Brazil, owners had identified the relationship between proper hoof-care practice and the impact to their economic well-being through reduced lameness of their horses.

This has not been the finding in other locations. In one study in Mexico (Aluja, 1998), a high percentage of donkeys and horses showed evidence of poor hoof care, with overgrown hooves being the most frequently observed problem. Furthermore, even in the stoic donkey, many were observed to be mildly to severely lame. The author's recent trip to Mexico indicated that proper hoof trimming was not being done (Fig. 17.23).

Lame equids need to be rested. Due to the difficulty in implementing this practice, SPANA clinics in Africa sometimes provide replacement donkeys for owners who are willing to relinquish their lame donkey. In this scheme, owners do not have to lose income, but their lame animal is provided a chance to rest, which is often the only therapy that is required.

Animals with high maintenance and chronic lamenesses need to be culled from working. There are no additional inputs for tolerating these types of lamenesses. The suffering is not only debilitating to the animal but also it is highly likely that family well-being is also affected.

Fair Treatment

Our experience in southern Brazil was that while the horses might be sadly neglected in terms of calorie inputs and water needs, it was extremely rare to see them purposely mistreated. In our weeks of observation, with hundreds of horses observed, we saw only two horses being goaded with the whip (and this was when driven by teenagers who appeared to be going for a joyride in their "car").

Our experiences in Mali were different. Many of the donkeys were guided with no bridle or halter,

Fig. 17.21. This thin donkey with numerous lesions is being rehabilitated by the SPANA clinic in Bamako, Mali. Eventually, this donkey will be healthy enough to return to work.

Fig. 17.22. Rope burns are also commonly observed in the carthorse population. On the positive side, it is good for horses to be tied out so they can eat grass whenever they are not working. On the negative side, some horses panic when they first get tangled in their tie-out rope and cause rope burns such as this.

Fig. 17.23. Poor hoof care observed in a remote Mexican village.

instead they were directed with a stick carried by the donkey handler. The stick was often used roughly and repeatedly. Numerous croup and loin lesions were observed that appeared to be a direct result of being struck with the stick (see Fig. 17.24). Some researchers have reported that owners in parts of Africa use sticks with nails protruding

from them (Herbert, 2006). Obviously, this contributes to the serious, and occasionally infected, lesions sometimes observed on the animals. Several studies have looked into the relationship between anthropogenic (human-caused) lesions and the behavior of the equid (Burn *et al.*, 2010; Popescu and Diugan, 2013). If owners can be trained to recognize the importance and benefits of compassionate handling of their animals, it will enhance greatly the welfare of these hard-working animals (Swann, 2006; McLean *et al.*, 2008). The authors

Fig. 17.24. Malian donkeys have been observed to have a high number of lesions in the croup area. It is believed this is because they get lacerations and deep tissue bruising from being hit with a stick. Some lesions ultimately become infected.

measured the heart rate of donkeys guided with either a stick or a halter and reins (McLean *et al.*, 2012). When either method was used correctly, there was no significant difference in heart rate.

Reproduction and Raising Young Stock

In truth, very few resource-poor equid owners are set up to raise young stock appropriately. However, replacement animals must be raised. In Brazil, we rarely observed young stock at any of the villages visited. When owners were asked about this, it appeared that they typically bought young stock from farmers who were raising horses for other purposes. We did observe a few young animals being raised, and in most instances it was obvious that the animals were receiving inadequate protein and inappropriate calcium:phosphorus ratios (NRC, 2007) (i.e. young stock were stunted and often had heads that were inappropriately large for their bodies).

In Mali, young donkeys were fairly often observed tagging along with their working dams. Again, as an attribute to the resiliency of the donkey, these young stock often looked surprisingly healthy. In the rural villages of Mexico that we have visited, overall, the health of young stock has been reasonably good.

As a side note, because castration is often hard to perform in resource-poor parts of the world, it is not uncommon for stallions to be the working animals nearly as often as geldings and mares. This is not without its problems, in terms of the potential for aggressive behavior or unplanned breeding. As with other behaviors, often these animals work so hard that there is little interest in causing some of the behavioral problems that would be considered typical to stallions.

Conclusion

More resources, in terms of time and money, are needed to enhance the welfare of working equids in resource-poor areas of the world. In addition, encouraging all opportunities for information dissemination between organizations and experts with the husbandry information and those who could derive benefit needs to take place (Dijkman *et al.*, 1999; Pearson and Krecek, 2006). We assume this is true for all species of draft animals. Although we have concentrated on relating our experience with working equids, other working livestock can benefit from some of these recommendations when applied in a unique species-specific manner (Fig. 17.25).

Fig. 17.25. Often, it is children who bring the draft animals to the veterinary clinics for treatment and/or immunizations. We have begun to implement educational programs on working equid husbandry that are geared specifically to a juvenile audience.

Helping working equids sometimes feels like the story of the boy and the starfish,[2] but it is that boy's words that encourage us to keep going.

Acknowledgement

We would like to thank Dr. Temple Grandin for inviting us to contribute this chapter to her book. We deeply appreciate her effort to broaden the awareness in the welfare science community about the plight of a high percentage of draft animals in the developing world.

Notes

[1]Body condition scoring will be discussed further in the section on nutrition.
[2]*The Starfish Story* (original story by Loren Eisley). "One day a man was walking along the beach when he noticed a boy picking something up and gently throwing it into the ocean. Approaching the boy, he asked, 'What are you doing?' The youth replied, 'Throwing starfish back into the ocean. The surf is up and the tide is going out. If I don't throw them back, they'll die.' 'Son,' the man said, 'don't you realize there are miles and miles of beach and hundreds of starfish? You can't make a difference!' After listening politely, the boy bent down,

picked up another starfish, and threw it back into the surf. Then, smiling at the man, he said ... 'I made a difference for that one.'"

References

AAEP Parasite Control Guidelines 2014. Available at: http://www.aaep.org/custdocs/ParasiteControlGuidelines Final.pdf. (accessed 19 May 2014).

Aluja, A. (1998) The welfare of working equids in Mexico. *Applied Animal Behaviour Science* 59, 19–29.

Arriaga-Jordán, C.M., Colunga-González, B. and Velázquez-Beltrán, L.G. (2006) The contribution of working equines to sustainable rural livelihoods. In: Pearson, R.A., Muir, C.J. and Farrow, M. (eds) *Proceedings of the 5th International Colloquium on Working Equines, Addis Ababa, Ethiopia*, pp. 264–275.

Bullard, R.W., Dill, D.B. and Yousef, M.K. (1970) Responses of the burro to desert heat stress. *Journal of Applied Physiology* 29(2), 159–167.

Burn, C.C., Pritchard, J.C., Farajat, M., Twaissi, A. and Whay, H.R. (2008) Risk factors for strap-related lesions in working donkeys at the World Heritage Site of Petra in Jordan. *The Veterinary Journal* 178, 261–269.

Burn, C.C., Dennison, T.L. and Whay, H.R. (2010) Relationships between behaviour and health in working horses, donkeys, and mules in developing countries. *Applied Animal Behaviour Science* 126, 109–118.

Clayton, H.M. (1991) Chapter 5, Thermoregulation. In: *Conditioning Sport Horses*. Sport Horse Publications, Michigan, USA, p. 66.

Diarra, M.M., Doumbia, A. and McLean, A.K. (2007) Survey of working conditions and management of donkeys in Niono and Segou, Mali. *Proceedings ASAS Annual Meeting* 85 (Suppl 1) (Abstract 59), p. 139.

Dijkman, J.T., Sims, B.G. and Zambrana, L. (1999) Availability and use of work animals in the middle Andean hill farming systems of Bolivia. *Livestock Research for Rural Development* 11(2), 1–12.

Doumbia, A. (2008) Fistulous withers: A major cause of morbidity and loss of use amongst working equines in West Africa. An evaluation of the aetiology and treatment of 33 cases in Mali. In: *Proceedings of the 10th International Congress of World Equine Veterinary Association, Moscow, Russia*. Available through International Veterinary Information Service (www.ivis.org), pp. 581–583.

Draught Animal News (DAN) Available at: http://www.vet. ed.ac.uk/ctvm/Research/DAPR/draught%20animal%20news/danindex.htm (accessed 20 May 2014).

Ellis, V., Ellis, R. and Claxton, J. (1980) *Donkey Driving*. J.A. Allen and Co. Ltd, London, pp. 115.

FAO (2002) Draught animals plough on. Accessed at: www.fao.org/ag/magazine/0009sp.1.htm (accessed 4 April 2015).

Geiger M. and Hovorka, A.J. (2015) Using physical and emotional parameters to assess donkey welfare in Botswana. *Veterinary Record Open*, doi:10.1136/vetreco-2014-000062.

Henneke, D.R., Potter, G.D., Kreider, J.L. and Yeates, B.F. (1983) Relationship between condition score, physical measurements and body fat percentage in mares. *Equine Veterinary Journal* 15, 371–372.

Herbert, K. (2006) At work in Morocco. In: theHorse. com, Article No 7001. Available at: www.thehorse. comarticles/16342/at-work-in-morocco (accessed 4 April 2015).

Jones, P. (2008a) Animal tolerance. *Draught Animal News* 46(1), 17–19. Available at: http://www.link.vet. ed.ac.uk/ctvm/Research/DAPR/draught%20animal%20news/danindex.htm (accessed 5 May 2015).

Jones, P. (2008b) Thoughts on harnessing donkeys for work, based on practical experiences in southern Africa. In *Draught Animal News* 46, 2, 64–70. Also accessible at http://www.link.vet.ed.ac.uk/ctvm/Research/DAPR/draught%20animal%20news/danindex.htm (accessed 27 March 2015).

Jones, W.E. (2004) Water, dehydration, and drinking. *Journal of Equine Veterinary Science* 24(1), 43–44.

Kitalyi, A., Mtenga, L., Morton, J., McLeod, A., Thornton, P., *et al.* (2005) Chapter 2, Why keep livestock if you are poor? In: Owen, E., Kitalyi, A., Jayasuriya, N. and Smith, T. (eds) *Livestock and Wealth Creation, Improving the Husbandry of Animals Kept by Resource-poor People in Developing Countries*. Nottingham University Press, Nottingham, UK, pp. 13–27.

Leeb, C., Henstridge, C., Dewhurst, K. and Bazeley, K. (2103) Welfare assessment of working donkeys: Assessment of the impact of an animal healthcare project in West Kenya. *Animal Welfare*, 12, 689–694.

McLean, A., Heleski, C.R. and Bauson, L. (2008) Donkeys and bribes … maybe more than just a cartoon! In: *Proceedings of the 42nd Congress of the International Society for Applied Ethology, Dublin, Ireland*, p. 72. Available at: http://www.applied-ethology.org/download 135.html (accessed 27 March 2015).

McLean, A.K., Heleski, C.R., Yokoyama, M.T., Wang, W., Doumbia, A., *et al.* (2012) Improving working donkey (*Equus asinus*) welfare and management in Mali, West Africa. *Journal of Veterinary Behavior* 7, 123–134.

NRC (2007) Chapter 13, Donkeys and other equids. In: *Nutrient Requirements of Horses*, 6th revised edn. The National Academies Press, Washington, DC, pp. 268–279.

Pearson, R.A. (2005) Nutrition and feeding of donkeys. In: Matthews, N.S. and Taylor, T.S. (eds) *Veterinary Care of Donkeys*. Available through International Veterinary Information Service (www.ivis.org).

Pearson, R.A. and Krecek, R.C. (2006) Delivery of health and husbandry improvements to working animals in Africa. *Tropical Animal Health Production* 38, 93–101.

Pearson, R.A., Nengomasha, E. and Krecek, R. (1999) The challenges in using donkeys for work in Africa. In: Starkey, P. and Kaumbutho, P. (eds) *Meeting the Challenges of Animal Traction*. Intermediate Technology Publications, London, pp. 190–198. Available at: www. atnesa.org (accessed 27 March 2015).

Pollock, P.J. (2009) Helping working equids in Egypt. In: theHorse.com, Article No 13537. Available at: www. thehorse.com/Print.aspx?ID=13537 (accessed 7 May 2015).

Popescu, S. and Diugan, E.-A. (2013) The relationship between behavioral and other welfare indicators of working horses. *Journal of Equine Veterinary Science* 33, 1–12.

Pritchard, J.C., Lindberg, A.C., Mail, D.C.J. and Whay, H.R. (2005) Assessment of the welfare of working horses, mules and donkeys, using health and behavior parameters. *Preventative Veterinary Medicine* 69, 265–283.

Pritchard, J.C., Barr, A.R.S. and Whay, H.R. (2007) Repeatability of a skin tent test for dehydration in working horses and donkeys. *Animal Welfare* 16, 181–183.

Ramswamy, N.S. (1998) Draught animal welfare. *Applied Animal Behaviour Science* 59, 73–84.

Riex (nee Broster) C.E., Burn, C.C., Pritchard, J.C., Barr, A.R.S. and Whay, H.R. (2014) The range and prevalence of clinical signs and conformation associated lameness in working draught donkeys in Pakistan. *Equine Veterinary Journal* 46, 771–777.

Sevilla, H.C. and León, A.C. (2007) Harnesses and equipment commonly used by donkeys (*Equus asinus*) in Mexico. In: Matthews, N.S. and Taylor, T.S. (eds) *Veterinary Care of Donkeys*. Available at: International Veterinary Information Service (www.ivis.org).

Starkey, P. (1997) Chapter 11, Donkey work. In: *The Professional Handbook of the Donkey*, compiled for The Donkey Sanctuary by Svensen, E.D. Whittet Books, Suffolk UK, pp. 183–206.

Svendsen, E. (1997) Chapter 13, Parasites abroad. In: *The Professional Handbook of the Donkey* compiled for The Donkey Sanctuary by Svensen, E.D. Whittet Books, Suffolk UK, pp. 227–238.

Swann, W.J. (2006) Improving the welfare of working equine animals in developing countries. *Applied Animal Behavior Science* 100, 148–151.

Upjohn, M.M., Shipton, K., Pfeiffer, D.V., Lerotholi, T., Attwood, G., *et al.* (2012) Cross sectional survey of owner knowledge and husbandry practices, tack and health issues affecting working horses in Lesotho. *Equine Veterinary Journal* 44, 310–318.

Upjohn, M.M., Pfeiffer, D.U. and Verheyen, K.L.P. (2014) Helping working Equidae and their owners in developing countries: Monitoring and evaluation of evidence-based interventions. *The Veterinary Journal* 199, 210–216.

Wickler, S.J., Hoyt, D.F., Cogger, E.A. and Hall, K.M. (2001) Effect of load on preferred speed and cost of transport. *Journal of Applied Physiology* 90(4), 1548–1551.

Wilgert, E. (2010) Impact of veterinary assistance on the health of working horses in Nicaragua. Thesis, Department of Nutrition and Management, Uppsala SLU, Sweden.

Zanella, R., Heleski, C. and Zanella, A.J. (2003) Assessment of the Michigan State University equine welfare intervention strategy (MSU EQWIS-ACTION) using Brazilian draught horses as a case study. In: *Proceedings of the 37th International Congress of the ISAE, Abano Terme, Italy*, p. 192.

Index